GUIDANCE AND CONTROL
1984

CONFERENCE COMMITTEE

CHAIRMAN:	Parker S. Stafford Martin Marietta Aerospace (303) 977-3226
CO-CHAIRMAN:	Zubin Emsley Ball Aerospace Systems Division (303) 441-4948
CO-CHAIRMAN:	Charles Cullian Ball Aerospace Systems Division (303) 441-4392
LOCAL ARRANGEMENTS:	Don Parsons Martin Marietta Aerospace (303) 977-7109
REGISTRATION AND FINANCE:	Bob Barsocchi Bendix Corporation (303) 979-5504
COMMITTEE MEMBERS:	Edward J. Bauman University of Colorado Colorado Springs
	Fred Bikle Martin Marietta Aerospace
	Robert D. Culp University of Colorado, Boulder
	Ed Dorroh Martin Marietta Aerospace
	Tom Fenner Martin Marietta Aerospace
	Carl Henrikson Ball Aerospace Systems Division
	John Montgomery Martin Marietta Aerospace
	Lou Morine Martin Marietta Aerospace

Thanks are due Teresa Wilson, Univelt, Inc., for final preparation of the manuscript for publication.

Front Cover Illustration:

Bruce McCandless floating free in the first space use of the Manned Maneuvering Unit (MMU). Photo courtesy of NASA and Martin-Marietta.

Frontispiece:

Transfer Orbit Stage (TOS). See paper AAS 84-045.

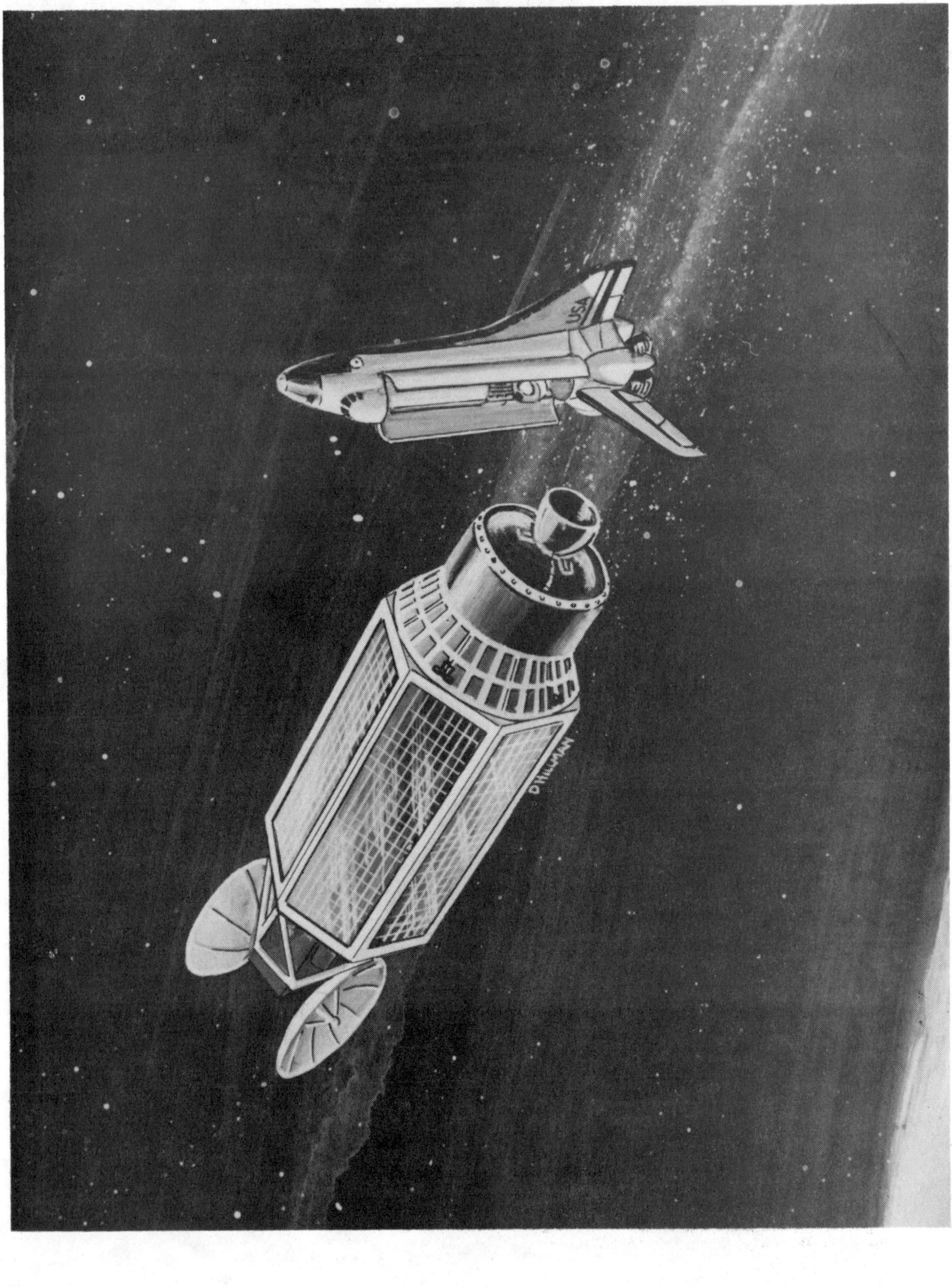
USA

AN AMERICAN ASTRONAUTICAL SOCIETY PUBLICATION

GUIDANCE AND CONTROL 1984

Volume 55
ADVANCES IN THE ASTRONAUTICAL SCIENCES

Edited by
Robert D. Culp
Parker S. Stafford

Proceedings of the Annual Rocky Mountain Guidance and Control Conference held February 4–8, 1984, Keystone, Colorado

Published for the American Astronautical Society by
Univelt, Incorporated, P.O. Box 28130, San Diego, California 92128

AAS Publications Office
P. O. Box 28130
San Diego, California 92128

Affiliated with the American Association for the Advancement of Science
Member of the International Astronautical Federation

First Printing 1984

Library of Congress Card No. 57-43769

ISSN 065-3438

ISBN 0-87703-199-1 (Hard Cover)

ISBN 0-87703-200-9 (Soft Cover)

Published for the American Astronautical Society
by Univelt, Inc., P. O. Box 28130, San Diego, California 92128

Printed and Bound in the U.S.A.

FOREWORD

HISTORICAL SUMMARY

The Annual Rocky Mountain Guidance and Control Conference began as an informal exchange of ideas and reports of achievements among local guidance and control specialists. Since most area guidance and control experts participate in the American Astronautical Society, it was natural to gather under the auspices of the Rocky Mountain Section of the AAS.

In 1977, Bud Gates, Don Parsons, Joe Spencer, and Bob Culp organized the conference formally and began the annual series of meetings the following winter. In March, 1978, the Annual Rocky Mountain Guidance and Control Conference, sponsored by the Rocky Mountain Section of the American Astronautical Society, met at Keystone, Colorado. It has met annually there ever since.

The Rocky Mountain Section of the Society established a broad-based Conference Committee, the Rocky Mountain Guidance and Control Committee, chaired ex-officio by the next Conference Chairman, to run the Annual Conference. The Conference has been a success from the start. Every year has brought increased attendance, as shown by these figures:

Year	Conference Chairman	Attendance
1978	Robert L. Gates	83
1979	Robert D. Culp	109
1980	Louis A. Morine	130
1981	Carl Henrikson	150
1982	W. Edwin Dorroh, Jr.	180
1983	Zubin Emsley	192
1984	Parker S. Stafford	201

In addition to providing for an annual exchange of the most recent advances in research and technology of astronautical guidance and control, the Conference brings in a guest speaker of special prominence to present a lecture series in one specific area of current interest. These speakers and their topics are:

1978 Professor Daniel Debra, Stanford University, "Navigation"

1979 Professor William L. Brogan, University of Nebraska, "Kalman Filters Demystified"

1980 Professor J. David Powell, Stanford University, "Digital Control"

1981	Professor Richard H. Battin, Massachusetts Institute of Technology, "Astrodynamics - A New Look at Old Problems"
1982	Professor Robert E. Skelton, Purdue University, "Interactions of Dynamics and Control"
1983	Professor Arthur E. Bryson, Stanford University, "Attitude Stability and Control of Spacecraft"
1984	Dr. William B. Gevarter, NASA Ames, "Artificial Intelligence and Intelligent Robots"

In 1983, Don Parsons, Joe Spencer, Ed Dorroh, and Bob Culp, at the request of the National Board of the American Astronautical Society, organized a charter which established the Annual Conference as the AAS Annual Rocky Mountain Guidance and Control Conference. The conference will be sponsored by the national organization of AAS, will meet annually at Keystone, Colorado, and will be run by the Rocky Mountain Section.

A national AAS Guidance and Control Technical Committee, with national representation, was established to provide oversight to the local conference committee. W. Edwin Dorroh, Jr. is the first chairman of the AAS Guidance and Control Technical Committee. This committee meets every year at the Keystone Conference, and also at either the summer Guidance and Control Conference, or the fall AAS Annual Meeting.

Thus, the AAS Rocky Mountain Guidance and Control Conference continues as the premier conference of its type anywhere. As a national conference sponsored by the AAS, it promises to be the preferred idea exchange for guidance and control experts for years to come.

The Conference Committee extends congratulations to the 1984 Conference Chairman, Parker Stafford, for an excellent conference and a job well done. His report follows.

Robert D. Culp
University of Colorado
Boulder, Colorado

PREFACE

THE 1984 AAS ROCKY MOUNTAIN GUIDANCE & CONTROL CONFERENCE

February 4 thru 8, 1984
at Keystone, Colorado

This was the seventh annual Rocky Mountain Guidance and Control Conference sponsored by the AAS. It began Saturday evening, February 4th, with an excellent banquet featuring Walter Boyne, Director of the Smithsonian Air and Space Museum, who gave an excellent presentation on the history of the museum, the future plans, and some fascinating details on how the national treasures are restored for display in the museum.

On Sunday we continued with our traditional Tutorial Session, having a six-hour lecture by Dr. Bill Gevarter summarizing the state-of-the-art of Artificial Intelligence, Expert Systems, and Robotics. Bill used as background NASA Technical Memorandum 85836 and National Bureau of Standards Report NBSIR 82-2479. Dr. Gevarter was an especially apropos speaker for this audience of Guidance and Control Engineers since his background is in Controls and he's practiced Guidance and Control Research for over 20 years while in various jobs with NASA.

There were five technical sessions at this conference highlighted by our first International Space Programs Conference chaired by Dr. Chris Markland, of ESTEC. All of the papers in this session were invited papers covering significant space program developments in Europe and Japan. Specifically, we looked at the attitude control systems and techniques used for the Olympus Satellite, the EUROSTAR, the Marine Observations Satellite, the Infrared Space Observatory, the Hipparcos Satellite, and the Planet-A Halley's Comet mission currently under development in Japan. It was clear to the audience that these programs are up to the current state-of-the-art and should provide some excellent scientific data.

Session II of the conference contained twelve papers in a storyboard format including several demonstrations of working hardware or video tapes of laboratory experiments. This was in conjunction with hors d'oeuvres and drinks so that for a three-hour informal session everyone could view the experiments and papers to any depth that they wished in a walk-around fashion. In addition to this, we were fortunate to have the STS 7 Mission film from NASA which was shown several times during the evening.

Our third session was chaired by Harry Buchanan of George C. Marshall Space Flight Center and Steve Szirmay of Jet Propulsion Laboratory entitled Rendezvous Docking and Orbit Servicing. In this session we had excellent papers on Space Station Rendezvous and Docking, a paper describing the role of the Manned Maneuvering Unit which coincidentally was given while the first orbital use of the Manned Maneuvering Unit was in progress by Astronaut Bruce McCandless.

In the fourth session we turned our attention to Vehicle Guidance and Control and its effects on payloads. The session was chaired by Col. Dave Younkin of Space Division and co-chaired by Bill Drake of Aerospace Corporation. In this session the Guidance and Control techniques used on the IUS, the CENTAUR, Orbit Maneuvering Systems, the Transfer Orbit Stage were described and options for integration of Launch Vehicle and payloads avionics were also discussed.

We closed out the conference on Wednesday, February 8, with our traditional session on recent experiences which focuses on the flights which have occurred in the previous year. In this session, we discussed in depth the IUS/TDRS Mission both from the IUS and TDRS points of view. We also looked at how the TDRS is operating now and doing its auto-tracking functions. We reviewed the Guidance and Control lessons learned from the first ten flights of the shuttle, and we were given an excellent overview of the effects of cosmic rays on modern digital electronic systems.

During the conference a meeting of the AAS Guidance and Control Technical Committee was held by Chairman Ed Dorroh and several constructive suggestions for next year's conference were provided by the committee. Next year's conference, the 1985 Annual Rocky Mountain G & C Conference, will be held again at Keystone, February 2-6, 1985. The chairman for that conference will be Charles Cullian, Ball Aerospace Systems, P.O. Box 1062, Boulder, co-chairman will be Parker Stafford, Martin-Marietta Aerospace, Box 179, Denver, CO 80201, area 303-977-3226. Both Chairman and Co-chairman would welcome any calls or written suggestions on improvements on next year's conference, and actively encourage those of you in the Guidance and Control community to plan to come next year to what we think is the premier Guidance and Control conference in the United States, if not the world.

I would like to take this opportunity to thank all of you who attended the 1984 conference at Keystone. It is with pleasure that I extend to you the invitation to attend the 1985 Keystone Conference on February 2-6, 1985. I hope to see you all again there.

Parker S. Stafford
Martin-Marietta Aerospace
Denver, Chairman,
1984 AAS Guidance & Control Conference
Keystone, Colorado

CONTENTS

CONTENTS (Cont'd)

CONTENTS (Cont'd)

CONTENTS (Cont'd)

Section I
INTERNATIONAL SPACE PROGRAMS

SESSION I

Chairman:	Dr. C. Markland European Space Agency, ESTEC, Noordwijk, The Netherlands
Local Chairman:	John Sand Ball Aerospace Systems Division

The following paper numbers were not assigned:

AAS 84-007 to -010

AAS 84-001

DEVELOPMENT OF THE ATTITUDE AND ORBIT CONTROL SUBSYSTEM FOR THE OLYMPUS SATELLITE

Michael Burton*

This paper describes the major trade-off studies performed during the development of the Attitude and Orbit Control Subsystem for a new generation of Communications Satellites, OLYMPUS. The resulting subsystem design is presented with descriptions of the main hardware elements and the modes of operation. Overall subsystem performance is summarised.

INTRODUCTION

OLYMPUS 1 is the precursor to the most powerful class of communications satellite in the Western World. The OLYMPUS class is compatible with both Space Shuttle and Ariane launcher systems and offers a platform on which may be based a whole range of communications packages varying from a low power payload of 2.5 KW up to a maximum of 7.8 KW. OLYMPUS 1 is designed to demonstrate the capability of this class of satellite to carry multi-purpose payloads that will offer new services to telecommunications authorities, TV companies and commercial and industrial corporations. OLYMPUS 1 will be launched on an Ariane 3 vehicle and the in orbit configuration is shown in Fig. 1.

The following major payloads are carried on OLYMPUS 1:

1. 20/30 GHz High Power Communications Package comprising 2 steerable spot beams (1° beam width) and a Pan-European, dual polarised beacon.

2. 12/14 GHz Specialised communications services, with a five-beam pattern steerable as a group and using Satellite-switched Time Division Multiple Access.

3. 12/18 GHz Direct TV Broadcast, providing one TV channel for Italy and one steerable channel for test transmissions by other European Countries.

* *Assistant Project Manager for the OLYMPUS Attitude and Orbit Control Subsystem. Mr. M. Burton is an employee of British Aerospace Dynamics Group, Space and Communications Division, Argyle Way, Stevenage, England*

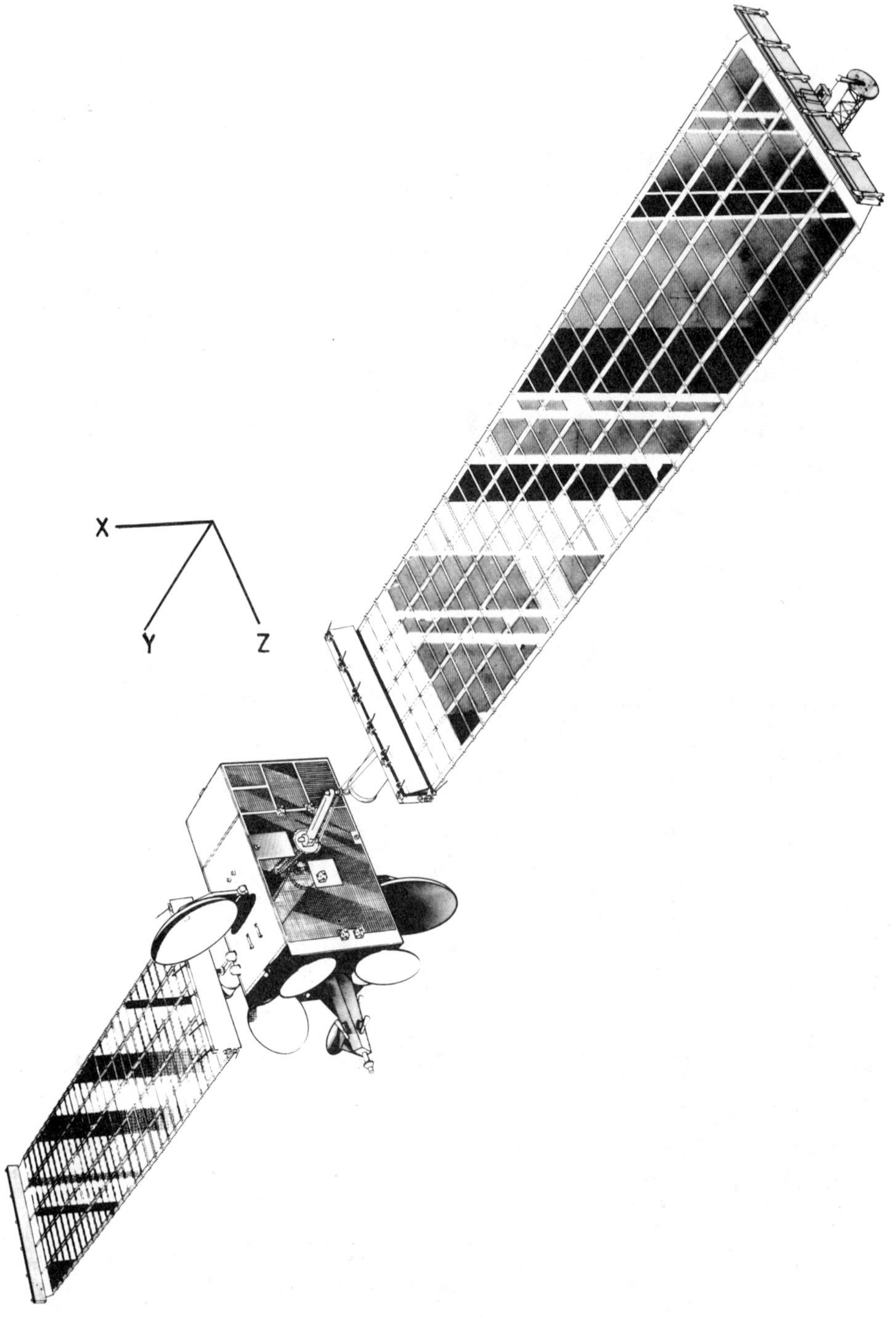

Fig. 1 OLYMPUS 1 In-Orbit Configuration

The dynamic characteristics of the wide range of antenna and solar array configurations envisaged for the OLYMPUS class presents the control system designer with a significant challenge, given the precision pointing requirements of high power payloads. Meeting this challenge has resulted in a new concept for European Communications Satellites which represents a major step forward from previous satellites such as the Orbital Test Satellite (OTS) and the European Communications Satellite (ECS). The resulting control system uses precision gyroscopes, Sun sensors and Earth sensors for attitude measurement, Reaction Wheels for control actuation and a microprocessor based Control Electronics Unit (CEU) for control law implementation.

The aim of this paper is to present the rationale which lead to the design concept for the Attitude and Orbit Control Subsystem (AOCS), to describe its constituent elements and to summarise its overall performance.

A synopsis of the evolution of the AOCS design is presented first. This includes the two key trade-off studies

1. The transfer orbit stabilisation concept - spinning or 3-axis stabilised

2. The operational control philosophy - momentum biased or zero momentum

which lead to the current design.

The main operating modes of the resulting design, during all mission phases, are then described and performance details given.

TRADE-OFF STUDIES

Transfer Orbit Stabilisation

At an early stage in the programme it was established that a liquid propulsion engine would be used for manoeuvres from the transfer orbit to geosynchronous orbit. A bi-propellant engine was selected for high fuel efficiency. The large mass of propellants required for the orbit manoeuvres, and subsequent attitude and orbit control functions, lead to a configuration in which the two propellant tanks (each about 1m in diameter and each with a capacity of 850 kg of propellants) were on the central axis of the spacecraft. This was taken as the starting point for the trade-off study, which was part of an overall design exercise aimed at determining the optimum Satellite Transfer Orbit configuration taking into account performance, mass, power and cost.

The AOCS Transfer Orbit design study evaluated the advantages and disadvantages of the two control philosophies

1. Spin stabilisation
2. Three-axis stabilisation

and concluded that the latter represented the optimum compromise between the conflicting requirements.

From the standpoint of relying on past experience, spin stabilisation is attractive since the design problems are well known, at least with solid propellants. However, the combination of the unstable satellite inertia ratio, the presence of liquids and the tank configuration cast doubt on the feasibility of spin stabilisation. During the early stages of Phase B a theoretical model of the propellant dynamics was developed. This model demonstrated instability due to the sloshing modes of the propellants. An active nutation damper, using gyros for rate measurement and thrusters for nutation control, was therefore developed to achieve stabilisation.

Fig. 2 shows the instability time constant (T_D) of the propellant dynamics as a function of tank fill ratio and satellite spin rate. The figure also shows the minimum controllable time constant (T_C), as a function of spin rate. The control margin was assessed as the ratio (T_D/T_C).

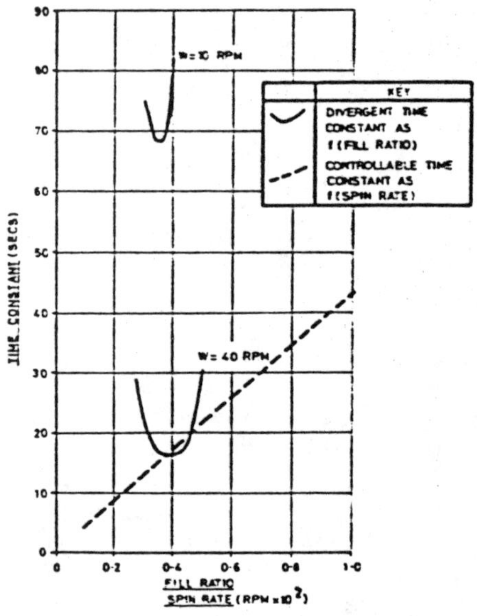

Fig. 2 Instability Time Constants

The fuel required to control the nutation was evaluated. The results are summarised in Table 1 below.

Table 1

STABILITY SUMMARY

Spin rate (rpm)	T_D / T_C	Fuel/Orbit (kg)	Control Fuel During Orbit Change (kg)
10	17.0	2.2	46
40	0.94	35.1	736

This table shows that the maximum spin rate to ensure stability ($T_D > T_C$) is just under 40 rpm. However the fuel penalty, even for a slow spin rate of 10 rpm, is high. In view of this conclusion, and also considering the approximations which had to be made to analyse sloshing in spherical tanks, it was concluded that 3-axis stabilisation was necessary during the transfer orbit for OLYMPUS satellites.

Further studies into power and thermal problems lead to the decision to deploy the solar arrays at an early stage in the transfer orbit.

The decision to adopt a three-axis stabilised control philosophy was only the first stage in the Transfer Orbit design optimisation. The selection of the attitude sensors resulted from a detailed evaluation of the available options in the light of constraints imposed by orbit selection and the need to minimise mass, power and cost.

To ensure Ariane and STS compatibility a nominal local midnight launch window was selected. This leads to a Sun-Earth geometry given in Fig. 3 and Fig. 4.

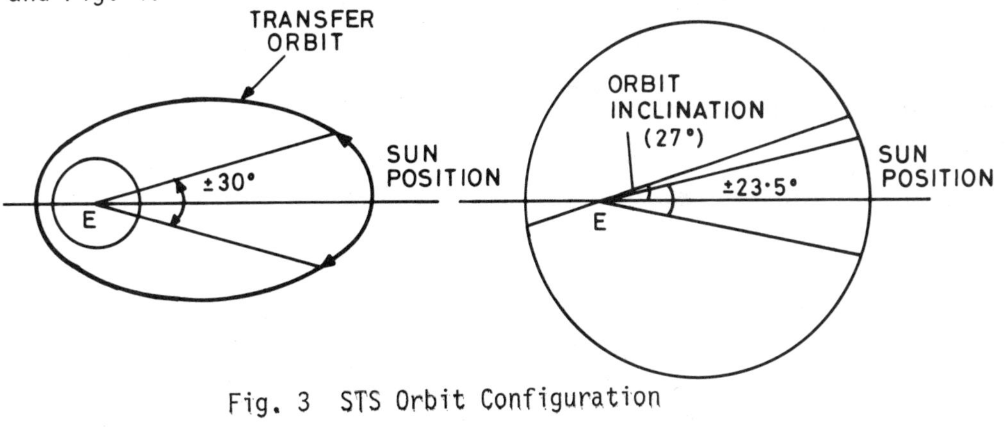

Fig. 3 STS Orbit Configuration

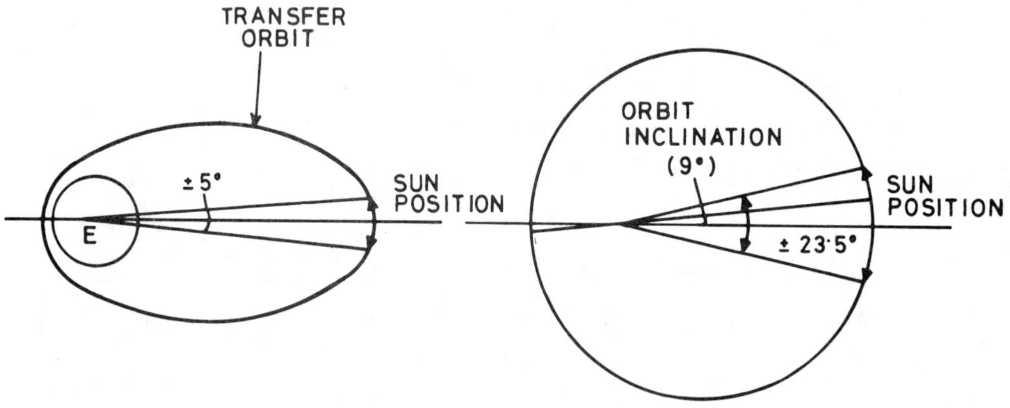

Fig. 4 Ariane Orbit Configuration

The obvious attitude references are the Sun and the Earth. However during the orbit change manoeuvre the Sun-Satellite-Earth configuration is close to colinearity. Thus inertial sensors would be required in addition to maintain continuous attitude measurement.

The alternative of using star sensors was considered but this was rejected on the basis of mass, cost and complexity. In order to minimise additional hardware, it was decided to use the Earth sensor required during the operational phase in the Transfer Orbit also. Since this could not be used during the whole of the transfer orbit (at altitudes below ∿ 20000 km the Earth's angular diameter is too large) the following sensing philosophy was adopted.

1. Gyroscopes are used for continuous sensing of rate and attitude

2. The gyroscopes are calibrated and periodically reset using Infra Red Earth Sensors (IRES) and Digital Sun Sensors (DSS).

The configuration of these sensors, in satellite axes, is shown in Fig. 5.

A summary of the main characteristics of the sensors is given in Table 2.

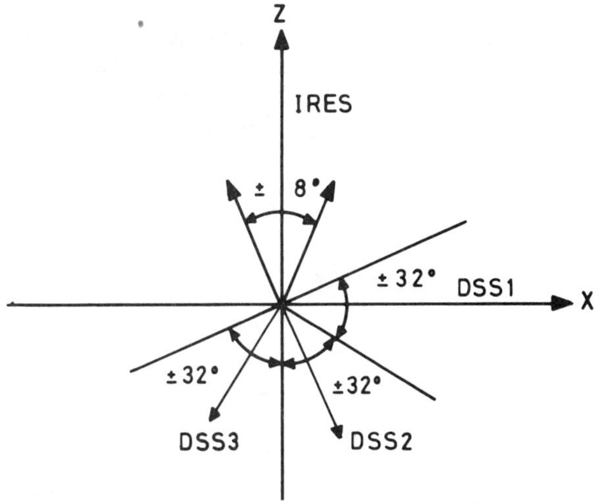

Fig. 5 Sensor Fields of View

Table 2

SENSOR CHARACTERISTICS

o	DIGITAL SUN SENSOR (ADCOLE)	- field of view	±31° x ±31°
		- accuracy	±0.017°
		- output	14 bit, LSB < 0.0048°
o	INFRA RED EARTH SENSOR (GALILEO)	- field of view	±14.5° x ±8° (wide) ±2.2° x ±4° (narrow)
		- bias	0.02°
		- noise	0.03° (3σ)
o	GYROSCOPES (FERRANTI)	- range	± 1° sec
		- output	16 bit, LSB 0.11°/hr
		- fixed drift	3°/hr
		- noise PSD	0.02 $(°/hr)^2$/hz
		- scale factor	± 1%

Operational Control

Previous European communications spacecraft (such as OTS and the MARECS and ECS series) have relied upon momentum stabilisation. The first of these, OTS, had a 25 NMS fixed momentum wheel. The concept was developed with the later communications satellites by having two wheels, symmetrically skewed about the nominal momentum axis (see Fig. 6).

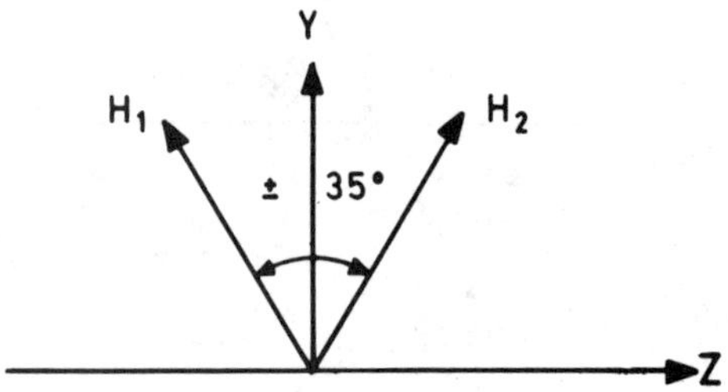

Fig. 6 Skewed Momentum Wheel Configuration

Control about the nominal axis was achieved by changing H_1 and H_2 in the same direction and control about the transverse axes was achieved by differentially controlling the two wheels.

For OLYMPUS 1 these two concepts were considered together with a zero momentum bias system, i.e. Reaction Wheels.

The momentum requirements for the three systems depend upon the level of environmental disturbance torques. Although some configurations within the OLYMPUS class will have significant gravity gradient disturbance torques, the principle source of environmental disturbance is due to solar pressure. The major contributions to the solar pressure torques are, in order

1. The large solar arrays, used for power generation

2. The deployed antennas

3. The satellite body

Detailed models of the disturbance torques were developed and a trade-off performed using two extreme satellite configurations.

1. The OLYMPUS 1 mission, comprising arrays of 3.5 kw power capability and two small (1.5m) deployed antennas

2. The OLYMPUS MAX mission, with arrays of 7.8 kw end of life power and two medium deployed antennas (3.0m).

The required angular momentum for a momentum biased system is determined mainly by the required yaw (z-axis) pointing accuracy

$$H = T_D / \psi \omega_o \quad (1)$$

where H is the angular momentum,

T_D is the disturbance torque,

ψ is the yaw accuracy, and

ω_o is the orbit rate

Reaction wheels have the capability to absorb cyclic disturbance torques and hence the angular momentum must be sufficient to absorb these and the constant disturbance torques over a period of one or more days. These constraints lead to much lower wheel angular momenta for a given level of environmental disturbance torque.

The trade-off addressed the following major areas:

1. Hardware implementation

2. Performance

3. Interfaces (Mass and Power)

4. Technical risk

5. Costs

The hardware implementation of a momentum bias system must consider techniques for disturbance torque compensation. A momentum bias system on its own must always be non-competitive for satellites in the OLYMPUS class due to the large angular momentum (and hence mass) required to achieve the yaw accuracy requirement and the mass of fuel required for periodic momentum desaturation due to the inability of this system to absorb cyclic disturbance torques. Techniques to compensate for environmental disturbance torques were considered in the trade-off study, including

1. varying the angle between the two solar arrays to produce a "windmill" torque to counteract some of the disturbance torques

2. periodic firing of thrusters, based on in-orbit estimation of disturbance torque levels, to achieve a more efficient cancellation of disturbance torques.

The performance factors considered were reliability and pointing accuracy. However these aspects, together with hardware implementation, eventually resolve into a comparison of mass and power, since the same overall performance can be achieved with sufficient hardware and adequate disturbance torque compensation techniques.

Detailed analysis of disturbance torque compensation techniques demonstrated that the zero degree of freedom momentum bias system was at least as mass efficient as the one degree of freedom system. Thus the final trade-off was between Reaction Wheels (RW) and Fixed Momentum Wheels (FMW).

Both systems required developments beyond existing European in-orbit experience. The requirement of 0.35 degree yaw accuracy leads to large angular momenta for the FMW system, especially at the top end of the OLYMPUS class. Prior in orbit experience was with wheels of 25 NMS, although 50 NMS wheels were being offered by European manufacturers.

It was decided, therefore, that the trade-off should be based on using multiple 50 NMS wheels for the FMW system. Magnetic suspension offers improvements in reliability (and hence reduced number of wheels). However this was rejected due to the higher risk involved in such a development.

Reaction wheels have not been flown on long life European missions previously. However, considerable in orbit and ground test experience exists in the United States and in addition, a ball bearing reaction wheel was under life test at British Aerospace and is currently in its fifth year of successful operation.

Although gyroscopes have been flown on previous European scientific and communications satellites, no manufacturer had established 10 years continuous operation. The OLYMPUS gyroscope manufacturer, Ferranti, have had 12 gyroscopes on continuous life test since early 1976. To date they have continued to operate without failure. This base of ground test results minimised the risk associated with a long life zero momentum control philosophy.

Thus, while both systems involve some degree of technical risk if they are to be capable of controlling the large and powerful configurations within the OLYMPUS class of satellites the level of risk has been reduced by relying upon on-going product development.

The final trade-off, therefore, was based solely on mass and cost. The differences in power consumption of the two systems was represented in terms of solar array mass required to provide the additional power. The costs considered were a balance of non-recurring and recurring costs. Figs. 7 and 8 show the final result, with mass and cost for the two systems plotted against surface area of arrays and antennas.

These charts infer that for satellites at the small end of the OLYMPUS class there is little to choose between the two systems. However, in order to meet the more demanding requirements of the larger payloads the Reaction Wheel system has a clear advantage in both mass and cost. Thus the Reaction Wheel system was selected for the OLYMPUS AOCS.

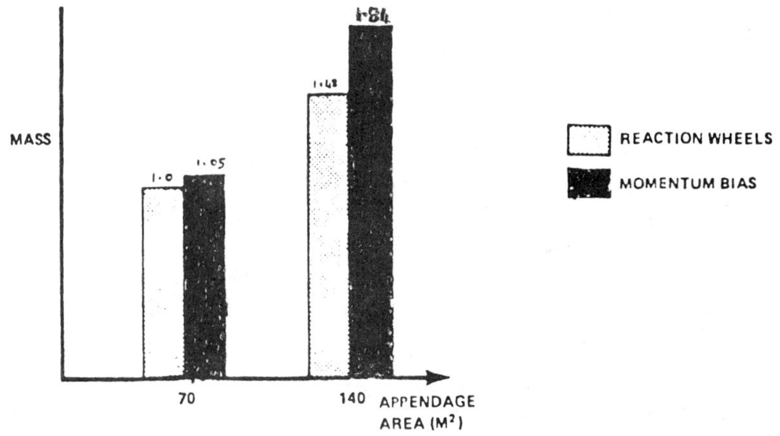

Fig. 7 Relative Mass

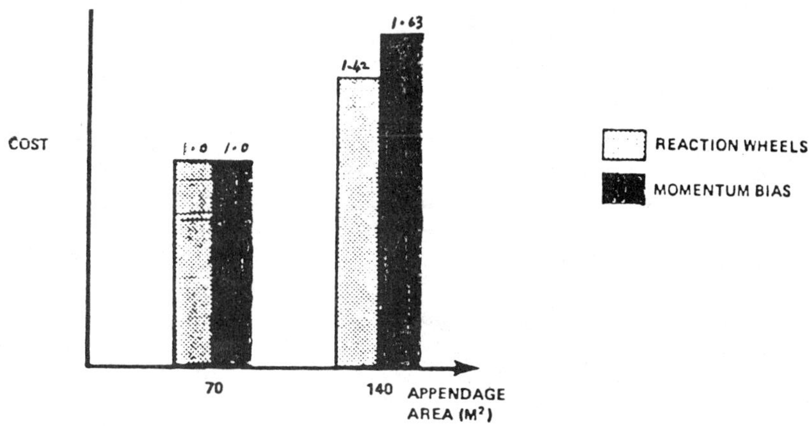

Fig. 8 Relative Cost

SYSTEM DESCRIPTION

The OLYMPUS AOCS is shown schematically in Fig. 9. It comprises the following main elements.

1. Control Electronics Unit (CEU), based on a Texas Instruments 9989 microprocessor communicating with internal and external units via a serial digital data bus.

2. Infra-red Earth Sensors (IRES), for attitude measurement during normal operation.

3. Digital Sun Sensors (DSS), for transfer orbit attitude determination and gyroscope calibration.

4. Sun Acquisition Sensor (SAS), used during the emergency back-up mode (Sun pointing).

5. Two Gyroscope Packages (GYP) together with their associated electronics (GYE), for inertial attitude measurement in the transfer orbit and yaw measurement on-station.

6. Four Reaction Wheels, together with the Wheel Drive Electronics (WDE) for operational control.

7. Thruster Drive Electronics (TDE), operating 16 bi-propellant Reaction Control Thrusters (RCT's) for Transfer Orbit Control, Station Keeping and Momentum dumping.

8. Two Solar Array Drive Mechanisms (SADM's), together with the Solar Array Drive Electronics (SADE) and four Solar Array Sun Sensors (SASS's). These are used to maintain the solar array sun pointing throughout the mission in order to maintain maximum power.

These elements are further described below.

Control Electronics Unit

The CEU is microprocessor based, using the Texas 9989 Microprocessor. The Executive, Sensor Processing, Attitude Control and Fault Handling are implemented in software using 16 K PROM and 16 K RAM. Whilst the PROM is almost fully utilised, the RAM contains spare capacity for in-orbit reprogramming operations.

Communication with the microprocessor module is via a Serial Digital data bus. This uses a 32-bit message structure operating at 100 K bits/sec and is linked to the CEU internal modules in addition to the DSS, IRES, GYE and WDE. Integrity of the data bus is maintained by a Bus Monitor, which automatically detects and clears bus errors.

The CEU also performs a central TM/TC function for the AOCS and controls the power switching of all AOCS units.

The CEU is internally redundant with all back-up modules operating in cold standby mode.

The unit is being designed and manufactured by British Aerospace.

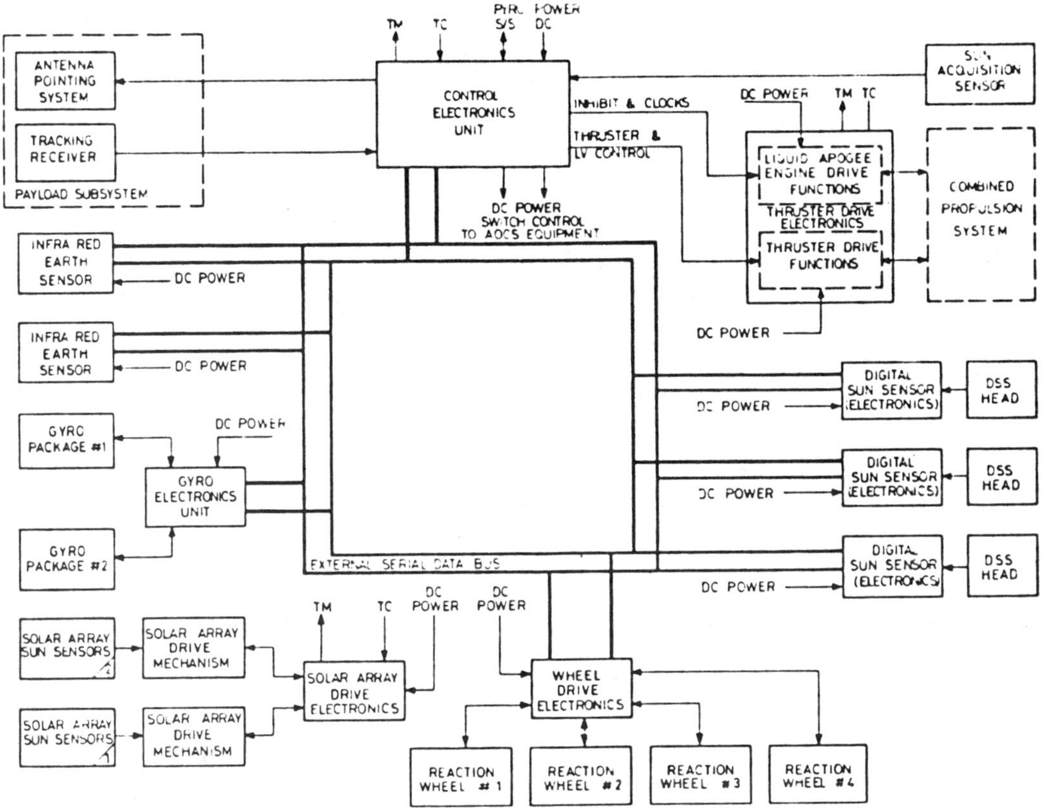

Fig. 9 AOCS Block Diagram

Infra-Red Earth Sensor

The IRES comprises 4 bolometers viewing the Earth via a mirror which is oscillated at 10 Hz, thus providing a scan of the Earth's Infra-red profile. It is manufactured by Galileo, Italy, and is based on a sensor design previously flown on OTS, ECS and MARECS.

The sensor determines the proportion of the scan which views the Earth for each bolometer and uses three of these four measurements to derive roll and pitch attitude data. Major performance characteristics are given in Table 2.

Digital Sun Sensor

This two-axis device, manufactured by ADCOLE, USA, has flown on many previous spacecraft. The sensor outputs two digital counts which are related to the sun angle by a complex trigonometric function. This function is evaluated by the CEU.

The sensor has two apertures for each axis. Behind the coarse aperture is a Grey Code pattern of solar cells which determines the sun angle to approximately 2 degrees. The fine aperture is used to give a more precise (0.017°) sun angle determination.

Sun Acquisition Sensor

Again, this is a space proven design, manufactured by TPD/TNO, Holland and MATRA, France. It is used to derive coarse sun position for use in the Initial Sun Acquisition and Emergency Acquisition Modes of operation. It comprises 4 (redundant) solar cells arranged in a pyramidal configuration. The SAS has a field of view in excess of 2 steradians. Two sets of outputs (prime and redundant) are produced, each giving an analogue representation of the Sun's position in two axes.

Gyroscopes

Ball bearing rate gyroscopes, manufactured by Ferranti, Scotland, have been selected for OLYMPUS. Two separate packages have been provided, the skew pack and the yaw pack. The skew pack is used in the transfer orbit phase and for roll and pitch rate measurement during station keeping. It comprises 3 orthogonal gyros with a fourth skewed equally to each of the other three. The yaw pack contains four gyros aligned with the yaw axis. It is used during the operation phase for yaw angle determination.

The gyroscope electronics outputs the average rate over 100 msecs for each operational gyroscope. This is then processed by the CEU to obtain both angle and rate data.

Reaction Wheels

Four reaction wheels have been provided, three aligned orthogonally and the fourth skewed equally to the other three. Each reaction wheel has an angular momentum capacity of 10 NMS and a nett output torque capability of 0.1 NM. The wheels are based on an existing British Aerospace design which has been on life test for 5 years.

Solar Array Drive

The large solar arrays envisaged for the OLYMPUS class require a powerful SADM, both in terms of torque capability and the power transfer capability. In order to minimise the disturbance to the satellite position a novel system has been developed by British Aerospace. The mechanism drive comprises a stepper motor and a gear box. A smooth drive is achieved by varying the power inputs to the stepper motor in a sinusoidal manner. Each SADM is capable of being driven independently at rates from 1 revolution/day to 0.2 degrees/second.

Thruster Drive Electronics

The TDE provides the interface between the CEU and the Reaction Control Thrusters. It is manufactured by Aeritalia, Italy. Additionally, the TDE provides the drive and timing circuitry for the Liquid Apogee Engine, used during the orbit change manoeuvre, and provides telemetry conditioning for the pressure transducers and latch valves.

MISSION SUMMARY

The AOCS comprises 11 modes of operation, all implemented in the CEU. These modes are dedicated to specific phases of the mission. With the exception of Emergency Sun Acquisition all modes are implemented in software. The OLYMPUS 1 mission and the associated control modes are summarised below.

Orbit Injection

When OLYMPUS is separated from the final stage of the launch vehicle, the arrays are stowed and the negative yaw axis is nominally pointing towards the sun direction. However, in the first 15 minutes the Sun is obscured by the Earth and so the spacecraft tumbles with the rates initiated by the separation. During this phase the CEU performs various initialisation functions, including priming the Propulsion subsystem. This mode is called the Automatic Initialisation Sequence (AIS).

Initial Sun Acquisition (ISA)

Fifteen minutes after separation the ISA mode is initiated. This mode first reduces the rates induced by separation and then points the negative yaw axis to the Sun using the SAS, gyros, and DSS's for measurement and the Reaction Control Thrusters (RCT's) for control. At the end of this manoeuvre the Solar Arrays are deployed by ground command,

Gyro Calibration

In order to perform the manoeuvres later in the transfer orbit it is necessary to calibrate the drift and scale factors of the skew pack gyros.

The drift rates of the roll and pitch gyros are calibrated whilst the negative yaw axis is pointing to the Sun. This mode is called -Z Sun Lock (ZSL) and is basically the same as ISA except that the rate reduction and manoeuvre logic is omitted. The SAS is not used in this mode for attitude measurement. The drift rate of the yaw gyroscope is calculated by pointing another spacecraft axis towards the Sun and maintaining this attitude. The roll axis is selected since a Digital Sun Sensor is positioned on this axis. The attitude is achieved and maintained using the Inertial Manoeuvres and Lock (IML) mode. This is a purely inertial loop which accepts demanded acceleration profiles to perform manoeuvres and subsequently lock onto the final position.

Gyro scale factors are calibrated by performing manoeuvres, within the fields of view of the various DSS's using IML. The difference between the manoeuvre angle determined by the DSS's and that determined by the gyroscope is a measure of the gyro scale factor.

Gyro Initialisation

The gyro integration algorithms require to be initialised periodically in order to remove the effects of residual gyro drift errors. This is performed when the spacecraft's yaw axis is pointing towards the Earth and the Sun is within the field of view of one of the DSS's. This attitude is acquired by a two stage process.

1. At an appropriate point in the orbit a coning motion is set up which should scan the IRES field of view across the Earth. This mode is called Earth Search (ERS).

2. When the IRES indicates the Sun is within its field of view control is switched to the Sun-Earth Acquisition and Lock (SEA) Mode which simultaneously locks onto the Earth and the Sun.

In this configuration the gyro integration algorithm may be initialised.

Apogee Engine Firing (AEF)

At the appropriate point in the orbit, the spacecraft is manoeuvred such that the LAE is aligned in the correct orientation. IML Mode is used to achieve this. A variation of IML is used to control during AEF. This variation allows the thrust direction to be steered at a uniform rate in two axes during the orbit change manoeuvre. This thrust vectoring allows a more efficient orbit change.

During AEF the Solar Arrays are in the deployed configuration. The acceleration produced by the LAE causes the array centre of mass to shift. This combined with misalignment of the LAE results in relatively large disturbance torques ($\approx$ 4 NM). In order to maintain the torques at an acceptable level, the Solar Array is aligned such that the thrust vector is in the plane of the solar array and maintained in this position throughout the manoeuvre.

Earth Acquisition and Lock (EAL)

Following AEF the operational Earth-pointing configuration must be achieved. This is implemented by IML using a sequence of manoeuvres. When the yaw axis is pointing to the Earth control is transferred to the EAL loops, using the IRES for roll and pitch measurement and a yaw gyro. Control is implemented using the RCT's.

Normal Mode (NRM)

The vast majority of the spacecraft life will be spent in the NRM. This is a fine Earth-pointing mode utilising the IRES for roll and pitch measurement and the gyros for yaw measurement. Control is effected by the reaction wheels.

Periodic updates of the yaw angle are derived from the DSS's.

Station Keeping Mode (SKM)

Every few weeks the spacecraft's orbital position is adjusted (either East-West or North-South) using the RCT's to apply a force through the centre of mass. Inevitably the disturbance torques during this mode are relatively high, requiring a separate control mode (SKM). This mode is similar to IML but the roll and pitch gyros are only used for rate measurement (the IRES providing angle measurement). The yaw gyro is used for both angle and rate measurement.

Back-up Operations

The CEU contains on-board logic to identify and contain a large number of possible failure modes. Whilst these are not the subject of this paper it is worth mentioning the ultimate back-up mode - Emergency Sun Acquisition (ESA). This uses the SAS to achieve a Sun pointing position when the operational modes (SKM, NRM or EAL) are unable to maintain an Earth pointing attitude. ESA is implemented in hardware and does not require the microprocessor for successful operation.

TRANSFER ORBIT

Scope

Space precludes a complete description of all the transfer orbit modes of operation, hence this section concentrates on two major aspects

1. Gyro integration

2. Inertial Manoeuvres and Lock

The function of the other transfer orbit modes has already been outlined. The loop structures of these other modes are almost identical (Fig. 10). They consist of position and rate information being combined and compared with a deadband. When this deadband is exceeded an RCT of the appropriate sign is fired for a fixed on-time. (Some modes contain logic to increase this on time during the acquisition stages).

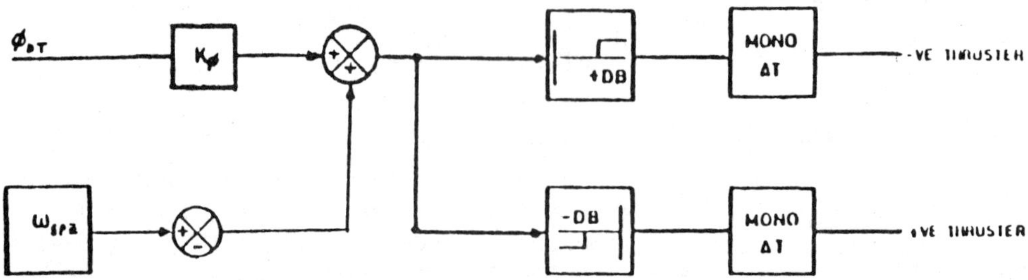

Fig. 10 Typical Transfer Orbit Control Loop

Gryoscope Integration

The gyroscopes measure the average rate over a precise period of 100 msecs. The choice of integration algorithm is trade-off between the accuracy of attitude determination and the complexity of implementation. The simplest method would be to multiply the gyro output by the update period to give angular increment. Thus the algorithm for gyro integration would become

$$\theta_{i+1} = \theta_i + \dot{\theta}_{g\,i+1}\,\Delta T \tag{2}$$

where θ_i is the angle estimate at the ith time point

$\theta_{g\,i}$ is the gyro output at the ith time point

ΔT is the gyro update period

This assumes that the second order terms in the Euler kinematic equations can be neglected. However, to achieve the required accuracy in setting up the initial AEF attitude it was concluded that this approach could not be adopted without excessive fuel consumption induced by maintaining very tight deadbands. More accurate integration techniques were considered, including

- o full Euler kinematics
- o direction cosines
- o quarternions

The Gibbs vector formulation of the quarternion equations was selected as it provided the best compromise between accuracy, complexity and flexibility of operation.

Gibbs Vector Implementation

The Gibbs vector, $\underline{g}$, consists of a scalar, s, and a vector, $\underline{r}$:

$$\underline{g} = \underline{e}/s \tag{3}$$

The kinematic equations are

$$\dot{s} = -(\underline{\omega}-\omega_s) \cdot \underline{r}/s \tag{4}$$

$$\dot{\underline{r}} = s(\underline{\omega}-\underline{\omega}_s)/2 + \underline{r} \wedge (\underline{\omega}+\underline{\omega}_s) \tag{5}$$

Equations (3) to (5) are combined to give

$$\dot{g} = [\underline{\omega}-\underline{\omega}_s + g \wedge (\underline{\omega}+\omega_s) + g(g \bullet (\underline{\omega}-\underline{\omega}_s))]/2 \tag{6}$$

where $\underline{\omega}$ is the angular rate vector

$\underline{\omega}_s$ is the demanded angular rate

Demanded rates, ω_s, are implemented via a demanded acceleration profile for each axis (see Fig. 11).

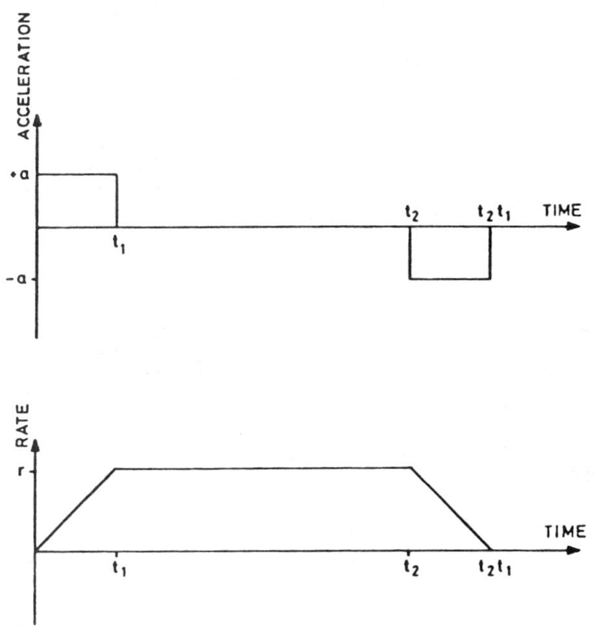

Fig. 11 Demanded Acceleration and Rate Profiles

Equation (6) is implemented in an incremental form

$$
\begin{aligned}
g_{i+1} &= g_i + \Delta t\, \dot{g}_{i+1} \\
&= g_i + [\underline{\delta\xi}_{i+1} - \delta\xi_{i+1} + g_i \wedge (\underline{\delta\xi}_{i+1} + \underline{\delta\xi}_{si+1}) \\
&\quad + g_i (g_i \cdot (\underline{\delta\xi}_{i+1} - \underline{\delta\xi}_{i+1}))]/2
\end{aligned}
\qquad (7)
$$

where $\underline{\delta\xi}_i$ is the angular increment measured by the gyros (see Eqn. (2) at the ith time point

$\underline{\delta\xi}_{si}$ is the incremental angular slew demand (derived from Fig. 11) at the ith time point.

This is shown diagramatically in Fig. 12.

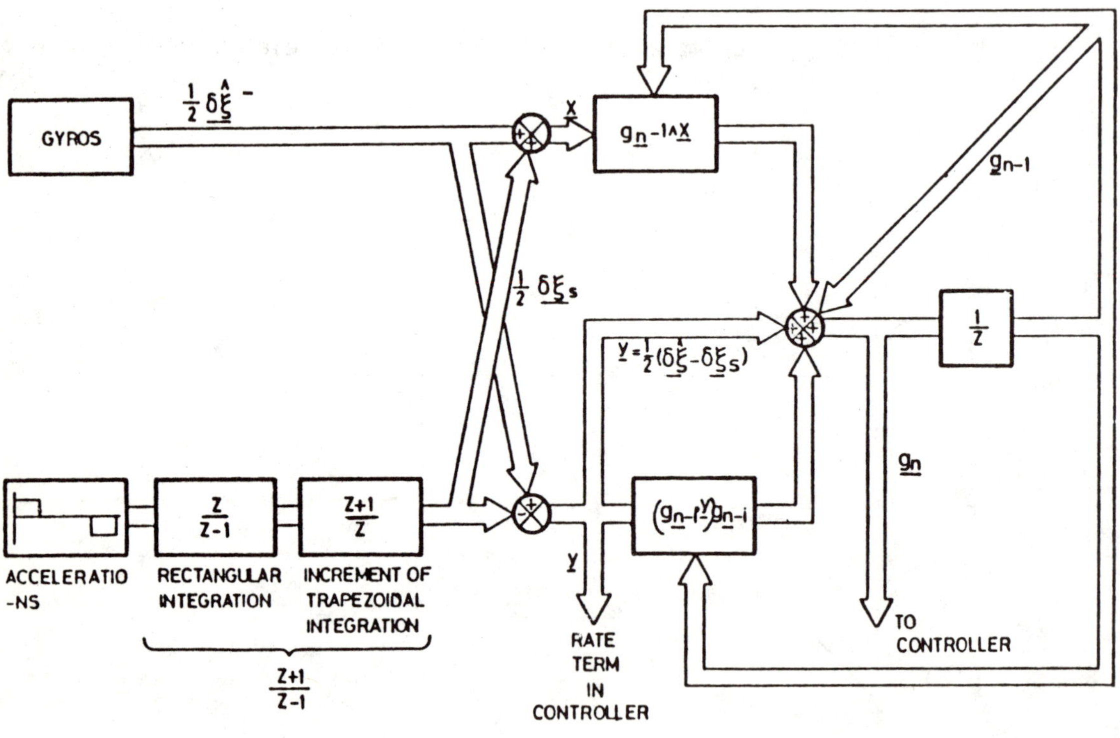

Fig. 12 Gibbs Vector Implementation

Inertial Manoeuvres and Lock

The majority of the transfer orbit will be spent in this mode (or a derivative of it used during AEF). The gyroscopes are the only sensors used in this mode and their data is processed using the Gibbs vector implementation described above.

The control loops are shown in Fig. 13. The inputs to the control loops are the rate vector $\underline{y}$ $(=(y_1, y_2, y_3)^T)$ and the Gibb vector $\underline{g}$ $(= (p, q, r)^T)$ shown in Fig. 12.

The form of control is proportional plus integral plus derivative. A second order filter is included because of the presence of high order modes in the solar array dynamics.

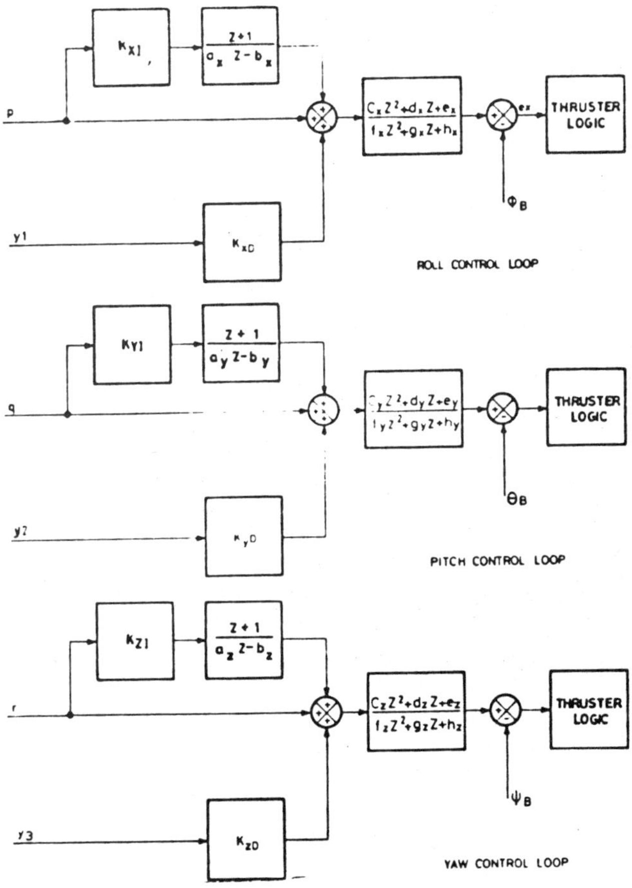

Fig. 13 IML Control Loops

The thruster logic has a two stage deadband. At the start of a manoeuvre the initial error will be large, and so a relatively long on-time is selected. When the errors have been reduced a smaller on time is used to achieve finer control. The logic also contains the ability to inhibit thruster firings for an integral number of loop update periods (100 msecs). This facility is included to avoid firing the RCT's at critical frequencies. Fig. 14 shows the logic for the roll axis. (The other axes are similar).

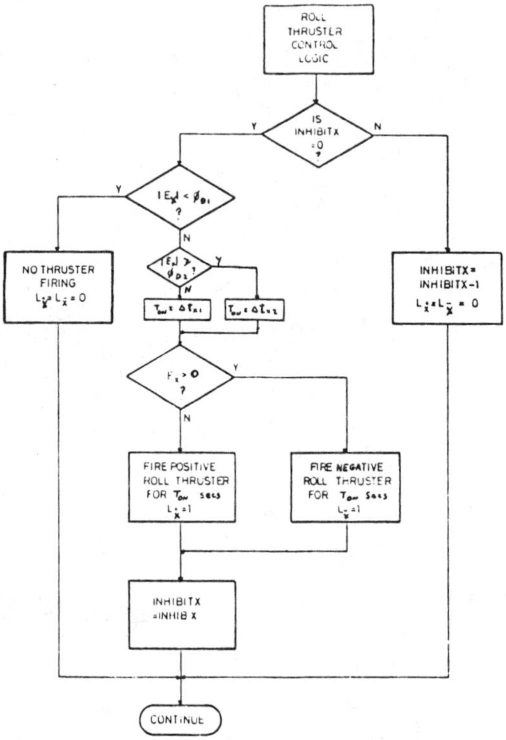

Fig. 14 IML Thruster Logic

Typical simulation results are given in Fig. 15 (Roll axis manoeuvre) and Fig. 16 (AEF). Fig. 15 shows the results of a demanded manoeuvre of 25 degrees about the Roll axis. It also shows the relative angular errors derived from the Gibbs Vector implementation. These can be seen to be well within the Transfer Orbit pointing accuracy requirement of ± 1 degree.

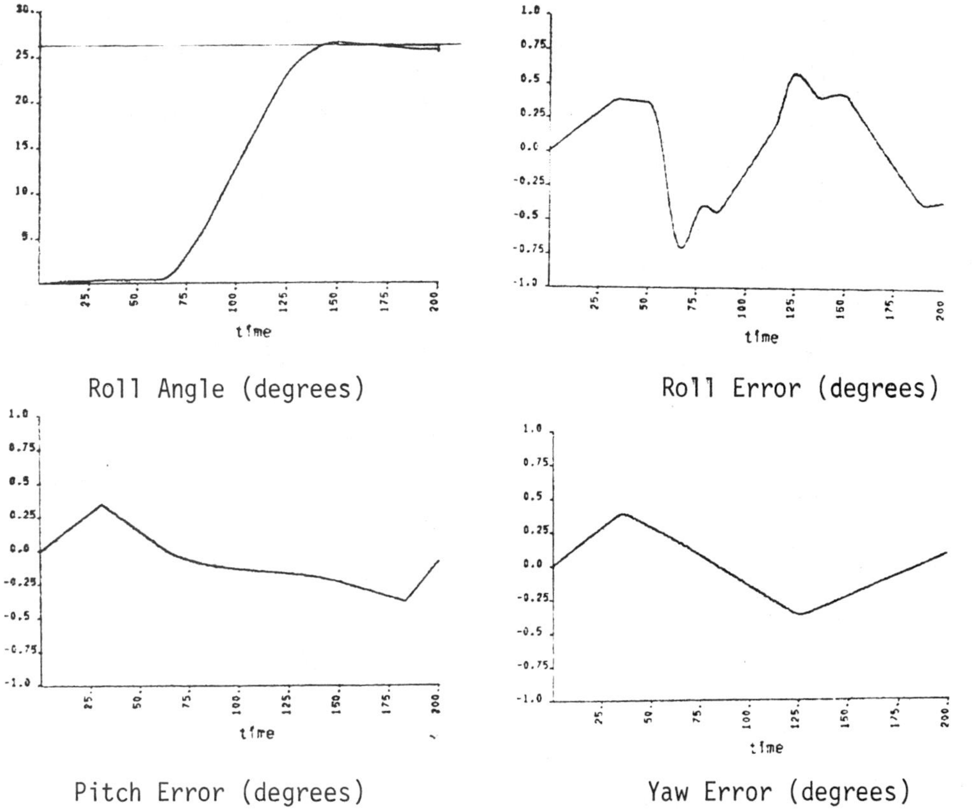

Fig. 15 Roll Manoeuvre Simulator Results

The Apogee Engine Firing Simulation (Fig. 16) represents a manoeuvre during which the thrust vector is steered about the pitch axis at a rate of approximately 9 degrees/hour. The large disturbance torques are seen to be well controlled after a small initial transient. The accuracy requirements during this mode are to set up and maintain the thrust vector within 0.8 degrees of its demanded orientation. The achieved accuracy, including errors associated with setting up the initial attitude, is better than 0.5 degrees, representing a significant saving in fuel during the manoeuvre. This accuracy is attributable to the use of the precision gyroscopes and the Gibbs vector integration method.

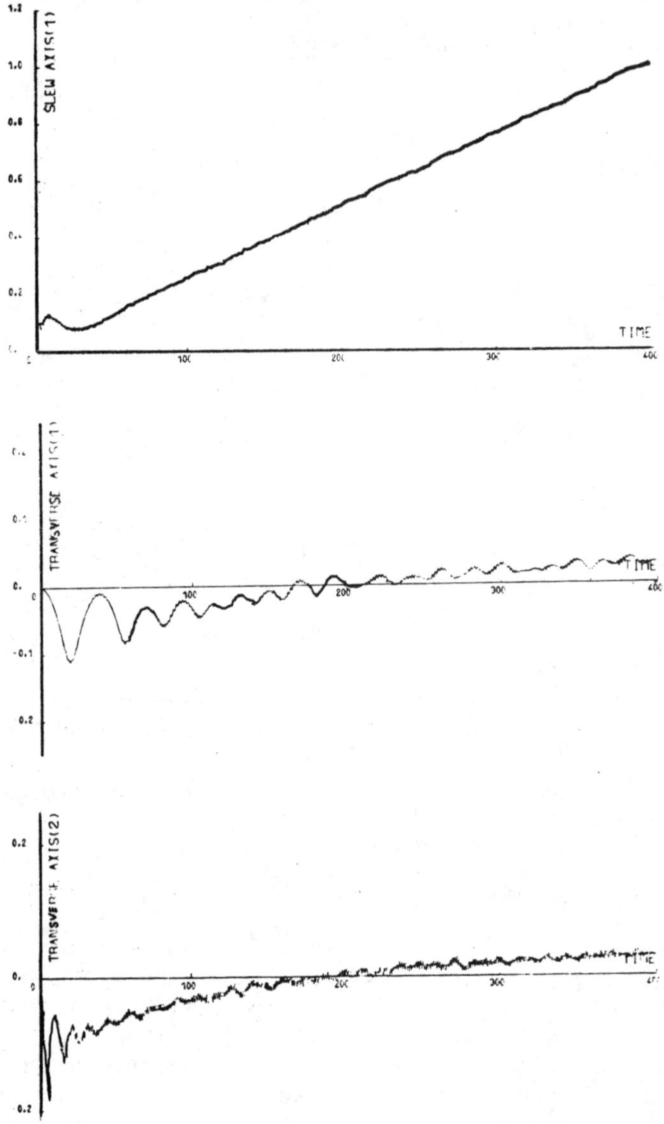

Fig. 16 AEF Simulator Results

GEOSYNCHRONOUS ORBIT

Earth Acquisition and Lock

This mode is used shortly after injection into the geosynchronous orbit, prior to the satellite reaching the required station (17° West for OLYMPUS 1). It is also used as a back-up mode whilst in-orbit failures are being rectified. Control is effected with the RCT's as the pointing requirements are relaxed, 1° in all axes being adequate.

Because of its dual role, the normal operational sensors are used (IRES and yaw gyroscopes). This means that rate information for roll and pitch must be derived. The control loops are shown in Fig. 17 (Roll and Pitch) and Fig. 18 (Yaw).

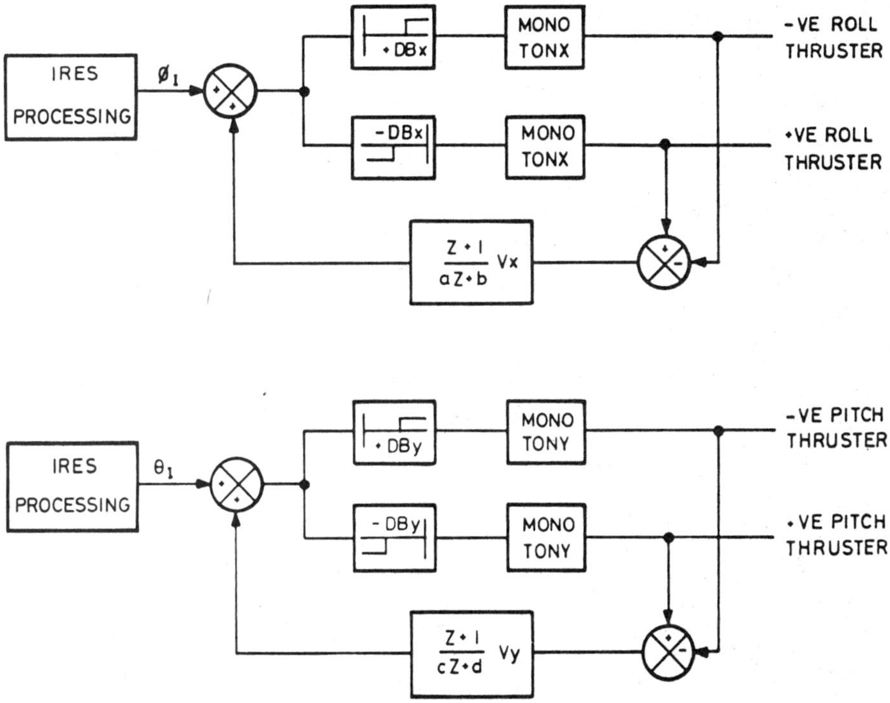

Fig. 17 Roll and Pitch EAL Control Loops

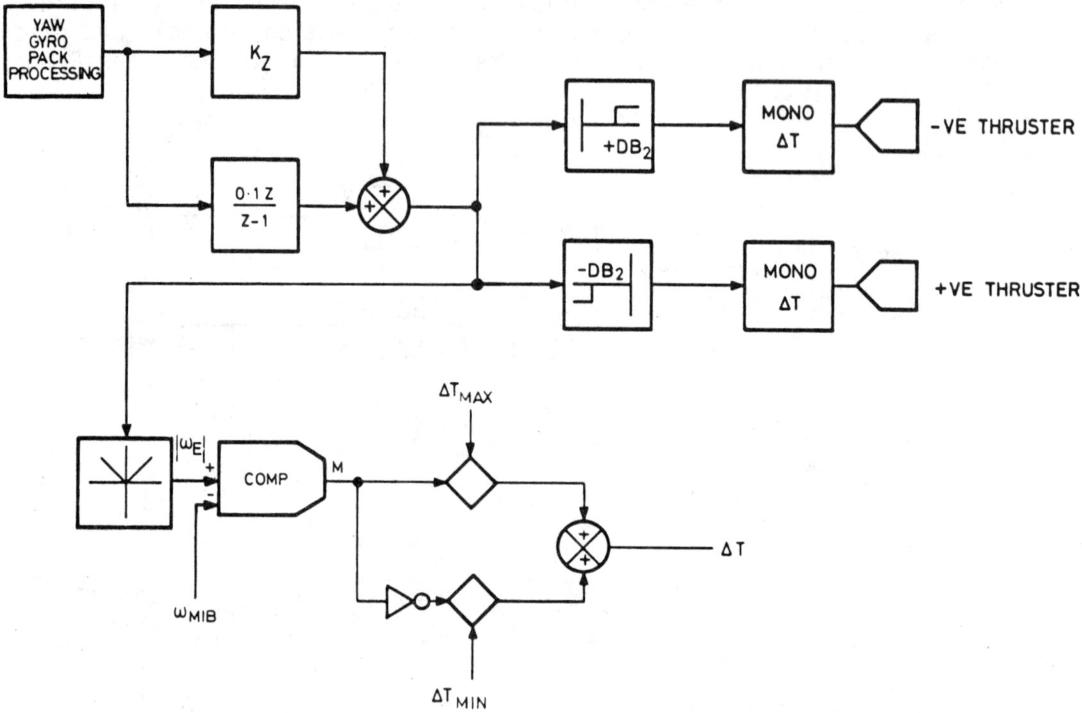

Fig. 18 Yaw EAL Control Loops

The pitch axis acquisition phase is shown in Fig. 19.

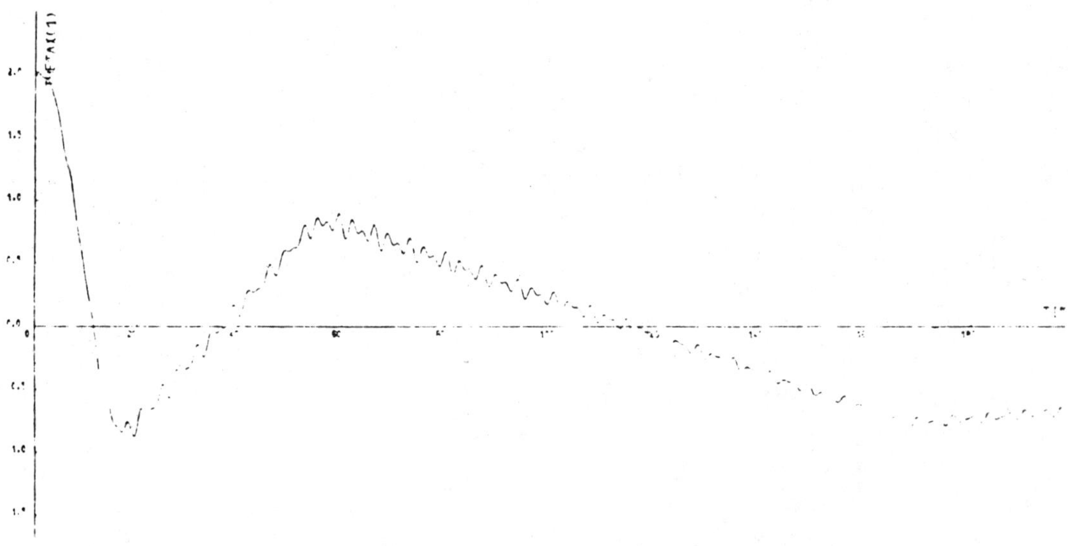

Fig. 19 Pitch Axis EAL Simulation Results

Normal Mode

The NRM control loops are shown in Fig. 20. The Roll and Pitch axes normally derive angular information from the IRES. However the OLYMPUS 1 direct TV broadcast payload includes a Radio Frequency Sensor (RFS). The AOCS has the facility to derive Roll and Pitch attitude either from the IRES or the RFS. Whichever source is used, the CEU uses a low pass filter to reduce noise level (not shown in Fig. 20) and has the capability to apply attitude offsets to the filtered sensor outputs. This latter capability could have applications during payload commissioning when offset pointing may be required.

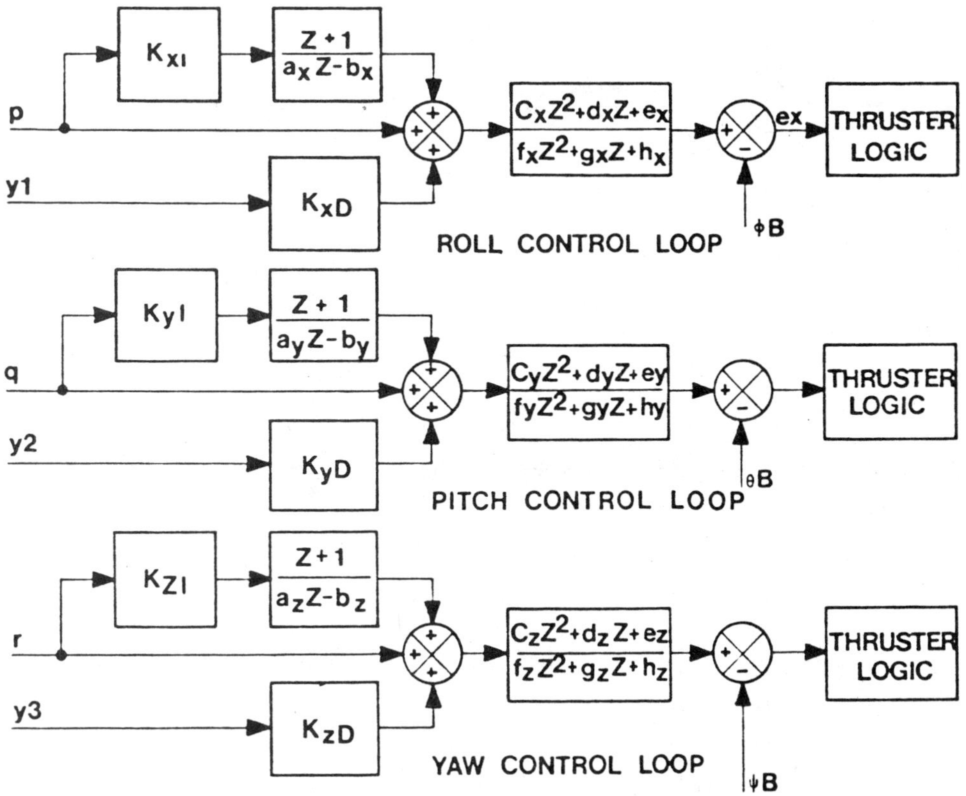

Fig. 20 Normal Mode Control Loops

The main control term is second order for the pitch axis and third order for the roll and yaw axes. An integrator is included to offset the reaction wheel friction torques. Since these torques are a function both of wheel speed and temperature an additional term, proportional to wheel speed, is included.

During momentum dumping, which is implemented using the RCT's, a feed forward term is included prior to the integrator. This reduces the attitude transient during dumping. A sequence of pulses equivalent to 0.3 to 0.4 NMS separated by several minutes is used to desaturate the wheel. This operation will be performed daily. Fig. 21 shows that the performance requirements are achieved during wheel desaturation.

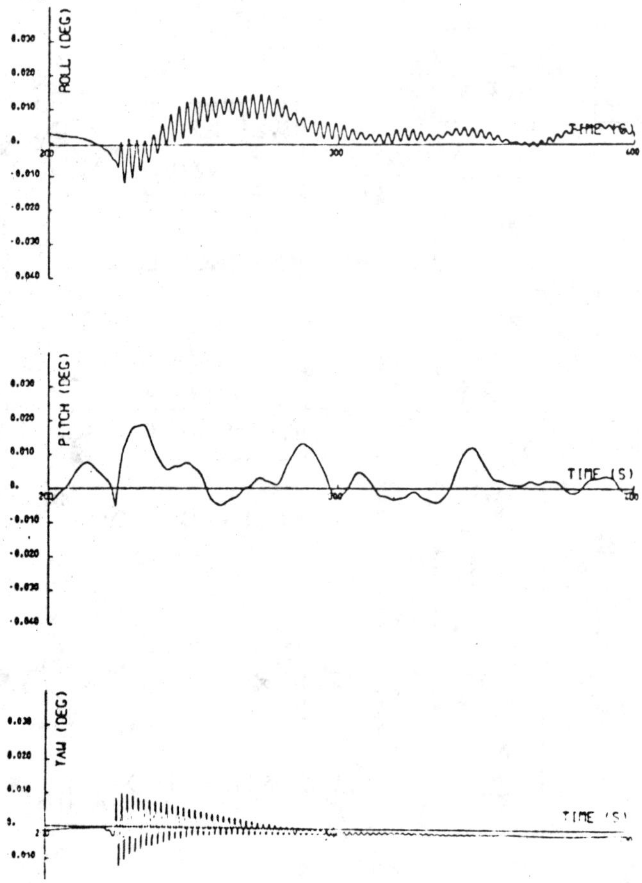

Fig. 21 Roll Reaction Wheel Desaturation Simulation

Since this loop has to operate in transient conditions (after a station keeping manoeuvre) it is necessary to limit the output of the main control term to avoid lengthy settling times and to increase the acquisition range. Fig. 22 shows the results of a transition from SKM to NRM.

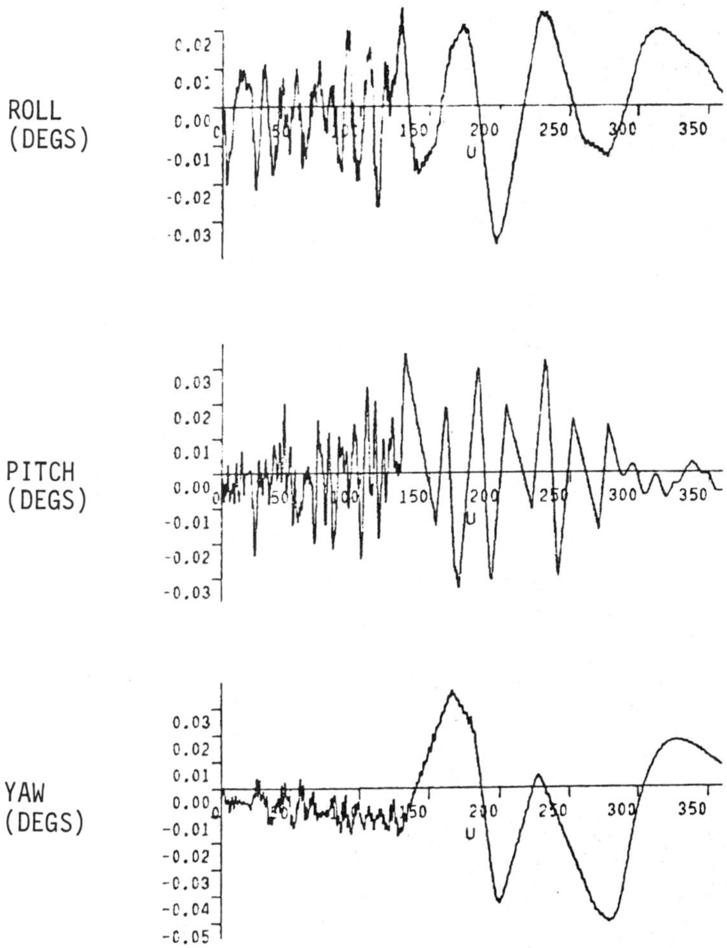

Fig. 22 Transition from SKM to NRM

Both Fig. 21 and Fig. 22 show that the roll and pitch accuracy requirements (±0.04 degrees) and the yaw requirement (±0.35 degrees) are achieved.

Station Keeping Mode

Station keeping has traditionally been the most difficult control problem for 3-axis stabilised communications satellites. The relatively high disturbance torques caused by misalignment of the RCT's and impingement of the RCT exhaust plume on the satellite structure (primarily the solar array) has dictated the need for a separate, high bandwidth control loop.

Early studies established that previous designs (e.g. as used on the Orbital Test Satellite and the European Communications Satellite) based on angular measurement alone would not be adequate to compensate the relatively low array bending frequencies (0.035 Hz) and high disturbance torques (4 NM) anticipated for the OLYMPUS class of satellites.

Initial trade-offs established the requirement for position and rate measurement and the first designs were based on a Luenberger Observer. The concept was to separate the rigid body dynamics from the flexible dynamics using the Observer and to actively control the rigid body mode, leaving natural damping to control the flexible array.

Detailed parametric analyses were performed which established the ability of the design to cope with a large range of array parameter uncertainties (± 25% in frequency and ±30% in modal gain) and rigid body inertia uncertainties (± 15%). However, the uncertainties in the RCT performance, and the associated disturbance torques, were such that the performance requirements (0.04° roll and pitch, 0.35° yaw) could not be achieved during the initial transient. After detailed comparative performance analysis the control loops shown in Fig. 23 (Roll and Pitch) and Fig. 24 (Yaw) were selected. These comprise a proportional plus integral plus derivative controller with a second order filter to remove higher order array mode interactions. The control loop structure is very similar to that used for IML and AEF. It should be noted that during North-South station keeping the yaw thrusters are used for the manoeuvre. Thus yaw control is achieved by switching off one of the thrusters for a preset time. Typical simulation results are given in Fig. 25.

These results demonstrate achievement of the pointing accuracy requirements of ±0.04 degrees in roll and pitch and ±0.35 degrees in yaw.

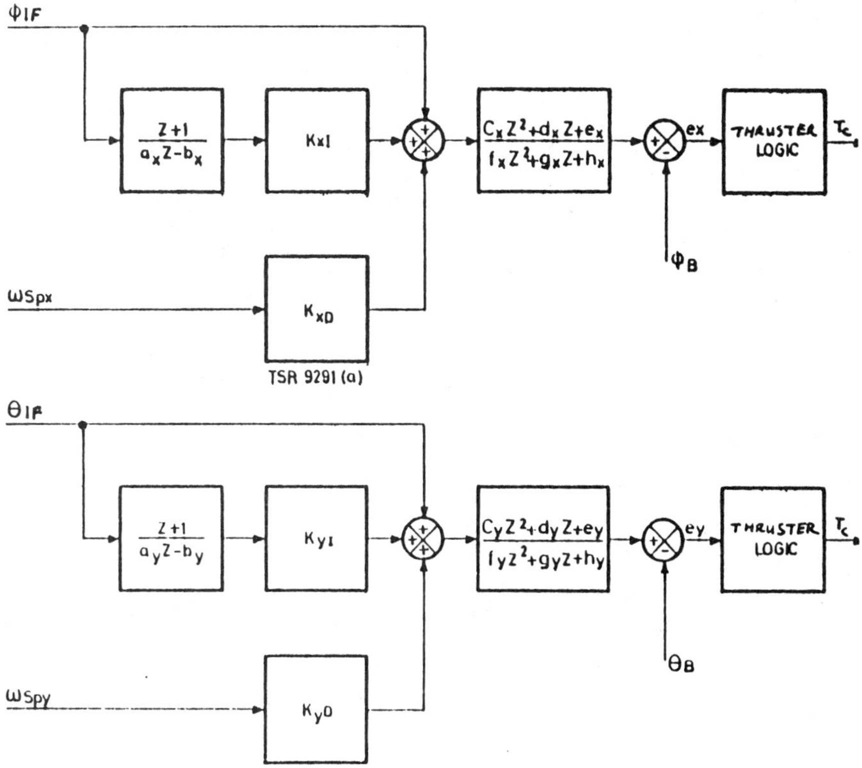

Fig. 23 SKM Roll and Pitch Control Loops

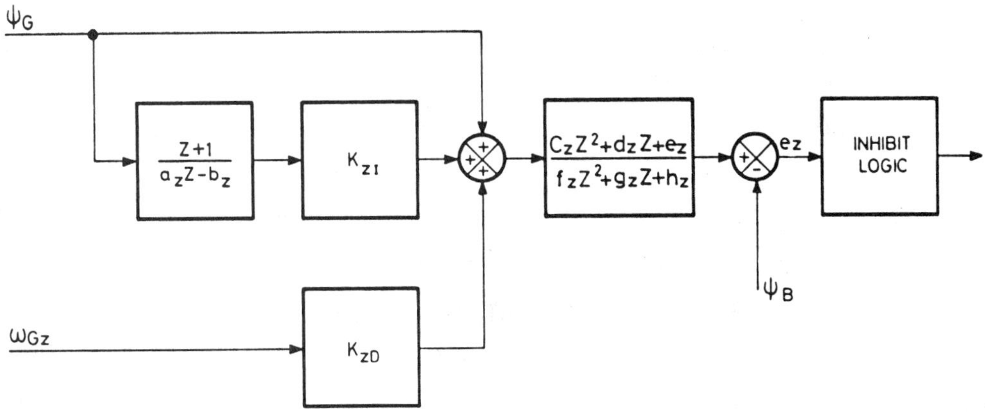

Fig. 24 SKM Yaw Control Loops

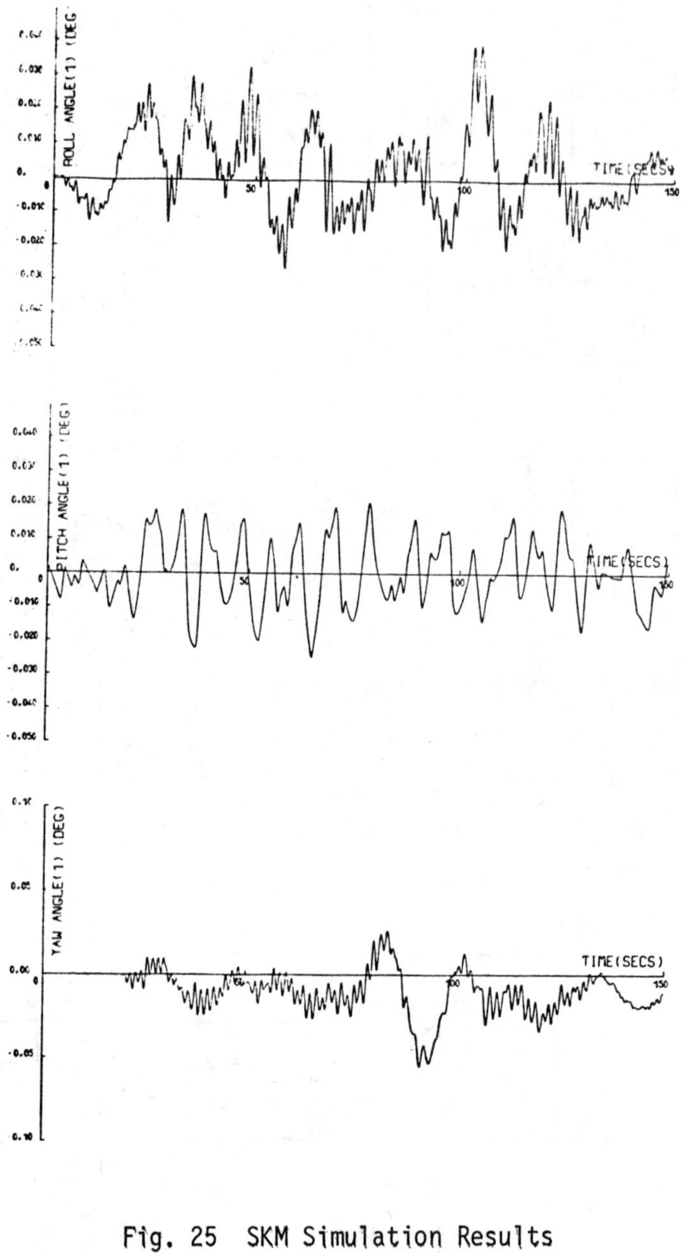

Fig. 25 SKM Simulation Results

CURRENT STATUS

At the time of writing Engineering Models for the AOCS units are in the process of manufacture and test. A combined subsystem performance test will be conducted in the third quarter of 1984. The purpose of this test will be to verify predicted performance using the actual AOCS units and a simulation of the spacecraft dynamics.

The subsystem design and analysis is therefore essentially complete and a major design review will be held in Spring 1984.

ACKNOWLEDGEMENTS

This paper is based on numerous technical notes and working papers produced by members of the OLYMPUS Control Subsystem design team. The author acknowledges the major contributions to this design made in particular by N. Bingham, S. Colling, J.A. Lewis, R. Marley and R.C. Rogers.

The author also wishes to acknowledge the helpful comments and criticisms made by members of the European Space Agency, in particular Dr. C. Markland and Mr. D. Wilson.

AAS 84-002

EUROSTAR MULTIMISSION PLATFORM ATTITUDE AND ORBIT CONTROL SUBSYSTEM

Roland Oskian*, Jean-François Poussin**
B. Govin+

The EUROSTAR program is a multimission platform for communication satellites, under development by SATCOM International, a joint BAe/MATRA company. It is an enhanced design, compatible with both Ariane and Shuttle launch, issued from the ECS generation flight proven spacecraft.

The AOCS requirements derived from the objectives cover a large range of applications, with minimum modifications and follow a design to cost approach. The AOCS utilizes spin stabilization with active nutation control during parking and transfer orbit operations; During all geosynchronous operations, the AOCS consists of a biased momentum three-axis control system. A digital decentralized architecture with permanent microprocessor and digital data bus is implemented. Growth capabilities can be provided.

An overview of the EUROSTAR platform, and associated requirements on AOCS is given. The main design features of the AOCS are introduced, a modes description and operation is provided, followed by a presentation of electrical architecture and equipment.

* *Engineer, Dynamics and Control Department, MATRA Space Division - B.P. N°1 - 78146 Vélizy Cedex, France*

** *Responsible for new generation Geostationary Attitude Control System team, MATRA Space Division*

\+ *MATRA Space Division*

OVERVIEW OF THE EUROSTAR PLATFORM

EUROSTAR is a multimission platform, designed as a standard vehicle for geostationnary telecommunication satellites. The development is under control of SATCOM INTERNATIONAL, a joint BAe/MATRA company. British Aerospace Dynamics Groups and MATRA have a large high technology background, and they did obtain contracts for a family of communications satellites: OTS, ECS, MARECS, TELECOM 1, SKYNET 4. Important contributions have also been made to all the Intelsat programs beginning with INTELSAT 3, and extending to INTELSAT 6.

EUROSTAR is a medium scale spacecraft which complements the highly successful OTS, ECS satellite series, first orbited in 1978, and the large scale Olympus series spacecraft.

The EUROSTAR platform is compatible with requirements of several missions requirements, among them:

- o high density trunk telephony on data. Cable TV. (4.11.12. GHz)
- o corporate communications at 11 - 12 GHz
- o direct television broadcast at 12 GHz
- o inter city links at 20 GHz
- o mobile communications at L. band.

The EUROSTAR maximum RF power from the payload is 700 Watts, with 10 year life and full eclipse capability.

The spacecraft is compatible both with STS, using PAM D or PAM D2 assist engines, and with ARIANE III and ARIANE IV, using the SPELDA double launch system.

The mass in transfer orbit will be from 1200 Kg up to 1800 Kg.

As a result of system trade off, the main design features are:

- o three-axis body stabilized on station
- o spin stable configuration in transfer
- o north-south solar arrays, with antennas on 3 sides: body, east-west walls
- o combined biliquid propulsion, with four large surface tension tanks, twelve (10 N) thrusters and a large apogee engine (400 N)
- o power subsystem is sun-regulated (nominally: 42.5 volts) and eclipse non-regulated (full regulation is proposed as an option)
- o a carbon-fiber structure
- o Electrical architecture is decentralized; The telemetry telecommand is distributed through a data bus (OBDH * type) controlled by a central interface unit.

The platform is designed for 10 year life duration in orbit, and 5 year on-ground storage.

EUROSTAR AOCS OBJECTIVES

The AOCS is under responsibility of MATRA, space products and technology division, who are already involved in AOCS design and development of previous European telecommunication programs (OTS, ECS, MARECS, TC1). Being designed for the commercial satellites market, the AOCS must use previously flight proven concepts, and critical equipment will be in qualified technologies. A cost efficient approach must be applied, both for equipment recurring costs, and for non-recurring mission adaptation costs.

AOCS must be designed so as to limit ground control to a minimum of telecommands, especially for normal mode during which it must be compatible with gound station unavailability for up to six hours.

It is requested that outages will be extremely limited. In case of a major failure, the AOCS will place the spacecraft in a safe condition,

---

* *OBDH: On board data handling, ESA standard.*

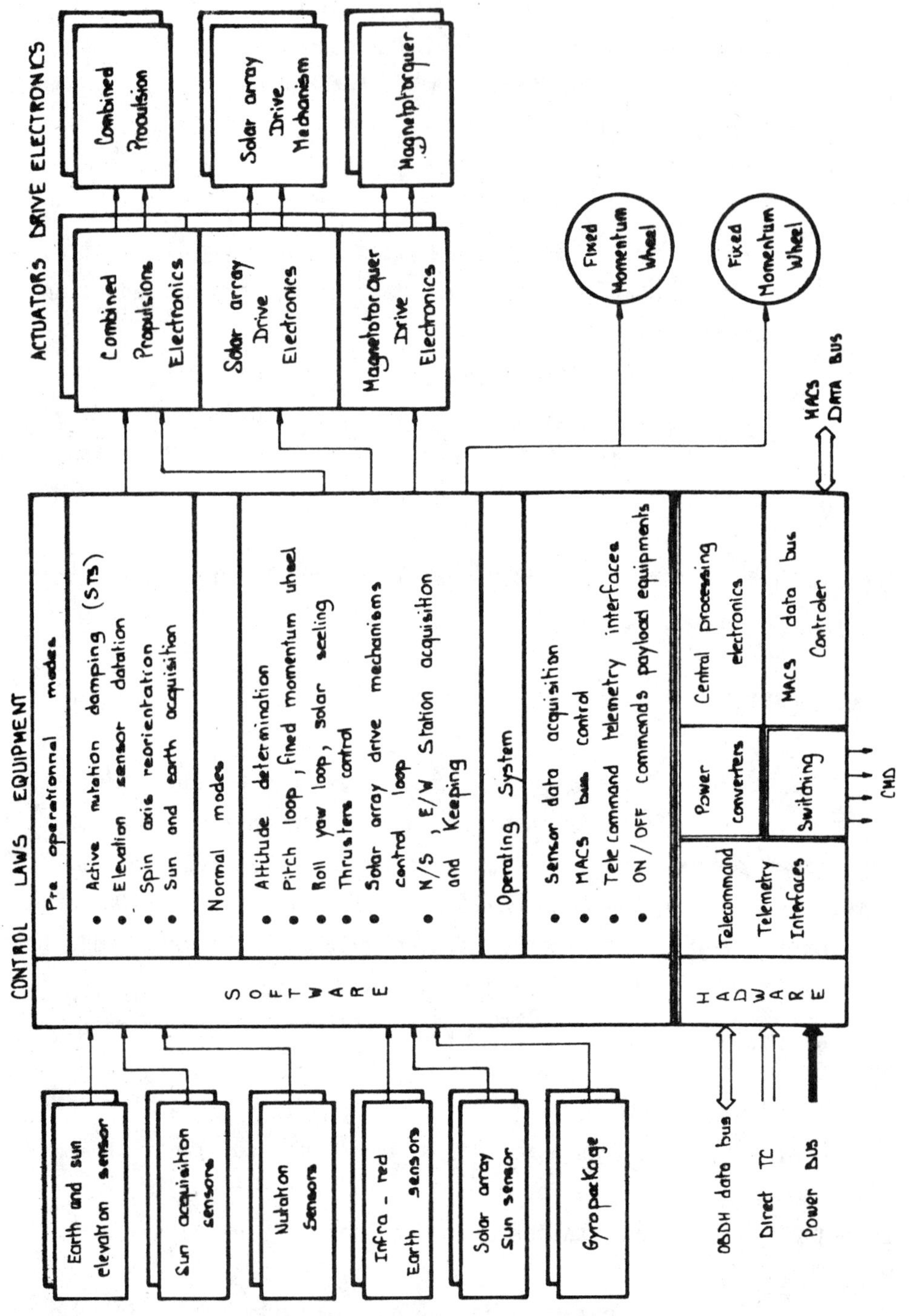

Figure 1 AOCS Block Diagram

so that normal operation will be recovered, under ground control, as soon as possible.

Mass and power consumption will be limited as much as possible and reliability will be better than 0.95 for ten years.

MODES DESCRIPTION AND OPERATIONS

A general block diagram of the AOCS is provided in figure 1. Major modes are:

- o parking orbit mode
- o transfer spinning mode
- o attitude acquisition modes (including despin, sun acquisition and earth acquisition modes)
- o operational modes (including normal mode and station keeping mode)
- o back-up modes.

Operational Modes

During all geosynchronous operations, the AOCS consists of a bias momentum three-axis control system.

In normal geosynchronous operation, roll and pitch errors are sensed by an accurate two-axis infrared earth sensor. When a preset error is exceeded in roll, solar array drive mechanisms are actuated in such a way to produce a differential bias angle between solar arrays, resulting in solar sailing torques. The earth sensor pitch error signal is processed and changes the momentum wheel speed as a function of the error. A gyroscopic stiffness is provided by a fixed-momentum wheel and a yaw sensor is not necessary.

In case highly accurate pointing is required, a radio frequency sensor and antenna pointing mechanism can be used for antenna beam pointing.

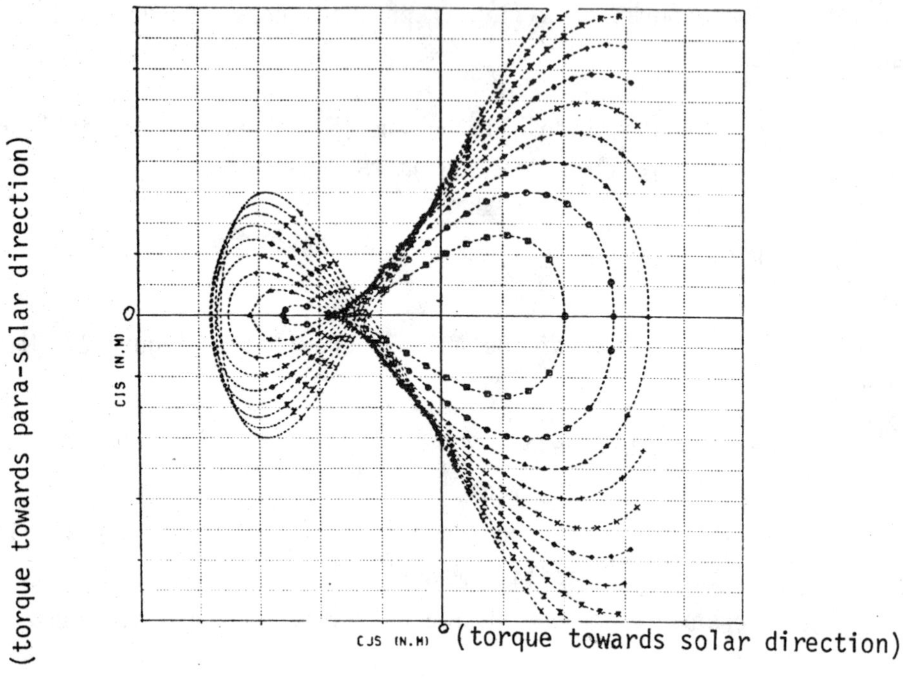

Figure 2 Solar Sailing Torques Capability

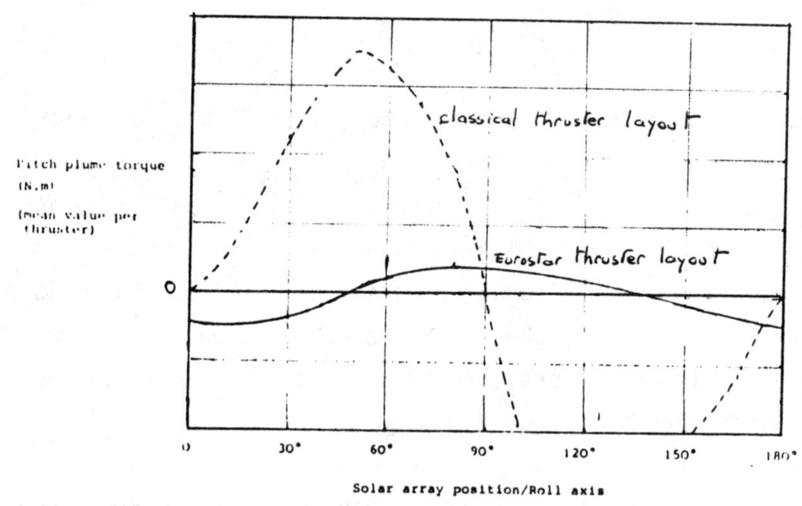

Figure 4 Optimization of Pitch Plume Torque

Due to spacecraft misalignments (solar arrays, center of gravity) and to the unsymmetrical antenna configuration, the solar pressure effect on the spacecraft creates the major disturbing torques which necessitates a control concept able to maintain the angular momentum in the required north-south direction. The selected design is based on solar sailing which offers a number of significant improvements with regard to classical concepts such as thrust control system (thrusters failure risk, fuel consumption) or full magnetotorquer actuation (sensitivity to magnetic storm).

As a result of this design, the normal mode will be very smooth and there is very limited risk of outages due to the thruster malfunctioning.

Derived from a concept successfully in flight, tested on OTS by the European Space Agency, the EUROSTAR solar sail has been significantly improved by use of fixed flaps and on-board loops. One of the main advantages of the EUROSTAR solar sail is the capability to compensate the disturbing torques by an adequate low bias on the solar array orientation, improving the yaw pointing performances.

Figure 2 presents an example of solar sailing torques capability for several array depointing angles. In case of high requirements for roll bias capability for a particular mission, an additional magnetic coil may be implemented to support solar sailing.

Due to the gravitational effects of the earth, moon and sun, and also the solar pressure effect on the large solar arrays, station keeping manoeuvres using 10 N biliquid thrusters are necessary.

Due to the important disturbing torques, a high band-width control loop using a yaw gyro sensor and thrusters actuation has been implemented. Control laws have been specially designed to maintain a high pointing accuracy in spite of low frequency flexible modes of the large solar arrays.

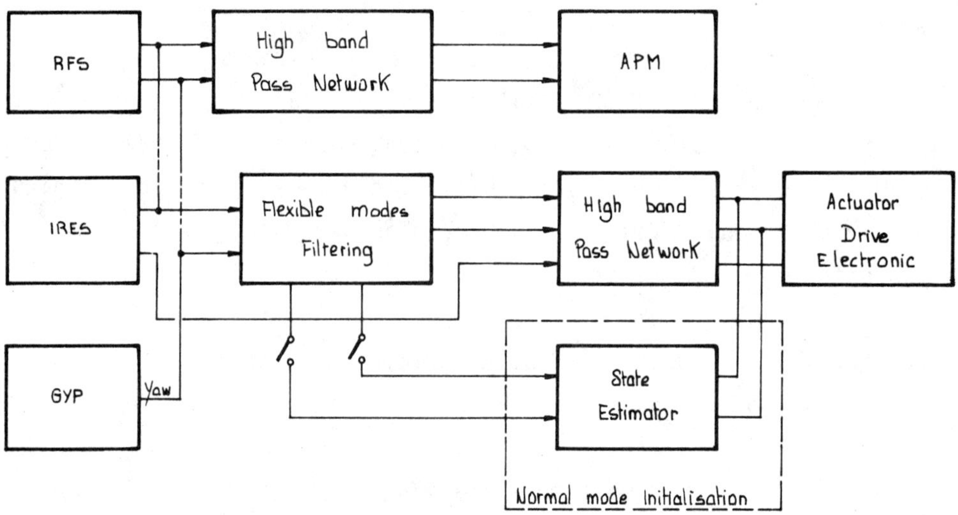

ANTENNA POINTING CONTROL DIAGRAM (OPTION)

Figure 3 Station Keeping Block Diagram

Furthermore, to damp the high rate level introduced by the thruster control, a special transition mode based on a state estimator is triggered for normal mode initialization (see figure 3).

Special care has been devoted to define a 12 thruster layout, taking into account platform configuration constraints and consumption minimization objectives. An example of optimization is shown in figure 4 concerning the parasitic pitch plume torque created during north-south station keeping manoeuvres by interaction between thruster jets and solar arrays.

Preoperational modes

In case of a Shuttle launch with a PAM D unstable spin configuration, active nutation damping using accelerometer detection is required to counterbalance the divergent motion due to the sloshing effect in the tanks. The AOCS will be able to maintain the body spin axis orientation within acceptable limits for PAM D burn efficiency. Control laws have been specially designed for such a requirement.

After injection in transfer orbit and PAM D separation, the spacecraft is dynamically stable; but the active nutation damping loop, with adapted parameters, may be switched on by ground in case of an unexpected nutation angle.

Due to the unified bi-propellant propulsion system, the impact of important liquid quantities during the spinning phases has been deeply analyzed. Stability conditions taking into account the sloshing effect have been determined (figure 5). Due to the stable configuration, the only required operations are the spin/despin and slew manoeuvers, initialized by ground after the reception of the sun and earth outputs from the Earth Sun Sensor.

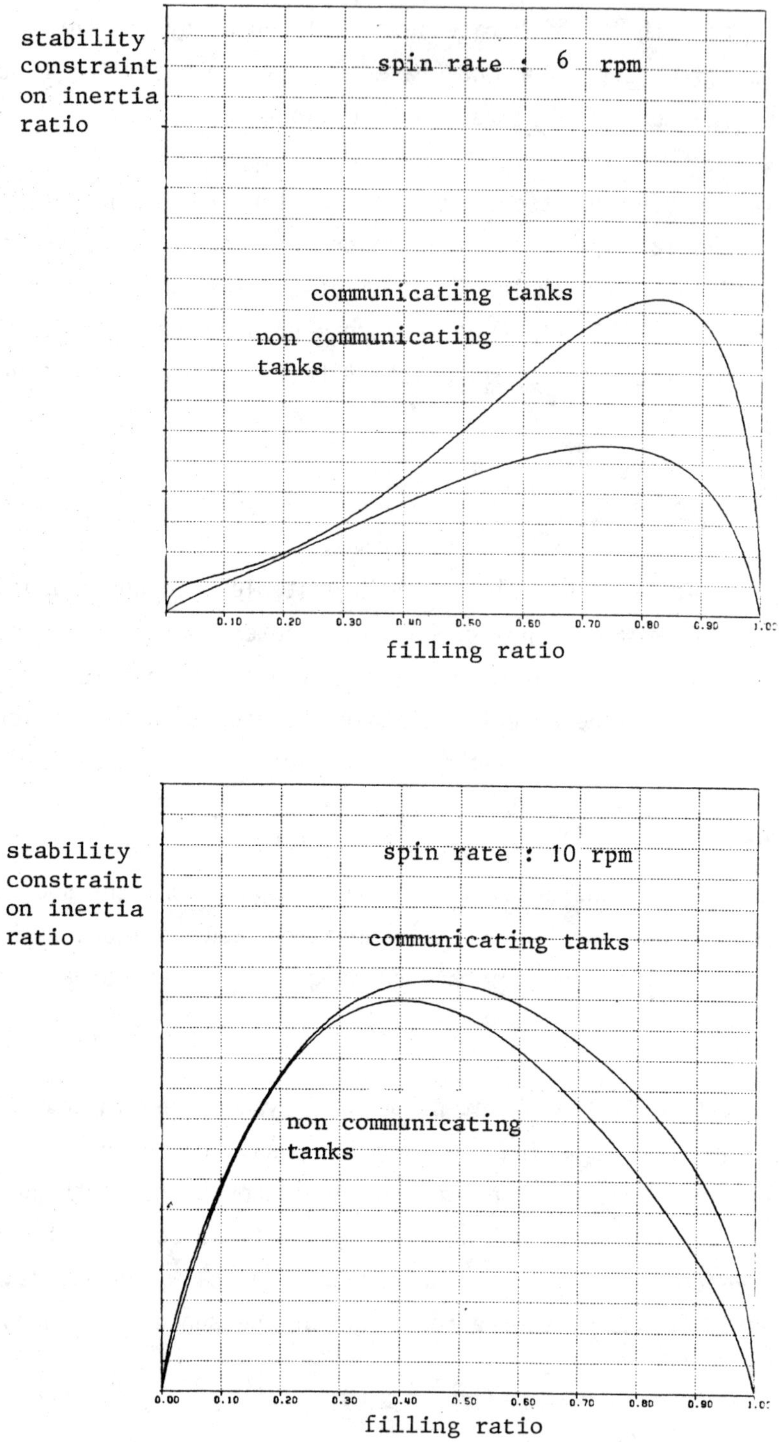

Figure 5 Rumyantsev Criteria

Subsequent to the last apogee burn, the spacecraft is despun, and a sun acquisition is performed using coarse analog sun sensors. Then an earth search mode is initialized by ground. The earth acquisition mode uses the infrared earth sensor in coarse mode. Fine station acquisition will be done by ground control, with station keeping control laws.

During the first days of operations, in-flight perturbing torque estimation will be performed to compensate for fixed repetitive ones.

Failure mode recovery strategy is as simple as possible, based on ground control for critical phases (like during station keeping), thanks to a specific design for failure criteria choice and detection implementation.

EUROSTAR AOCS ELECTRICAL ARCHITECTURE

The EUROSTAR AOCS design was driven by flexibility and modularity objectives. Mass and cost requirements were the main constraints. The electrical architecture is shown in figure 6.

The AOCS design is organized around a central permanent microprocessor:

- 16 bit word length
- integrated injection logic technology
- radiation tolerant.

Design of the central processing unit was carefully performed, to avoid heavy ion and cosmic-ray soft errors, and transient voltage disturbances. Reliability was another factor in the design.

An error detection correction (EDC) circuit is implemented, with hard radiation CMOS random access memories. Control laws software is in PROM, with some capabilities for in-flight parameters modifications. Telemetry is done through a direct memory access.

Interfaces with equipment is done through a digital real time data bus: MACS TELECOM, derived from the MACS European Space Agency standard.

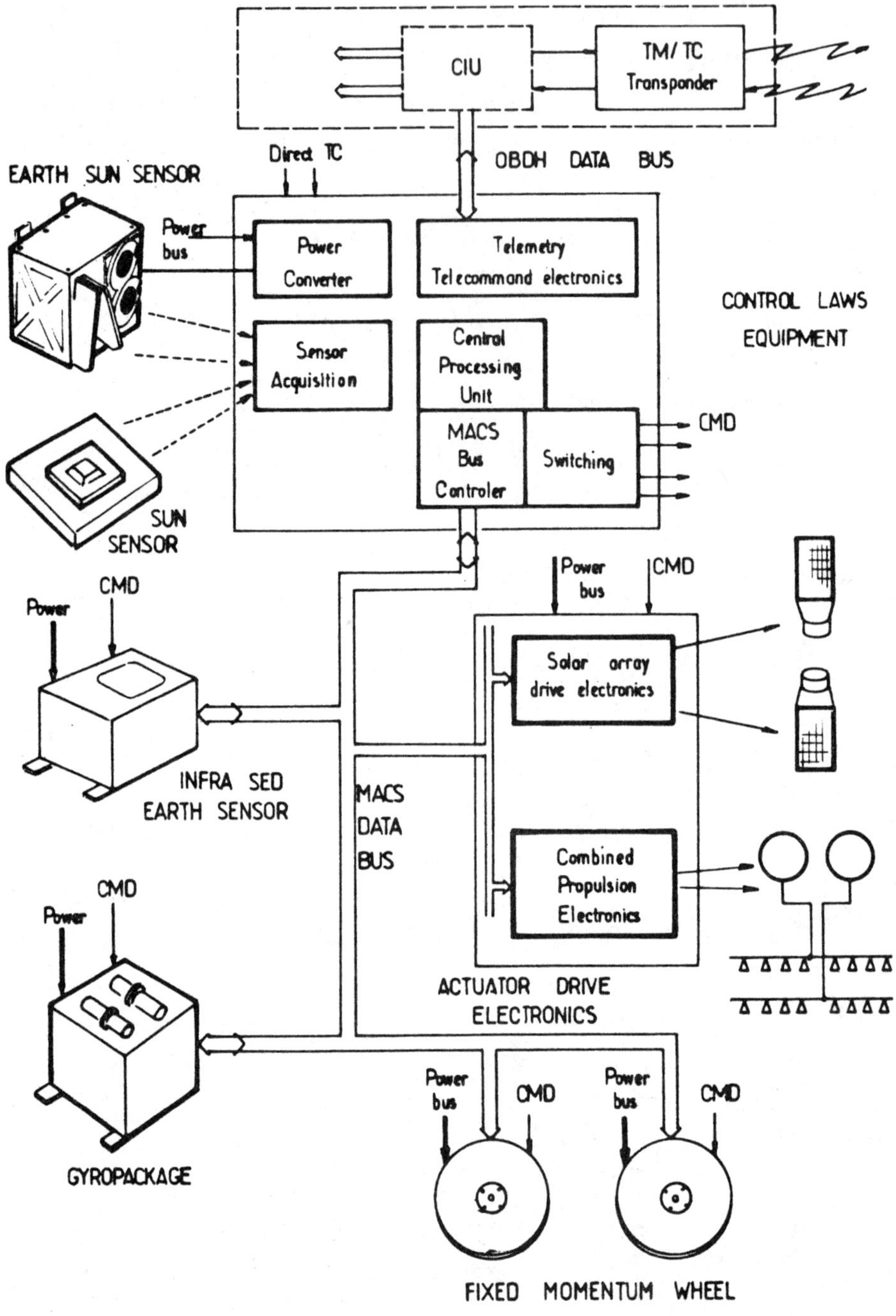

Figure 6 EUROSTAR AOCS Electrical Architecture

The main features of the MACS bus are:

- o serial data links, with two signals (clock and data)
- o contention protocol for access to data bus control, allowing fast response for interruptions
- o clock frequency up to 500 KHz. Communications are half duplex, each word consisting of 24 bits. Every message consists of one instruction word and one data word (Fig. 7).
- o Exchanges are secured with parity bit error and acknowledge answer bit. A supervisor level is provided.
- o Instructions are data transmission or request command distribution protected or not against errors, broadcasting. Block data exchanges are possible.
- o 32 subscribers can be connected, with 32 subaddresses for read or write in each of them.
- o User interface was especially optimized in such a way to reduce dedicated peripheral circuits: A DMA capability is provided for microprocessor type equipment. For simple equipment, all protocol control and decoding is done inside the semi-custom interface circuit.
- o Hardware interface is reduced to hybrid circuit for fault tolerant receivers-drivers, and one semi-custom CMOS, radiation hardened for protocol control and decoding (Fig.8).

The use of the data bus inside the AOCS results in harness mass saving and provides a robust EMC interface.

Power converters are distributed in equipment as soon as their power consumption is large enough, or when they require specific performances (sensors). Converters were designed on a standard basis, using recent LSI technology and high frequency pulse wave modulation. Power converters are compatible with both American (28 V bus) and European (50 volts) systems.

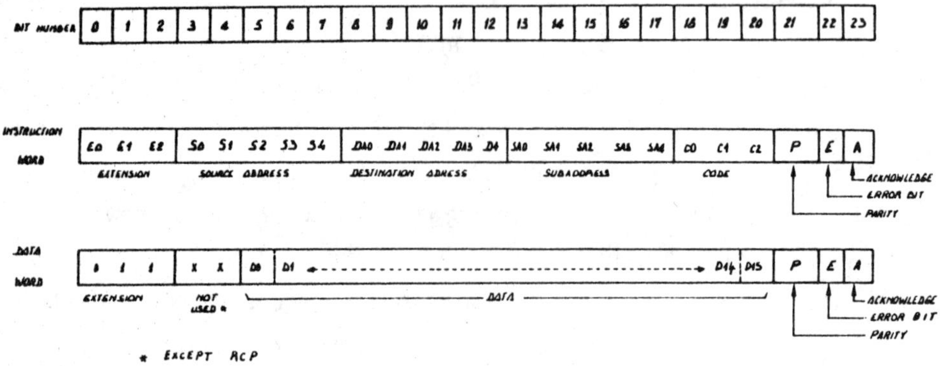

Figure 7 MACS Word Size

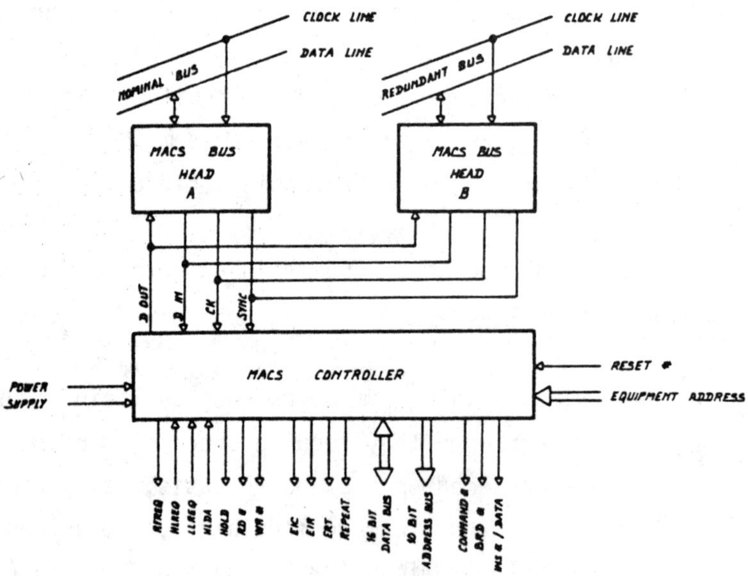

Figure 8 MACS Hardware Interface

Switching is done in a decentralized hierarchical approach; the AOCS Kernel is switched by the control interface unit, through direct telecommands. Then switching is done by the control processing unit, with information coming through the OBDH data bus. (128 telecommands are provided for functions or equipment).

Equipment is organized around the AOCS central processing unit, which is named "Control laws equipment". It includes:

- TM/TC, OBDH data bus interfaces
- power converter
- central processing unit
- MACS data bus controller
- analog to digital converters for some light sensors like analog sun sensors.

Technology is large pcb, multilayer type with intensive use of hybrids, circuits, and ceramics.

Sensors for a basic mission include:

Analog coarse sun sensors provide a full field of view of 2 pi steradian. It is composed of a central truncated pyramid of square section, with nominal and redundant silicon cells on each face. The equipment is used for sun and earth acquisition.

Infrared earth sensor is a scanning type based on the mechanical modulation of the energy coming from the earth horizon. This modulation is obtained by a mirror fitted on a Bendix flex pivot driven by a torque motor. The infrared detections are four immersed bolometers placed in the focal plane of the input objective. The roll and pitch attitude errors are derived from the deviation between reference pulse and earth/space transition for each bolometer measured by clock counting. The sensor is used in coarse and fine mode for earth acquisition and on station control. Pointing accuracy in fine mode is 0.03°.

Earth and sun sensor is used for spacecraft attitude determination in spin mode during the transfer phase. It consists of two types of detection:

- o two independent infrared beams, with bolometers
- o two redundant V-beam sun sensors, using silicon cells.

Nutation sensor package is used for nutation sensing during the spin phase, especially during STS launch with PAM D. Three independent strap down accelerometers are implemented within the package.

Gyropackage provides yaw rate and angle information from two rate integrating gyros used in closed loop. It is used for initial sun acquisition and orbit corrections.

Actuators are driven by two types of equipment:

Actuator drive electronics contain electronic circuits to drive:

- o the solar array drive mechanism, using a stepping motor
- o propulsion drivers compatible with a 16 thruster layout, latching valves and monitoring of pressure transducers and latching valves status.
- o magnetic coil power drivers.

Fixed momentum wheel is used during normal mode to provide a gyroscopic stiffness to the satellite around the roll yaw axes, and to provide pitch control. A rotating mass ensures a momentum of 25 Nms up to 50 Nms, with a typical wheel speed of 4100 rpm; the unit can provide a reaction torque up to + 0.1 Nm. within a speed range of - 20 % up to + 10 % from nominal speed.

The unit is designed with space qualified oil-lubricated ball bearings, with oil storage. The motor is a brushless DC type, with three phases and coil angle position detections for commutation. The equipment was flight proven on many missions and the electronics was largely improved, using recent technologies, especially for the power driver. This results in a very efficient electrical and mechanical design.

For all equipment, an important effort of technology was performed. Components were choosen from a new preferred parts list including a lot of new hard radiation tolerant components. A semi-custom was designed and manufactured for the MACS data bus, and several hybrid circuits were developed (up to 15 different types). An enhanced design for packaging was implemented, resulting in up to 20% mass saving from classical equipment design.

The electrical architecture, largely decentralized, results in a very modular and flexible design; use of a data bus will especially permit to add or suppress equipment without modifications in hardware at the control law equipment level.

EUROSTAR MULTIMISSION GEOSTATIONARY AOCS

While it is not the baseline for the EUROSTAR platform, the architecture of the AOCS can be easily modified to implement:

- o three-axis transfer, adding gyropackage and digital sun sensors for example
- o 1-DOF with a V-wheel configuration and a transverse momentum wheel.

Both case provisions for switching are included in the control laws equipment, and new peripheral sensors or actuators will be connected on the MACS data bus. Provisions in software, data processing power, and memory extensions are provided to implement adequate control laws in software.

REFERENCE

MACS - The ESA Standard for Guidance/Control and Robotics - IAF 83 - 346 - XXXIV Congress, Budapest 1983.

AAS 84-003

MARINE OBSERVATION SATELLITE-1 SYSTEM AND CONTROL CONCEPTS

Y. Ishizawa and M. Kusanagi*
T. Shimamura, G. Shirako and E. Nakagawa+

The MOS-1 (The first Japanese Marine Observation Satellite) is planned to be launched by N-II Launch Vehicle in the summer of 1986. The MOS-1 program started in April 1980 with the following major objectives:
(i) Establishment of fundamental technologies which are common to both marine and land observation satellites,
(ii) Observation of the state of sea surface and atmosphere using visible, infrared and microwave radiometers and verification of these sensors.
To achieve those objectives the MOS-1 carries three radiometers. The MOS-1 hardware configuration is discussed and some results of the mission analysis and operational aspects are explained. Attitude and orbital control concepts and the critical points of control subsystem design are also explained. Computer simulations and relevant engineering test results of control subsystem are cited to show the validity of the design.

INTRODUCTION

National Space Development Agency of Japan (NASDA) is developing a marine and land observation satellite series for Japan's remote sensing activities in space. The first satellite of this series, MOS-1, is under development with a launch target date in 1986. The basic design and critical design were completed in 1981-1983, and the MOS-1 program is now in the integration process of Prototype Model.

In the MOS-1, four payload equipments are installed. They are Multispectral Electronic Self-Scanning Radiometer (MESSR), Visible and Thermal Infrared Radiometer (VTIR), Microwave Scanning Radiometer (MSR) and Data Collection System Transponder (DCST). MOS-1 is a modular construction with both bus and mission modules. The mission module carries all of the payload equipments, while the bus module provides the vital electrical and mechanical components. Both modules are approximately cubic and a deployable solar paddle which is driven around the pitch axis is attached to one side of the body. The total weight is expected to be less than 750 kg, and the design life is two years.

* *1st Satellite Design Group, National Space Development Agency of Japan, Tokyo, Japan*
\+ *Space Development Division, NEC Corporation, Yokohama, Japan*

The normal orbit is a sun-synchronous and near-recurrent orbit. The altitude is about 909 km and the inclination is about 99 deg. The local time at descending node is kept between 10:00 AM and 11:00 AM. The recurrent period of the orbit is 17 days and the satellite covers about 14 revolutions per day. The orbit ground track is controlled via Reaction Control Subsystem (RCS) that can either add or remove orbit velocity through the controlled firing of 1 Newton thrusters. This control is necessary for removing launch vehicle injection errors and for compensating energy losses due to atmospheric drag decay.

The Attitude and Orbit Control Subsystem (AOCS) is a three-axis stabilization type with a kind of bias momentum scheme (controlled bias momentum). Two of four wheels store angular momentum and compensate the disturbance torque, e.g., that caused by solar radiation. Magnetic torquers unload the wheels when necessary. For attitude control during acquisition phase and orbit correction phase, RCS is employed.

SATELLITE SYSTEM SUMMARY

The outside view of MOS-1 is illustrated in Fig. 1. The main body of MOS-1 is composed of mission module and bus module. Both modules are approximately cubic and give the satellite dimension of about 1.3m (width) × 1.5m (depth) × 2.4m (height). A three-segment deployable solar paddle is attached to one side of the mission module. The size of deployed solar paddle is 2.0m (width) × 5.3m (length). Three types of payload sensors, MESSR, VTIR and MSR are mounted on the mission module. The total weight of MOS-1 is expected to be about 750 kg. The design life is two years.

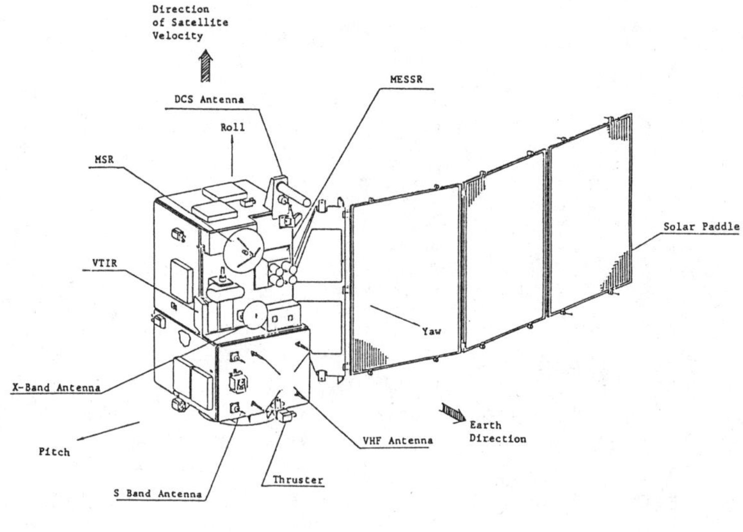

Fig. 1 MOS-1 Configuration

Payload equipments of MOS-1 are MESSR, VTIR, MSR and DCST.
MESSR operates in four wavelength between 0.51 and 1.1 microns. By measuring the relative brightness in these four bands a sea surface color can be discriminated. MESSR has two optical systems. Each optical system is composed of the Gauss type telescope, the prism to divide the incident ray into two parts depending on wavelength and two sets of CCD having 2048 elements. Two sets of MESSR, signal processor and 8 GHz transmitter are installed to increase the reliability. The viewing directions of MESSR's are canted slightly (2.7 deg) each other. Using two sets of MESSR, signal processor and 8 GHz transmitter simultaneously, the swath width becomes 200 km.

Sea surface temperatures are measured by VTIR. VTIR has one visible channel and three infrared channels. Each channel has two detector elements in order to increase reliability. The optical part of VTIR is Ritchey-Chretien type telescope which focusses incident radiation onto the detectors. Scanning is achieved by rotating the reflection mirror 7.3 times a second.

MSR is used to measure the water content of the atmosphere. MSR is composed of two Dicke type radiometers in frequency of 23 GHz and 31 GHz. Scanning is performed by rotating an offset parabolic reflector with about 18 rpm. The scanning coverage of MESSR, VTIR and MSR is illustrated in Fig. 2.

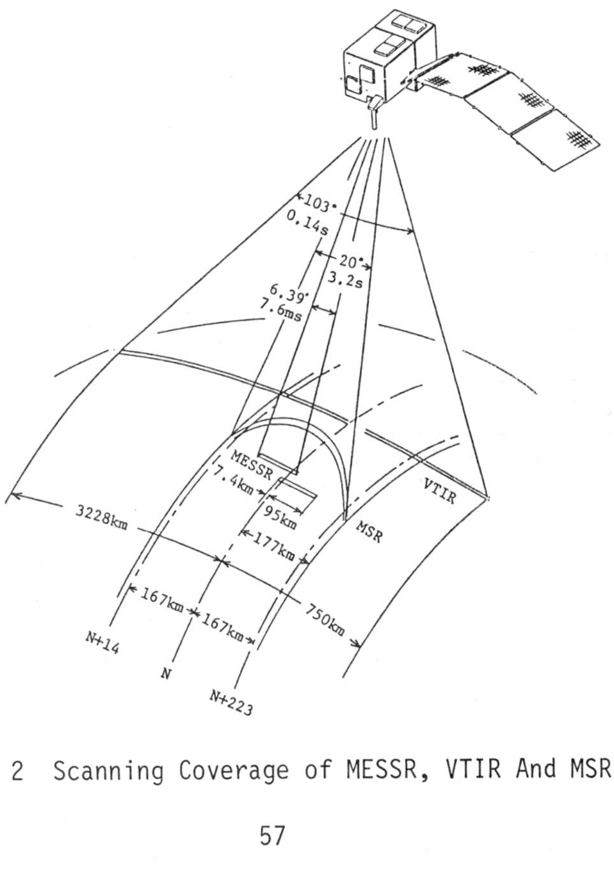

Fig. 2 Scanning Coverage of MESSR, VTIR And MSR

DCST relays data from remote instruments connected to Data Collection Platform (DCP) to ground station whennever MOS-1 is within transmission and reception range of both DCP and the ground station.

These payload equipments and other vital equipments are integrated into MOS-1. The vital equipments of MOS-1 are categorized with Telemetry Tracking and Command Subsystem (TTC), Electrical Power Subsystem (ESP), Solar Paddle Subsystem (PADL), Attitude and Oribt Control Subsystem (AOCS), Reaction Control Subsystem (RCS), Thermal Control Subystem (TCS), Structure Subsystem (STR), Mechanical Integration (MINT) and Electrical Integration (EINT).

ORBIT AND ATTITUDE REQUIREMENTS

Required orbit is sun-synchronous and near-recurrent orbit. The orbital parameters are shown in Table 1.
Sun angle history and ground track based on these orbital parameters are shown in Fig. 3, Fig. 4 and Fig. 5. Fig. 3 shows the history of sun angle for a year, where sun angle is defined as the angle between the sun direction and pitch axis of the satellite. The rotation of the orbital plane synchronized with the sun direction provides a constant sun pitch angle. Fig. 4 shows the ground track for a day and Fig. 5 shows the ground track for a week (7 days) in the vicinity of Japan. The satellite circles the earth every 103 minutes, completing 14 orbits per day and viewing the entire earth every 17 days.

Table 1
THE ORBITAL PARAMETERS

ALTITUDE	909 km
INCLINATION	99.1 deg
ECCENTRICITY	less than 0.004
PERIOD	103 min
RECURRENT PERIOD	17 days
GROUND TRACE ACCURACY	20 km (deviation at the equator)
LOCAL TIME AT DESCENDING NODE	10:00 AM to 11:00 AM

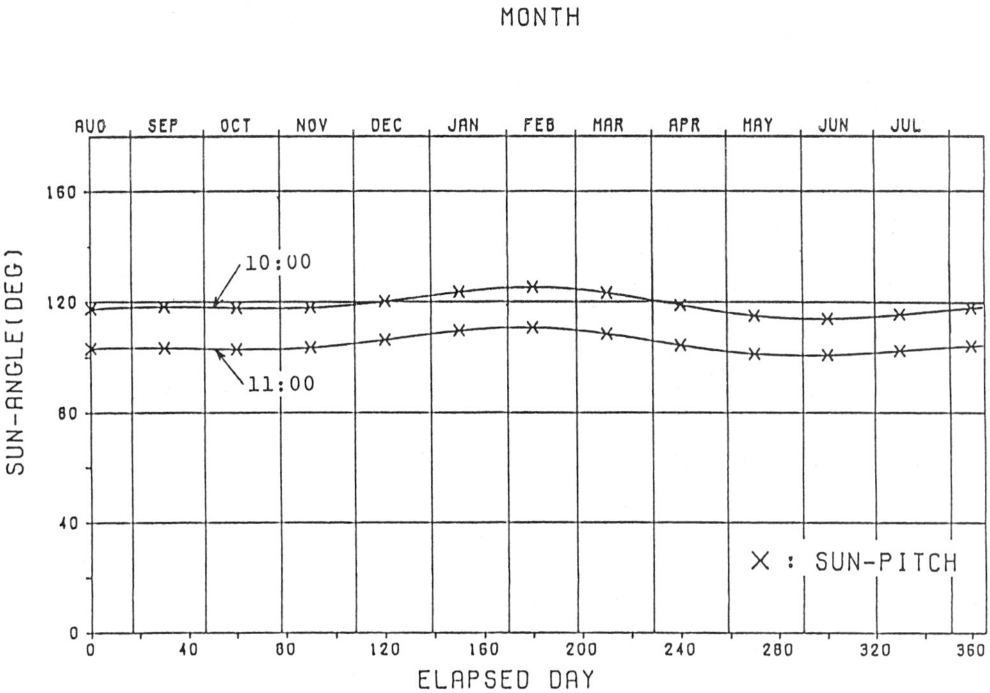

Fig. 3 Sun Angle History for a Year

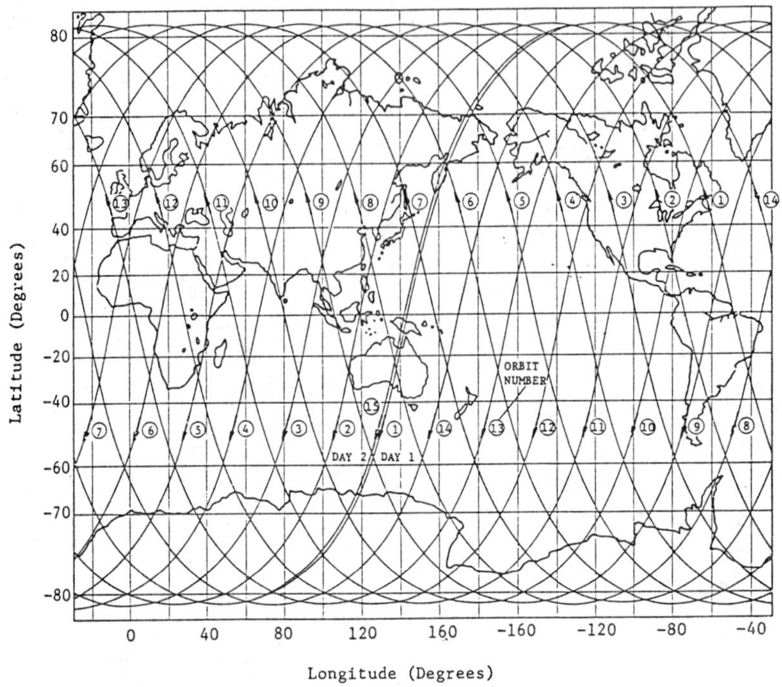

Fig. 4 Orbital Ground Track for a Day

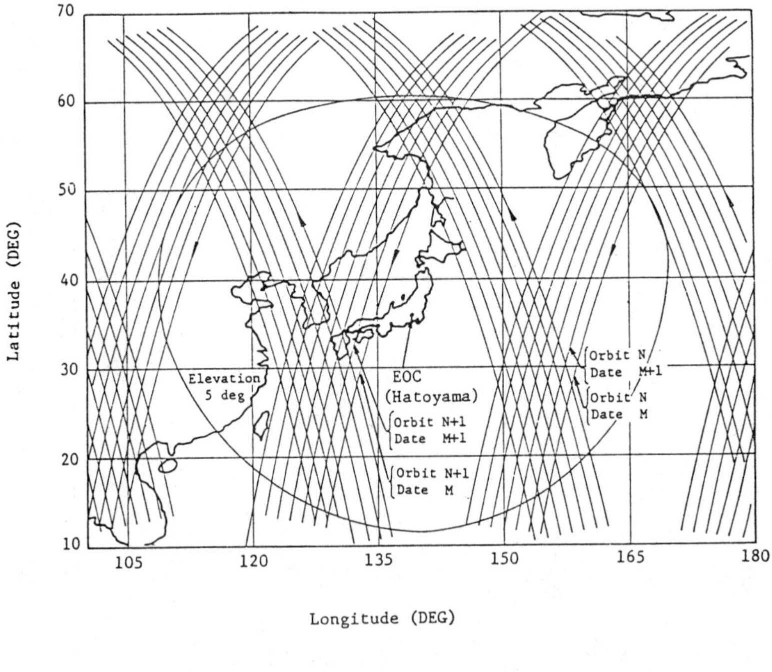

Fig. 5 Orbital Ground Track for a Week

Several perturbing forces (atmospheric drag, the gravitational attraction of the sun and the moon and radiation pressure of the sun) act upon the satellite after the desired orbit has been attained. These forces cause changes to the orbit that compromise the desired coverage and repeatability characteristics. If the orbit correction is not applied, the ground trace deviation from initial orbit during a period of two years will be estimated to be 1500 km. Orbit correction is necessary to keep the ground trace deviation within 20 km (required accuracy). Fig. 6 shows the estimated ground trace deviation during mission life (two years) when orbit control with respect to altitude is put into practice every 8 recurrent cycles (136 days). Altitude rises about 95 meters on each orbit control time (expected to be performed 5 times during mission life), and maximum ground trace deviation is estimated to be 13.4 km.

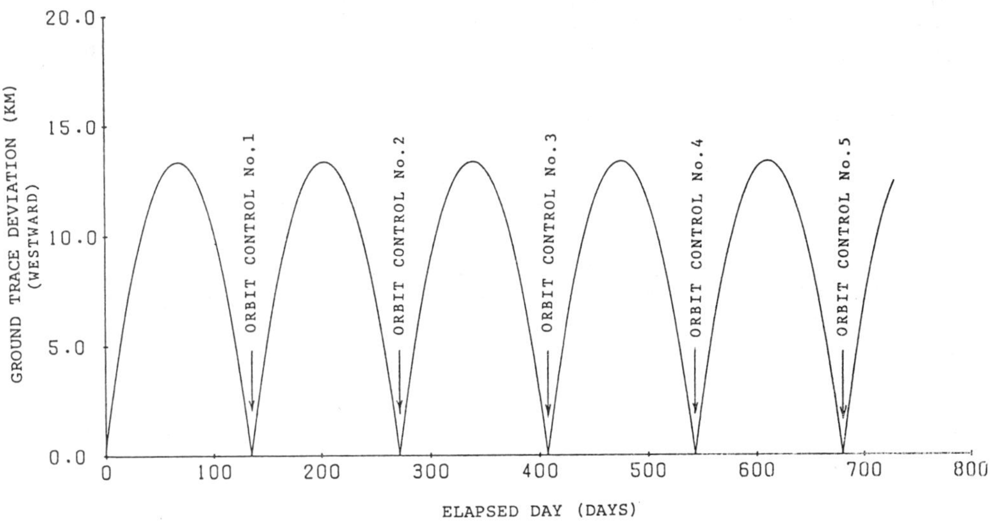

Fig. 6 Ground Trace Deviation and Orbit Control

Attitude drift error of each axis causes the image distortion. The allowable drift error of each axis during one scan time was assumed to be less than 4% of one pixel FOV of each payload sensor. The drift rate requirement for the satellite which is based on the allowable drift error of each payload sensor is summarized in Table 2.

Table 2
DRIFT RATE REQUIREMENT

Item \ Axis	Roll	Pitch	Yaw
Drift rate requirement (deg/sec)	±0.016	±0.016	±0.05

The allowable pointing error of sensors was assumed to be less than 7% of 1 scene (1 FOV) of MESSR. Because image resolution of MESSR is the highest of the three sensors. The requirement for pointing error around each axis is summarized in Table 3. Pointing error shown in Table 4 is apportioned to each compornent. Error components are 1) AOCS control performance, 2) Structural distortion, 3) Thermal distortion and 4) Alignment error of each payload sensor. Error component which can not predicted are considered under margin. These components are shown in Table 4.

Table 3
POINTING ERROR REQUIREMENT

	Roll	Pitch	Yaw
Pointing error (3σ) (deg)	±0.45	±0.45	±1.0

Table 4
APPORTIONMENT OF POINTING ERROR

Component \ axis	Roll(deg)	Pitch(deg)	Yaw(deg)
1) AOCS Control performance	0.4	0.4	0.9
2) Structural distortion	0.1	0.1	0.1
3) Thermal distortion	0.1	0.1	0.1
4) Alignment error of radiometer	0.11	0.11	0.4
5) Margin	0.1	0.1	0.1
Total (RSS)	0.45	0.45	1.0

FUEL BUDGET

Reaction Control Subsystem (RCS) is utilized to correct the orbit injection error, and the ground trace of orbit. It is also utilized to control the attitude of the satellite during acquisition phase, orbit correction phase and reacqisition phase (in case of Safety mode). Seven one Newton thrusters are selected to activate by Attitude and Orbit Control Subsystem. Impulses of the thrusters which are necessary for the attitude and orbit control during mission life (two years) are estimated and fuel (hydrazine) consumption is derived in accordance with the characteristics of RCS, i.e. Impulse and Isp. The estimated Impulses and Fuel Consumption are shown in Table 5. 18.5 kg of fuel is to be filled in the Tank of RCS considering this estimate.

Table 5
ESTIMATED THRUSTER IMPULSES AND FUEL CONSUMPTION

		Thruster Impulse (Ns)	Fuel Consumption (kg)
Correction of Orbit Injection Error	Δa	7445	3.6
	Δi	6700	3,3
Ground Trace Correction		190	0,1
Attitude Control	Initial Acquisition and Orbit Control	6880	3.9
	Reacquisition	3995	3.6
Total		25210	14.5

AOCS SUBSYSTEM

Outline of AOCS

Attitude and Orbit Control Subsystem (AOCS) consists of a so called biased momentum type three axis control analog system as a main part. As aquisition phase sensors, it utilizes rate integrating gyros, sun sensors and earth sensors sucessively. Most of those transient phase, conbination of one Newton thrusters are employed as an actuator.

Reaction Control Subsystem (RCS) consists of two fuel tanks with 18.5 kg hydrazine and N_2 pressurized gas. Seven one Newton thrusters are arranged as Fie. 7, to generate roll, pitch and yaw control torques and X and Y axial thrusts for orbit correction maneuver. Each thruster is clustered to pair to form redandant system. Those thrusters are driven by Valve Drive Electronics (VDE) which receives the control signals from Attitude and Orbit Control Electronics (AOCE).

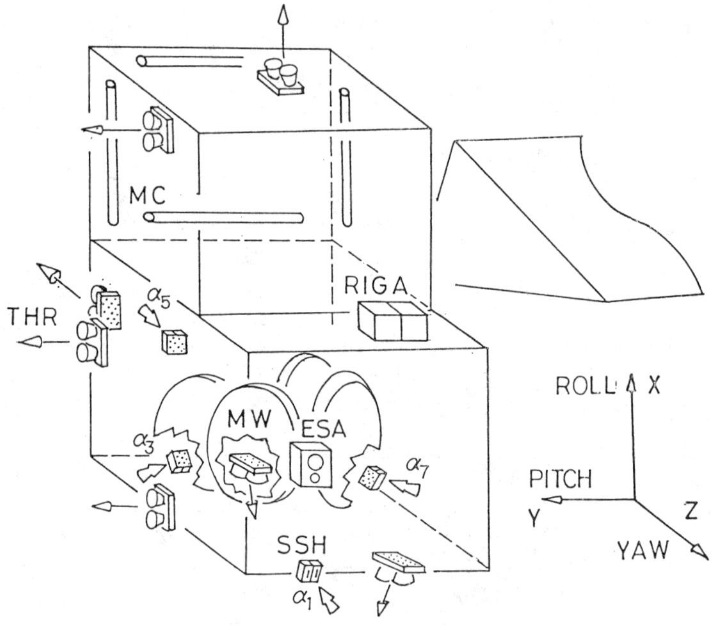

Fig. 7 Allocation of Sensors and Actuators

As Normal Mode (NM) or operational mode, skewed V-shaped pair of 50 Nms Momentum Wheel (MW), driven by Wheel Drive Electronics (WDE), generates the main driving force. Besides its built in stabilizing effect due to the 50 Nms bias momentum, it can control pitch and roll angles by changing the speed of each wheel independently. Yaw angle is not actively controlled throughout the operational mode, therfore the pointing accuracy for yaw angle is the worst. However, roll-yaw coupling in a long run will help to converge the yaw error to zero too. Since the wheels have saturation limit when their speed reached either below 4140 rpm or over 5060 rpm, the momentum wheels are controlled by wheel unloading system utilizing the earth magnetic field by switching the magnetic coils several time in every revolution. Also Reaction Control Subsystem is considered to be used as a back up for the wheel unloading.

AOCS components are connected as illustrated in Fig. 8. Some of the characteristics of AOCS components are listed in Table 6.

Table 6
AOCS COMPONENTS CHARACTERISTICS

COMPONENTS	PERFORMANCE AND CHARACTERISTICS	
Rate Integrating Gyro Assembly	±2 deg/sec ±12 deg/hr	Range Non G sensitive drift rate
Sun Sensor Assembly	±20 deg/±45 deg ±70 deg	Linear/Operating rang Off axis operating rang
Earth Sensor Assembly	±4 deg/±60 deg	Linear/Operating range
Momentum Wheel & WDE	50 Nms±10% (4600 RPM ± 10%	Angular momentum range Rotational speed)
Magnetic Coil & MCE	0.07 Nm 50 Atm^2	Maximum Output torque Maximum magnetic dipole moment
Reaction Control Subsystem & VDE	±0.6 Nm ±1.0 Nm	Roll torque Pitch and Yaw torque

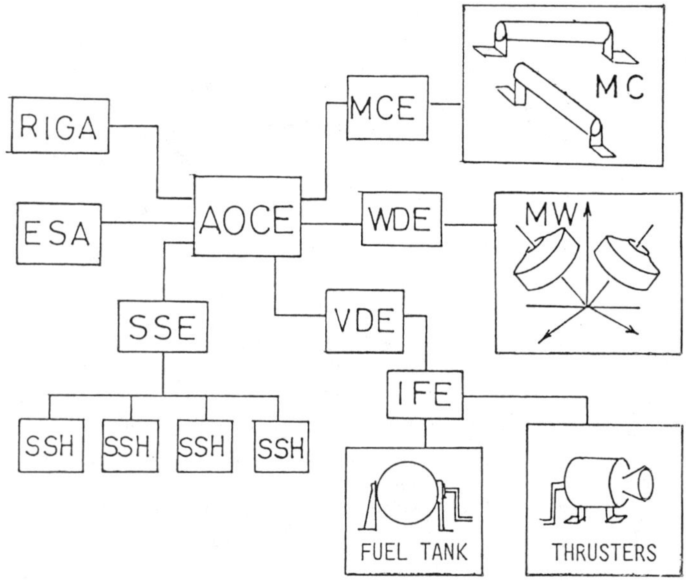

Fig. 8 Layout Diagram of AOCS Components

Function of AOCS

The attitude control is divided into two parts, i.e. acquisition modes and a normal mode. Acquisition modes are transient phase modes up to the establishment of an operational attitude. Normal mode is equivalent to operational mode, where pointing and stability requirements are the severest. The functions of each modes will be described along with the sequence of flight (Fig. 9).

N-II launch vehicle's flight path is so selected as not to let the payload sensors see the sun directly and also as to avoid the extreme local heat gradient.

The satellite will be separated at the eclipse zone with 10 deg and 0.4 deg/sec injection allowance. Two minutes after the separation, when the rate integrating gyro is fully warmed up, the Rate Damping Mode (RDM) starts to settle the satellite motion to the low constant yaw rate (Fig. 9, ① - ②).

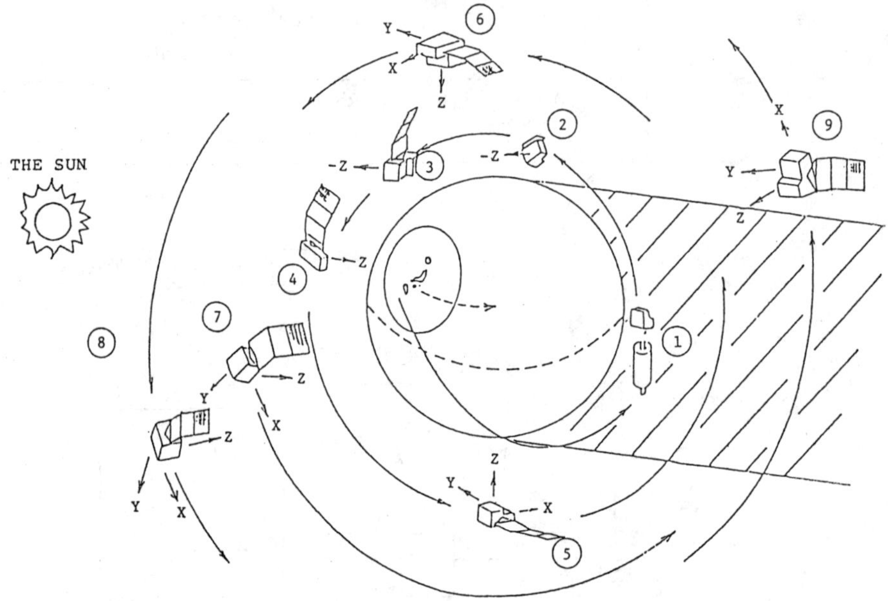

① SEPARATION AND RATE DAMPING
② SUN ACQISITION
③ SOLAR PADDLE DEPLOYMENT
④ EARTH ACQUISITION
⑤ COARSE YAW ACQUISITION
⑥ FINE YAW ACQUISITION
⑦ PADDLE DRIVE START
⑧ ⑨ NORMAL MODE

Fig. 9 MOS-1 Orbital Sequence

As soon as the satellite comes out of the eclipse zone, sun sensors are ready to work and hence Sun Acquisition Mode (SAM) starts. Through three steps -Z axis (anti-earth pointing side) captures the sun, remainning 0.7 deg/sec yaw rate around the Z axis. About 30 minutes after the sun acquisition starts, paddle deployment is taken place (Fig. 9,③), so as to assure the event from the tracking station in Japan, and it proceeds to the next event. Earth acquisition sequence will be initiated by earth command (Fig. 9,④). Earth Acquisition Mode (EAM) completes when pitch and roll deviations are maintained within 5 deg while 0.7 deg/sec yaw rate remains. Affirming the EAM state, the command, which initiate the Coase Yaw Acquisition Mode (CYAM), is sent from a tracking station (Fig. 9,⑤). In this phase the yaw angle is kept within 15 deg from the orbit plane with the aid of sun sensors. The momentum wheels are energized in this phase. CYAM is followed by Fine Yaw Correction Mode (FYCM), where pitch and roll errors are controlled within 4 deg (Fig. 9,⑥). Because of roll-yaw coupling caused by momentum wheels, not only the roll angle, which is controlled constantly, but also the yaw angle is controlled. The speed of wheels are kept constant throughout this mode, hence the deviation of angular momentum is controlled by RCS using an earth sensor and a rate gyro.

After the all acquisition sequences are completed, Orbit Control Mode (OCM) will be taken place to correct the orbit injection errors in both inclination and altitude. Later, in the mission duration, once in every 8 recurrence period or so, altitude correction will take place in order to compensate orbit decay due to the solar pressure, the aerodynamic drag, etc. In those mode momentum wheels are kept at a constant speed and used only to augment the stability, and active control is done by RCS.

Besides all those modes mentioned so far, there is one additional recovery mode or Safety Mode (SAFETY). If RIGA output, wheel speed, thruster operational duration or AOCE converter output exceeds the preset limits or if ESA loses its earth presence signal; safety controller detects them and switches to the back up elements. After wheel run down, the controller select the sun acquisition status, where thorough trouble shooting will take place. Once the abnormal element is identified, the controller will isolate the element from the others, and resume the acquisition sequence from EAM. Fig. 10 and Table 7 show all those transient modes control block diagram.

Operational mode or Normal Mode is controlled by biased momentum wheel system with wheel unloading by magnetic coil. Unloading by thruster is also available as a back up system in case of large disturbance torque. The momentum wheels are arranged in V shape with 30 deg half cone angle and the plane is skewed with 20 deg offset angle. About 90 Nms nominal bias momentum is available by two wheels around pitch axis. Due to the built in offset angle effect, the momentum difference of the two wheels create roll and yaw control torque at the same time. The Normal Mode control block diagram is shown in Fig. 11.

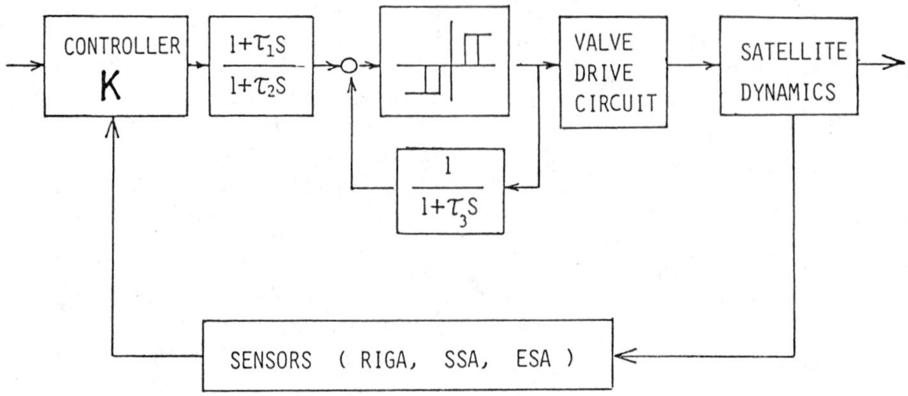

Fig. 10 Transient Momdes Control Block Diagram

Table 7
TRANSIENT MODES SENSORS AND CONTROL LAWS

MODE	SENSORS	CONTROL LAWS ROLL	PITCH	YAW
RDM	RIGA(w_x, w_y, w_z)	$K_D w_x$	$K_D w_y$	$k_D(w_z - w_{zc})$
SAM Step1	RIGA(w_x, w_y, w_z)	$K_D w_x$	$K_D w_y$	$K_D(w_z - w_{zc})$
Step2	SSA($\alpha_{13}, \alpha_5, \alpha_7$)	$K_D w_x$	$K_D(w_y - w_{yc})$	$K_D w_z + K_P \alpha_{13}$
Step3		$K_D w_x + K_P \alpha_{13}$	$K_D w_y + K_P \alpha_5$	$K_D(w_z - w_{zc})$
EAM	RIGA(w_x, w_y, w_z) ESA(ϕ, θ)	$K_D w_x + K_P \phi$	$K_D w_y + K_P \theta$	$K_D(w_z - w_{zc})$
CYAM	RIGA(w_x, w_y, w_z) ESA(ϕ, θ) SSA($\alpha_{13}, \alpha_5, \alpha_7$)	$K_D w_x + K_P \phi$	$K_D(w_y + w_o) + K_P \theta$	$K_D w_z + K_P(\alpha_7 - \alpha_c)$
FYCM	RIGA(w_x, w_y, w_z) ESA(ϕ, θ)	$K_D w_x + K_P \phi$	$K_D(w_y + w_o) + K_P \theta$	$K_D w_z$
OCM	RIGA(w_x, w_y, w_z) ESA(ϕ, θ)	$K_D w_x + K_P \phi$	$K_D(w_y + w_o) + K_P \theta$	$(K_D + K_P/s) w_z$

Note: Suffix "c" denotes constant bias, and "O" denotes initial value.

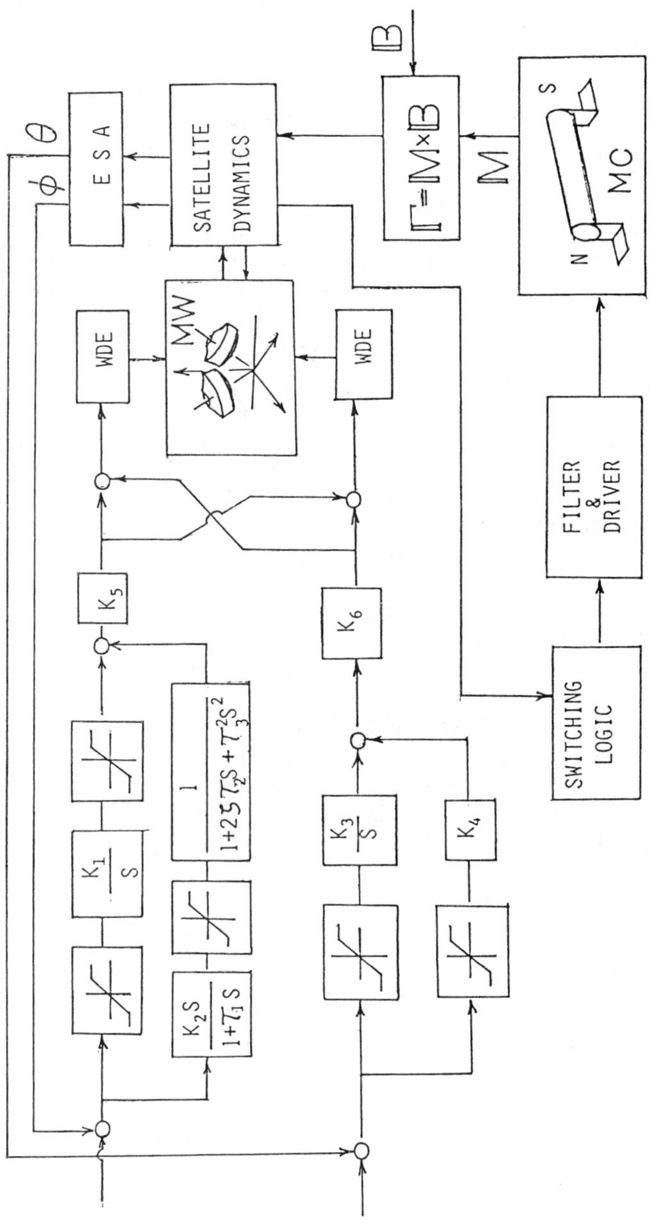

Fig. 11 Normal Mode Block Diagram

Development Status of AOCS

AOCS subsystem is designed and fublicated at Mitsubishi Electric Corporation, and installed into the satellite by NEC Corporation after integration tests. Electronic components such as AOCE, MCE and WDE are started with Bread Boad Models, while other components are started with Engineering Models.

Stability of the system is assured by Nyquist method etc. and total logics are checked its acceptability by software simulation programs. Some of those results are shown in Fig. 12.

The test facility consists of solar and earth simulators, three degree of freedom flight tables and digital computer which simulates the actuator and satellite dynamics. Prototype Model components are tested to stand the flight environment, and now final function tests are to be completed.

One of the most critical points during development was in the verification of the stability against the paddle flexibility. Deployed paddle itself has relatively low resonant frequencies. They are 0.37 Hz (Out of plane 1st bending mode), 0.87 Hz (In plane 1st bending mode), 2.01 Hz (1st torsion mode), 2.33 Hz (Out of plane 2nd bending mode) so on. Those lower frequencies are still higher enough to avoid the interference with the satellite nutation frequency caused by momentum wheels. The stability of AOCS is satisfactorily checked by digital simulation and the Nyquist method in the frequency domain. However, if the damping factor is extremely low, it is known by analysis that the limit cycle could exist, in case of the coupling between paddle 1st torsion mode and thruster pulse trains. Even if this limit cycle is the stable one, it is not desirable from the observation point of view and also the mechanical point of view. To make clear this problem, the damping factor and flexible mode of solar paddle are tested in the vacuum chamber to verify the sufficiency of the structure damping of the Solar Paddle Subsystem and to estimate the effect of air mass upon the shift of frequencies.

(1) Sun Acquisition Mode

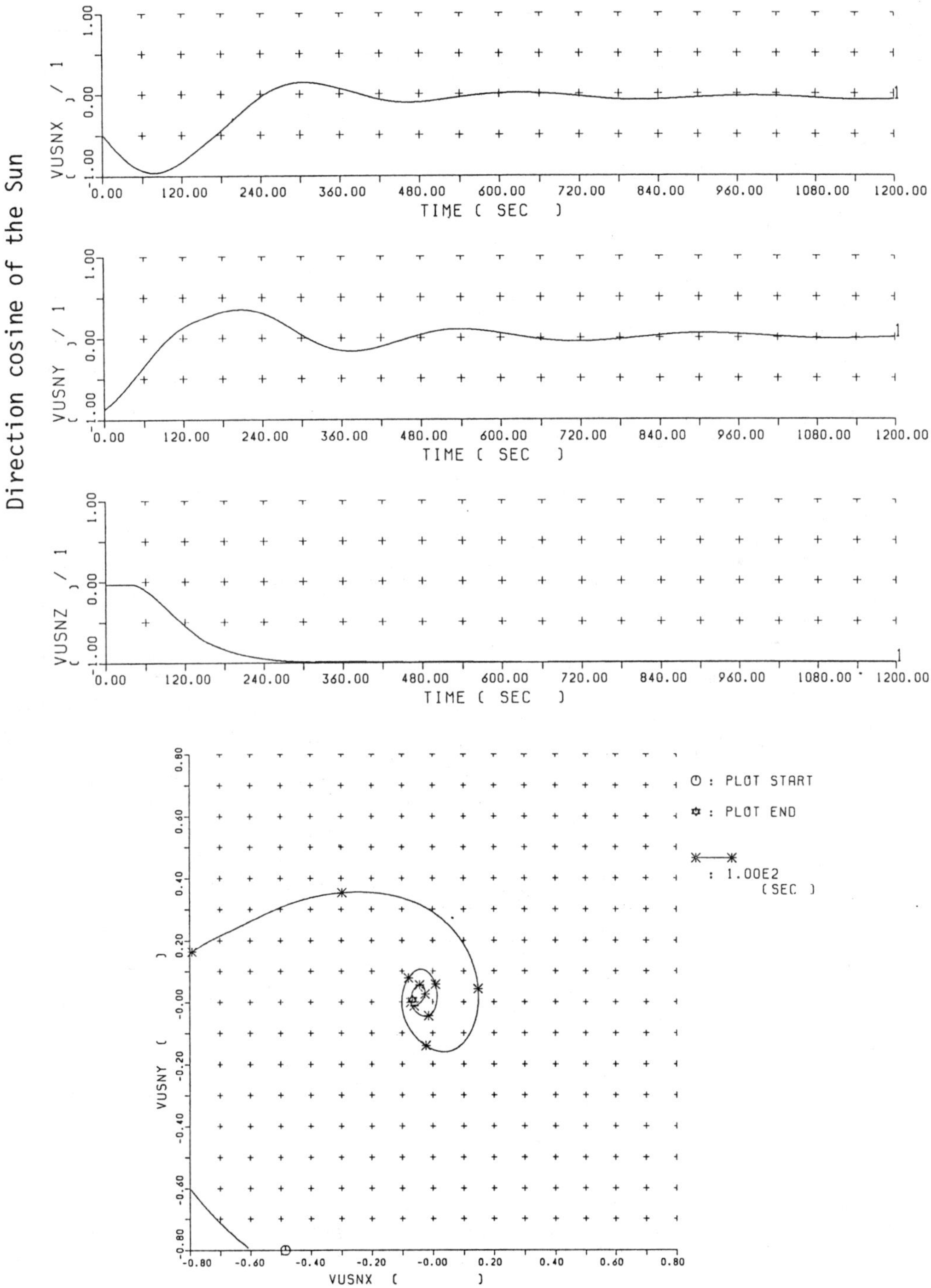

Fig. 12 Simulation Results of Attitude Control (1/2)

(2) Fine Yaw Correction Mode

Euler Angle

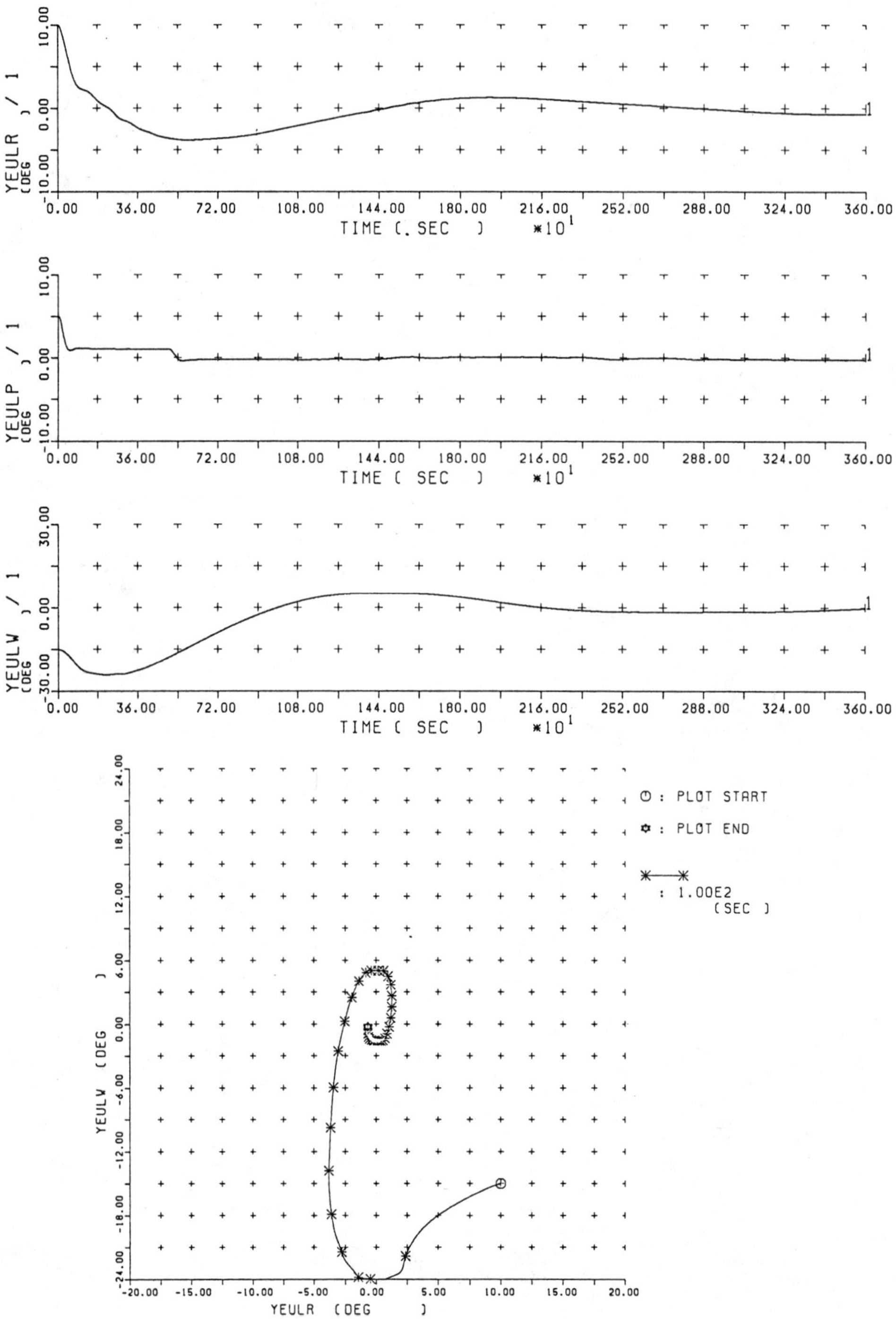

Fig. 12 Simulation Results of Attitude Control (2/2)

CONCLUDING REMARKS

Flight worthiness of the MOS-1 satellite in a simulated flight environment was tested by thermal, static and dynamical test models. Overall system functions and electrical performance tests were conducted by using the Engineering Model. Based on these results the critical design review was completed and the design finalized.

As for the AOCS subsystem, the design was verified by analyses. The performance of hardware components at a subsystem level was tested by using the Engineering Mode (EM) in all modes and the validity of the subsystem components design was demonstrated. AOCS subsystem components of the Prototype Model (PM) went through the subsystem quarification level environment tests successfully. Performance tests results of PM AOCS subsystem were the same as those of the EM subsystem, hence the design was verified twice.

Qualification tests of most of MOS-1 subsystems have been completed and no major design changes are foreseen. However, some flexibility in design still remains as detailed operational plans for the MOS-1 are not yet finalized.

The development of the MOS-1 AOCS subsystem was undertaken by Mitsubishi Electric Corporation Kamakura Works as a subcontractor to NEC, the prime contractor of MOS-1 satellite, and primarily monitored by the Guidance and Control Division of NASDA Tsukuba Space Center. In conclusion, we would like to express our appreciation for the excellent cooperation and assistance of the many personnel who contributed to the success of the MOS-1 project.

AAS 84-004

ATTITUDE CONTROL OF THE INFRARED SPACE OBSERVATORY AND ITS PREDECESSORS

R.J. Hamann
Fokker B.V., Space Division, Schiphol-Oost, The Netherlands

ABSTRACT

This paper describes the design of the ISO attitude and orbit control subsystem and shows the features it has in common with the two highly successfull national Dutch projects ANS and IRAS. The continuous development of a computer centered, highly autonomous and flexible concept is illustrated by discussing the role of the computer, the a priori determination of accuracies and the control algorithms employed.

INTRODUCTION

The Infrared Space Observatory (ISO) has been selected in 1983 as the next scientific satellite of the European Space Agency. Phase B of the project will start in the beginning of 1986 and launch is planned for end 1992. During the industrial phase A study, performed in the period from October 1981 to November 1982, Fokker B.V. has been responsible for the Attitude and Orbit Control Subsystem as a subcontractor to MBB.

The concept of the ISO AOCS draws heavily on experience gained during the two national Dutch projects ANS (Astronomical Netherlands Satellite, launched in August 1974) and, together with the US and the UK, IRAS (Infrared Astronomy Satellite, launched in January 1983). In order to illustrate the commonality of the concepts and the development over more than a decade, a short description is given of the attitude control concept of both ANS and IRAS.
The evolution of the ISO AOCS design is best illustrated by comparing the three concepts (ANS, IRAS and ISO) according to the following characteristics:

- The role of the On-Board Computer in the Attitude (and Orbit) Control System.
- The specified accuracies and the methods to determine them a priori.
- The control algorithms employed.

During the IRAS mission a number of in-orbit experiments have been executed especially aimed to obtain a better knowledge on some areas of the ISO design. Those related to

- identification of slosh phenomena
- autonomous on-board modelling of misalignment change

are described and a summary of the results is given.

A(O)CS DESCRIPTION

Astronomical Netherlands Satellite (ANS)

ANS, the first satellite designed and built in the Netherlands, has been launched in August 1974 in a sunsynchronous twilight orbit with a perigee of 266 km and an apogee of 1175 km, instead of the intended 540 km circular orbit. The mission required pointings of limited duration (± 15 minutes), for which a star sensor integrated in the telescope of the major instrument was used. In addition a slow scan has been incorporated to study extended sources. The mission duration has been 1½ years. The satellite was 3-axis stabilized and the attitude was defined such that the instruments, aligned with the satellite X-axis, were pointing away from the earth. The Z-axis was directed towards the sun. Operational fine pointing modes were

- slew at rates upto 2 arcdegrees/second,
- scan with orbital speed and 2 arcminutes/second accuracy,
- slow scan at 0.1 times the orbital speed,
- star pointing with 1 arcminute accuracy,
- off set pointing.

Sensors in the fine pointing modes were the Fine Sunsensor (FSS) and the Horizon Sensor (HSE, slew and scan) or Starsensor (SSE, pointing). Actuators were three orthogonally placed reaction wheels (RWL).

In addition the following acquisition and back-up modes were present

- Despin after separation by means of a yo-yo (YY) system.
- Solar detumble under guidance of six coarse sun sensors (CSS) and with three magnet coils (MC) as actuators.
- Magnetic detumbling, reducing residual angular momentum to near zero; sensor was a three-axial magnetometer (MGM), actuators were the coils (MC).
- Sun acquisition. Sensors: CSS, actuators the RWL's.
- Intermediate control to keep the Z-axis pointed to the sun. Sensors: two intermediate sun sensors (ISS), actuators the RWL's.
- Emergency mode, a minimum power mode keeping the Z-axis pointed to the sun.
- A passive eclipse mode, during which only CSS and ISS data are taken to determine the satellite illumination status.

ACS unit	ACS functions									
	despin	*solar detumble*	*magnetic detumble*	*sun acquisition*	*intermediate control*	*slew/scan*	*star pointing*	*angular momentum unload*	*emergency mode*	*eclipse mode*
MGM			X		X			X	X	
CSS		X		X					X	X
ISS					X				X	X
FSS						X	X			
HSE						X				
SSE							X			
YY	X									
MC		X	X					X		
RWL				X	X	X	X		X	
ACL		X	X	X	X			X	X	
OBC						X	X			

Table I ANS Attitude Control functions

- An angular momentum dump mode, to transfer the excess satellite angular momentum stored in the reaction wheels by means of the coils to the earth magnetic field.

The ACS is grouped around the central On-Board Computer, which contains the algorithms for the fine pointing modes. The other control modes are contained in the hardwired Attitude Control Logic (ACL). A block diagram of the subsystem in shown in figure 1, while the ACS control functions are listed in table I.

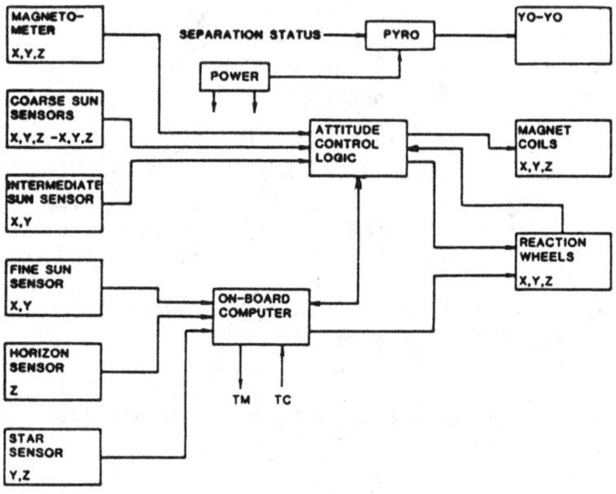

Figure 1 ANS block diagram

<u>Infrared Astronomy Satellite (IRAS)</u>

IRAS has been launched in January 1983 into a circular sunsynchronous twilight orbit of 900 km altitude. The mission has ended in November 1983 upon depletion of the Helium cooling fluid for the telescope. The mission objectives were

- to produce a complete survey of the sky in the infrared,

- to perform special observations of preselected sources.

Corresponding requirements on the Attitude Control Subsystem asked for

- scanning the sky at a constant speed and at solar aspect angles between 60 and 120 arcdegrees,
- performing pointings and rasterscans of 15 minutes duration maximum.

In addition the telescope aperture is to be kept away from heat sources, primarily sun and earth, under all conditions, a requirement which demands extensive attitude safeguarding functions.

A block diagram of the 3-axis stabilized IRAS is shown in figure 2. Its Attitude Control Subsystem is a 2 Hz sampled data system. It is of a modular structure and relies on a sensor configuration which is partly redundant and which allows several back-up possibilities in case of failures in hardware or software. All ACS processes are controlled by a redundant central

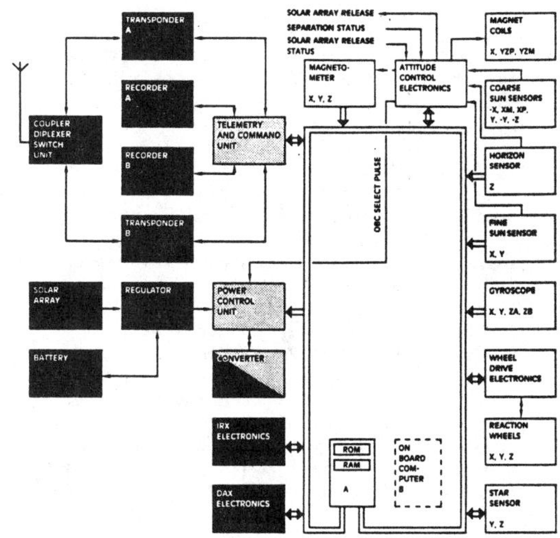

Figure 2 IRAS block diagram

ACS unit	detumble	search (sun acquisition)	intermediate control	back-up intermediate ctl.	MGM eclipse mode	GYR eclipse mode	coarse control	MGM coarse control	fine control	coarse attitude calibration	fine attitude calibration	angular momentum unloading	attitude safeguard (ROM) 1)	attitude safeguard (RAM)	all gyro	eclipse survival
CSS		X														
FSS			X	X			X	X	X	X	X		X	X		
HSE			X				X			X			X	X		
MGM	X			X	X			X		X 2)		X				
GYR						X		X	X	X	X		X	X	X	X 3)
SSE											X 2)					
RWS		X	X	X	X	X	X	X	X	X	X	X				X
MCL	X											X				
ACE	X	X										X				
OBC	X	X	X	X	X	X	X	X	X	X	X	X	X	X	X	X

1) hardwired (computer switch) safeguard in ACE
2) not in control loop
3) Z-axis only

Table II IRAS Attitude Control functions

on-board computer (OBC). The mission critical functions are contained in a non-volatile Read Only Memory (ROM) while the fine control algorithms are contained in a Random Access Memory (RAM) which is reprogrammable from the ground.

As prime actuators a set of three orthogonal reaction wheels is used. Attitude references are formed by the sun, the earth, the earth magnetic field, inertial space, and, for gyro calibrations and reconstruction purposes, by stars. The nominal attitude is when the X-axis, which is the telescope axis, points away from the earth while the Z-axis points towards the sun for illumination of the solar panels. The attitude with respect to the sun can be measured directly by a fine sun sensor: FSSX for rotations about the X-axis, FSSY for the Y-axis. The attitude about the Z-axis is obtained from an earth-horizon sensor - HSE - or, when performing the observations, using a rate integrating gyro. The three reaction wheels are controlled by central wheel drive electronics.
The two redundant on-board computers each have a 3 K ROM and a 29 K RAM.

The ROM contains all the functions which are essential for the safety of the satellite. These are summarized hereafter.

After separation from the launcher a SEARCH algorithm rotates the satellite so that the sun enters the field of view of the FSS after which a transition to Intermediate Control is effectuated. In the search mode attitude information is provided by five coarse sun sensors.

The objective of the INTERMEDIATE CONTROL mode is to bring and keep the satellite in an attitude where the Z-axis is directed towards the sun and the X-axis away from the earth. Sensors are the fine sun sensor and the horizon sensor, actuators the reaction wheels.

In case of a malfunctioning HSE, the BACK-UP INTERMEDIATE CONTROL mode, using the Z-axis gyro as sensor, rotates the satellite with orbital speed in order to maintain a safe attitude with respect to the earth. Attitude updates are performed over the earth's poles by alignment of the X-axis with the earth magnetic field.

For an anomalous separation inducing an amount of momentum into the satellite that exceeds the capacity of the reaction wheels, a DETUMBLE algorithm has been implemented, where the access angular momentum is transfered to the earth magnetic field by the magnet coils.

Two eclipse modes have been implemented. In the MGM ECLIPSE MODE the experiment axis is to be aligned with the local earth magnetic field because the normal, earth induced, eclipses occur near the north pole where the field is nearly vertical. During moon induced eclipses, which are expected to occur near the equator, the GYRO ECLIPSE MODE guarantees a safe attitude with respect to the sun and the earth by keeping the attitude inertially fixed. In case of malfunctioning of one of the sensors used, both eclipse modes can serve as a back-up to the other.

The ATTITUDE SAFEGUARDING function, active under RAM control only, anticipates dangerous attitudes. When observing attitude/rate combinations from which attitudes outside the field of view of FSS or HSE could develop, control is transfered to ROM.
The operational fine control modes are

- scan
- pointing
- rasterscan,

and are executed under guidance of the FSS and the gyro's. At regular intervals the attitude error build up around the Z-axis due to gyro drift is updated by a calibration by means of the focal plane integrated starsensor. A summary of ACS control functions is given in table II.

The Infrared Space Observatory (ISO)

Observations during the ISO mission call for inertial pointing over extended periods of time during a 600 days satellite lifetime. AOCS tasks are:

- Imparting a velocity increment of 85 m/s to the satellite in order to acquire and maintain the operational orbit.
- To control the experiment package attitude during observations and to determine the real attitude such that the mission requirements are met.
- To execute slews up to 270 degrees at a rate of 6 degrees per minute.
- To safeguard the satellite attitude with regard to the sun, earth and moon and to perform subsystem reconfiguration based on internal mode transition criteria or health check results.
- To counteract disturbances generated by environment and payload.

The operational ISO orbit is an equatorial 12 hours orbit with a perigee of 1000 km and an apogee of 39395 km. The attitude is defined such that the telescope line of sight (X-axis) is pointing in an inertially fixed direction, maintaining the Y-axis nominally normal to the sunvector. The solar aspect angle is constrained between 60° and 120°, while the satellite may have any attitude about the sunvector within the safeguard constraints.
To obtain sufficient accuracy both absolute (stars, sun, earth magnetic field) and relative (inertial space)

attitude references are required.

Besides the fine pointing mode, other AOCS modes are designed to ensure a safe operation of the satellite at all times during the mission:

- Acquisition mode, reducing separation induced angular momentum to near zero, and ensuring sun and earth acquisition.
- Safe mode, both fall-back mode if fine pointing is lost, and transition mode after initial acquisition.
- Intermediate mode, a transition mode between safe mode and the fine pointing mode, controlling angular rates such that initial star acquisition can take place.
- Fine pointing mode, the 3 axes inertially fixed attitude mode for scientific observations, including a dedicated mode to calibrate the misalignment between telescope line of sight and the absolute attitude reference.
- Slew mode, required to point the spacecraft to a new target.
- Velocity increment manoeuvre mode, closely resembling the fine pointing mode, to execute ΔV-manoeuvres.

The predicted accuracy of the fine pointing mode for observations is (AOCS contributions only):

	half cone angle (arcsec)
Absolute Pointing Error	6.9
Absolute Pointing Drift	1.6
Relative Pointing Error	2.1
Absolute Measurement Accuracy	4.5

In addition to the control modes a function is implemented to dump excess angular momentum by means of the reaction control system. This function is only activated during perigee pass in order to

- minimize orbit perturbations,
- minimize telescope contamination hazard,
- prevent unacceptable attitude excursions during operational pointing.

Major ISO mission characteristics and actuator algorithm performance are for operational pointing slightly more demanding than those for the IRAS mission, whereas the orbit control requirement can be treated as another subsystem building block. Consequently the IRAS system has been taken as the baseline for ISO and upgraded or expanded (orbit control function) where appropriate.

The AOCS for ISO is of a modular structure; sensors and actuators being grouped around an Attitude Control Computer, which contains both read-only and reprogrammable memory, and which is slaved to the Data Handling Control Terminal Unit to the extent that autonomous AOCS decisions may be inhibited if required. A block diagram of the subsystem is given in figure 3.

The AOCS sensors are two coaligned startrackers, a two-axes fine sun sensor, a three-axes gyro package, a quadrant star sensor alignment reference integrated in the telescope focal plane, five coarse sun sensors and a three-axes magnetometer. The actuators are reaction wheels (three orthogonal, one skewed) and a hydrazine reaction control system (c.f. figure 4) for orbit control and excess angular momentum dump. Most hardware has been

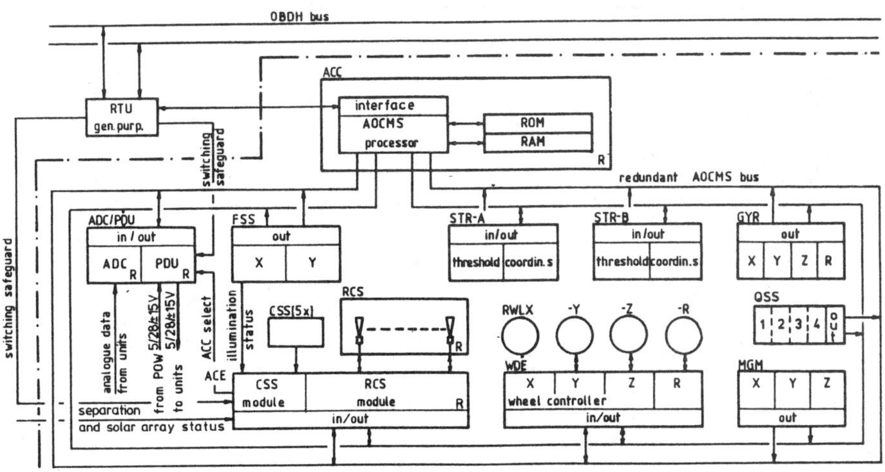

-E Electronics (-)R Redundant -S Sensorhead RTU Remote Terminal Unit (-)X,(-)Y,(-)Z X,Y,Z-axis

Figure 3 ISO AOCS block diagram

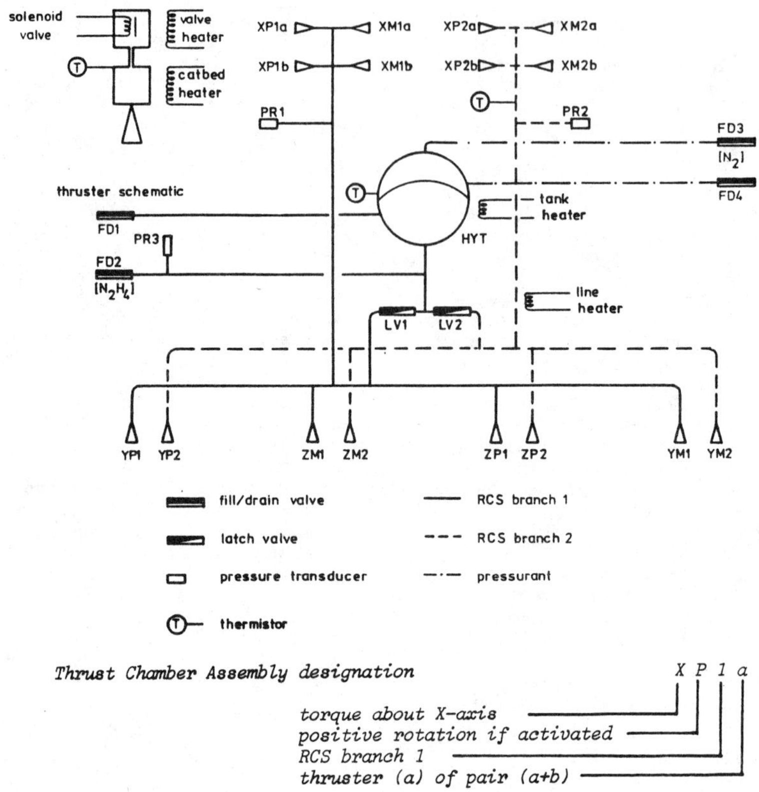

Figure 4 ISO reaction control system block diagram

unit	origin	modification	supplier
Fine Sun Sensor (FSS)	IRAS	minor (electronics)	Adcole (USA)
Coarse Sun Sensor (CSS)	IRAS	none	TPD (NL)
Gyro Package (GYR)	IRAS	minor (electronics)	Ferranti (GB)
Magnetometer (MGM)	IRAS	minor (electronics)	Schonstedt (USA)
Startracker (STR)	EXOSAT	minor (electronics)	Sodern (F)
Quadrant Star Sensor (QSS)	–	new design	–
Reaction Wheel (RWL)	IRAS	none	Teldix (D)
Wheel Drive Electronics(WDE)	IRAS	minor	Teldix (D)
Attitude Control Electronics (ACE)	IRAS	major (RCS module)	Fokker (NL)
Attitude Control Computer (ACC)	–	new design, based on IRAS subsystems	Philips (NL) Saab (S)
Analogue Digital Converter/Power Distribution Unit (ADC/PDU)	–	new design, based on well known technology	Philips (NL) ETCA (B)
Reaction Control System(RCS)	ECS	minor (lines)	ERNO (D)

Table III ISO hardware status and availability

taken from previous programs with only minor modifications in a number of cases (c.f. table III). The AOCS control functions are summarized in table IV.

Total AOCS mass is 216 kg, of which 118 kg is hydrazine and 26 kg is dry reaction control subsystem mass. Subsystem power consumption in the various modes is:

• Launch	68W
• Safe mode (initial acquisition)	78W (109W)
• ΔV manoeuvres	129W
• Operational modes	
– sunlight	76W
– eclipse	78W
– momentum dump	108W

AOCS verification and testing is based upon the philosophy

- verification of S/S performance for both in-orbit conditions and for specific ground test conditions by extensive computer simulations,
- execution of static closed-loop tests of all modes and transitions during S/S acceptance testing,
- execution of dynamic closed-loop tests of operational pointing modes on a single axis air bearing.

For all units requiring minor or no modifications a limited model philosophy (flight and spare models only) is chosen; for the remaining units also an engineering model will be produced.

ACS unit	ACS functions: *launch* 1)	*detumble*	*sun acquisition* 2)	*earth acquisition* 2)	*safe/coarse mode* 2)	*intermediate mode*	*alignment calibration*	*operational pointing*	*angular momentum unload*	*attitude manoeuvres*	*ΔV-manoeuvres*	*attitude safeguard (ROM)* 3)	*attitude safeguard (RAM)*
ACC	X	X	X	X	X	X	X	X	X	X	X	X	X
ADC/PDU	X	X	X	X	X	X	X	X	X	X	X	X	X
RWS			X	X	X	X	X	X	X	X			X
GYR	X	X				X	X	X		X	X		X
MGM				X	X	X 4)							
STR							X 4)	X			X	(X)	X
ACE	X	X	X						X		X		X
FSS			X		X	X	X	X		X	X	X	X
CSS			X										
RCS	X	X							X		X		
QSS							X						

1) stand-by
2) during T/O perigee: actuator RCS
3) hardwired (computer switch) safeguard in ACE
4) not in control loop

Table IV ISO Attitude and Orbit Control functions

CHARACTERISTIC A(O)CS ASPECTS

The A(O)CS aspects of ANS, IRAS and ISO reveal their commonality and evolution in a clear way, if they are compared on the characteristics

- role of the On-Board Computer in the A(O)CS,
- accuracies,
- control algorithms.

Table V summarises the most important aspects of these characteristics.

The role of the On-Board Computer

The ANS central On-Board Computer performed the following tasks:

- To control the attitude of the satellite according to various fine control modes as required by the experiments.
- To store and execute observation programs of twelve hours duration for the three on-board scientific experiments.
- To collect, reduce and store experiment data, attitude data and a selection of housekeeping data.
- To present time tagged commands via Telecommand Unit.
- To provide clock signals to the telemetry, the experiments and other subsystems.

All remaining attitude control modes, i.e. those other than the fine control modes, were accommodated in the hardwired Attitude Control Logic. The OBC memory was a core memory, which conserves the stored information also if the unit is not powered. The size of the attitude control program was 0.5 Mbits; a twelve hours

	ANS	IRAS	ISO
On Board Computer			
interface memory capacity	serial star 0.46 Mbits core	serial bus 2x0.05 Mbits ROM 2x0.59 Mbits COMS RAM	serial bus 2x0.05 Mbits ROM 2x0.22 Mbits RAM
AOCS program size	0.05 Mbits	0.05 Mbits ROM 0.12 Mbits RAM	0.05 Mbits ROM 0.14 Mbits RAM
addition time attitude control program	128 µs acquisition and safe modes in hardware; operational fine modes in firmware;	16 µs acquisition and safe modes in firmware (ROM) operational fine modes in software; hardware watchdog	16 µs acquisition and safe modes in firmware (ROM); operational fine modes in software; hardware watchdog; extensive on-board healthcheck and reconfiguration
autonomy	12 hours	12 hours	2 hours around perigee
A(O)CS accuracies			
boresight requirement	control accuracy . pointing 1' . scan 2'/s	control accuracy* . scan 603"/67" . pointing 27"/23" reconstruction accuracy* 18"/23"	Abs. Pointing Error 17" Abs. Pointing Drift 10" Rel. Pointing Error 4" Abs. Measurement Acc. 9"
accuracy definition	-	-	ESA Handbook of Satellite Pointing Errors
accuracy analysis misalignments effects	RSS prime attitude sensor focal plane integrated	RSS frequent attitude calibrations with focal plane integrated reference sensor	pseudo-statistical on-board modelling of alignment change and update of model by frequent calibrations with focal plane integrated reference sensor
Control algorithm	averaging of noisy sensor output (Horizon Sensor); PID	mixing filter of Fine Sun Sensor and Gyro outputs; PID	mixing filter of Startracker and Gyroscope outputs; state estimator; PID

* Fist value: in scan; second value: cross scan

Table V Summary of A(O)CS characteristics

observation sequence, including attitude changes, was commanded by a Satellite Operations Plan, consisting out of 1K16 bit words. Program timing was obtained by an one-second hardware restart.
The reprogrammability of the OBC has shown to be indispensable for mission success. Examples are

- the adaptation of on-board attitude control algorithms for the highly anomalous orbit,
- continuation of observations during eclipses (not foreseen).

The IRAS OBC performs the same tasks as the ANS OBC, and in addition

- ground command handling,
- down link data formatting,
- all ACS safe and acquisition modes,
- housekeeping monitoring and control.

The computer is a derivative of the Philips P850 mini-computer series, and employs the same instruction set.

The stringent attitude safeguard requirements for the cryogenic telescope have introduced many additional back-up modes, while the higher accuracy demands more sophisticated on-board attitude calculation and control laws. The programs for the six safe modes and the attitude safeguard are located in a ROM, while the operational programs are stored in RAM. A simple hardwired watchdog device (the ACE) supervises OBC actions by executing a check on the actual attitude of the satellite. If the attitude test is failed, it commands a switch over to the redundant computer.

The computer system is of a modular structure, and has a standard serial bus. The software program is subdivided into 4 msec slots, a feature which simplifies trouble shooting and even allows a bypass of (faulty) ROM located functions.
Again the reprogrammability of the computer has been a useful characteristic. It has allowed the operations team to optimize some operation modes and has eliminated an otherwise fatal hardware anomaly (erroneous fine sun sensor outputs). In total 18 program changes have been made in orbit.

The Attitude Control Computer (ACC) in the ISO AOCS is set up along the same lines as the IRAS OBC. However, the ACC is slaved to the ISO data handling system, and controls only AOCS related tasks.

Software organisation is identical to IRAS and the bus concept is maintained. In addition to the hardwired simple watchdog function the ISO Data Handling computer may overrule the (otherwise autonomous) configuration management of the ACC. The more severe redundancy requirements of the ISO project enable extensive on-board unit healthcheck, in which way more functional back-up modes could be implemented.

AOCS accuracies

From the conceptual design phase of a satellite system onwards it is of utmost importance to generate and maintain a coherent set of accuracy requirements. Two aspects play a role

- definition of the accuracies,
- combination of individual error sources into an overall system accuracy.

A good understanding of these aspects is essential to ensure the matching between mission objectives and actual system performance and the assignment of subsystem and unit accuracy budgets.

In practice both aspects are not always treated with sufficient attention. Accuracies are defined in terms of control accuracy, stability and jitter, the definition of which is specific to a project. Error combination rules vary between root-sum-square to worst case adding, and individual error sources are often not explicitly classified. Below we will indicate the development in establishing accuracy performance in the three projects under discussion. Emphasis will be given to predicted accuracies.

In the ANS project accuracy definition has been limited to the fine control pointing modes. In practice the analysis has been executed for the (focal plane integrated) star sensor only. Only absolute errors in rate and position were defined, and individual error sources were combined in a root-sum-square (RSS) fashion, irrespective of their frequency domain. As a result out-of-spec performance has been found only during computer simulations and hardware tests, sometimes requiring crash actions to eliminate discrepancies.

System level accuracies for IRAS have been defined and allocated in the same fashion as for ANS, however some specification in the frequency domain has been added. Telescope and ACS error sources were combined in a RSS fashion, without distinguishing between systematic and other errors. Analytical verification of the ACS accuracy in a later stage of the project showed some potential problem areas related to the chosen error combination rules, as the RSS and worst case ACS accuracies showed a large difference upto one order of magnitude. This led to a more refined approach, where individual error sources were asigned to three categories:

- systematic (constant),
- long-term pseudo-systematic,
- short-term random.

The error sources within a category were combined in a RSS fashion, while the resulting accuracies per category were linearly combined to achieve overall ACS accuracies. The error distributions were assumed to be Gaussian.
Although the approach outlined above still leaves room for subjective judgement, it is felt to be more realistic than the RSS method (too optimistic) or the worst case method (too pessimistic, and hence leading to overdesign).

In the ISO project the subject of accuracy has been treated according to well established definitions and error combination rules from the very beginning. Accuracy definitions were taken from the ESA Handbook of Satellite Pointing Errors.

Table VI shows the unified terminology framework presented in the handbook. The relevant accuracies selected for ISO are:

- Absolute Pointing Error (APE)
- Absolute Pointing Drift (APD) over one hour
- Relative Pointing Error (RPE) over 30 seconds
- Absolute Measurement Accuracy (AMA).

Definitions of these accuracies are:

- The Absolute Pointing Error is the angular separation between the commanded or desired pointing direction and the actual or achieved instantaneous direction of a spacecraft fixed axis, in the plane containing the two axes.
- Absolute Pointing Drift is the angular separation between a reference axis or axis frame during some time interval T_1 and a like reference axis or axis frame defined during another time interval T_2, in the same observation period.
- Relative Pointing Error is the angular separation between the instantaneous absolute orientation of a spacecraft fixed axis or axis frame at some time t and a reference axis or axis frame defined over a specified time interval T, where t belongs to T.
- Absolute Measurement Accuracy is the angular separation between the actual and the measured:
  - location of a spacecraft fixed axis or axis frame
  - location of a celestial phenomenon defined instantaneously or over a time interval.

The following individual error sources have been taken into account:

dS_1	absolute sensor internal error
dS_2	absolute sensor internal error change
dS_3	absolute sensor internal noise
dA_1	absolute sensor static misalignment
dA_2	absolute sensor misalignment change
dD	gyro random drift
dB	gyro bias-drift change
$dC_{1,a}$	limit cycle
$dC_{1,b}$	state estimate error
dC_2	limit cycle residue
dE_1	accumulating error in on-board calculating proces
dE_2	random error in on-board calculating proces

Classification of these sources is:

systematic	dS_1, dA_1
long term pseudo-systematic	dS_2, dA_2, dB, dE_1
short term random	dS_3, dD, $dC_{1,a}$, $dC_{1,b}$, dE_2, dC_2

For the computation of the system accuracies, the actual ACS configuration and operation mode has to be taken into account, e.g. for the APE during observations the accuracy resulting after

ATTITUDE CONTROL REQUIREMENT	ATTITUDE MEASUREMENT REQUIREMENT
– POINTING • Absolute Pointing • Absolute Pointing Error • Attitude Reproducibility	– POINTING • Absolute Measurement • Absolute Measurement Accuracy
– STABILITY • Relative Pointing • Relative Pointing Error • Relative Pointing Rate • Absolute Pointing Drift • Absolute Rate Error	– STABILITY • Relative Measurement • Relative Measurement Accuracy

Table VI Unified framework for pointing and measurement terminology

a misalignment calibration has to be taken into account.
Accuracies are first calculated per axis, and are then converted to half cone angles. Resulting accuracy values are:

	requirement	predicted performance
coarse/safe mode RPE (arcdeg)	5	2.4
ΔV-manoeuvres		
APE (arcmin)	60	36.3
APD "	30	0.7
RPE "	15	5.1
AMA "	60	35.8
fine control (observations)		
APE (arcsec)	13.4	6.9
APD "	8.7	1.6
RPE "	5.0	2.1
AMA "	7.6	4.5

These values include only AOCS contributions. The difference between RSS, worst case and "mixed" error combination rules is illustrated by the following numbers for the fine control APE:

- RSS 4.2 arcsec
- worst case 12.7 arcsec
- "mixed" 6.9 arcsec

The configuration of the error cones is illustrated in figure 5.

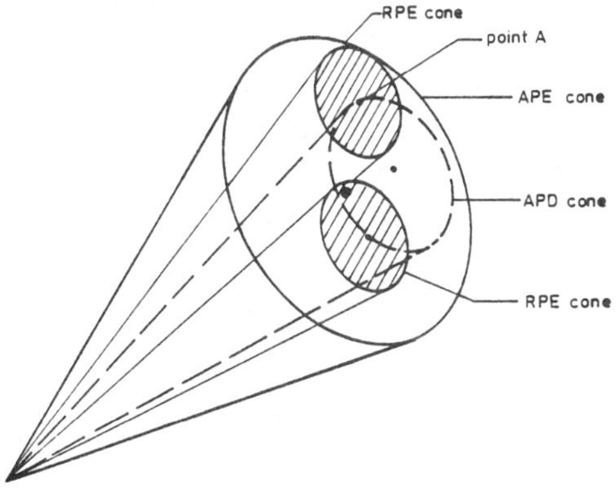

Figure 5 ISO error cone configuration

Point A indicates the worst case position of the alignment reference line of sight immediately after target acquisition, preceded by an alignment calibration, corresponding to a half cone angle

APE − RPE = 6.9 − 2.1 = 4.8 arcsec.

Over one to ten hours pointing duration point A will wander (as a maximum) over the APD cone, full distance = 2 x half cone APD = 3.2 arcsec.

Overall system accuracies are obtained by repeating the accuracy calculation including the thermo-structural misalignment (changes).

Control algorithms

The control algorithms employed in the Attitude Control Systems of ANS, IRAS and ISO show much commonality, but indicate a clear development towards more refined solutions.
In this section the algorithms for

- sun acquisition,
- safe mode (ANS and IRAS: intermediate control),
- fine (operational) modes and attitude manoeuvres

are described.

Sun acquisition

All three satellites have a fixed attitude with regard to the sun. Coarse information on the position of the sun w.r.t. the satellite is provided by five or six coarse sun sensors (CSS) of which the f.o.v.'s cover together the total celestial sphere.

Based on attitude information provided by the CSS, the satellite is rotated about the X- and Y-axis such that the Z-axis is pointed towards the sun. The torque request generation, TR, is defined in table VII as function of the illumination status of the CSS.
The Z-axis torque request is set equal to zero because the only purpose is sun acquisition.
The algorithm provides a fast and reliable sun acquisition and has remained essentially unchanged since ANS. Figure 6 illustrates a simulated sun acquisition sequence for ISO.

Safe mode

The objective of this mode is to bring and keep the satellite in an attitude where the Z-axis is directed towards the sun and the X-axis away from the earth. The criticality of the algorithm for the IRAS and ISO missions is obvious when one keeps in mind that once the cover of the experiment has been ejected, heat input from either the sun or the earth will destroy the telescope system.
The control laws used are basically of the bang-bang type. They effectuate minimum time arrivals in the center of the field of view of the fine sun sensor or at predefined output values of the sensor controlling the third axis.

The ANS algorithm is illustrated in figure 7, and is a bang-bang algorithm amplified with a damping term to limit overshoots.

illumination	CSS+X	CSS−X	$\overline{\text{CSSX}}$
CSS+Y	TR_x = − TR_y = +	TR_x = − TR_y = −	TR_x = − TR_y = 0
CSS−Y	TR_x = + TR_y = +	TR_x = + TR_y = −	TR_x = + TR_y = 0
$\overline{\text{CCSY}}$ and/or CSS−Z	TR_x = 0 TR_y = +	TR_x = 0 TR_y = −	TR_x = 0 TR_y = −

Table VII Sun acquisition algorithm. TR is maximum positive (+), maximum negative (−) or zero.

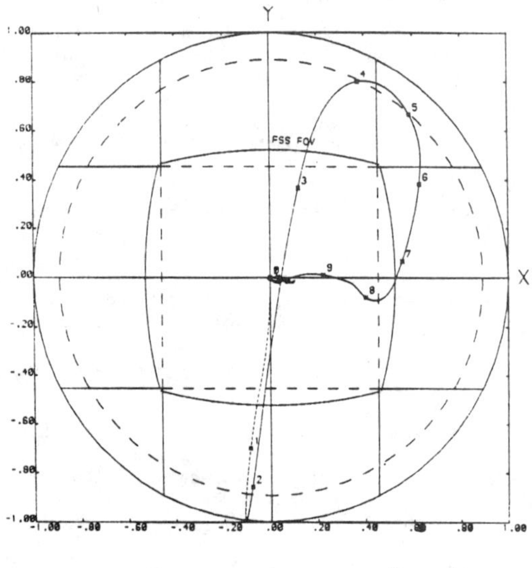

***Figure 6** ISO simulated sun acquisition performance*

It shall be kept in mind that the algorithm has been implemented hardwired, and that the control law has been realized partly in the control logics, partly in the sensor design itself. The algorithm demands always full torque, and hence full power, from the actuators.

In case of IRAS and ISO the bang-bang algorithm has been adapted for the central resolution elements in order to limit the power consumption by the reaction wheels. As a result of this modification constant maximum torqueing is avoided in the limitcycle. In figure 8 the complete control law is explained. The parabola parts in the outer regions discriminate between maximum positive and maximum negative torque.
Due to disturbances (crosscoupling effects mainly) the character of the actual limitcycle will be different from the one indicated in figure 8 which has been obtained analytically.

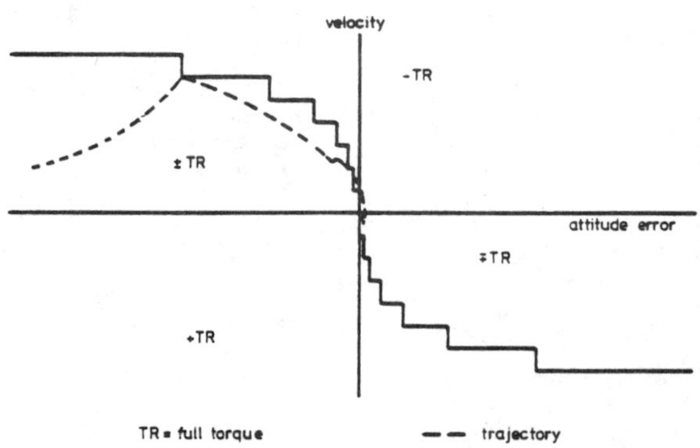

***Figure 7** ANS safe mode algorithm*

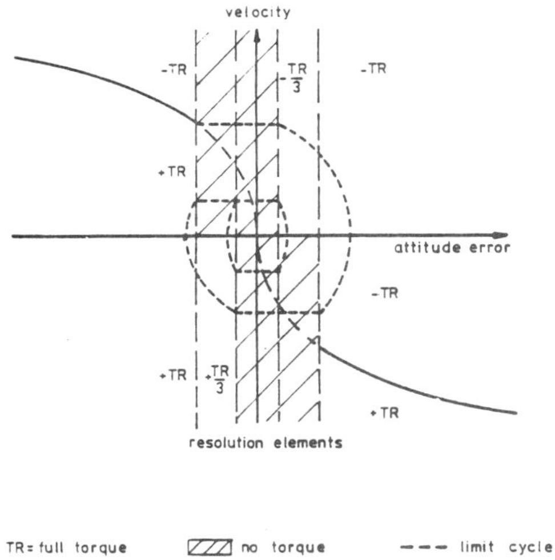

Figure 8 IRAS/ISO safe mode algorithm

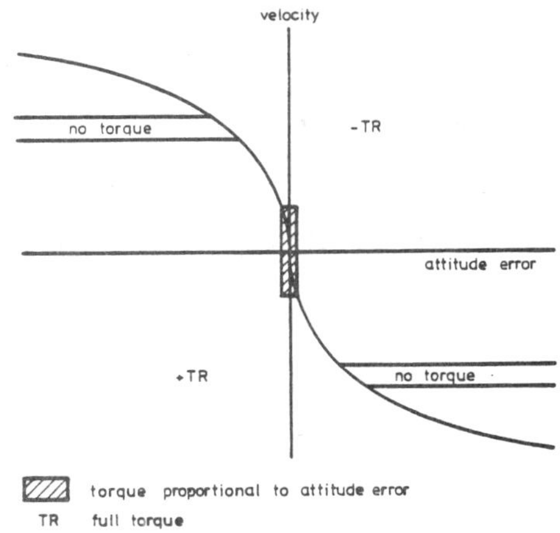

Figure 9 ANS slew algorithm

Whereas the average control torque using an unadapted bang-bang control law would be the maximum value .2 Nm per axis, the here obtained control effort is .2/6 Nm on an average. Further it is emphasized that the modification does hardly increase the amplitude of the limitcycle which, when implementing the mere bang-bang algorithm, would be far from perfect because of the coarse attitude information used and the bad estimation of the angular velocity.

Fine (operational) modes and attitude manoeuvres

The control algorithm implemented for ANS is subdivided into:

- slew algorithm for manoeuvres,
- three sets of PID algorithms for pointing and scan.

The slew algorithm is illustrated in figure 9. Satellite rate is obtained by subtracting two subsequent absolute sensor outputs. The algorithm effectuates minimum time slews between two attitude setpoints and is of a bang-bang type with two modifications.

- The satellite rate is limited to a maximum value to avoid crosscoupling between the axes (regions with torque request zero in figure 9).
- To enable a smoother transition to steady state pointing or scan, the torque request in the center part of the switching parabola varies proportional to the attitude error.

The pointing or scan algorithms are of a conventional PID type

$$T= P\, e_i + D(e_i - e_{i-1}) + I \sum_{j=0}^{i} e_j$$

Figure 10 shows the subdivision of the phase plane into areas with different sets of PID-parameters. The I-term is deleted in the outer regions in order to obtain a faster system response to large errors.

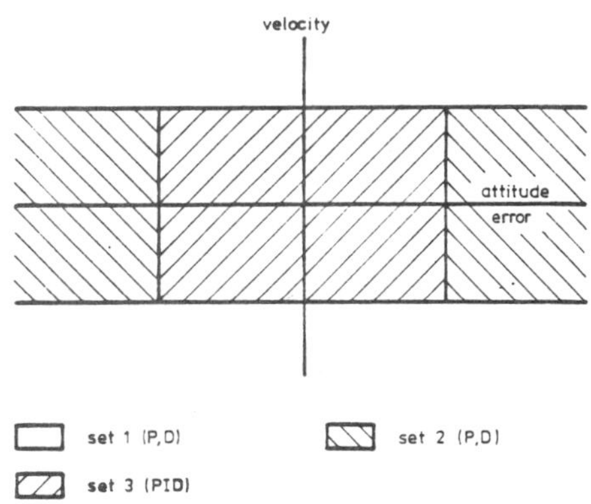

Figure 10 ANS fine pointing laws

The IRAS control law consists out of two building blocks:

- a filter (X- and Y-axis only) to combine the gyro and fine sun sensor outputs in order to limit the effects of noise (fine sun sensor) and drift (gyro),
- a control law, which is used both for slew and pointing or scan.

Through primarily incorporated for safety reasons and to allow observations during eclipse, the gyros have profitably been combined with the FSS in the calculation of the attitude errors. The error is obtained using:

$$e_i = \alpha(\theta_{required} - \theta_{fss}) + (1-\alpha)(e_{i-1} + \Delta_{gyro})$$

in which $\alpha \rightarrow 0$, through which the main contribution to the calculated error is provided by the angular increment $\Delta gyro$ as obtained from the rate integrating gyro. The FSS contribution now consists in continuously updating the attitude error provided by the inertially measuring gyro.

For quick changes of the required attitude a bang-bang control law is used in which the parabola $e + \beta\dot{e}|\dot{e}| = 0$ discriminates between maximum positive and maximum negative torque and so effectuates a minimum time arrival at the newly required attitude. Fast updates of the attitude error are achieved by setting $\alpha = 1$ in the error computation defined before. In order to restrict time consuming overshoots, a smooth transition from bang-bang to PID is desired. This is acomplished by the socalled dual mode control law - DMCL - which combines both in one algorithm as follows:

$$\text{Torque} = Pe + \beta\, Pe|\dot{e}| + D\dot{e} + I\Sigma e$$

in which the integration term is set to zero for large errors.
By inspection of this law the following observation can be made. For large errors it appears that $\beta\, P|\dot{e}| >> D$, which reduces the DMCL to the ordinary bang-bang law. When the errors are small the steady state PID-law is the result.

ISO requires even better fine control accuracy performance (c.f. table V) than IRAS, while the absolute attitude reference, the startracker, exhibits a considerably more noisy behaviour than the fine sun sensor. Application of the IRAS algorithm (α-filter data smoothing and dual mode control law) improves the accuracy by a factor of 1.8, which is amply within specification for the prime observation modes.
However, science return will increase considerably, if the accuracy would be improved further. Hence the application of a state estimator has been investigated. It appears that in that case an improvement of at least a factor of 2.25 may be expected. The resulting structure of the algorithm is shown in figure 11.

The three operational fine control modes

o Observation pointing mode
o Calibration mode
o Attitude manoeuvres

are all handled by the same algorithms, be it that some algorithm parameters may be

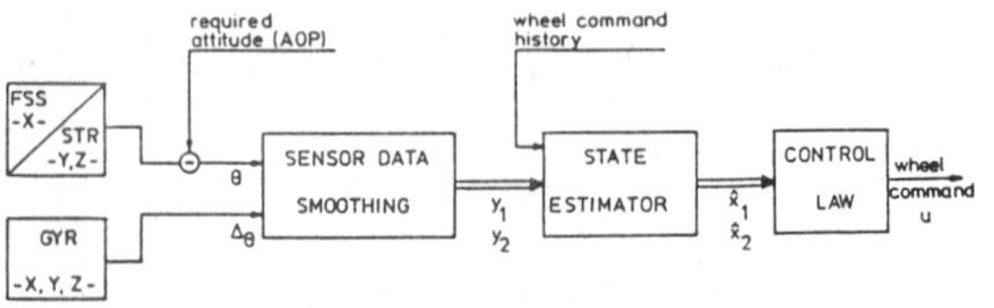

Figure 11 ISO control algorithm structure

adapted to the actual sensor configuration. The algorithms contained in the control blocks are

- Sensor data smoothing:

$$y_1(k) = \alpha\,\theta + (1-\alpha)(y_1(k-1) + \Delta_\theta)$$

$$y_2(k) = (y_1(k) - y_1(k-1))\,\frac{1}{\Delta}$$

where α is filter gain; Δ is sampling period; θ is FSS, STR or QSS output; Δ_θ is GYR output; y_1 attitude and y_2 rate.

This block is essentially identical to the IRAS algorithm.

- State estimator (Luenberger observer):

The discrete time single axis dynamic description for ISO is given by

$$\underline{x}(k) = A\underline{x}(k-1) + Bu(k-1)$$

$$\underline{y}(k) = C\underline{x}(k)$$

in which: $\underline{x}$ is the state vector (attitude, velocity),
u is the control effort,
$\underline{y}$ is the measurement vector (attitude, velocity),
A,B,C are (closed loop) system matrices.

For further reduction of the measurement errors the attitude and velocity estimates $\underline{y}$ ($=y_1$, y_2), can be improved by application of a Luenberger (full order) observer. The observer consists of a prediction part, based on the dynamic description of the plant, and of an update part, based on the difference between measurement and prediction.
The observer equations for the system are formulated as

$$\text{Predictor: } \hat{\underline{x}}^-(k) = A\,\hat{\underline{x}}^+(k-1) + B\,u(k-1)$$

$$\text{Update: } \hat{\underline{x}}^+(k) = \hat{\underline{x}}^-(k) + K(\underline{y}(k) - C\,\hat{\underline{x}}^-(k))$$

where:

$\hat{\underline{x}}$ is the state vector estimate (attitude and velocity),

$\underline{y}$ is the measurement vector,
$\underline{u}$ is control effort plus estimated disturbance torque,
K is the observer gain matrix.

- Control law (bang-bang + PID):

$$u = P\,\hat{x}_1^+ + P\,\beta|\hat{x}_2^+|\,\hat{x}_2^+ + D\,\hat{x}_2^+ + (I\Sigma\hat{x}_1^+)$$

(where $\hat{x}_1^+$ is the attitude error and $\hat{x}_2^+$ is the velocity error)

Again this block is essentially identical to the IRAS algorithm.

Disturbance torques in steady state may be coped with either by the integration term $I\Sigma x_1$ in the PID-controller or by direct feedback of a disturbance torque estimate obtained with an extended observer (extension by one state: the disturbance torque).

IRAS IN-ORBIT EXPERIMENTS

Two In-Orbit Experiments, especially ISO oriented, have been executed with IRAS, and have been funded by ESA.

The experiments have been defined and executed in order to

- try and provoke and identify liquid Helium slosh phenomena,
- measure and model on-board thermo-structural misalignment changes between ACS prime sensors and telescope boresight.

Slosh experiment

The objective of the experiment was the identification, modelling and reproduction of slosh phenomena, interacting with the attitude control. Previous studies indicated that bulk motion of helium could produce a maximum disturbance torque of 1.8×10^{-3} Nm, with a period of 300 seconds while the disturbance torques due to local vortices in the fluid were expected to be much less than 0.001 Nm; such local effects would occur as a result of bulk fluid motion as well. Torques of this magnitude can only be monitored by means of the integrator term in the PID control algorithm. A provision has been made to take these data during the experiment.
To provoke the bulk motion of the fluid three manoeuvres have been carried out (see figure 12):

- Two one-axial slews about Y- and Z-axis.
- One two-axial conical motion about the X-axis, which is the axis of the toroidal tank.

Examination of the disturbance torque and the control behaviour after the manoeuvres did not reveal any evidence of interaction between fluid motion and attitude control. Further identification of possible presence of slosh phenomena will be attempted by means of a frequency analysis of the data taken.

Thermo-structural misalignment

The prime absolute attitude sensor of IRAS could not be accommodated in the cryogenic telescope, as will be the case for ISO. Consequently the relation between telescope boresight and ACS coordinate frame is not continuously known and varies in time due to thermo-structural deformations. Both IRAS and ISO carry a focal plane integrated alignment calibration sensor to update at set intervals the actual misalignments. However, the time devoted to these calibrations has to be limited, as it consumes observation time and as suitable calibration stars are not abundantly available.

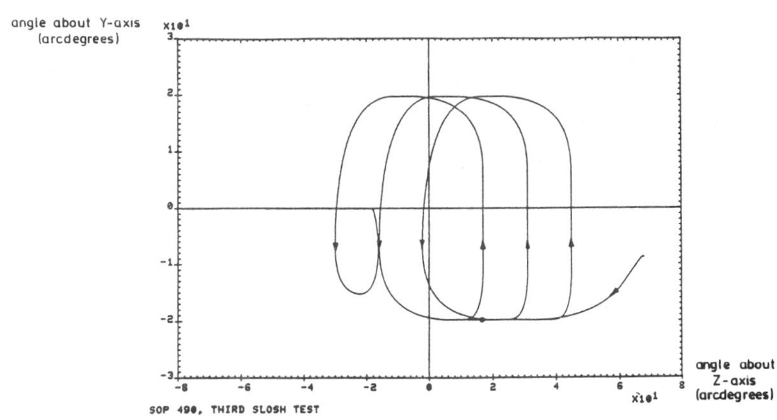

Figure 12 Conical motion to provoke slosh

In order to achieve high pointing accuracies, a suffiently accurate on-board model to predict misalignment changes and update the attitude accordingly is mandatory.
An in-orbit experiment has been executed with IRAS in order to demonstrate the feasibility of such a solution. In-flight measured data on the misalignment changes have been analyzed and two effects have been identified.

- The transfer function of the fine sun sensor has changed with regard to the pre-launch measured data.
- A thermo-structural deformation mechanism exists, which can be reasonably modeled by means of a second order differential equation, expressing the misalignment change as a function of time and solar aspect angle.

The model describes the misalignment error as

$$F(t) = A + Be^{-Ct} + De^{-Et}$$

where C and E are the two major time constants identified, and A, B and D are representing the thermal input to the system as a response to a step change in the solar aspect angle.

Table VIII shows the results of a simulation exercise for a number of SOP's (12 hours observation period). It appears that the model improves the pointing accuracy by 2.5 to 15 arcseconds. The magnitude of the 1σ residual errors is acceptable in view of the magnitude of the "noise" sources (fine sun sensor and star sensor accuracy, limit cycle amplitude), which amounts to root-sum-squared 1σ value of 5 arcseconds.
The algorithm has been implemented on-board, and flight data are evaluated at the time of writing.

	standard deviation of residual errors (simulation-observation) arcsec.	
	SOP 426-429	SOP 490
measured data	27.3	18.3
+ corrected FSS transfer function	21.0	11.0
+ misalignment model	6.4	8.5

Table VIII Results of misalignment model experiment

CONCLUSIONS

The concept of a computer centered Attitude (and Orbit) Control Subsystem for scientific satellites, as has been developed during more than a decade for ANS, IRAS and ISO, has shown to be a flexible, easily adaptable design, which allows for stepwise upgrading of its performance both in hardware and software. Throughout the development the system exhibits a continuous trend towards a modular structure, to which functions and units can be added according to specific mission needs.
The continuous development of such a modular system stimulates the use of standard design tools, an example of which is the systematic approach to accuracy budgetting and a priori performance determination.
The nucleus of the system, a reprogrammable computer, allows recovery from otherwise fatal mission anomalies and enables also the use of the satellite as a flying testbed for future projects.

References

1. ANS Functional Description; November 1972.

2. Arink, G.J.A.; The On-Board Computer of ANS; Philips Technical Review; Vol. 34, No. 1; 1974.

3. Bloemendal, W.; The Netherlands Astronomical Satellite (ANS); Ph.Techn. Rev.; Vol. 33, No. 5; 1973.

4. Van Otterloo, P.; Attitude Control for the Netherlands Astronomical Satellite (ANS); Ph. Techn. Rev.; Vol. 33, No. 6; 1973.

5. Wakker, K.F.; The Astronomical Netherlands Satellite; Ruimtevaart; Vol. 19, No. 1; April 1970.

6. Blomeyer, L.S.; IRAS Mission Joint Project Requirements; 623-6 rev. A; JPL, USA; March 1981.

7. Van Holtz, R.C. and Hameetman, G.J.; In flight software reconfiguration for ANS and IRAS; NLR, the Netherlands.

8. A Handbook of Satellite Pointing Errors; ESS/SS 97; British Aerospace Dynamics Group, UK; March 1980.

9. Teule, F.J.P.; Control Algorithms On board IRAS; Proceedings VIII IFAC Symposium on Automatic Control in Space; Oxford UK; July 1979.

10. Hamann, R.J.; ISO phase A AOCMS Fokker report RV-R-82-043; the Netherlands; September 1982.

11. Teule, F.J.P., IRAS Attitude Control - Principles and Verification -; Proceedings IAF Congress; München; 1979.

12. Teule, F.J.P., and Karsten, L.; Slosh experiment no. 1; Fokker note TR-N-83-026; the Netherlands; Nov. 1983.

13. Karsten, L.; IRAS misalignment modelling; Fokker note TR-N-83-IOX-027; the Netherlands, Nov. 1983.

14. Pouw, A; System design and operational experience with the IRAS Attitude Control Subsystem; AIAA paper 84-0152; AIAA 22nd Aerospace Sciences Meeting; Reno, Nev., USA; Jan. 1984.

AAS 84-005

ATTITUDE DETERMINATION AND CONTROL OF THE HIPPARCOS SATELLITE

Dominique P. Vilain* and Robert S. Harris**

The Hipparcos satellite is an ESA program dedicated to astrometry. The AOCS requirements derived from the scientific mission specifications lead to an AOCS concept based on an opto-inertial attitude estimator of 1 arcsec class and a cold gas actuator system with control action optimized to minimize the number of actuations. The AOCS design is described and the mission phases and operational modes are detailed. The simulation results demonstrating the ability of the baseline design to comply with the most critical requirements are presented.

PRESENTATION OF THE HIPPARCOS MISSION

The Hipparcos project is the European Space Agency's (ESA) astrometry satellite, scheduled for launch by Ariane 3/4 in 1988, for which MATRA (France) has been selected as Prime Contractor. For the B2 phase which ended in December 1983, the AOCS studies have been jointly performed by MATRA (attitude estimation) and BAe (UK) (control laws).

The space astrometry mission Hipparcos aims at measuring the positions, proper motions, and trigonometric parallaxes of about 100,000 selected stars, down to magnitude levels of about B = 13 mag, most of them being brighter than about B = 10 mag.

The end product of the mission is constituted by the Hipparcos catalogue containing these five astrometric parameters with an expected average precision of 0.002 arcsec for the positions and parallaxes and 0.002 arcsec/year for the proper motions.

* *MATRA Space Division - B.P. N°1 - 78146 Vélizy Cedex, France*

** *British Aerospace, Space and Communication Division, Bristol BS12 7QW, England*

The basic principle of observation is to scan continuously and systematically the entire sky with a telescope capable of accurately measuring the angles between stars separated by a large angle (58°). Measurements are obtained by comparing, in the focal plane of a single telescope, the positions of stars imaged from two different fields of view (FOV's) superimposed by means of a beam combiner. By numerically combining several million such angular measurements collected over a time span of 2.5 years, it is possible to derive the astrometric parameters with the required accuracy.

The reference system frame defined by positions and proper motions observed by the satellite, although homogeneous, will not in itself be inertial. However, provided that a sufficient number of stars can be linked to extragalactic objects whose positions are measured through other techniques (e.g. interferometry for radio-sources or Space Telescope for quasars), it can be related to an inertial frame.

For continuous payload operation, a permanent RF link between the satellite and the ground is required either to uplink the next star observation program or to receive the measurement data generated by the payload. As a result, the satellite will be placed in a geosynchronous orbit at 12°W.

The scanning of the whole celestial sphere is obtained by the combination of two motions :

- The satellite is spinning at a constant rate of about 11 revolutions per day around the axis normal to the 2 telescope lines of sight.
- The direction of the spin axis revolves about the direction of the sun, describing a cone of semi-angle = 43° at a mean rate of 6.4 revolutions per year.

The Hipparcos configuration is shown in fig. 0.

DERIVED REQUIREMENTS FOR THE AOCS

- As the satellite has a prolate configuration, the AOCS must provide an active nutation damping (AND) capability for transfer and near synchronous orbit when the satellite is spin stabilized.
- During normal mode the key requirements for AOCS design are the following :
  - To control the satellite attitude in order to follow the scanning law previously defined with a maximum error cone semi-angle of 10 arcmin on the spin axis orientation and a maximum error angle of $\pm$ 10 arcmin around the spin axis. This control has to be performed without generating prohibitive jitter on the optical axes of the payload.
  - To estimate the satellite attitude in real time with an accuracy such that the position of the telescope optical axes with respect to their ideal position can be derived with an accuracy better than 1 arcsec RMS (error cone semi-angle).

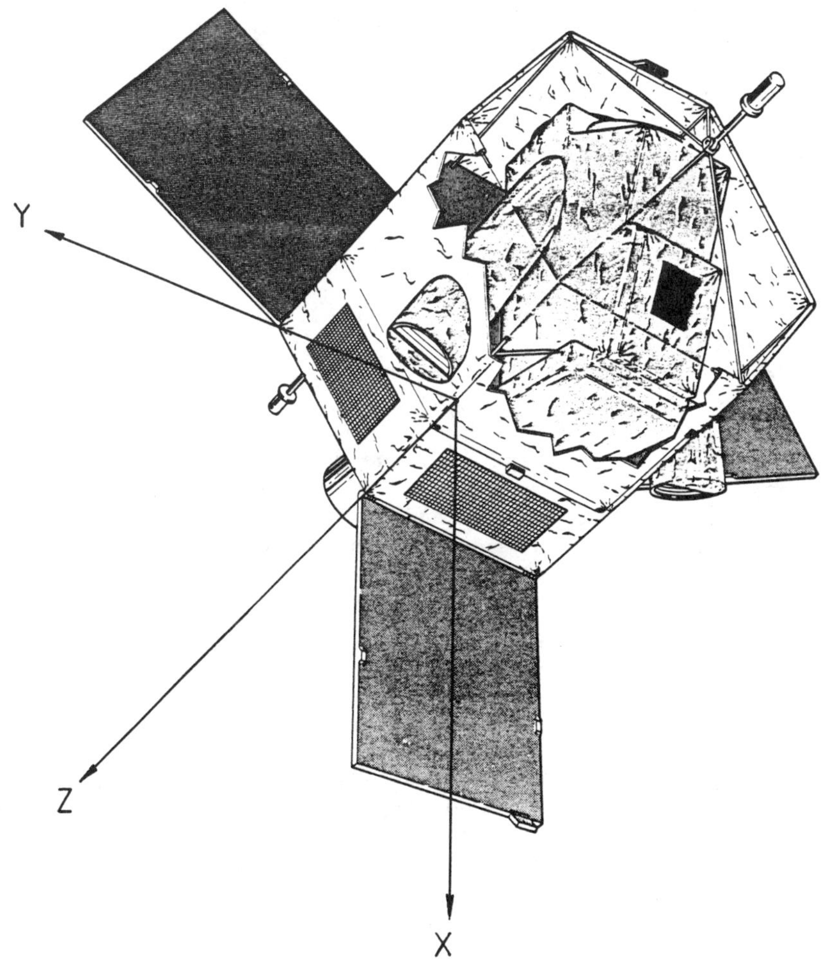

Fig. 0 - Hipparcos Satellite View

This requirement comes from the necessity for the payload to pilot the photomultiplier sensitive area on the star to be observed with a very high accuracy.

From these requirements it is clear that a 3 axis stabilized control concept is necessary. In addition, as the reference scanning law leads the spacecraft to highly variable inertial orientations, a bias momentum wheel concept cannot be applied.

During the feasibility study phase, three main control concepts were considered and traded-off :

a - Reaction wheel continuous control

b - Cold gas intermittent control

c - Cold gas and magneto-torquing

Since the attitude control precision in itself is not very demanding providing the attitude estimation is performed accurately, the selection criterion was the jitter amplitude generated at the payload optical axis level.

For the continuous control concept (a), the constraint was expressed in terms of vibration power spectral density (PSD) envelope. In particular, between 50 and 200 Hz, the PSD had to be lower than 3×10^{-3} (arcsec/ Hz) ; using a NASTRAN model of the satellite excited by the white noise torque associated with a 2 Nms reaction wheel (the smallest capacity off-the-shelf wheel), some flexible mode peaks were found with an amplitude 10 times higher than this upper limit ; therefore the reaction wheel concept had to be rejected.

Cold gas control (b) provides a more attractive alternate solution. In fact, at the expense of more sophisticated control laws, the satellite has quiet periods without any pertubation, unlike the wheels which continuously generate vibrations. In turn, when an actuation occurs, its effect is very strong and can impair for a certain time the payload operations ; thus the validity of the concept fully relies on the capability of the structure to damp out the jitter generated by the cold gas actuations.

A NASTRAN analysis combined with a refinement of the jitter specification expressed in the time domain leads to a set of constraints specifically defined for the cold gas concept :

- maximum thrust level : 0.02 N
- maximum on-time : 0.5 s
- synchronisation between actuations on the 3-axes.
- maximisation of the separation time between actuations with a minimum interval of 100 s.

Magneto-torquing (c) complementing the cold gas actuators was envisaged in order to minimize the number of actuations around Z. However, due to the scanning law which is refered to the Sun, the position of the satellite with respect to the Earth's magnetic field varies over a large range ; this leads, at certain periods of the mission, to a very poor torquing efficiency associated with additional perturbance torques on the transverse axis, inducing compensating gas actuations countering the previous advantages.

As a result, concept (b), using pure cold gas control for the operational phase of the mission, has been retained.

Concerning the attitude estimation, the required high level of accuracy (1 arcsec) necessitated the use of an opto-inertial concept using an Inertial Reference Unit (IRU), constituted by high performance gyros, and star-mappers located in the payload focal plane in order to eliminate mechanical distortion effects.

AOCS BLOCK DIAGRAM AND SUBSYSTEM DESCRIPTION

A block diagram of the AOCS is shown in fig. 1. The Attitude Control Electronics (ACE) is the core of the subsystem and utilises a microprocessor to run the attitude control software. It interfaces with the rest of the spacecraft via the on-board data handling (OBDH) bus ; within the AOCS, it interfaces with the Control and Actuation Electronics (CAE), the Gyro Measurement Electronics (GME) and the Power Supply Electronics (PSE). Its main functions are subsystem telecommand and telemetry management, closed loop attitude control, open loop attitude manoeuvres, real time attitude determination (RTAD) and subsystem configuration and equipment monitoring. During on-station operation, it receives intermittent attitude information via the OBDH bus, which is derived within the on-board computer from star mapper data.

Gyros provide continuous three axis rate and incremental attitude data during final despin and the three axis stabilized mission phases. The gyro package (GYP) consists of five inertial class rate integrating gyros, 2 aligned with the satellite Z axis, one with the X axis, one midway between the X and Y axes and one midway between the X and -Y axes for adequate redundancy. The rate and attitude information is provided by the GME to the ACE via the internal AOCS bus.

The CAE also interfaces with the ACE via the internal bus. Its functions are to condition the outputs of the earth and sun sensor (ESS), the sun acquisition sensors (SAS), the attitude anomaly detector (AAD) and the accelerometer package (ACP), to drive the reaction control subsystem (RCS) latch valves, flow control valves and catalyst bed heaters in response to commands from the ACE, to provide sensor power switching, to provide hardwired emergency attitude control modes during on station operation overriding the ACE, and to provide automatic fault detection and recovery during the spin stabilised phase initiating automatic reconfiguration of the ACE and RCS.

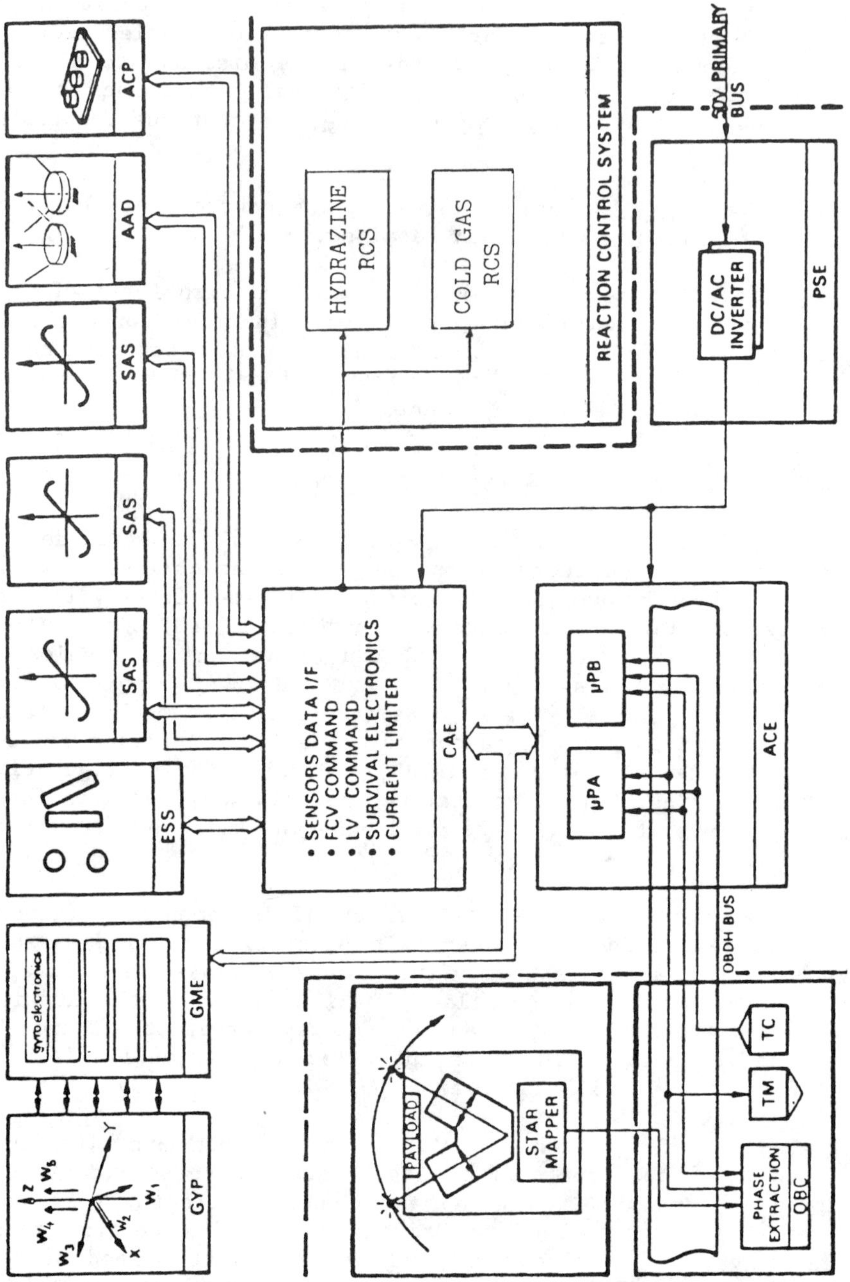

Fig. 1 - AOCS Block Diagram

The ESS is a combined dual beam infrared earth sensor and X-beam sun sensor, used to provide satellite attitude information during the spin stabilized phase. The ACP houses three linear accelerometers which are aligned with the satellite spin axis, to measure the nutation angle. The SAS is a wide angle sun sensor which is used during sun acquisition and sun pointing modes to give, in two axes, the deviation of the sun line from the sensor boresight. It is internally redundant and three are required to give the full 4π steradian coverage. The AAD is a simple sun presence indicator with a cone shaped field of view (46° half angle), which is aligned with the satellite Z axis. During on station operation, a zero output from this sensor may initiate an emergency mode.

The AOCS is powered from the main 50 V dc bus. Loads with varying power demands are taken directly from this bus ; constant loads are powered from the AC bus provided by the PSE.

MISSION PHASES AND OPERATIONAL MODES

Preoperational Spin Phase

On separation from Ariane, the AOCS is initialised into AND mode. The spin rate will be in the range 5 to 10 rpm and the satellite nutation angle may be up to 7°. The AND system senses nutation using a linear accelerometer (3 of which are present in the ACP for automatic fault detection and recovery) and controls nutation using thrusters providing +Y and -Y torques. At 5 rpm the nutation is controlled to within 3°, at 10 rpm to less than 1°.

Some minutes after separation, once the ground link has been established, the AND is inhibited and the spin rate increased to 60 rpm to provide the required stiffness for apogee motor firing (AMF). Spin monitoring is performed on ground from the datated output of the X beam sun sensor (part of the ESS). The spin-up has the effect of reducing the nutation angle present at 5-10 rpm, in spite of the dedamping energy dissipation. On completion of spin-up, the AND is reactivated, the nutation being controlled to below 0.05°.

Prior to AMF the spin axis must be slewed through about 10° for correct alignment of the ABM. The magnitude and direction of the slew are calculated on ground using the datated outputs of the ESS. The manoeuvre is performed by disabling the nominal AND and by pulsing the +Y and -Y thrusters synchronised to the ESS sun reference pulse. The pulses last 250 msec. Nutation is maintained below 0.05° by inhibiting pulses that would increase nutation, as indicated by the sign of the accelerometer output signal. The nominal AND is enabled at the end of the slew and the new satellite attitude calculated on ground. A further fine slew of typically 0.5° will be required in order to achieve the required accuracy for spin axis alignment.

At initiation of AMF, (a Mage 2 motor is baselined) the AND is disabled. Due to the satellite inertia characteristics, nutation angles in the range 9° to 12° are present at the end of AMF and the final spin rate is between 44 and 72 rpm. A preset time after AMF initiation (about 40 secs) the AND is enabled to remove this nutation . Subsequently, the spin rate is measured on ground and restored to 60 rpm by an open loop Z-axis thruster pulse. Normal AND operation then maintains the nutation angle below 0.05°.

East-west thruster actuations will be required for orbit adjustment and station acquisition. As a result, the spin axis must be reorientated normal to the orbit plane. This is performed in the same way as the pre-AMF slew. Subsequently station acquisition is performed using simultaneous +Z and -Z thuster pulses phased relative to the ESS sun reference pulse to produce the required ΔV, the pulse train having been calculated on ground. The nominal pulse length for each thruster is 250 msec. The AND system operates normally during station acquisition.

This near synchronous orbit phase lasts about 12 days, during which, as the hydrazine RCS is utilised in blow-down mode, the nominal thrust level dropping from 6N to 2N, the AND control parameters are updated from ground.

Once on station, the satellite spin rate is reduced from 60 rpm to 10 rpm in three steps. Each spin change is performed in the same way as the initial change from 5-10 rpm up to 60 rpm by disabling the AND, performing an open loop despin and re-enabling the AND. The despin is performed in three steps to avoid excessive nutation build-up. At 10 rpm, the remaining hydrazine is drained off by firing +Z and -Z thrusters continually. The spin rate is monitored on ground from the datated ESS sun reference pulse and any spin rate change due to thruster mismatch is countered by off modulating the appropriate Z thruster for a period which is evaluated on ground. During the hydrazine drain the AND remains active. Subsequently when no further thrust is produced - detected on ground from telemetred tank pressure and thruster temperature data - all the hydrazine thrusters are opened to permit any residual hydrazine to evaporate.

Throughout the spin phases, the capability exists for flat spin recovery, i.e. recovery from a position of spin about the major (Y) axis. Flat spin will only occur due to the production of abnormal nutation during AMF (> 20° at a spin rate of 60 rpm) or a failure outside the AOCS which disables the AND for a significant period. All other faults within the AOCS during the spin phase are detected and corrected automatically on-board. The recovery manoeuvre is performed using the following sequence. When the nutation about the Y axis has decayed to below 10°, the direction of the spin rate about Y is determined by commanding a +Y thruster pulse to change the spin rate by about 5 % and measuring the change in nutation frequency on ground from the telemetred ACP data ; a decrease in frequency indicates a negative spin rate. The relevant Y thruster is then fired to reduce the spin rate to 10 rpm.

At this time, the same Y thruster is pulsed 750 ms on, 1250 ms off, to continue the Y spin rate reduction and the -Z thruster fired continually. This has a similar effect to the use of a skewed thruster (e.g. as on METEOSAT) and results in a flip-over to a spin about the Z axis once the Y rate has been reduced to around 0.6 rpm. Once a spin rate of about 8 rpm is achieved, the nutation angle will be about 20°, so the spin up is stopped and the AND is reinitialised to kill the nutation. This point is identified by a particular signal level at the output of the ACP and the switch on of AND is automatic. The sequence of operations from 8 rpm to 60 rpm is identical to that for the initial spin up.

Preoperational Acquisition Phase

During this phase, the satellite is despun from 10 rpm using the cold gas RCS and is sun-pointed on the +Z axis using three SAS's and gyros for rate information. The despin starts open loop, with spin rate monitoring via the SAS outputs. At about 1 rpm, all three axes are placed in closed loop rate control with a demand of 168.75°/hr on the Z axis and zero rate on the X and Y axes (to remove the residual nutation). The total despin takes about 5.6 hours.

Sun acquisition is performed once the rate has reached 168.75°/hr. Three SAS's are oriented on the spacecraft to give full 4π steradian cover (apart from a small blind spot of less than 0.2 steradians). The outputs of the three SAS's are tested to determine control action, the gyro rate outputs being used to provide damping. Acquisition is performed inside 27 minutes, the +Z axis being sun pointed with a half cone accuracy of 1.55°. The Z axis rate control is maintained at 168.75°/hr $\pm$ 2 %. Solar array deployment is performed when the Z axis is sun pointed.

Commissioning Phase

During this phase which lasts 40 days, the payload is commissioned and the satellite is positioned at the start of the nominal scanning law. Initially the payload covers are removed and, with the satellite rotating as described above, star pattern recognition is performed on the ground from the datated star mapper outputs in order to determine the phase of the rotation and the actual Z axis pointing to within the normal RTAD accuracy. By telemetring the gyro data in addition, the gyros are calibrated on ground, after which the RTAD is initialised and the normal mode controller, as used in the operational phase, switched in. This controller maintains the Z axis pointing within 600 arc sec (half cone) of the desired direction with a Z rate within $\pm$ 2 % of 168.75°/hr.

Upon completion of payload commissioning, the Z axis must be slewed 43° from the sun line to the start of the scanning law, the slew being performed about the X axis. Attitude information for the control loops is provided by the calibrated gyros through the RTAD. Initially, at a predetermined phase of the satellite rotation, when the X axis is in the correct inertial orientation for the subsequent 43° slew, the Z axis rate demand is set to zero.

The Z axis position control system then stops the rotation at the required X axis inertial orientation, after which the 43° slew is performed about the X axis. Due to gyro drift and scale factor errors, the resulting Z axis pointing error will be about 8 arc minutes, within the required normal mode pointing accuracy of 10 arc minutes. At this time, the Z axis rate demand is reset to 168.75°/hr, X and Y axis rate demands are applied to achieve the slow Z axis coning motion about the sun line and the normal mode control loops are enabled. Both gyros and star mapper are used in the RTAD to provide attitude information.

Operational Phase

This last sequence of events initiates the operational phase. Throughout this phase, the program star file for use within the RTAD is regularly updated. In nominal operation, the payload scientific data and the star mapper updates to the RTAD are only interrupted when (a) the earth or moon are about to enter the field of view of either telescope - payload shutters are used to avoid damage to the focal plane instrumentation, or (b) stationkeeping is performed. Only East-West stationkeeping is required ; the disturbance torques are much larger than those present in normal mode as a result of which the normal mode performance requirements cannot be met so that the payload data must be interrupted. The basic control law structure is the same for stationkeeping as for normal mode, although the parameter values are modified.

Throughout the three axis stabilized phase, i.e. from sun acquisition onwards, five back-up modes are usable, four in sunlight and one in eclipse.

(i) Sun Reacquisition Mode

This typically is initialised by telecommand following a fault detected on ground. It uses software control loops driven from the three SAS's only. It could also be triggered automatically if during the first sun acquisition the sun is not in the AAD field of view 1600 sec from the start of the sun acquisition.

(ii) AOCS Free Running Mode

This mode results from star mapper/OBC data being rejected by the RTAD. Attitude control is maintained using the normal mode control loops with attitude measurements being provided by the gyros alone through the RTAD.

(iii) AOCS Standby Mode

This mode is automatically implemented if the ACE watchdog detects a software failure. It causes ACE A to switch off and ACE B to switch on in standby mode i.e. all thruster commands inhibited and telemetry/telecommand management capability present.

(iv) Emergency Sun Reacquisition Mode

This is a hardwired control loop located in the CAE. It is the final back-up mode in sunlight which overrides the software back-up modes and is included to ensure that the sun does not approach too closely the payload boresights. If the sun goes outside the AAD field of view, this mode is triggered. It uses attitude measurement data from the Z axis SAS with the redundant SAS conditioning electronics. The main RCS branch A is switched off by means of the latch valves and branch B is used to provide the control torques. The hardwired control laws are simple phase advance plus deadband controllers on the X and Y axes, there being no control about the Z axis.

(v) Emergency Spinning Mode

This is also a hardwired control loop located in the CAE. It is the final back-up mode in eclipse overriding the software back-up modes, and is used to ensure that, if a failure occurs at this time, then, upon emerging from eclipse, the sun will not have approached the payload boresights too closely. The accumulated thruster control action on each axis, during eclipse, is compared with the expected accumulated control impulse ; if the difference exceeds a preset threshold, this emergency mode is triggered. RCS branch A is disabled by latch valves and branch B enabled. A Z thruster pulse of 200 sec is fired to increase the satellite spin momentum from about 0.26 Nms to about 3.5 Nms, after which the AOCS is put in standby mode. This ensures that the disturbance torques present in eclipse do not cause the Z axis to deviate by more than 55° from the sun line. On emerging from eclipse, a despin to 168.75°/hr is commanded by firing the opposite Z thruster for 200 sec and, simultaneously, emergency sun reacquisition is enabled.

MAIN STUDY RESULTS VS REQUIREMENTS

AND Controller

The nutation angle caused by AMF is used to size the AND thrusters. Maximum nutation angles of around 10° at 60 rpm are obtained by simulation of AMF. Based on a dedamping time constant, τ, of 300 sec and a spin inertia of 350 kgm^2, the minimum torque required to control this nutation is 4 Nm. As a result, the use of two 5 N ERNO thrusters at moment arms of $\simeq$1 m is baselined.

A conventional design for the AND has been adopted. The transverse angular momentum causing nutation is sensed as it rotates in body axes and, at a particular phase of the rotation, the appropriate thruster is fired to reduce this momentum component. The sensing device is a linear accelerometer mounted on the Y axis close to the edge of the spacecraft, nominally in the cg plane, with the input axis aligned with the spin axis.

The output is, to a good approximation, a sinusoid proportional to the transverse satellite Y rate, together with a bias term due mainly to alignment errors. When the AND is enabled, the mean value of the sinusoid, i.e. the bias, is evaluated over the first few nutation cycles. The bias is then subtracted from the ACP signal and the resultant passed through a first order high pass filter (time constant of 100 sec) to remove any residual bias.

The AND system operates in a cycle lasting 1.5 nutation periods, the cycles overlapping by 0.5 nutation periods. The complete cycle is shown in fig. 2. The ACP output occuring 1/4 period after a positive zero crossing is stored and checked against three thresholds to determine the magnitude of control required, i.e. the duration of the +Y and -Y thruster pulses. This duration determines the interval (< 1/4 nutation period) after the negative zero crossing before the -Y thruster is fired ; 1/2 nutation period later the +Y thruster is fired. In between these pulses, the positive zero crossing indicates the start of the next cycle.

In normal operation, only the lowest threshold will be exceeded. In this case, only the -Y thruster is fired for control purposes, to avoid spurious zero crossing caused by the ACP detecting linear accelerations. The size of the inner threshold, θ_3, is kept to a minimum to minimise fuel consumption, within the constraints imposed by noise in the ACP signal and thruster minimum impulse bit. At 60 rpm, the threshold is set at 0.05° the thruster on-time being defined such as to cause a nutation change $\Delta\theta_3$ of $\theta_3 = 0.05°$ in one pulse. The outer threshold, θ_1, is set at a value corresponding to the nutation angle removed by two 1/2 nutation period pulses (one -Y followed immediately by one +Y pulse) ; the +Y and -Y thruster on-times when this threshold is exceeded is 1/2 nutation period. The intermediate threshold, θ_2, is set at the geometric mean of θ_1 and θ_3 ; the +Y and -Y thruster on-times when this threshold is exceeded is such that the two pulses cause a total nutation change $\Delta\theta_2 = \theta_2$.

Obviously as the hydrazine is depleted and the thrust level drops, the on-times and thresholds must be regularly updated from ground. In addition, these parameters must also be updated at the different spin rates.

Fig. 3 shows the motion of the momentum vector in satellite X, Y axes caused by AMF followed by AND initiation, to remove the ACP bias, and AND operation. AMF lasts 33 sec leaving a nutation angle of about 10° (defined at the time the X body rate is zero) ; AND initiation takes a further 40 sec during which the nutation angle increases to 11° ; AND then remove the nutation, the angle being reduced to below 5° in a further 120 sec.

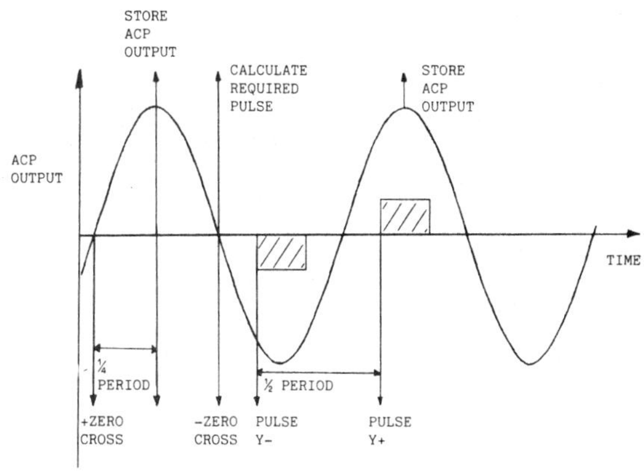

Fig. 2 - AND Controller Sequence of Events

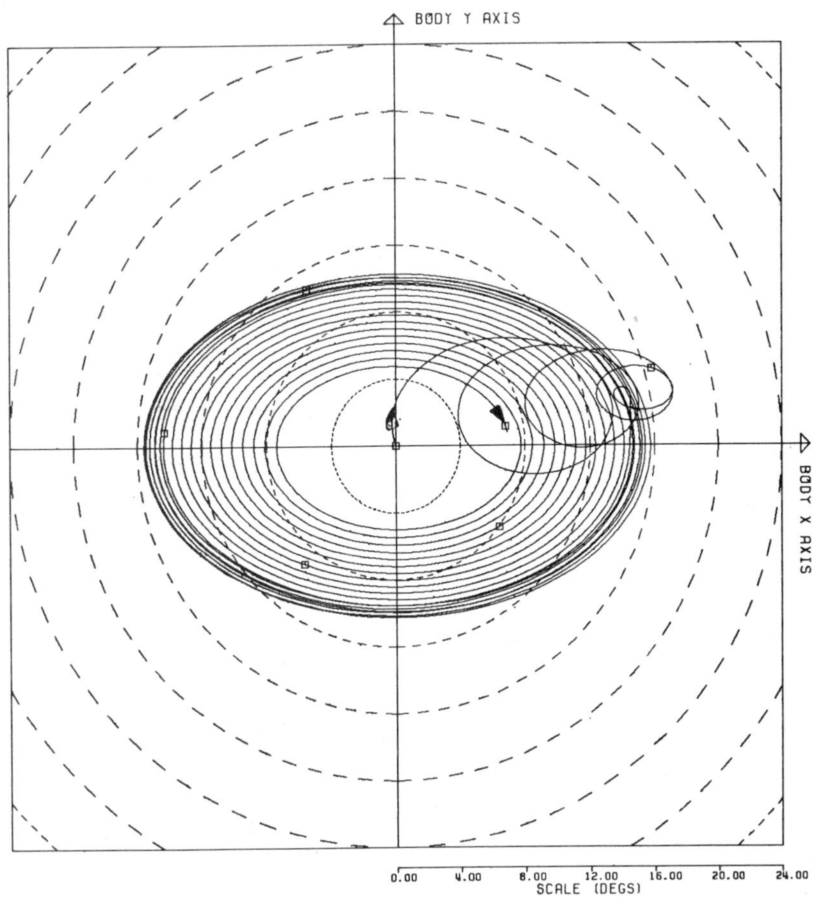

Fig. 3 - Nutation Angle Caused by AMF and the Subsequent AND

Flat Spin Recovery

The method of performing flat spin recovery has been outlined above. Simulation results showing the motion of the momentum vector in body axes, starting from a Y spin rate of 10 rpm, with 5° of nutation about Y, and ending with a Z spin rate of 8 rpm with 20° of nutation about Z are shown in fig.4.

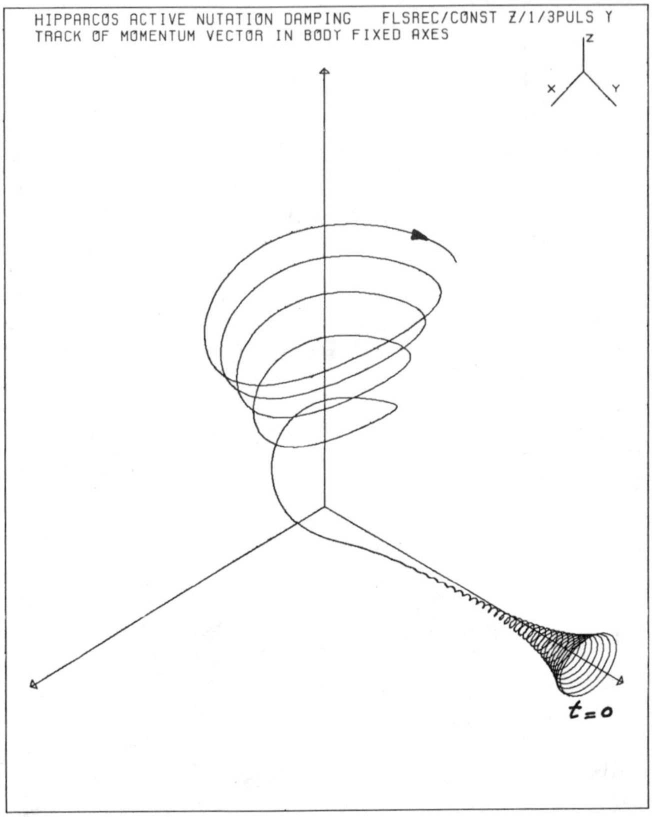

Fig. 4 - Flat Spin Recovery

Normal Mode Controller

Requirements

The requirements on the control system, derived from the mission science objectives, are to follow the scanning law within these constraints :

Rotation rate around Z : 168.75°/hr ± 2 %

Pointing error of Z axis : < 10 arc min of the nominal scan law

Pointing error about Z axis : < 10 arc min (for adequate control of the phase of the Z axis rotation)

Thruster magnitude = 0.02 N with moment arm $\leqslant$ 1 m.

Minimum thruster on-time $\geqslant$ 50 ms (minimum impulse bit considerations)

Maximum thruster on-time $\leqslant$ 500 ms. (jitter considerations).

Minimum time between any thruster firings $\geqslant$ 100 s and should be maximised within the above constraints.

Controller Design

The last requirement infers synchronising thruster pulse actuations on the three axes. In order to achieve this minimum separation time, whilst maintaining the required pointing accuracy within the thruster on-time constraints, use is made within the controller of the prelaunch estimate of the disturbance torques. With the satellite rotating at 168.75°/hr about the Z axis, which is inclined at a constant 43° to the sun line, the solar radiation pressure torques are deterministic and predictable, with a high degree of accuracy, as they are mainly caused by differing reflectivities, shadow effects and centre of pressure - centre of mass offsets.

Random misalignment effects, e.g. unpredictable solar array twist, give rise to very much smaller torque components. It is anticipated that the error in determining the solar disturbance torques is less than 10 %. The only other major disturbance torque source is due to the Z axis gyro flywheel. This gyro has its wheel axis perpendicular to the satellite rotation axis, resulting in a precession torque which is constant in body axes and is aligned with the satellite X axis.

The total disturbance torques, in μNm, are :

$$T_x = -8.25 \sin\Omega_d t + 1.79 \sin 2\Omega_d t - 0.49 \sin 4\Omega_d t + 0.37 \sin 5\Omega_d t + 10.6$$

$$T_y = -8.27 \cos\Omega_d t - 1.81 \cos 2\Omega_d t - 0.49 \cos 4\Omega_d t - 0.35 \cos 5\Omega_d t$$

$$T_z = -1.38 \sin 3\Omega_d t - 0.38 \sin 6\Omega_d t - 0.11 \sin 9\Omega_d t \qquad (1)$$

where Ω_d = 168.75°/hr and t = 0 when the sun is in the satellite XZ plane. Other disturbance sources, such as gravity gradient or magnetic torques, are in excess of an order of magnitude smaller and may therefore be neglected. During eclipse, the disturbance torques are given to a good approximation, by

$$T_x = 10.6 \ \mu\text{Nm (gyro precession torque)}$$

$$T_y = T_z = 0 \qquad (2)$$

The controller is driven by three variables, one for each axis :

$$\varepsilon_x = \varphi + K_x \dot{\varphi}$$

$$\varepsilon_y = \theta + K_y \dot{\theta}$$

$$\varepsilon_z = \psi + K_z \dot{\psi} \qquad (3)$$

where Kx, Ky, Kz are gain factors to take account of rate errors (a suitable value was found to be 50 s for each) ; the attitude errors, φ, θ, ψ, and their derivatives are produced by the RTAD. Thruster actuation is demanded as soon as one of the following conditions occurs :

$$|\varepsilon_x| \geqslant \mu_x \text{ or } \quad |\varepsilon_y| \geqslant \mu_y \quad \text{or} \quad |\varepsilon_z| \geqslant \mu_z \tag{4}$$

where $\mu x, \mu y, \mu z$ are deadband thresholds. These are maximised, within the pointing constraints, to minimise the frequency of actuations. As a result

$$\mu x = \mu y = 420 \text{ arc sec (so that } \sqrt{\varphi^2 + \theta^2} < 10 \text{ arc min)}$$

$$\mu z = \quad 550 \text{ arc sec (so that } \quad \psi < 10 \text{ arc min)} \tag{5}$$

Once one of these thresholds is crossed, control action is taken on all three axes. The control impulse on each axis may be varied by means of the commanded on-time within the range 50 to 500 ms.

The on-times are calculated using the attitude and rate errors, at the time the control action is demanded, and a forecast of the disturbance torques, such that the attitude error becomes zero at a time tp after the control action is applied. The calculation is as follows for the X axis but applies also to the Y and Z axes.

If the attitude and rate errors immediately after the thruster firing, are φ_o and $\dot{\varphi}_o$, then, in the presence of a constant disturbance torque $\bar{T}x$, the attitude error is reduced to zero in a time tp given by

$$\dot{\varphi}_o = -\varphi_o/t_p - \bar{T}_x t_p / 2I_{xx} \tag{6}$$

where Ixx is the X axis inertia. This gives the desired rate after the thruster firing to achieve a particular value of tp. The change in rate $\Delta\dot{\varphi}$ is given by

$$\Delta\dot{\varphi} = \dot{\varphi}_o - \dot{\varphi}(t) \tag{7}$$

where $\dot{\varphi}(t)$ is the rate error at time t, just before the thruster firing. the thruster on-time is then given by

$$\begin{aligned} t_{on} &= |\Delta\dot{\varphi} . I_{xx} / T_{xc}| \\ &= \left| \frac{I_{xx}}{T_{xc}} \left(\dot{\varphi}(t) + \frac{\varphi(t)}{t_p} + \frac{\bar{T}_x t_p}{2 I_{xx}} \right) \right| \end{aligned} \tag{8}$$

where Txc is the X axis control torque and $\varphi(t)$ is the attitude error just before the thruster firing, which is assumed not to change during firing. The constant disturbance torque in the above expression is taken as

$$\bar{T}_x = \frac{1}{2} \left(\hat{T}_x(t) + \hat{T}_x(t + t_p) \right) \tag{9}$$

where $\hat{Tx}(t)$ is the estimate of the disturbance torque, Tx(t), at the time the control action is demanded. The on-time is restricted to lie within the range 50 to 500 ms. If a firing of less than 50 ms duration is requested, the on-time is forced to 50 ms ; if no control action were taken, the threshold could be exceeded. If an on-time in excess of 500 ms is requested, only 500 ms is allowed.

Simulation Results

In the simulation, due to the ACE cycle times, the on-times are quantised by rounding down to an integer number of 13.3 ms quanta. The simulation results given in fig. 5 are based on a value for tp of 400 sec ; only the mean and fundamental disturbance torque components are included in the torque estimate used for calculating the on-times, although the complete disturbance torque model is used in the simulation to drive the satellite dynamics. The results show a minimum separation time between thruster actuations of 420 sec, with a mean separation time of 629 sec over a complete great circle. The minimum value is governed by the maximum on-time allowed of 0.5 sec and the size of the disturbance torques.

The controller is fairly insensitive to errors in the disturbance torque estimate. A 20 % error in the estimate of the fundamental component of solar radiation torque causes a worst case minimum separation time of 400 sec with little impact on the mean value. A 50 % error causes a worst case minimum value of 270 sec and a mean value of 576 sec.

Controller Adaptability

The main advantage of this controller is that it is adaptable to eclipse conditions, to payload commissioning and to stationkeeping. In eclipse, the disturbance torque estimate is simplified to a constant X-component of 10.6 μNm and zero Y and Z components (eq. 2). During payload commissioning when the Z axis is sun pointed, solar radiation torques are again negligible, i.e. less than 1 μNm, so that the same estimate as for eclipse is used. From simulation results, the mean and minimum separation times may be increased to 680 sec and 550 sec respectively at this time by using a value for tp of 500 sec.

During stationkeeping, the disturbance torques are much higher than in normal mode and are caused by thruster mismatch, thruster misalignment and displacement of the thrust vector from the centre of mass. Stationkeeping is performed by firing both the +Z and -Z thrusters over an arc of the satellite rotation ; attitude control about Z is achieved by off modulating the appropriate Z thruster, whereas X and Y axis control is achieved in the normal way. Anticipated maximum disturbance torque levels are 60 μNm, 370 μNm and 830 μNm about X, Y and Z respectively. In view of these magnitudes, the maximum on-time of 0.5 sec no longer applies. In addition, the gain factors Kx, Ky, Kz must be reduced as the rates will be much higher. A suitable value for each is 2 sec. The disturbance torque will not be known prior to launch so they will be computed in orbit.

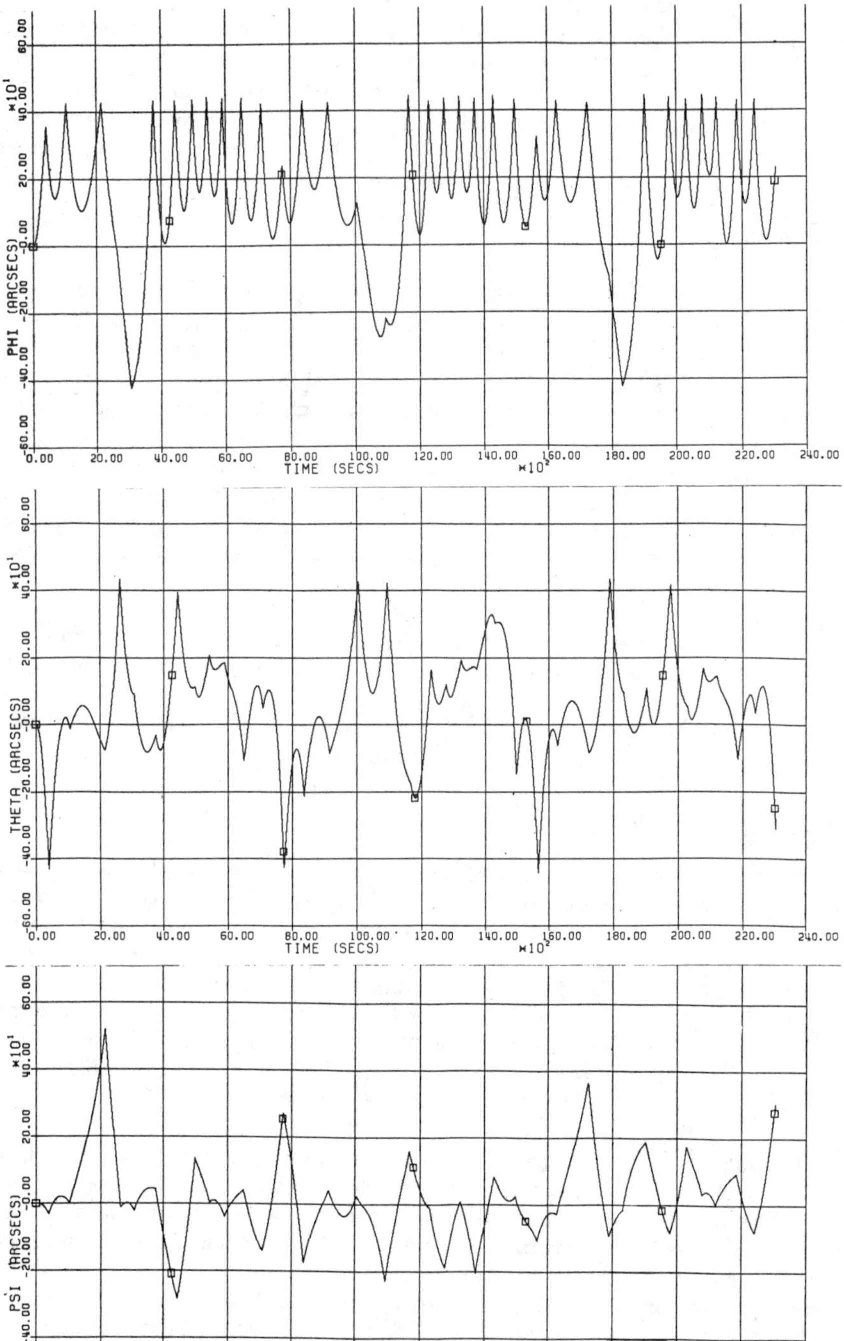

Fig. 5 - On-station Operation with Normal Mode Attitude Control

As they are reasonably constant during the manoeuvre, they may be computed from the gyro outputs according to :

$$\hat{T}_x = I_{xx}\left(\dot{\varphi}(t_1) - \dot{\varphi}(t_2)\right)/(t_1 - t_2) \tag{10}$$

(Similarly for Y and Z).

The look-ahead time, tp must also be reduced, in order to remain within the 10 arcmin pointing. A reasonable value is in the range 50 to 100 sec. With these changes included, the calculation of the thruster on-times and the firing logic is exactly the same as for the normal mode. As a result, simulation results indicate regular thruster firings at a separation of 74 sec for tp = 50 sec and 106 sec for tp = 100 s.

Real Time Attitude Determination

Requirements

In order to drive correctly the payload detector the real time attitude determination (RTAD) function has to provide the On-Board Computer (OBC) with a three-axis attitude φ, θ, ψ such that :

- the current optical axis orientation of the two telescopes with respect to their ideal position defined by the reference scanning law is known with an accuracy better than 1 arcsec (RMS).
- The three attitude rate $\dot{\varphi}$, $\dot{\theta}$, $\dot{\psi}$ are estimated with an accuracy better than 0.1 arcsec/sec (RMS).

These requirements do not apply during cold gas actuations.

Estimator Design

Defining the reference frame (Xo,Yo,Zo) as the frame which perfectly follows the scanning law required by the scientific mission, the current attitude of the satellite is described by the three shift angles φ, θ, ψ defined between the reference frame and the satellite frame (Xs, Ys, Zs) as shown on fig 6.

Basically, the RTAD filter uses a Kalman structure where the process model is given by the kinematic equations, receiving the gyro data as a command vector and updated by the measurements issued from the different star mappers.

The kinematic equations are derived from the transformations of fig 6.

$$\begin{aligned}
\dot{\varphi} &= p\frac{\cos\psi}{\cos\theta} - q\frac{\sin\psi}{\cos\theta} - \omega_x - tg\,\theta\left(\omega_y \sin\varphi - \omega_z \cos\varphi\right)\\
\dot{\theta} &= p\sin\psi + q\cos\psi - \left(\omega_y \cos\varphi + \omega_z \sin\varphi\right)\\
\dot{\psi} &= r + tg\,\theta\left(q\sin\psi - p\cos\psi\right) + \frac{1}{\cos\theta}\left(\omega_y \sin\varphi - \omega_z \cos\varphi\right)
\end{aligned} \tag{11}$$

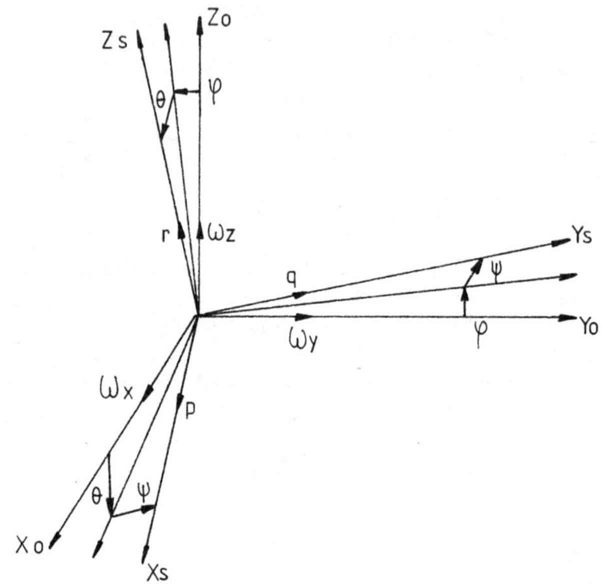

Fig. 6 - φ, θ, ψ Angle Definition

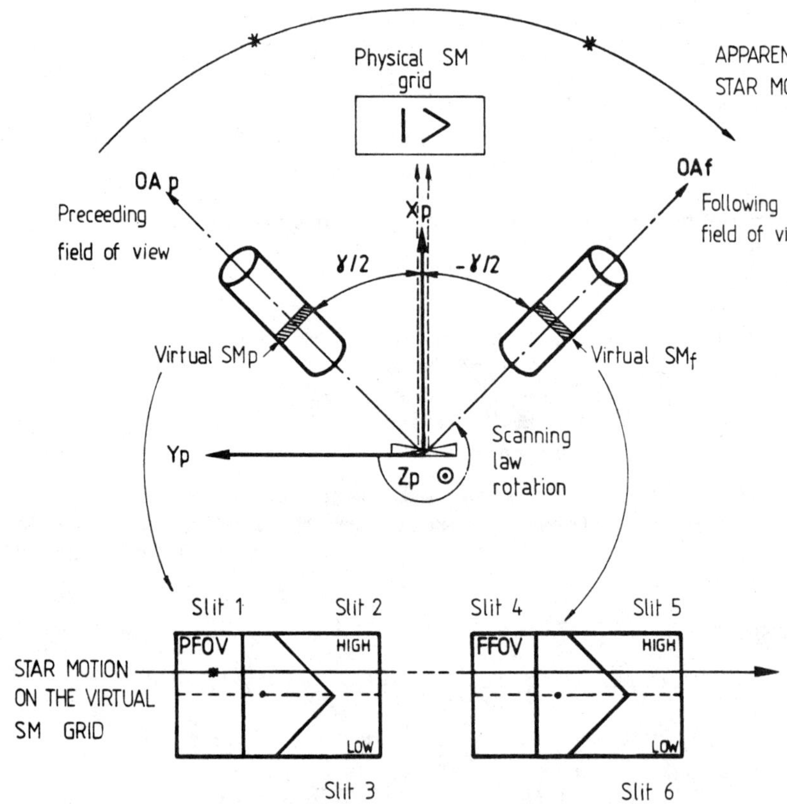

Fig. 7 - Definition of the SM Virtual Slits

where p, q, r are the absolute rotation rates around Xs, Ys, Zs and $\omega x, \omega y, \omega z$, are the absolute rotation rates around Xo, Yo, Zo.

$\omega x, \omega y, \omega z$ are derived from the scanning law definition and therefore are fully deterministic. According to the control specification φ, θ, ψ never exceed 10 arcmin, therefore the kinematic equations can be simplified and linearized to obtain the model to be used in the filter.

$$\begin{aligned} \dot{\varphi} &= \omega z\,\theta + p - q\psi - \omega x \\ \dot{\theta} &= -\omega z\,\varphi + q + p\psi - \omega y \\ \dot{\psi} &= r - p\theta - \omega z \end{aligned} \tag{12}$$

The p, q, r rates are obtained from the IRU, after geometric transformation to take into account the position of the gyro input axes with respect to the satellite axes. These measurements are corrupted by noise and the complete equations are :

$$\begin{aligned} gx &= p + dx + Vx \\ gy &= q + dy + Vy \\ gz &= r + dz + Vz \end{aligned} \tag{13}$$

gx, gy, gz are the virtual gyro measurements after the geometric transformation, dx, dy, dz are the virtual gyro drifts and Vx, Vy, Vz the virtual gyro rate noise. In order to improve the final performance, the drifts are modelled as a Wiener process and incorporated in the process model, and the rate noises are modeled as white noises according to the IRU gyro noise PSD :

drift noise : $\dot{d} = W$; with PSD (W) = Qw = 10^{-21} $(rd/s^2)^2/Hz$

rate noise : V ; with PSD (V) = Qv = 10^{-3} $(°/h)^2/Hz$

Then the classical state equation for the process model is obtained :

$$\begin{bmatrix} \dot{\varphi} \\ \dot{\theta} \\ \dot{\psi} \\ \dot{d}x \\ \dot{d}y \\ \dot{d}z \end{bmatrix} = \left[\begin{array}{ccc|ccc} 0 & \omega_z & 0 & -1 & 0 & 0 \\ -\omega_z & 0 & 0 & 0 & -1 & 0 \\ 0 & 0 & 0 & 0 & 0 & -1 \\ \hline & 0 & & & 0 & \\ & & & & & \end{array}\right] \cdot \begin{bmatrix} \varphi \\ \theta \\ \psi \\ dx \\ dy \\ dz \end{bmatrix} + \begin{bmatrix} (g_x-\omega_x)-(g_y-d_y)\psi \\ (g_y-\omega_y)+(g_x-d_x)\psi \\ (g_z-\omega_z)-(g_x-d_x)\theta \\ 0 \\ \\ \\ \end{bmatrix} + \begin{bmatrix} -V_x \\ -V_y \\ -V_z \\ W_x \\ W_y \\ W_z \end{bmatrix} \tag{14}$$

The filtering process is performed digitally in the ACE microprocessor and the continuous state equation has to be written in a recursive form ; a sampling period of T=1 s has been retained for the state computation and the gyro acquisition.

In fact, these gyro acquisitions are angular increments defined by :

$$G_i(t_k+T) = \int_{t_k}^{t_k+T} g_i(\tau)\,.\,d\tau \tag{15}$$

Hence, the command vector of the process becomes :

$$\Gamma_\varphi(k+1) = G_x(k+1) - \omega_x(k).T - \psi(k).G_y(k)$$

$$\Gamma_\theta(k+1) = G_y(k+1) - \omega_y(k).T + \psi(k).G_x(k)$$

$$\Gamma_\psi(k+1) = G_z(k+1) - \omega_z(k).T - \theta(k).G_x(k) \qquad (16)$$

the variation of $\omega x, \omega y, \omega z$ and φ, θ, ψ being negligible over T.

Finally, the discrete model of the process is given by :

$$X_{k+1} = \begin{bmatrix} \varphi \\ \theta \\ \psi \\ dx \\ dy \\ dz \end{bmatrix} = \left[\begin{array}{ccc|ccc} 1 & \omega_z T & 0 & -T & 0 & 0 \\ -\omega_z T & 1 & 0 & 0 & -T & 0 \\ 0 & 0 & 1 & 0 & 0 & -T \\ \hline & & & 1 & & 0 \\ & 0 & & & 1 & \\ & & & 0 & & 1 \end{array}\right] . \underbrace{\begin{bmatrix} \varphi \\ \theta \\ \psi \\ dx \\ dy \\ dz \end{bmatrix}}_{X_k} + \begin{bmatrix} \Gamma_\varphi(k+1) \\ \Gamma_\theta(k+1) \\ \Gamma_\psi(k+1) \\ 0 \\ 0 \\ 0 \end{bmatrix} \qquad (17)$$

Thus, this model is used to estimate, every second, the spacecraft attitude when no star mapper measurement is available. The attitude error rates, $\dot{\varphi}, \dot{\theta}, \dot{\psi}$ are derived from two consecutive state estimates.

Star mapper measurement insertion

Each of the two redundant star mappers (SM's) located in the focal plane of the payload is made of 1 vertical and 2 inclined slits ; since each receives the sky image from the two telescopes, this real SM is equivalent to two virtual SM's located on each optical axis resulting in 6 virtual slits as shown in fig. 7.

One measurement in slit number i is defined as $Zi = \omega z\,(tR-tN)$ where tN is the nominal crossing time of a given star in slit i if the spacecraft were perfectly following the scanning law and tR is the observed crossing time of the same star in the same slit. The quality of the star mapper measurements depends on the star characteristics (magnitude, color) and on the star position a-priori knowledge which is used to build the tN file. In the present case the RMS error of SM measurements ranges from 0.21" to 1.04". The measurement Zi is computed by the on board computer (OBC) from the SM raw data and the tN catalogue and sent to the ACE after non linearity correction, along with the slit number.

However, due to the software task synchronisation in the OBC and the duration of the computation of tR, the measurement Zi is always received with a variable delay in the ACE. The maximum delay separating tR from the moment when the measurement is available in the ACE is estimated at 3.25 sec.

Over such a period, the variation of φ, θ and ψ can be of several arcsec ; thus it is necessary to extrapolate the received measurement in order to make it consistent with the current state issued from the prediction algorithm.

Defining Hi as the observation matrix between the measurement in slit i and the state vector (Zi = Hi.X) and Zi (extr) as the extrapolated measurement, the current state vector is updated in

$$Xk^{+} = Xk^{-} + Ki \left[Zi(extr) - Hi.Xk^{-} \right] \quad (18)$$

where Ki is the gain vector associated with slit i.

Gain selection

One of the major issues of the RTAD study has been the selection of the gain vector. In the first place, the use of a real adaptive Kalman gain has been rejected for the performance improvement is not worth the computational burden increase for the ACE and the extension of the star file in the OBC. This study case showed the main parameter driving the variation of the Kalman gain was the characteristics of the observed stars in terms of position a-priori error - and not the duration between star updates during which the gyro noise is integrated. An intermediate case, using two sets of fixed gains, one for each star class (small or large a-priori error), has also been tested and showed similar results with the full Kalman case. However this solution has been rejected at system level because it required storage of the class in the star file. Finally, a unique gain set has been selected and, for each slit, the gain vector is defined as the mean value of the Kalman gain series associated with that slit over one great circle ($\simeq$ 7700s). The Table 1 shows the RMS error over one great circle, of this estimator for those three strategies.

Table 1
RTAD PERFORMANCES VS UPDATE GAINS

RMS error over one G.C	Optimal Kalman gains	Double set of gains	Fixed average gains
$\sigma(\varphi)$	0.533"	0.534"	0.587"
$\sigma(\theta)$	0.353"	0.350"	0.401"
$\sigma(\psi)$	0.274"	0.271"	0.304"

The influence of the word length has been tested and demonstrated the necessity to make the filtering computation over 24 bits (2 X 12 bit word) - a 12 bit processor, the Harris 6100 was baselined during the study phase.

Performance statistical assessment

The RMS error computed over one great circle is a suitable performance index to check the behaviour of the filter against variations of deterministic parameters such as the gain vector or word length. However the performance of the filter cannot be assessed against the stochatic perturbation sources - SM noises, a priori position errors, gyro noises - in the same way, because these random processes are not stationary in the wide sense and thus not ergodic.

For instance the random walk of the gyro drift or the measurement error variance which depends on the star characteristics (magnitude, color, a-priori knowledge) are typically time dependent.

In other words, the attitude RMS error defined in the requirements is time-varying and has to be computed at each instant as an ensemble average. This is done by simulating the behaviour of the filter over one great circle, 10 times in a row and making the RMS of the 10 observed errors at every instant. The general flowchart of this simulation is given in fig. 8.

Simulation results :

The star pattern is built by spreading, with a uniform probability, about 1000 stars over one great circle ; however due to the track and size of the telescope field of view, only 220 detections per vertical slit and 100 detections per inclined slit occur along the 7700s of 1 great circle - that is to say the mean separation time between 2 detections on inclined slits of one FOV is 38s.

Fig. 9 shows for each telescope, the variation of the RMS value $\sigma(t)$ of the angle separating the estimated optical axis from the real one and the separation time during which no star detections occur on an inclined slit. The filter performance is summarized in table 2.

($\overline{\sigma(t)}$= mean value of $\sigma(t)$ over one great circle).

Table 2
FILTER PERFORMANCES

error on	preceeding FOV	following FOV
$\sigma(t) > 1''$	2.77 %	0.00 %
$\overline{\sigma(t)}$	0.628 "	0.566 "
$\sigma(t)$ max	1.238 "	0.976 "

RMS error	$\dot{\varphi}$	$\dot{\theta}$	$\dot{\psi}$
on rates	0.059"/s	0.058"/s	0.053"/s

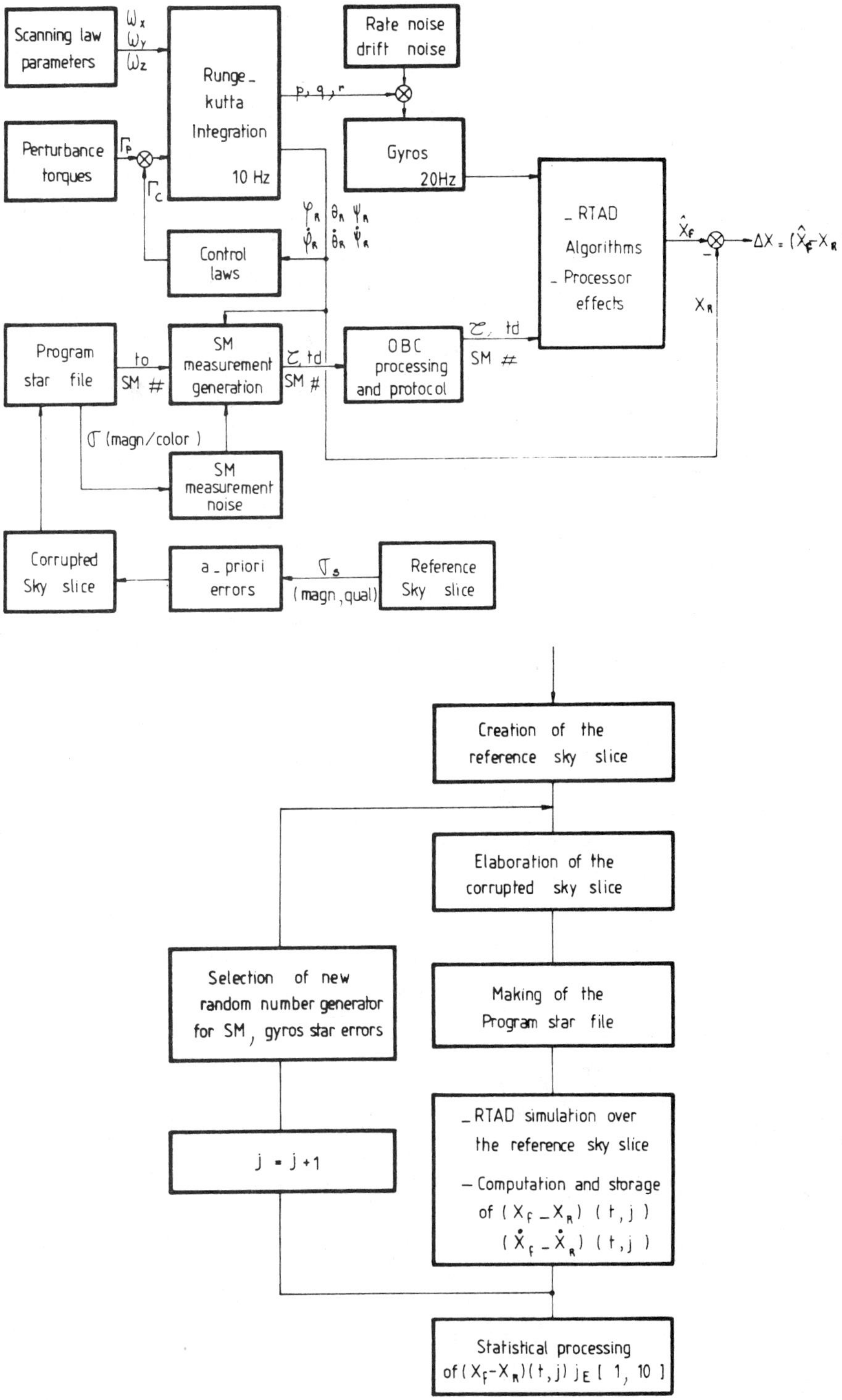

Fig. 8 - Simulation Flowchart

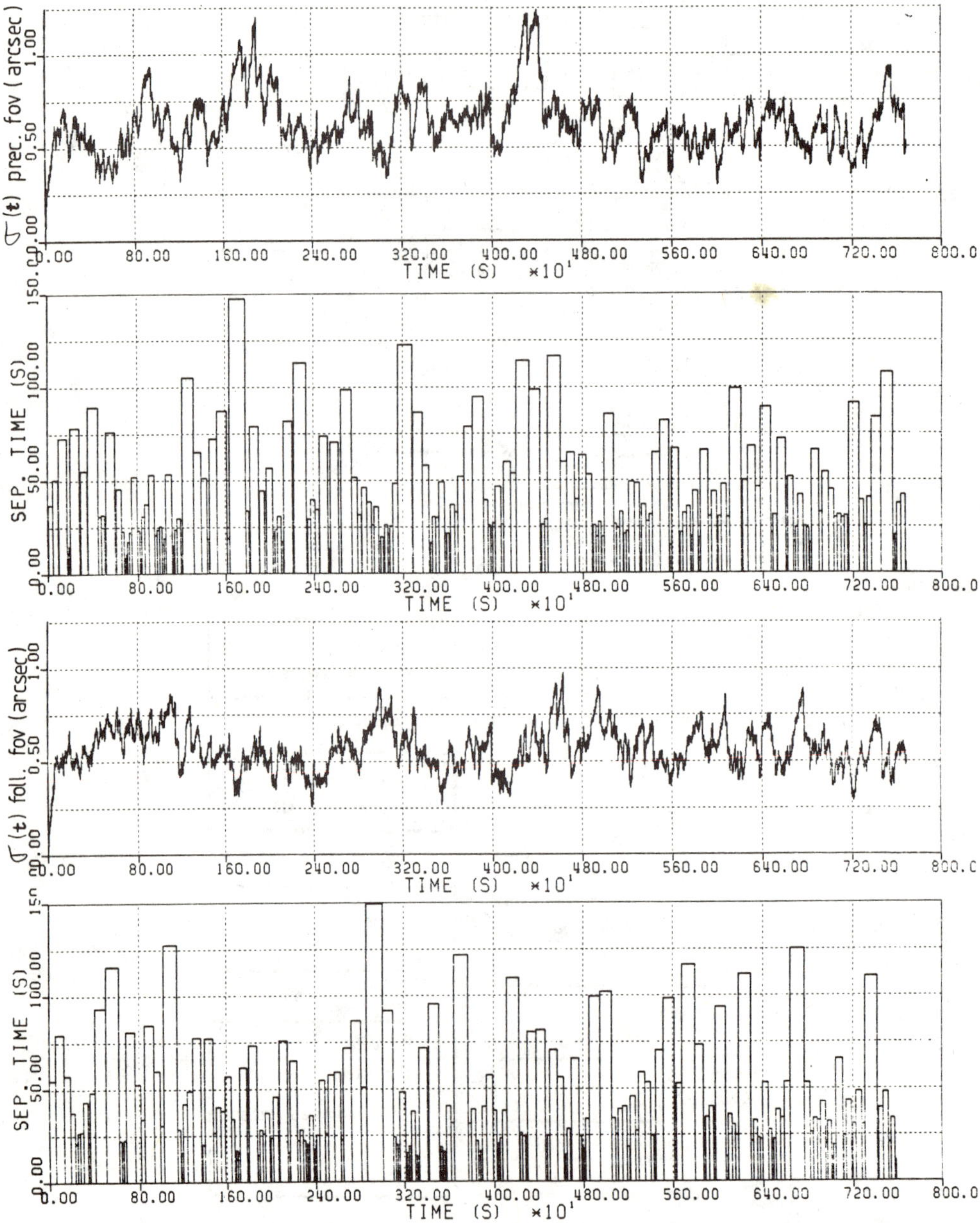

Fig. 9 - Variation of σ(t) and Separation Time Histogram

On-Ground Attitude Reconstitution (OGAR)

The last critical issue of the study phase was to demonstrate the possibility to reconstitute a posteriori and on-ground the three axis attitude of the spacecraft from the data continuously telemetred - gyro samples, star mapper raw data, actuation instant and characteristics - within an accuracy on each axis of 0.1", time averaged RMS over one great circle.

Since this task is performed a posteriori, there is no constraint on the computational complexity, and backward processing of the received data is possible ; thus optimal smoothing techniques can be fully applied. For each great circle, this smoothing is performed in three steps :

- Forward filtering using the optimal Kalman gains on SM updates.
- Backward filtering using the optimal Kalman gains on SM updates.
- Optimal combination at every instant of the results of the two previous steps.

In the first place, this optimal smoothing has been applied with the kinematic filter where equations (12) are replaced by a second order development of (11).

The gyro noise characteristics remain identical whereas the SM measurement noise has been reduced in the range 0.03" to 0.08". This SM noise reduction is due to a more sophisticated ground processing of the SM raw data taking advantage of the available payload data. Also, the number of stars constituting a great circle is extended to 1400 as the ground processing permits the use of additional stars to those included in the on-board catalogue. The statistical performance assessment method, previously applied to the RTAD, is reused and leads to the results of Table 3.

Table 3
OGAR PERFORMANCES WITHOUT PRE-FILTERING

error on	φ	θ	ψ
$\overline{\sigma(t)}$	0.113"	0.093"	0.055"
$\sigma(t)>0.1"$	62 %	37 %	4 %

In order to improve this result, the dynamical properties of the spacecraft may be introduced in the process model to reduce the impact of the gyro rate noise. However, to avoid handling large dimension matrices, this is done by means of pre-filtering of the virtual gyros, based on a one axis Kalman filter :

$$\begin{bmatrix} \dot{p} \\ \dot{\gamma}_x \end{bmatrix} = \begin{bmatrix} 0 & 1/I_x \\ 0 & 0 \end{bmatrix} \cdot \begin{bmatrix} p \\ \gamma_x \end{bmatrix} + \begin{bmatrix} 0 \\ b_\gamma \end{bmatrix} \quad \text{process model} \qquad (19)$$

$$\frac{G_x}{T} = \begin{bmatrix} 1 & 0 \end{bmatrix} \cdot \begin{bmatrix} p \\ \gamma_x \end{bmatrix} + V_x \quad \text{observation equation} \qquad (20)$$

with p = absolute rotation rate around the X axis
γx = perturbance torque modeled as a Wiener process : $\dot{\gamma}_x = b_\gamma$
T = gyro sampling period = 1 s
Ix = inertia of the relevant axis
Gx = virtual gyro measurement about X axis
Vx = virtual gyro rate noise.

The pre-filtered values are used to feed the kinematic filter replacing Gx by G'x = p.T. It must be noticed that the prefilter is disconnected during actuations in order to recover the full gyro bandpass during the pulse duration.

Table 4 shows that the dynamic pre-filtering of the gyro significantly improves the final performance of the OGAR smoother :

Table 4
OGAR PERFORMANCES WITH GYRO PRE-FILTERING

error on	φ	θ	ψ
$\overline{\sigma(t)}$	0.083"	0.056"	0.033"
$\sigma(t)$>0.1"	26 %	6 %	0.4 %

CONCLUSION

The study phase has shown the most demanding performances required from the AOCS could be met with the selected design and with sufficient margins to enter the C/D phase confidently.

AAS 84-006

PLANET-A ATTITUDE AND ORBIT CONTROL SUBSYSTEM

K. Ninomiya*, K. Uesugi*, H. Hara**
H. Yamamoto**, N. Muranaka**, A. Katoh**

PLANET-A is a scientific space probe to be sent to Halley's comet that will approach the sun early 1986. Objectives of the PLANET-A are to take the ultraviolet image of Halley's comet and to measure the solar wind in its vicinity. In this paper the design and the expected performance of attitude and orbit control subsystem (AOCS) are described in relation to the scientific requirements. The approach of testing and verifying the performance of the AOCS are also described. The mission of PLANET-A, its system design, and Japanese deep space station configuration are briefly presented.

INTRODUCTION

Outline of Mission

In the beginning of 1986, Halley's comet will again make a closest approach to the sun after its 76 years' trip in the solar system. The comet is described as dirty ice ball containning some primordial solar matters. As the commet nears the sun, their frozen water and gases sublimate and form a halo of gas, called the coma, around the nucleus. Sunlight ionizes its gas to form a plasma. Driven back by the solar wind the plasma and dust particles form a long tail pointing away from the sun.

The Institute of Space and Astronautical Science (ISAS), Japan, is planning to send a scientific space probe to investigate the evolutional history of the comet's coma and interactions between the comet and solar wind. The improved type of M-rocket, called M-3SII that is now under development at ISAS, is to be used as a launch vehicle. The rocket will have a capability of launching a spacecraft (S/C) of 140 kg weight into an interplanetary orbit.

* *Institute of Space and Astronautical Science, 4-6-1 Komaba, Meguro-ku, Tokyo 153, Japan*
** *Space Development Division, NEC Corporation, 4035 Ikebe-cho, Midori-ku, Yokohama 226, Japan*

The spacecraft, called "PLANET-A", equipped with vacuum ultra-violet imaging camera and energy spectrum analyzer of the solar wind plasma will be launched in August of 1985, and intercept Halley's comet near its descending node.

The ISAS is also developing a test spacecraft, called "MS-T5", which will be launched 8 months earlier to prove the launching capability of M-3SII rocket and the technologies applied to PLANET-A, as well as to verify the Japanese deep space communication network. However, MS-T5 is equipped with three scientific experiments. It will be launched into an interplanetary orbit also approaching Halley's comet.

Scientific Instruments

The scientific objectives of PLANET-A are to take ultraviolet images of Halley's comet and to measure the properties of the solar wind in its vicinity. The following two scientific instruments are aboard PLANET-A:

1) Ultraviolet Imaging Camera (UVI); UVI comprises an optical system to collect the ultraviolet radiation, a charge coupled device (CCD), a CMOS-microprocessor, and a hybrid-memory to store the images. Time delay and integration (TDI) method is applied to enhance the S/N ratio of images and to avoid smears due to the spinning of the S/C. The field of view of UVI is $2.5^\circ \times 2.5^\circ$ and effective pixels of CCD are 153 ×122. The instrument requires slow and stable spinning of S/C as low as about 0.2 rpm.

2) Energy Spectrometer (ESP); ESP comprises two spherical electrostatic analyzers for ion and electron respectively, each provided with microchannel plate (MCP). Energy and angular distributions of the solar wind particles will be measured by utilizing the S/C spin.

On the other hand, the following three scientific instruments are aboard the test S/C, MS-T5:

1) Interplanetary Magnetic Field experiment (IMF); IMF consists of three ring core-type magnetometers deployed by a 2m boom from the S/C and measures the three components of interplanetary magnetic field. Magnetic cleanliness is required for the equipments aboard MS-T5.

2) Plasma Wave Probe experiment (PWP); The instrument measures the electric and magnetic field components of the plasma wave by dipole antenna (10 m tip to tip) and loop antenna respectively.

3) Solar Wind experiment (SOW); Retarding potential trap type analyzer is used to measure the plasma densities and bulk velocities of the solar wind.

Spacecraft Design

MS-T5 and PLANET-A will be launched in January and August 1985 respectively from Kagoshima Space Center (KSC), the launch site of ISAS. Both S/C will be directly injected into interplanetary space to save the fuel and equipments required on parking orbit. In Fig-1 are shown their interplanetary orbits along with that of Halley's comet. Both S/C will approach to Halley's comet in early March, 1986 near at its descending node.

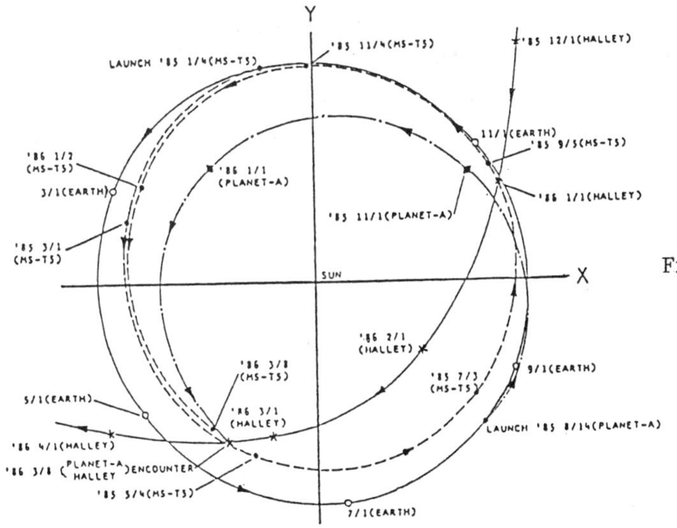

Fig. 1

Interplanetary Orbits of PLANET-A and MS-T5

MS-T5 and PLANET-A are depicted in Fig-2. The spacecraft are cylindrical in shape, about 0.7 m tall and 1.4 m in diameter. PLANET-A and MS-T5 weigh 135 kg and 142 kg respectively including 10 kg of initial fuel weight. Basically both spacecraft are of the same design except for the scientific instruments.

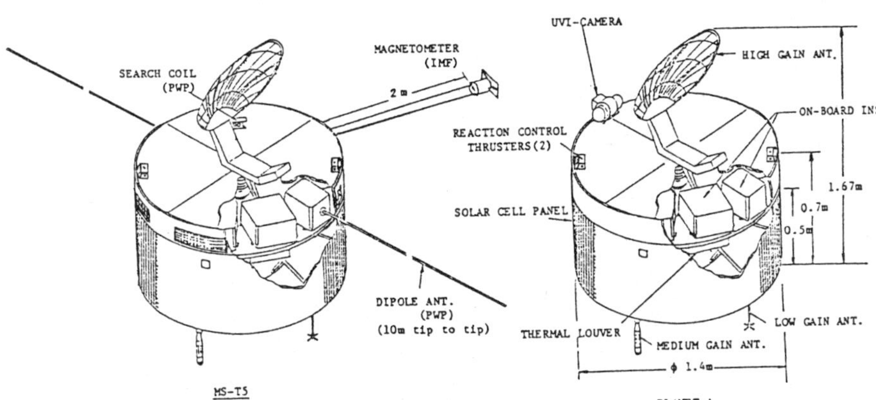

Fig. 2 MS-T5 and PLANET-A Spacecraft Configurations

Solar cells around the spacecraft's cylinder will supply 90 watts of power during the closest approach, at a distance of 0.83 AU from the sun. Ni-Cd battery of 2 AH capacity will supply the power to the S/C at launch and injection phase where sufficient power is not available from the solar cells.

The spacecraft is spin stabilized. The mechanically despun high-gain antenna (HGA) points toward the earth. HGA is an offset parabola (0.8 m in diameter) with a pencil beam radiation pattern of ±5° beam width. The spacecraft is also equipped with a medium-gain antenna (MGA) and a low-gain antenna (LGA) for early phase operation after the injection of S/C. MGA has a flat disk-like pattern that is perpendicular to the spin axis and LGA has a wide conical pattern which almost covers the lower hemisphere in S/C coordinate, as shown in Fig-3.

S-band signals are used for telemetry, command and S/C tracking. The transmission power of 5 watts or 70 mw is selectable by a ground command depending on the distance of S/C from the earth. S-band transmitter has non-coherent and coherent modes. The latter mode enables two-way doppler measurements of the velocity component of the S/C along its line of sight. The command codes and ranging codes at a rate of 16 bps will be transmitted alternately for the S/C control and S/C range measurements.

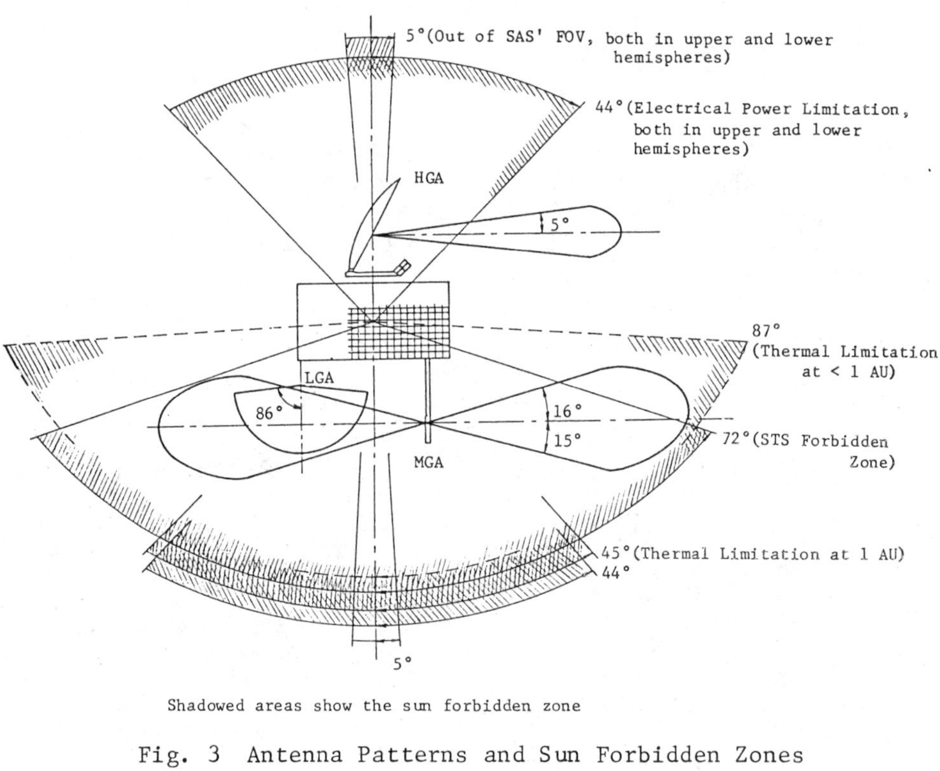

Shadowed areas show the sun forbidden zone

Fig. 3 Antenna Patterns and Sun Forbidden Zones

Data Processing Unit (DPU) has functions of data handling and command distribution to the equipments aboard the S/C. Five types of commands are implemented; Discrete Commands with direct execution (EDC), Discrete Commands without direct execution (NDC), Block Commands (BC), Program Commands (PC), and Execution Command (EC). NDC's and BC's need EC for their executions. The command system is shown in Fig-4. BC's are used to define the operational modes of plural number of specified equipments like UVI, ESP and AOCE by a block of command codes. A PC which is a set of DC's stored sequencially on DPU memory is executed step by step every 128 sec, 2048 sec or 8192 sec. One of these time steps is selected by DC. PC's enable the autonomous control of the S/C for a time spun up to 24 days.

Telemetry data edited at DPU are transmitted at a rate of 2048 bps or 64 bps depending on the S/C distance from the earth. Telemetry data at 64 bps are convolutionally coded with R=1/2, K=7, Q=8. Newly developed data recorder of 1 M bits capacity utilizing magnetic bubble memory is employed to store the data to be obtained when the S/C is out of sight from the earth station.

Thermal louvers maintain the S/C temperature within a reasonable range against the solar energy input variation ranging from 1 to 2.2 solar constant. Heaters are also used for the thermal control of equipments such as RCS.

For more detailed information of the spacecraft and the project, see references 1 to 3.

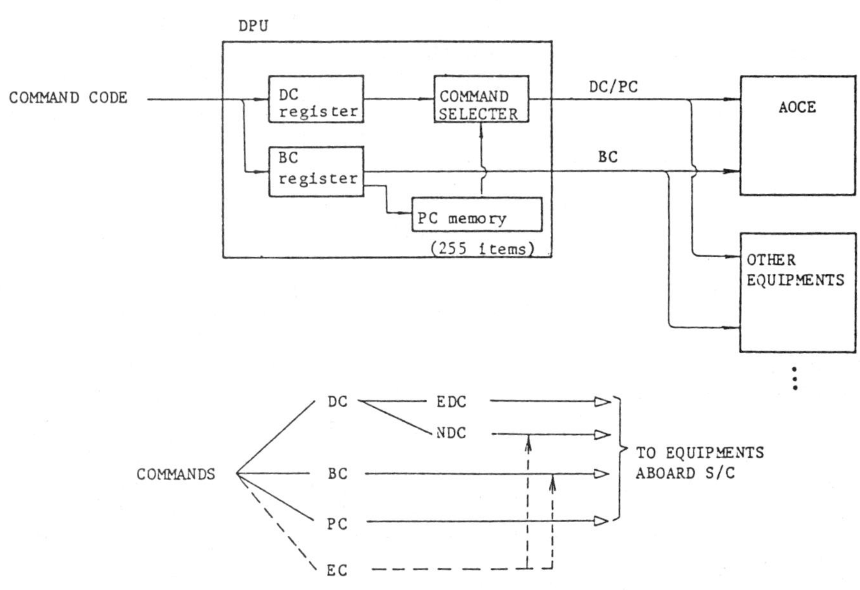

Fig. 4 Command System of PLANET-A/MS-T5

AOCS DESCRIPTION

AOCS Subsystem

From the thermal and scientific requirements, S/C is designed as a spinner. To meet the requirement from the UVI imagery S/C is equipped with momentum wheel. For taking images of Halley's comet, momentum wheel is rotated and spin rate of the S/C body is reduced from its nominal rate of 6 rpm to 0.2 rpm, which is a limiting spin rate for the thermal condition of the S/C without heat pipes or any other sophisticated thermal control design. Spinnig design of S/C has an advantage for UVI in that only the rotational phase for TDI initiation with reference to the sun's direction need be specified by command because azimuthal scanning of Halley's comet is automatically performed by the spin.

In a cruising and observation phase the spin axis is oriented perpendicular to the ecliptic plane with the allowable deviation of $\theta_s = 90^\circ \pm 3^\circ$ and $\theta_e = 90^\circ \pm 5^\circ$ (θ_s = spin-sun angle, θ_e = spin-earth angle). Constraints on the S/C attitude are shown in Fig 3 along with antenna patterns.

The moment of inertia of the S/C around its spin axis is about 32 $kg.m^2$ and its inertia ratio to transverse moment is about 1.4. This results in the highly stable dynamical configuration of S/C.

Attitude and orbit control subsystem is composed of spin-type sun sensor (SAS), star scanner (STS), momentum wheel assembly (MWA),nutation damper (ND), hydrazine reaction control system (RCS), and attitude and orbit control electronics (AOCE). The block diagram of AOCS along with despin control subsystem (DCS) is shown in Fig-5 and configuration of these components on PLANET-A is shown in Fig-6.

AOCS provides the means of
1) Attitude determination
2) Spin axis maneuvers
3) Spin rate control with RCS
4) Orbit correction
5) Slow spin control of S/C body-rate by MWA

SAS and STS provide the S/C attitude data with which attitude determination will be accomplished by a software at ground station's computer. RCS thrusters provide the means of spin axis maneuvers, spin rate control, and orbit correction that are required mainly during the early phase of orbit. All of the spin axis maneuvers are performed according to the rhumb line method using the sun as the reference. In a cruising and observation phase, attitude touch-up and unloading of angular momentum are required to compensate for the disturbance torque due to the solar radiation pressure.

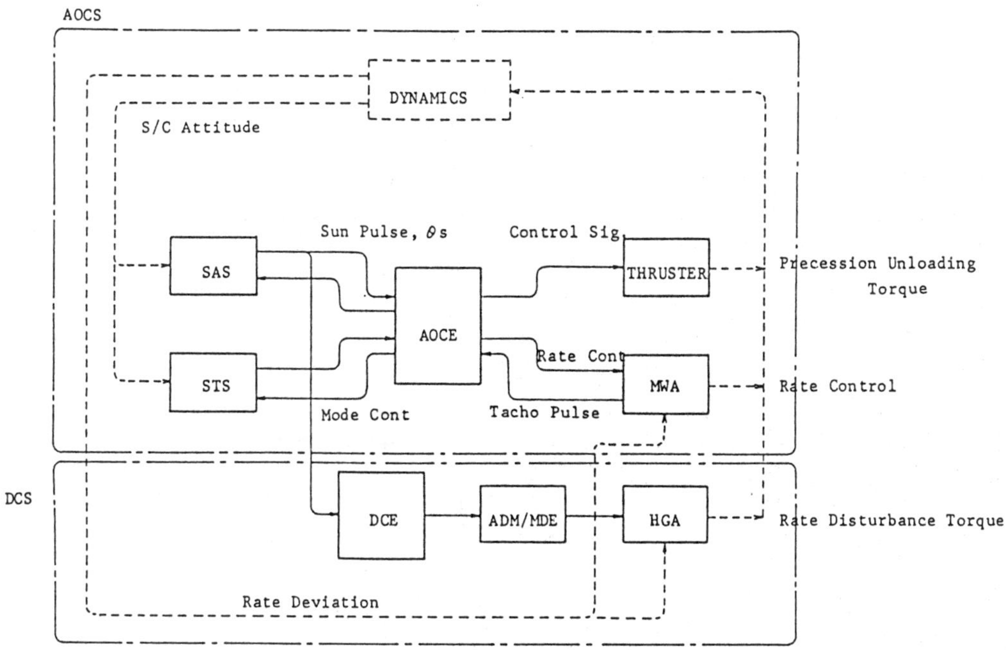

Fig. 5 AOCS Subsystem Block Diagram

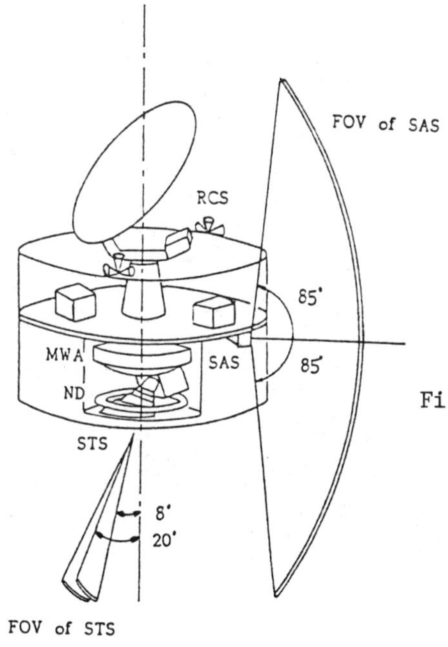

Fig. 6 Location of AOCS Components on spacecraft

AOCS Components and Their Functions

Sun Sensor (SAS)

Spin-type ADCOLE sun sensor provides spin axis-sun angle (θ_s) and sun pulse information. SAS consists of two sensor heads plus two corresponding electronics completely independent functionally. Each sensor has a thin fan beam shaped field of view (FOV) of 128° and tilt-mounted to each other by 42° . In this way two sensor heads cover FOV of ±85° with respect to the equatorial plane of S/C coordinate. The ±43° central portion of FOV is covered by both sensors.

These sensors give "Gray coded" coarse sun angle (7 bits) with accuracy of 1° , fine analogue sine and cosine outputs and the sun pulse, when FOV of sensors has just scanned the center of the sun's disk. Combined accuracy of the coarse and fine sun angles, and the accuracy of sun pulse timing are 0.1° at the central portion of FOV and 0.25° at the peripheries.

The sun pulse is utilized for, i) the timing of thruster firing in RCS pulse mode, ii) synchronizing the TDI of UVI, iii) the reference for despin control, and iv) the reference for the measurement of timings of star coincidence pulses.

Star Scanner (STS)

The star scanner, developed by BASD, is a passive scan type with two stripe-shaped silicon photodiodes placed on the focal plane of the optical system foming a "V"-like shape. The sensor optical axis is tilted at an angle of 14 degrees from the minus Z axis of the S/C so as to use Canopus (α-Carinae) as main target star for attitude determination. Two photodiode-elements subtend fan beam shaped FOV of 12° , one FOV is nominally parallel to spin axis and the other inclined 20° with respect to the first. The measurement accuracy between two star-pulses as expressed in azimuthal angle is 0.1 deg.

A threshold command sets a minimum star signal level. STS responds to stars brighter than this level and produces star coincidence pulses. A spin rate command defines one of two operation modes of STS; high spin mode at 6 rpm and slow spin mode at 0.2 rpm. These commands are incorperated in one BC (see Fig-10).

Sun shade rejects sunlight and earth albedo so long as their incidence angles are greater than the specified values. The sensitivity of STS is nominally 1.2 silicon magnitude.

Momentum Wheel (MWA)

TELDIX momentum wheel driven by DC brushless motor is used to stabilize the spacecraft attitude during the UVI's operation. The momentum wheel stores the angular momentum of 20 Nms at 2000 rpm rotor speed. Tacho pulses picked up magnetically is used to control and assure the very stable rate of the wheel rotation. The dynamic range for wheel speed is 2000 ± 300 rpm. The run up time for the wheel from 0 to 2000 rpm is about 5 min and free coast-down time is nominally 40 min.

Reaction Control System (RCS)

The monopropellant hydrazine propulsion system, developed by MHI , for attitude control and orbit correction consists of two tanks, manifolds, and six body fixed 3 N thrusters. N_2 pressurant gas and hydrazine (N_2H_4) fuel are stored in the tanks and blow-down system is adopted. Total fuel weight of 10 kg is to be filled up into the tanks. The schematic of the propulsion system is shown in Fig-7.

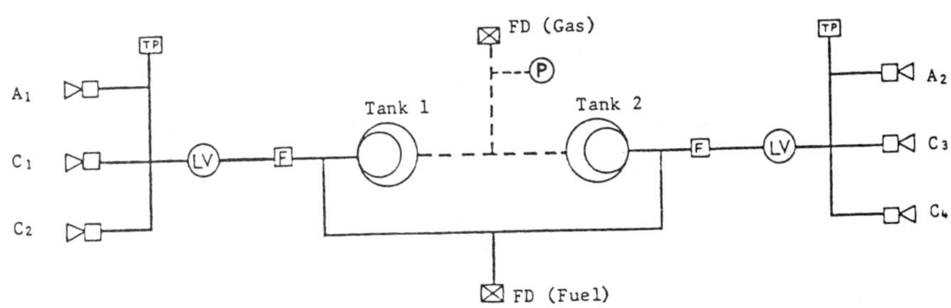

P = Pressure Transducer
FD = Fill and Drain Valve
LV = Latching Valve
F = Filter
TP = Test Port

Fig. 7 RCS Block Diagram

Two of thrusters are mounted pararell to the spin axis (A_1, A_2) and four are cant-mounted on the upper surface of the S/C (C_1, C_2, C_3, C_4). Thruster operation has redundant modes as shown in Fig-8 and Table-1.

Nutation Damper (ND)

To attenuate a nutation motion of the S/C a passive hoop-type nutation damper is used. Scilicon oil fills to about 1/3 of the annular pipe volume. The time constant for the attenuation of the nutation motion is estimated to be about 5 min at 6 rpm spin rate.

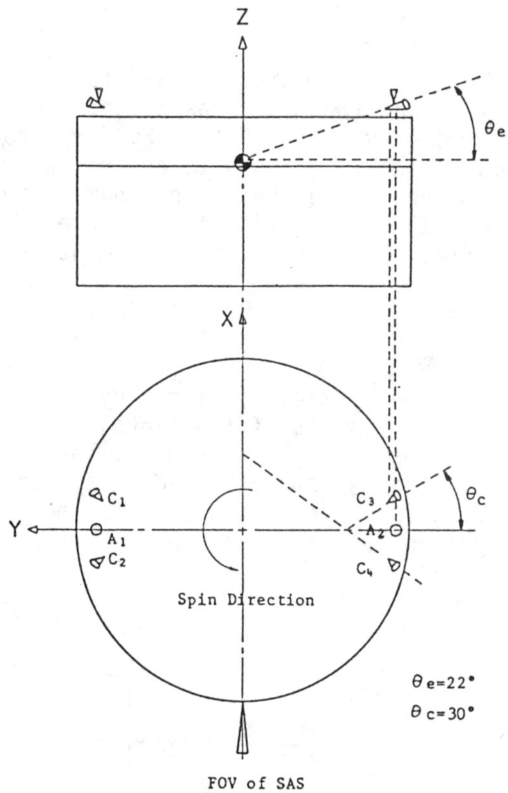

Fig. 8 Thruster Location

Table 1 THRUSTER UTILIZATION FOR VARIOUS MANEUVERS

Maneuver		Thruster Combination	Firing Mode
Angular Momentum Control	Initial Despin	C_2+C_3, (C_2), or (C_3)	Continuous
	CCW Unloading (Spin Up)	C_1+C_4, (C_1), or (C_4)	Continuous
	CW Unloading (Spin Down)	C_2+C_3, (C_2), or (C_3)	Continuous
Spin Axis Maneuver	Coarse Precession	A_1 or A_2	Pulse
	Fine Precession	C_1+C_3 or C_2+C_4	Pulse
Orbit Correction	Axial Mode	A_1+A_2, (A_1), or (A_2)	Continuous
	Canted Mode	C_1+C_2 or C_3+C_4	Pulse
	Vector Mode	Combination of A,C Thrusters	Continuous/Pulse

Thruster combinations within brackets are used only in a backup mode.

Attitude and Orbit Control Electronics (AOCE)

AOCE has following functions; 1) power ON/OFF control of SAS, STS and MWA, 2) temperature control of RCS, STS, and ND, 3) firing control of RCS thrusters, 4) control of wheel speed, 5) preprocessing of the attitude data of SAS and STS. In Fig-9 functional block diagram of AOCE is shown schematically. All these functions are realized by hardwired logics.

Control of RCS thrusters: Four kinds of thruster control logics are implemented in AOCE; i) thruster control by BC, ii) 90°-sun-angle acquisition control, iii) $\Delta\theta$ control (incremental change of spin axis around solar vector) and iv) tilt-and-scan control (used in contingency where S/C to ground communication is accidentally lost).

In the thruster control by BC, a control law is once stored by BC into one of four 5-byte memories each having triple redundancy. By the DC or PC the control law is read out with 2 out of 3 selection logics and exectuted. An adequate thruster combination out of 6 thrusters (A_1, A_2, C_1, C_2, C_3, C_4) is selected according to the contents of BC (see Fig-10). For continuous firing mode, jet on-time can be set from 125 ms to 512 sec by a step of 125 msec. For pulse firing mode number of pulses to fire can be set from 1 to 1023 pulses with increment of one(1) pulse. In this mode jet on-time can be selected from 100, 200, 400, 800 msec. The timig to fire the thrusters with respect to the sun pulse can be set either by 128 Hz timer clock or by "angle clock" which is generated by phase lock loop (PLL) synchronized with sun pulse train. Angle clock divides the full rotation of S/C (360°) into 2^{12} pulses and therefore 1 bit of this clock corresponds to 0.088 deg.

In the 90°-sun-angle acquisition control, coarse sun angle from the SAS is utilized to control the S/C spin axis to and within 90°± 3°. The control logic is shown in Fig-11. When the coarse sun angle (Gray code) shows θ_s less than 87°, A_2 thruster is fired, whereas when the coarse sun angle shows θ_s greater than 93°, A_1 thruster is fired synchronously with the sun pulse detection. Pulse width is automatically changed between 100 ms (for S/C spin rate lower than 7.5 rpm) and 200 ms (for S/C spin rate higher than 7.5 rpm).

As for the $\Delta\theta$ control, C_1+C_3 or C_2+C_4 thrusters are fired at the instant of the sun pulse detection for 32 pulses consecutively. By this control S/C spin axis is rotated around sun-S/C line in CW or CCW direction by about 4.5 deg, the value corresponding to the beam width of HGA. Jet on-time in this control is selected from 100 ms or 200 ms depending on the S/C spin rate as in the case of autonomous sun acquisition logic.

The tilt-and-scan control in contingency is implemented for the reacquisition of communications with the S/C in case S/C is accidentally lost from the earth station. By the DPU signal which is generated when no commands are received on the S/C for one week, AOCE

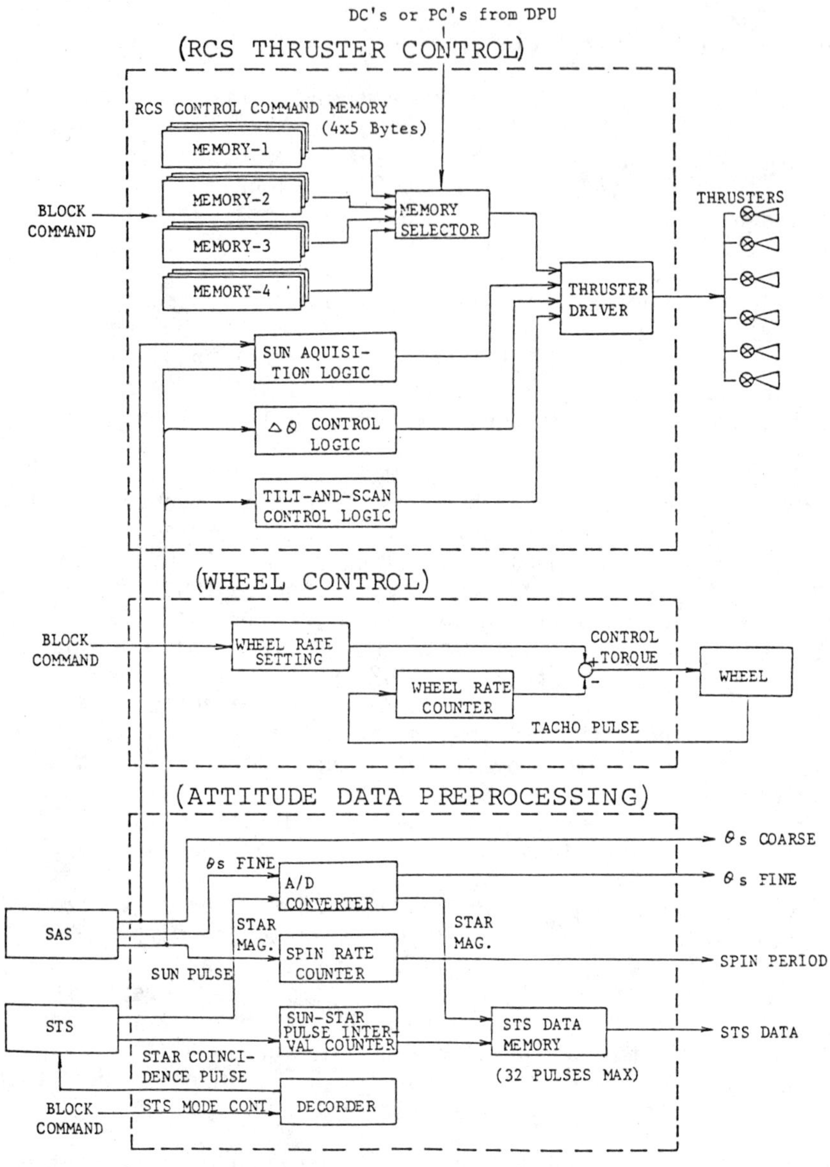

Fig. 9 AOCE Functional Block Diagram

automatically starts this attitude maneuver. First, S/C spin axis is reoriented to $\theta_s = 90° \pm 3°$ by the sun-angle acquisition logic. Then, by the $\Delta\theta$ control logic with $C_1 + C_3$ thrusters the spin axis of S/C is precessed in step of 4.5 degrees at the interval of 22.8 hr. During this interval HGA is slowly rotated at a rate of 0.7 /32 sec. In this manner, the beam of HGA will eventually complete all sky scan. At the time when the signal from the S/C is received at the earth station, this earth acquisition maneuver is completed and the maneuver is stopped by the command from the ground.

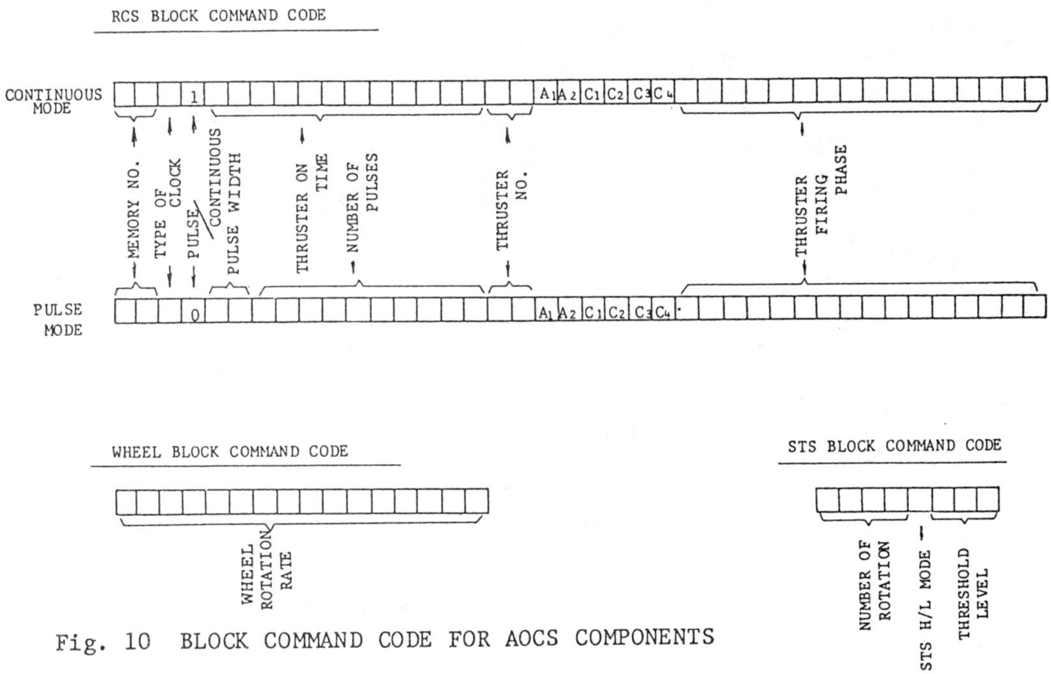

Fig. 10 BLOCK COMMAND CODE FOR AOCS COMPONENTS

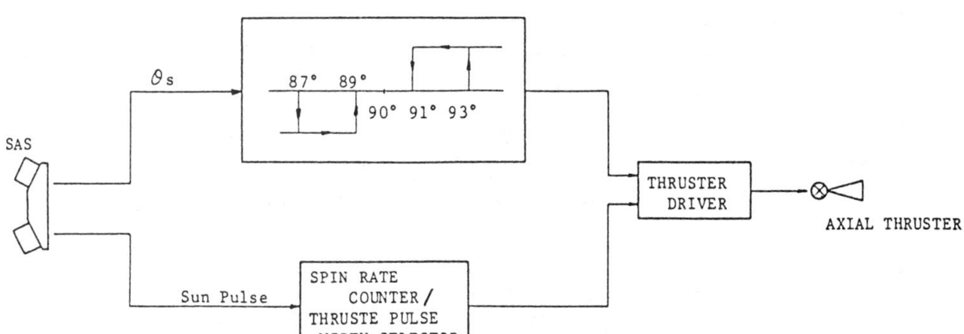

Fig. 11 Sun-Angle Acquisition Logic

To fire the thrusters, the sun pulse from one of sun sensors designated by command is used as the timing reference. However, when the sun goes out of FOV of the selected sensor, AOCE switches to another sensor automatically.

Control of wheel: wheel speed can be set by the BC shown in Fig-10 by the increment of 0.07 ± 0.02 rpm. Wheel rate (f_w) is defined by

$$f_w = 62,914,560/N \text{ (rpm)}$$

where N is the digit number represented by BC. In 0.2 rpm mode smooth and stable rotation of 0.2 rpm ± 0.4% is required to prevent the UVI's image from smearing. This means that the deviation of the wheel rotation at around 2000 rpm should be less than 0.19 rpm in a time spun of 5 min. In wheel control logic, the difference between the rate set by the BC and actual rate measured by the tacho pulses is filtered by the first order linear filter and then the torque which is proportinal to the filter output is applied to the wheel rotor. In this control loop, the stability of less than 0.018 rpm (3σ) is confirmed as a result of open loop test. To shorten the communication loss between the S/C and ground station due to the transition time of S/C spin, the wheel run down time is reduced to 14 min by actively decelerating the wheel.

Attitude Data Preprocessing: The analogue sine and cosine signals from the sun sensors, and the analogue star intensity signal are A/D converted into 8 bits. Spin period from one sun pulse to another and time intervals from sun pulse to star pulses, as shown in Fig-12, are measured by 4096 Hz clock at 6 rpm mode and by 128 Hz clock at 0.2 rpm, each resulting in 16 bits digital data. The quantization error corresponds to 9.5×10^{-3} deg in spin angle for both cases. Star scanner data (pulse timings and its intensities) are stored at the AOCE memory. The number of S/C rotations during which star data is stored in the memory is selectable between 1 to 15 rotations by BC, shown in Fig-10, so long as the number of total star pulses is less than 32 pulses.

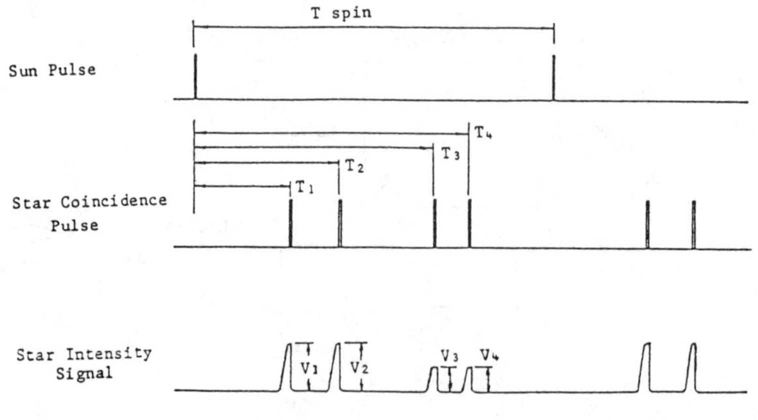

Fig. 12 Timing Chart of Attitude Data Measurement

AOCS OPERATION

Event sequence after the S/C injection is divided into the following four phases

1) initial acquisition phase
2) orbit correction phase
3) cruising phase
4) observation phase

The event sequence is shown in Fig-13. The initial acquisition phase includes reduction of the S/C spin rate from 120 rpm to final 6 rpm and spin axis maneuvers for sun and earth acquisition. Orbit correction phase follows at 4 to 5 days after injection. After orbit correction, S/C cruises toward the descending node of Halley's comet. Observations of Halley's comet will be conducted en route to the comet. The operational sequence from initial acquisition to cruising phase is depicted in Fig-14.

Initial Acquisition

S/C spin rate after injection is as high as 120 rpm, then by firing C_2 + C_3 thrusters spin rate of the S/C will be reduced to intermediate spin rate of 30 rpm to save the fuel comsumption during the following sun acquisition maneuver. When S/C spin-rate has reached 30 rpm, AOCE

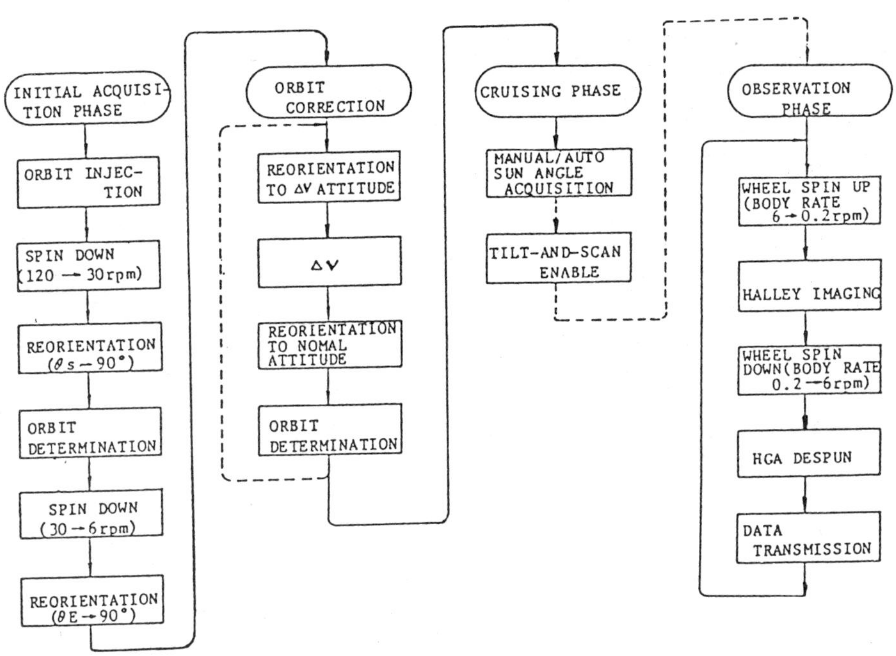

Fig. 13 AOCS Operation Sequence

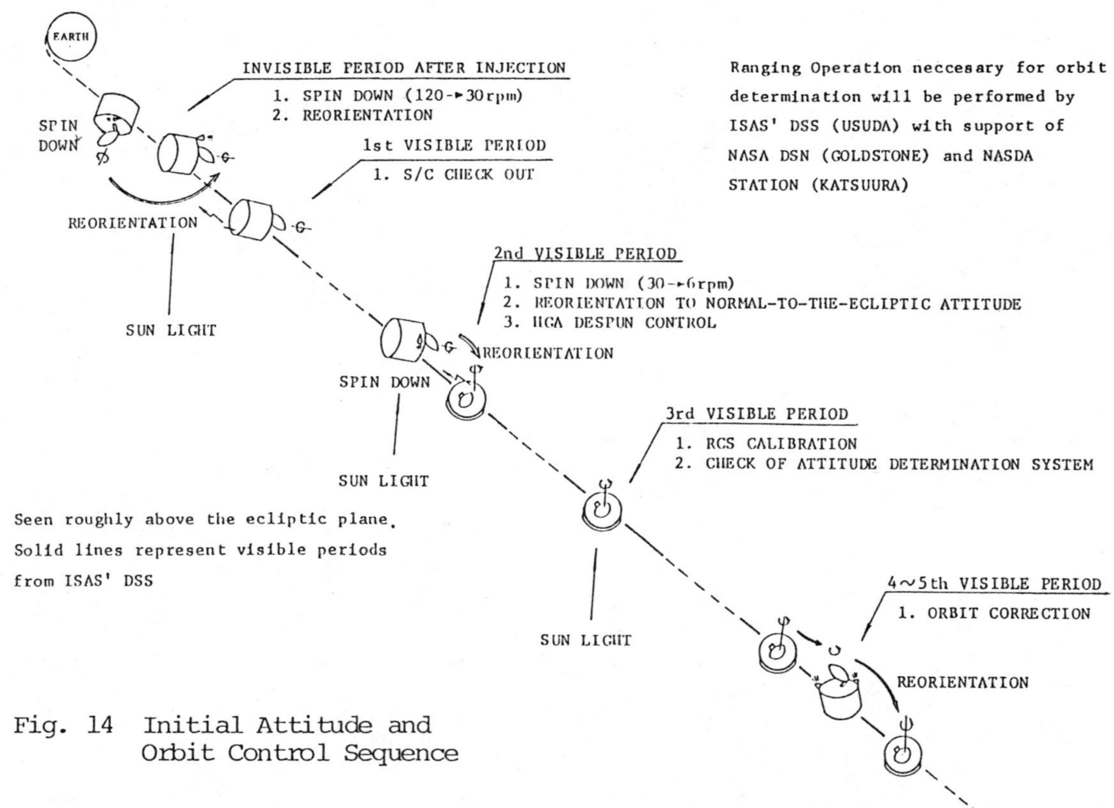

Fig. 14 Initial Attitude and Orbit Control Sequence

automatically stops the firing C_2+ C_3 thrusters. After that, by the PC from DPU AOCE starts the autonomous sun-angle acquisition control in which spin axis is maneuvered to be perpendicular to the sun-S/C line by using A_1 or A_2 thruster. Low gain antenna will be used for the communication between the S/C and the ground station. During the first visible period from the Japanese ground station onboard instrument will be checked out then S/C spin rate will be reduced further to 6 rpm by C_2+ C_3 thrusters. By firing the axial thrusters A_1 or A_2 the earth acquisition sequence will follow which reorients the spin axis neary normal to the ecliptic plane. The trajectory of the direction of minus Z axis projected on sky is shown in Fig-15. Prior to the earth acquisition maneuver attitude determination system starts to determine the direction of the S/C spin axis with SAS and STS data.

In the initial acquisition phase, NASA deep space station at Goldstone, Calif. and NASDA satellite station at Katsuura, Japan are being scheduled to support the orbit tracking to provide the orbital elements used to initiate the tracking at ISAS' Deep Space Station (at Usuda, Japan).

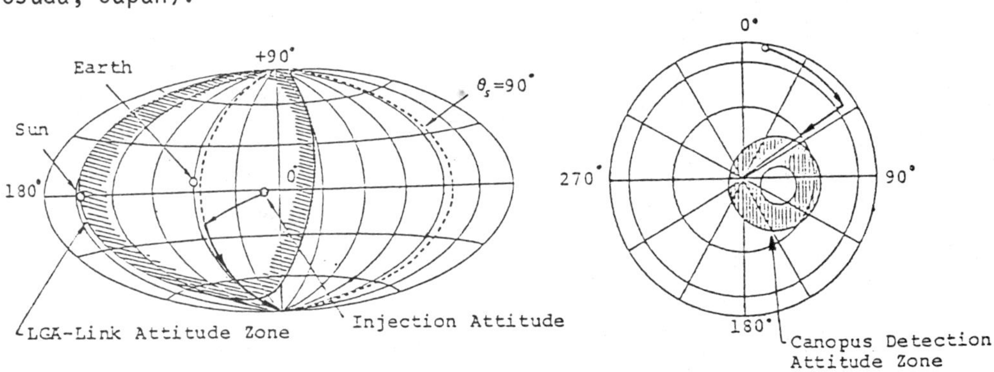

Fig.15 Trajectory of S/C Attitude (Minus Z-Axis of S/C) in Initial Acquisition Phase

Orbit Correction

4 to 5 days after the injection, the orbit correction maneuvers are to be performed to trim the injection error. Initial orbit injection error is estimated as 100 m/s (3σ). Orbit correction capability is limited to about 50 m/sec. The sequence of the orbit correction maneuver is as follows.

i) orbit determination with range and range-rate data
ii) ΔV calculation (by the ground software)
iii) search for the optimum attitude for the correction in terms of fuel consumption and required time for thruster operation.
iv) spin axis maneuver to the one decided in step iii)
v) orbit correction firing
vi) spin axis maneuver to the initail attitude (spin axis normal to the ecliptic plane)

ΔV-attitude must satisfy the thermal and communication's requirements, permit sufficient power generation from solar cells and it also should allow the attitude determination by star and sun sensors. As shown in Fig-16, ΔV-attitude, satisfing these requirement, is confined within dashed narrow zones on a celestial sphere. In spite of this, by choosing the adequate thrusters for operation it is possible to generate ΔV in almost any direction as shown in Fig-17.

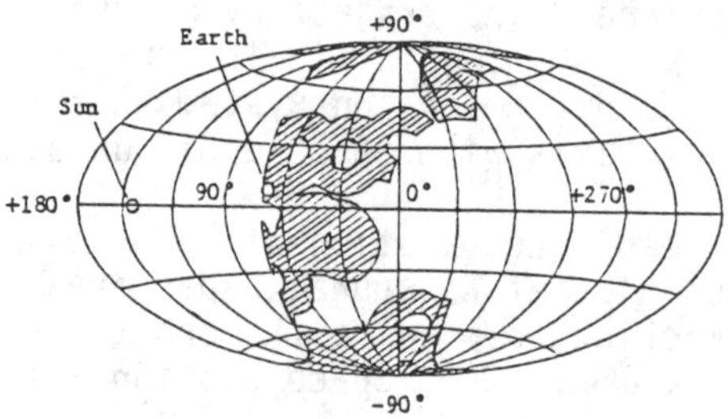

Dashed zones show the possible ΔV-attitude (direction of minus Z-axis of the S/C) in a spacecraft-centered ecliptic coordinate system.

Fig. 16 Possible ΔV-Attitude

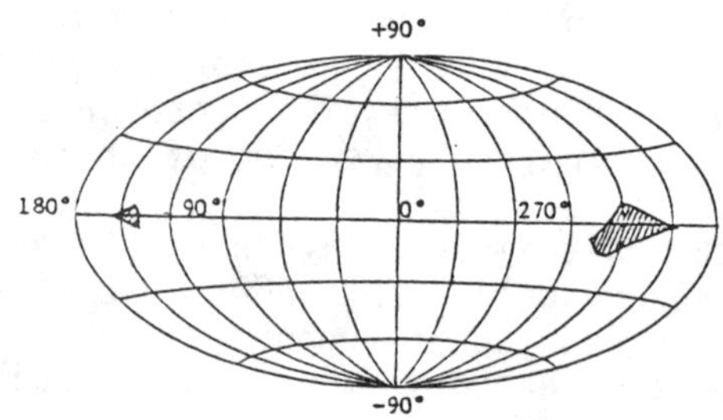

Dashed zones show celestial directions into which S/C cannot be accelerated with possible ΔV-attitude (within left dashed zones) and through only one sequence of orbit-correction maneuvers.

Fig. 17 Attainable ΔV-Directions

Cruising Phase

In cruising phase, spin axis of the S/C is kept normal to the ecliptic plane. The main disturbance torque to the attitude is caused by the solar pressure in PLANET-A case. Unbalanced distribution of the sunlight receiving area around the center of mass of S/C produces the disturbance torques which cause the precessional motion of the attitude around the sun-S/C line as well as the angular momentum accumulation. In PLANET-A case, once per three weeks attitude touch-up maneuver will be required to keep the HGA beam within $\pm 5°$ toward the earth. The unloading operation of the excess angular momentum is estimated to be less than once per two months.

Observation Phase

To take images of Halley's comet by UVI camera, momentum wheel is started to rotate. About five minutes later wheel rate reachs to 2000 rpm and S/C spin rate reaches to 0.2 rpm. UVI camera adjusts the elevation angle of its reflecting mirror to catch the Halley's comet at the center of FOV. Images of the comet thus obtained are once stored at UVI memory, compressed to 1/3 of the initial amount of image data, and then stored at 1 Mbits Data Recorder (DR). Up to several images thus obtained are storable on the DR. Wheel will then be stopped. This returns the S/C spin rate to its normal cruising rate of 6 rpm. Despin control electronics, also by referencing the sun pulse, starts to control the HGA to point toward the earth to transmit the UVI image data. Solar wind observation by ESP will be performed only at 6 rpm spin rate. Thus, observations by UVI and ESP are performed alternately.

AOCS GROUND SUPPORT SOFTWARE/DEEP SPACE NETWORK

Ground support softwares around deep space network is outlined in Fig-18. Four software groups which support the AOCS operations run on a central computer (FACOM M-380) at ISAS' Komaba campus. The central computer is linked to both of KSC and DSS at Usuda. Orbital data (range and range rate) and telemetry data received at KSC and DSS are transmitted to Komaba. House-keeping data will be processed on Satellite Control Equipment, data from scientific instruments will be processed by Scientific Instrument (SI) Data Processing System, and attitude and orbit data are transfered to and are processed at central computer M-380.

Orbit and ΔV-Determinations

Orbit of the S/C will be determined by the program called "ISSOP" using the ranging and doppler data. In this program, the recursive least square method with differential correction is utilized as an orbit estimation algorithm. Orbit will be determined with the accuracy around few hundred kilometers at the time of encounter with Halley's comet (Reference 4). After the orbit determination by "ISSOP", ΔV needed to encounter with Halley's comet will be determined by program "HOMAN".

Attitude Determination Software

S/C attitude determination are based upon the sun-angle data (θ_s), star elevation angle data (δ) with respect to the S/C coordinate, and rotation angle (ψ) of S/C from sun pulse detection to star pulse detection. Star elevation angle (δ) is derived from the time interval of star coincidence pulses based upon STS slit map. The attitude determination process is as follows.

i) "Definitive" star pulse pairs are searched among star pulses in pulse train for which it is definite that they belong to the same star.

ii) For the "tentaitive" stars obtained through the process of step i) the sun-star and star-star angular distances are calculated and compaired with those of the star catalogue.

iii) If the matching of stars is good, stars are considered identified and coarse attitude will be determined.

iv) If the above mentioned process fails, all possible pulse pairs are formed and then steps ii) and iii) are repeated. If stars are

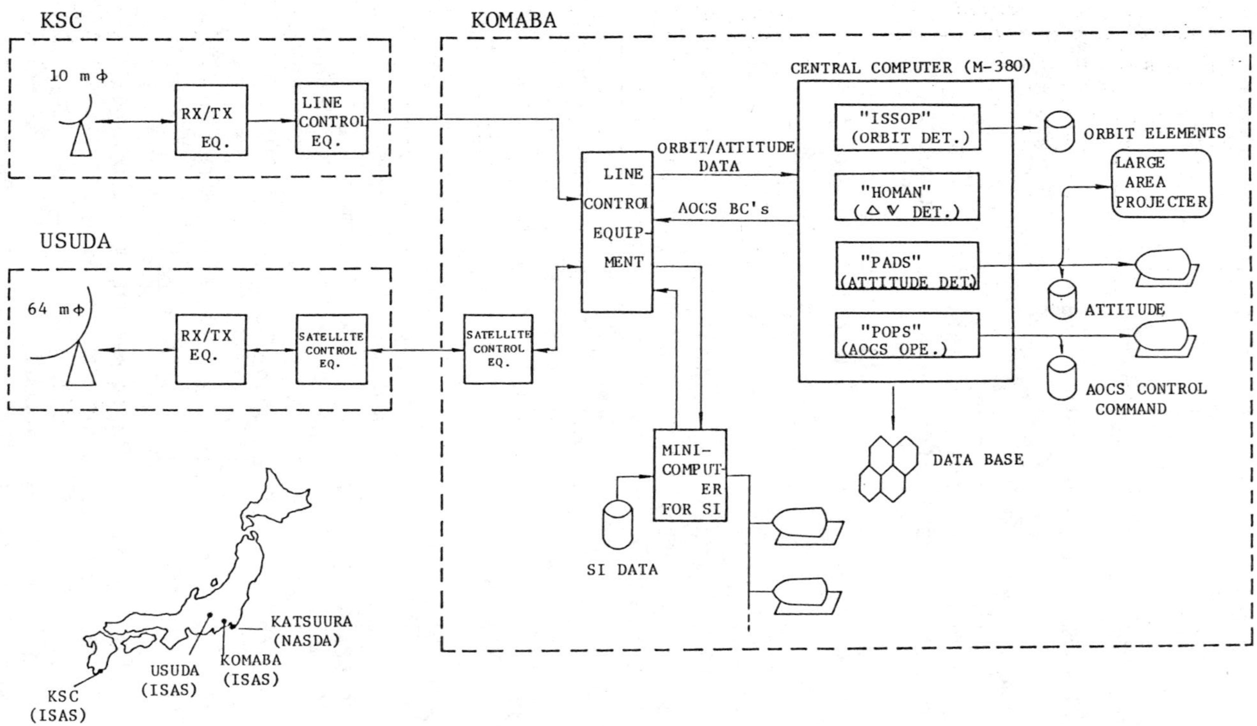

Fig. 18 Ground Support Software and Japanese Deep Space Network

not identified even in this case, alarms to change the STS threshold level will be issued on the display.

v) Based on the coarse attitude determined in step iii) simulated star pulse-train is generated for confirmation. The difference of the pulse positions between the STS data and those of simulated pulses is the measure of fitness of the estimated attitude.

vi) If the fitness is smaller than the specified threshold, data is used in the subsequent fine attitude determination phase.

The attitude determination flow is shown in Fig-19.

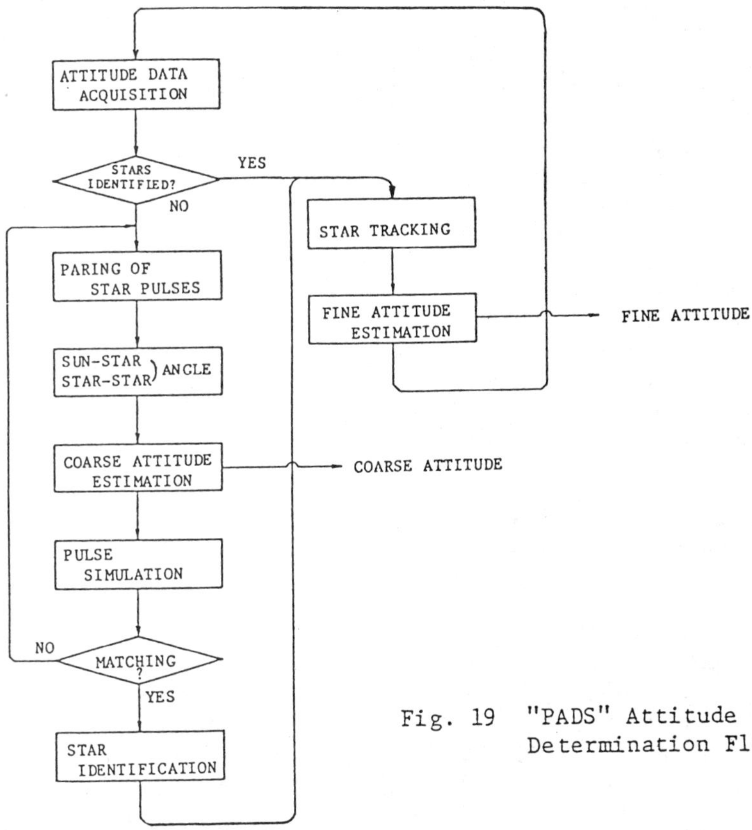

Fig. 19 "PADS" Attitude Determination Flow

In the fine attitude determination phase, sequential least square algorithm is employed on the consecutive spin data. The tilt due to the dynamic inbalance of S/C and the sensor misaligument are to be estimated along with the S/C attitude parameters. The attitude determination software works both in real-time mode and in batch mode. Overall attitude determination accuracy is 0.1 deg. Simulation software ("PSDG") generates the sun and star pulse train as will be obtained in space and is used to check the performance of the "PADS". In "PSDG", the mathematical model of SAS, STS and S/C dynamics are constructed and used to generate the output.

AOCS Operation Software

AOCS operation program, called "POPS", generates the block commands for the RCS control for attitude maneuvers and $\Delta \mathbf{v}$ operation as well as the block command for the HGA despin control.
This software has following functions

i) The generation of BC's for RCS and despin control, the main function.
ii) Simulation of the S/C and HGA dynamics under the BC's generated.
iii) Estimation of mechanical parameters of the S/C such as fuel weight and S/C weight, moment of inertia, and RCS pressure, all of which will change due to the fuel consumption.
iv) Calibrations of the thruster parameters.

The BC generation for RCS control are based upon the performance tables or formulae of thruster characteristics (such as total impulse, centroid location, and specific impulse) for the continuous and pulse mode firing. Based upon the knowledge gained from previous maneuvers in orbit, calibration factors (η's) for the individual thruster are obtained to be used to modify the performance data. The sun-S/C-earth angle (α) and its time derivative ($\dot{\alpha}$) will be calculated based upon the orbital elements of the S/C and converted to BC codes for the control of despun antenna. The BC for α controls HGA to point toward the earth and that for $\dot{\alpha}$ permits the autonomous pointing control of HGA for a month or so (Reference 5).

For item ii) above, the S/C dynamics or pointing motion of the HGA under the BC thus generated are simulated by the numerical integration of the dynamic equations of motion to confirm BC's effects and to reject any incorrect BC's that may be generated by inadequate inputs.

As to the estimation of mechanical parameters, a current fuel weight is estimated in three ways in PLANET-A case; i) from the state equation for ideal gas using the telemetered tank pressure, ii) by integration of fuel consumption at each maneuver, and iii) from the change of the S/C moment of inertia, the change being detected by wheel turn-on and-off. The last method seems to provide the most accurate estimation of fuel weight. Other paramaters such as S/C weight and RCS pressure are calculated on the base of the current fuel weight thus estimated.

AOCS VERIFICATION

AOCS system verification has been planned through following three steps.

1) Spin-system verification-test with actual subsystem hardwares on a rate table.
2) Performance verification-test for initial attitude acquisition.
3) Verification test for attitude determination system involving flight and ground segments.

Spin-System Verification-Test

To confirm the stability of the S/C spin rate the verification test for the spin rate control system was performed by using 1-axis rate table mounted with momentum wheel and despun HGA. The control system is depicted in Fig-20 and its test configuration is depicted in Fig-21. Based upon the measurement of wheel rate and rate of HGA with respect to S/C body, an analogue computer simulating S/C dynamics is used to drive the rate table at the corresponding rotation rate of the S/C. The test results confirmed that there is no interferences between wheel and despin control system. The test result in Fig-22 shows that spin rate deviation of S/C is about 2×10^{-4} rpm (3σ) and smaller than the UVI requirement (less that 8×10^{-4} rpm).

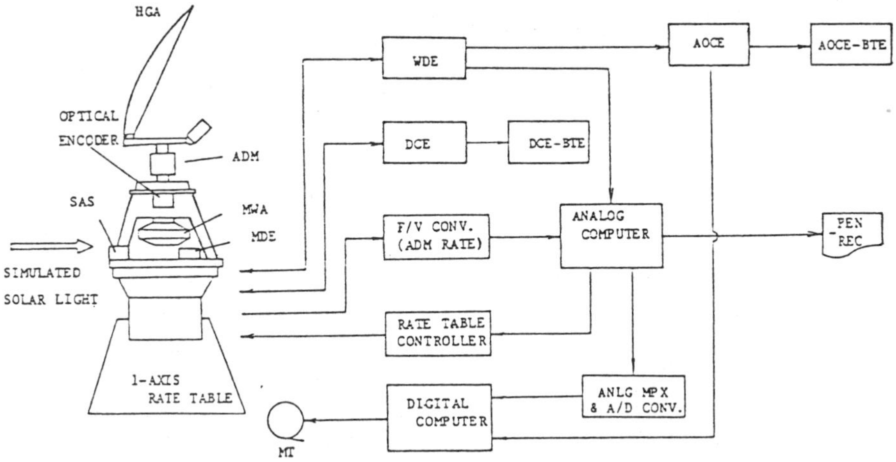

Fig. 21 Spin Control System Test Layout

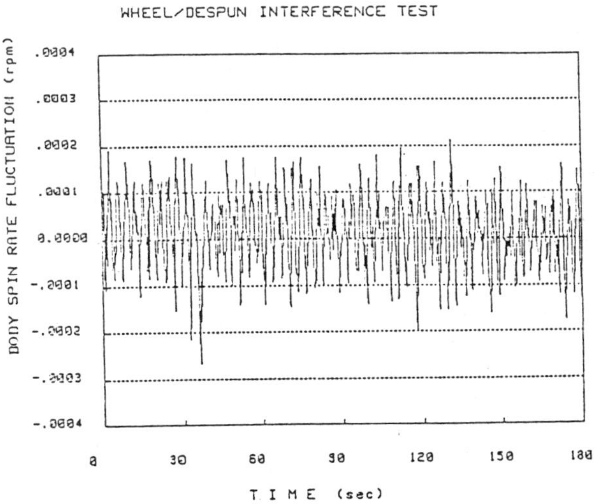

Fig. 22 Spin Control System Verification-Test Result

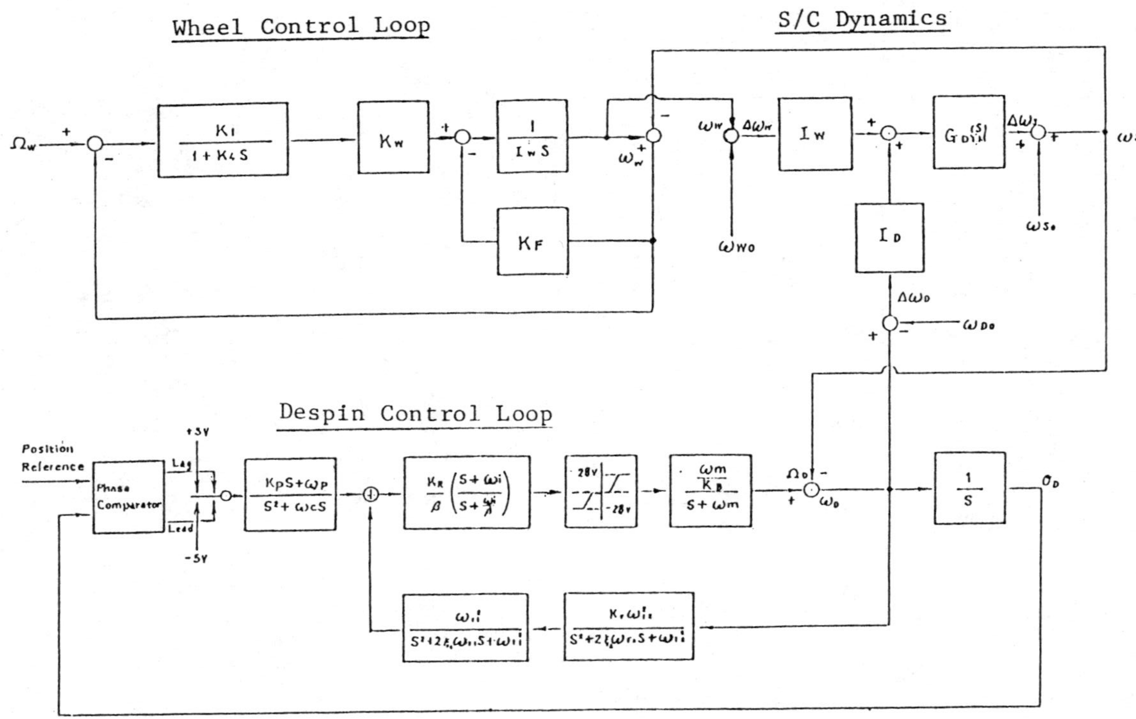

Fig. 20 Spin Control System Functional Block Diagram

Performance Verification-Test for Initial Attitude Acquisition

This test is planned to check the functions of control logics such as sun-angle acquisition, $\Delta\theta$ control, and the thruster control by BC for initial phase operation. Test layout is shown in Fig-23. S/C dynamics and the output from the detectors of SAS are simulated by digital minicomputer. The latter signals drive sun sensor electronics which supplies the sun pulse and θ_s signal to the AOCE. According to the logic imbeded in AOCE, AOCE activates the thruster valves. Thruster on/off times are measured by minicomputer to calculate the S/C motion. History of the S/C attitude and spin rate are displayed and monitored for the comfirmation of the operation.

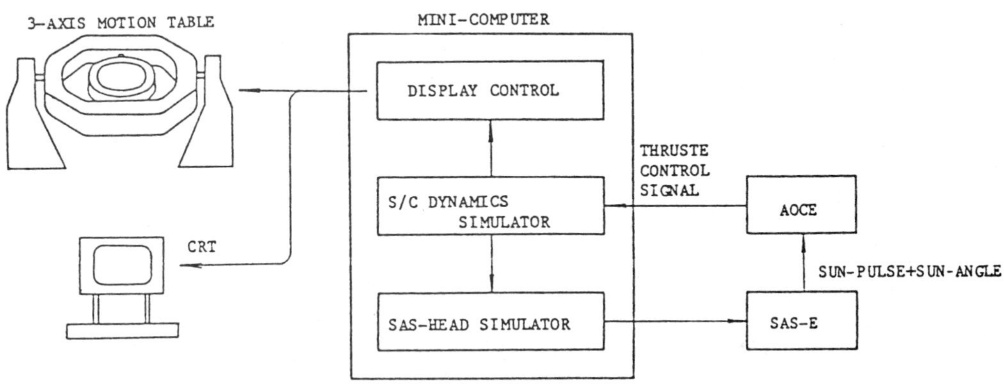

Fig. 23 AOCS Dynamic Functional Test

Verification Test of Attitude Determination System

This test aims to check the performance of the attitude determination system (SAS, STS, AOCE and PADS). Test layout is shown in Fig-24. Data recorded on an analogue magnetic tape is used to stimulate the SAS electronics and the LED attached on STS hood cover. These analogue data simulating the sun sensor detector output and the light input to STS are generated and recorded based upon the mathematical model of SAS, STS optics, star catalogue, S/C dynamics, and ephemeris. LED is used to stimulate the STS. Telemetry data from the S/C are processed by ground support equipmant and transferred to the central computer on which "PADS" works. In this way the attitude determination software is validated.

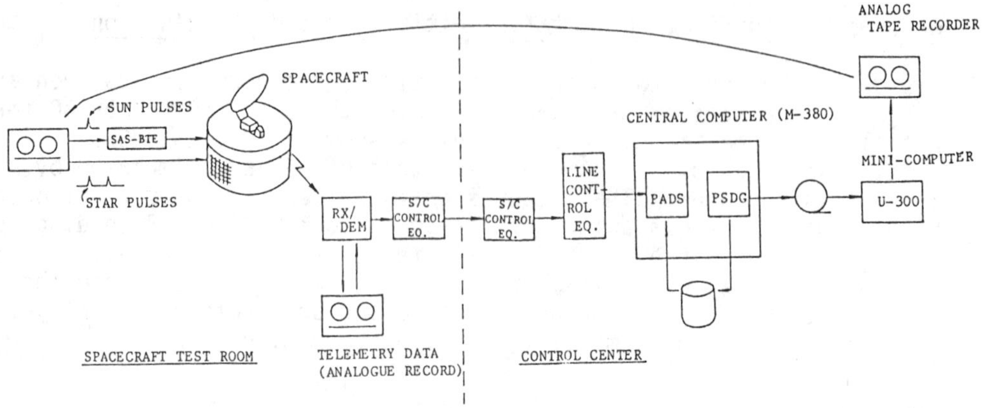

Fig. 24 Attitude Determination System Test

CONCLUDING REMARKS

The AOCS of the first Japanese interplanetary mission, PLANET-A, has been described. It has functions of attitude determination, attitude maneuvers and orbit correction. A unique point is the slow spin control of S/C body using a momentum wheel, which permits the ultraviolet imaging of Halley's comet. The adequacy of the design has been proven by the computer simulation and the tests using the engineering-type model hardwares. The test spacecraft MS-T5 is now under the mechanical and electrical interface test at ISAS and is scheduled to start the system performance test from May 1984 including the test of the Japanese deep space network. After the launch of MS-T5 in January 1985, PLANET-A is scheduled to start the system performance test for the forthcoming launch in August 1985. AOCS system test involving flight hardwares and ground support softwares (PADS) will complete the verification of AOCS functions and performances.

REFERENCES

1. Nomura, T., Hayashi, T., Hirosawa, H., Ninomiya, K., Ichikawa, M., "Deep Space Project in Japan," AIAA/NASA Symposium on Space Tracking and Data Systems, Pentagon City/Virginia 16-18 June 1981.

2. " PLANET-A Japanese First Interplanetary Flight Status Report," ISAS Report SES-TD-82-006 2nd Edition, dated June 1982.

3. Hirao, K., Matsuo,H., "The Japanese Halley Exploration Project," 33rd Congress of IAF, Paris, 9/27-10/2 1982 (IAF 82-191).

4. Nishimura, T., Kato, T., Wakagi, T., " Orbit Determination Problem of the Planet-A Mission, " Proceedings of The Thirteenth International Symposium on Space Technology and Science, Tokyo 1982, pp 957-962.

5. Kawaguchi, M., Masumoto, Y., Yamamoto, H., Hara, H., Hayashi, T., " Despin Antenna Control System for PLANET-A," Proceedings of The Thirteenth International Symposium on Space Technology and Science, Tokyo 1982, pp 975-980.

Section II
STORYBOARD PAPERS/DISPLAY

SESSION II

Chairman:	R.D. Neal Martin Marietta Denver Aerospace
Co-Chairman:	William E. Frazier Ball Aerospace Systems Division

The following papers were not available for publication:

AAS 84-014	Model of Hysteretic Properties of a Piezo-electric Crystal Stack, P. Dahl, R. Wilder, The Aerospace Corporation
AAS 84-019	High Performance Actuated Equipment, K. Silveira, Lockheed Palo Alto Research Laboratory

The following paper numbers were not assigned:

AAS 84-022 to -030

AAS 84-011

DIRECT QUATERNION DETERMINATION FROM BARS MEASUREMENTS

Dr. Stephen A. Skaron†

The Band Angle Readout System (BARS) inertial measurement unit is under consideration for use in missile guidance systems. The output data from the instrument includes 6 so called "band angles", 4 of which can be used to determine the missile's attitude. This can be done by computing the 9-parameter direction cosine transformation matrix from a reference set of axes to the missile body set of axes. Since most flight programs now use quaternion mathematics, the equivalent 4-parameter quaternion can then be extracted by use of a suitable, singularity-free extraction routine. However, this procedure has been shown to be costly in terms of ROM and RAM memory and in execution speed.

In the presentation, it will be demonstrated that the attitude band angles are intimately related to Euler angles, and that quaternion determination is a relatively straight-forward matter. Since 12 conventions exist for defining Euler angles (i.e., 313 sequence, 213 sequence, etc.), several direct determination algorithms can be developed. These algorithms shall be briefly compared and one routine discussed in detail.

A. Introduction

The Band Angle Readout System (BARS) all-attitude inertial measurement unit (IMU)[1] was developed by the C.S. Draper laboratory and is currently under consideration for use in missile and submarine guidance systems. The IMU contains an inner sphere which is the inertially stabilized element and an outer sphere. The inner sphere is motion isolated from

† *Member of the Technical Staff, TRW Electronics and Defense, Spacecraft Engineering Division, One Space Park, Redondo Beach, California 90278*

the rest of the IMU by flotation in a hydraulic fluid. The outer sphere supports the fluid-suspended inner sphere and acts as a data interface for the inner sphere.

The output from the instrument includes 6 so called attitude "band angles", 4 of which can be used to determine the vehicle's attitude. Since most operational computer programs now utilize quaternion mathematics[2] (discussed briefly in Appendix I), the attitude described with the band angles must somehow be converted to the attitude described with quaternions. This can be done by first computing the attitude direction cosines from the band angles, and then extracting a quaternion from the direction cosines by use of a suitable, singularity-free quaternion extraction procedure.[3,4,5] This method, however, can be costly in terms of memory allocation and execution speed.

Attitude band angles are closely related to Euler angles (a set of three angular parameters which can be used to specify an attitude). From a particular set of Euler angles (there exist 12 uniquely defined sets), quaternion extraction is quite straightforward. This technique shall be referred to as direct quaternion determination and shall be discussed in detail.

B. The Band Angle Readout System[6]

The geometry of the BARS is depicted in Figure 1. The inertially fixed inner sphere contains three mutually orthogonal driver bands mounted such that the driver band planes 1, 2, and 3 are normal to the inner sphere axes IS1, IS2, and IS3 respectively. The outer sphere contains a single circular receiver band in the plane of the axes OS1 and OS2. A third axis OS3 is perpendicular to the receiver band plane such that OS1, OS2, and OS3 form an orthogonal set of outer sphere axes.

The three driver band angles ϕ_1, ϕ_2 and ϕ_3 are measured about axes IS1, IS2, and IS3 respectively in a counterclockwise fashion. Their magnitudes are defined as follows

- ϕ_1 The angle from axis IS2 to the appropriate intersection (subsequently defined) of driver band 1 with the receiver band.
- ϕ_2 The angle from axis IS3 to the appropriate intersection (subsequently defined) of driver band 2 with the receiver band.
- ϕ_3 The angle from axis IS1 to the appropriate intersection (subsequently defined) of driver band 3 with the receiver band.

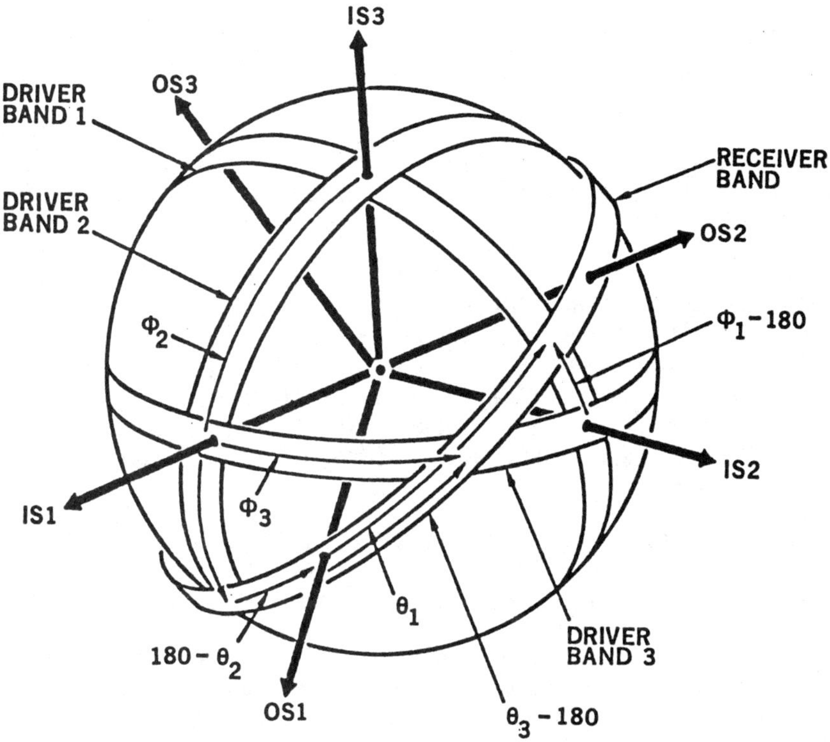

Figure 1. BARS geometry.

The three receiver band angles Θ_1, Θ_2, and Θ_3 are all measured along the receiver band in a counterclockwise fashion. Their magnitudes are defined as follows

- Θ_1 The angle from axis OS1 to the appropriate intersection (subsequently defined) of driver band 1 with the receiver band.
- Θ_2 The angle from axis OS1 to the appropriate intersection (subsequently defined) of driver band 2 with the receiver band.
- Θ_3 The angle from axis OS1 to the appropriate intersection (subsequently defined) of driver band 3 with the receiver band.

Note that driver band i intersects the receiver band at two points. The appropriate intersection of driver band i with the receiver band when measuring an arc ϕ_i is the one in which ϕ_i does not continue below the hemisphere with OS3 normal at the top, yet stops in *this* hemisphere. Similarly, the appropriate intersection of the receiver band with driver band i when measuring an arc Θ_i is the one in which Θ_i does not continue below the hemisphere with ISi normal at the top, yet stops in *this* hemisphere.

It can be shown that three independent parameters are needed to uniquely specify an attitude.[7] Since the BARS measurements include 6 band angles, a redundancy exists. In practice, two band angles Θ_i and ϕ_i associated with a single driver band are rejected[6,8] in order to avoid computational inaccuracies in the determination of the appropriate quaternion or direction cosine transformation from the inner sphere to the outer sphere attitude. Clearly, if the receiver band overlaps a particular driver band, the angles associated with this driver band are not defined. Similarly, if the angle of intersection between the receiver band and a particular driver band becomes too small, the measurements of the associated band angles are inaccurate and they must be rejected. The 4 band angles selected (Θ_i, ϕ_i, Θ_j, ϕ_j) therefore represent an optimal set to be used for attitude determination.

C. Attitude Direction Cosines

The direction cosine transformation matrix from the inner sphere to the outer sphere attitude, $\mathbf{A}$, can be shown to be[6]

$$\mathbf{A} = \frac{1}{K} \begin{pmatrix} C_{11} & C_{12} & C_{13} \\ C_{21} & C_{22} & C_{23} \\ C_{31} & C_{32} & C_{33} \end{pmatrix} , \qquad (1)$$

where

$$
\begin{aligned}
K &= \sin(\Theta_i - \Theta_j) \\
C_{1i} &= -\sin\phi_j \sin\Theta_i \\
C_{1j} &= \cos\phi_i \sin\Theta_j \\
C_{1k} &= -\cos\phi_j \sin\Theta_i + \sin\phi_i \sin\Theta_j \\
C_{2i} &= \sin\phi_j \cos\Theta_i \\
C_{2j} &= -\cos\phi_i \cos\Theta_j \\
C_{2k} &= -\sin\phi_i \cos\Theta_j + \cos\phi_j \cos\Theta_i \\
C_{3i} &= -\cos\phi_i \cos\phi_j \\
C_{3j} &= -\sin\phi_i \sin\phi_j \\
C_{3k} &= \cos\phi_i \sin\phi_j
\end{aligned}
\qquad (2)
$$

In equation 2, i, j, and k are related to the rejected driver band. The indices (i, j, k) equal (1, 2, 3) if driver band 3 is rejected, (2, 3, 1) if driver band 1 is rejected, and (3, 1, 2) if driver band 2 is rejected.

D. Quaternion Extraction from Direction Cosines[3,4,5]

The inner sphere to outer sphere transformation quaternion $\tilde{q}$ with components (q_1,q_2,q_3,q_4) is related to the direction cosine matrix **A** as follows[7,9]

$$
\mathbf{A} = \begin{bmatrix} q_1^2-q_2^2-q_3^2+q_4^2 & 2(q_1q_2+q_3q_4) & 2(q_1q_3-q_2q_4) \\ 2(q_1q_2-q_3q_4) & -q_1^2+q_2^2-q_3^2+q_4^2 & 2(q_2q_3+q_1q_4) \\ 2(q_1q_3+q_2q_4) & 2(q_2q_3-q_1q_4) & -q_1^2-q_2^2+q_3^2+q_4^2 \end{bmatrix} . \qquad (3)
$$

Note that $-\tilde{q}$ with elements $(-q_1,-q_2,-q_3,-q_4)$ results in the same transformation matrix as $+\tilde{q}$ with elements $(+q_1,+q_2,+q_3,+q_4)$. This sign ambiguity is resolved by choosing the quaternion element q_4 to always remain positive.

In terms of the direction cosine matrix elements A_{11}, A_{12}, etc., the magnitude of the quaternion elements q_1, q_2, q_3, q_4 can be seen to be

$$
\begin{aligned}
|q_1| &= +\frac{1}{2}\sqrt{1+A_{11}-A_{22}-A_{33}} \\
|q_2| &= +\frac{1}{2}\sqrt{1+A_{22}-A_{11}-A_{33}} \\
|q_3| &= +\frac{1}{2}\sqrt{1+A_{33}-A_{11}-A_{22}} \\
|q_4| &= +\frac{1}{2}\sqrt{1+A_{11}+A_{22}+A_{33}}
\end{aligned}
\tag{4}
$$

The signs for the quaternion elements can be determined from the following set of equations

$$
\begin{aligned}
q_1q_2 &= \frac{1}{4}(A_{12}+A_{21}) & \qquad q_1q_4 &= \frac{1}{4}(A_{23}-A_{32}) \\
q_1q_3 &= \frac{1}{4}(A_{13}+A_{31}) & \qquad q_2q_4 &= \frac{1}{4}(A_{31}-A_{13}) \\
q_2q_3 &= \frac{1}{4}(A_{23}+A_{32}) & \qquad q_3q_4 &= \frac{1}{4}(A_{12}-A_{21})\ .
\end{aligned}
\tag{5}
$$

Note that an individual quaternion element q_1, q_2, q_3, or q_4 appears in only 3 of the equations listed above. In order to avoid computational inaccuracies, the quaternion element with the largest magnitude is selected from equation 4. The 3 equations in which this element appears are then selected from the list above in order to determine the remaining parameters. If q_4 is found to be negative, the sign on all the elements is reversed so that $q_4>0$. For example if q_4 is the quaternion element of largest magnitude, $\tilde{q}$ can be extracted from $\mathbf{A}$ as follows

$$
\begin{aligned}
q_4 &= +\frac{1}{2}\sqrt{1+A_{11}+A_{22}+A_{33}} \\
q_1 &= \frac{1}{4q_4}(A_{23}-A_{32}) \\
q_2 &= \frac{1}{4q_4}(A_{31}-A_{13}) \\
q_3 &= \frac{1}{4q_4}(A_{12}-A_{21})\ .
\end{aligned}
\tag{6}
$$

In the case where $q_4 = 0$, no convention exists for uniquely determining the sign on $\tilde{q}$. However, since $-\tilde{q}$ performs the same transformation as $+\tilde{q}$, no inconsistencies result.

There are several disadvantages in utilizing this algorithm for real time flight software. The complete direction cosine matrix must be computed first before the transformation quaternion can be extracted. This can be costly in terms of ROM memory allocation as well as throughput. Furthermore, RAM memory allocation suffers a bit as a result of storing 9 direction cosine matrix elements instead of just 4 quaternion elements.

E. Direct Quaternion Extraction from BARS Measurements

The transformation from one Cartesian axis system to another can be performed by means of three successive rotations done in a specified sequence. The three angles of rotation are referred to as Euler angles.[7] However, since 12 conventions exist for defining the Euler angles (i.e., 213 sequence, 313 sequence, etc.) care must be taken to avoid confusion. The inner sphere to outer sphere transformation describable by the generalized Euler sequence ijk is begun with a counterclockwise rotation of the IS axis system through an angle ϕ about axis ISi resulting in the OS' frame. This step is followed by another counterclockwise rotation of the OS' axis system through an angle Θ about the intermediate axis OSj' resulting in the OS" frame. Finally, the OS" system is rotated counterclockwise through an angle ψ about axis OSk" resulting in the required OS coordinate system. It is imperative in the above sequence that $i \neq j$ and $j \neq k$ in order that three *independent* rotations be performed. The case where $i=k$ poses no problem. Once an Euler angle convention has been established, the transformation quaternion elements q_1, q_2, q_3, q_4 can be easily expressed as trigonometric functions of the Euler angles. Furthermore, the attitude band angles are closely related to the Euler angles (within a specific convention) so that the quaternion elements can also be expressed as relatively simple trigonometric functions of the band angles.

Of the 12 possible Euler sequences, only 4 are immediately applicable for direct quaternion determination from band angle measurements. They are the 123 sequence which requires band 1 not be rejected, the 213 sequence which requires band 2 not be rejected, the 313 sequence and the 323 sequence which both require that band 3 not be rejected. Since all 4 applicable Euler sequences require that a particular band not be rejected, at least two sequences (any two except the 313 and 323 sequences) are needed in order to directly extract the attitude quaternion for all possible attitudes. The details for the 313 sequence will be presented and the results for the others simply summarized.

Figure 2 illustrates the various stages of the 313 sequence of Euler rotations. The sequence is begun by rotating the IS axis system by an angle ϕ counterclockwise about the IS3 axis, resulting in the OS' coordinate system. In the second stage, the intermediate OS' axis system is rotated by an angle Θ counterclockwise about the OS1' axis, resulting in the OS" axis system. Finally, the second intermediate OS" axis system is rotated by an angle ψ counterclockwise about the OS3" axis resulting in the OS coordinate system. (Note that in Figure 2c, the angle ψ is depicted as a clockwise rotation. There is no inconsistency since ψ, as shown, has a negative value.)

The quaternion elements can in turn be shown to be trigonometric functions of the Euler angles for the 313 sequence as follows

$$
\begin{aligned}
q_1 &= \cos \frac{\phi-\psi}{2} \sin \frac{\Theta}{2} \\
q_2 &= \sin \frac{\phi-\psi}{2} \sin \frac{\Theta}{2} \\
q_3 &= \sin \frac{\phi+\psi}{2} \cos \frac{\Theta}{2} \\
q_4 &= \cos \frac{\phi+\psi}{2} \cos \frac{\Theta}{2}
\end{aligned}
\qquad (7)
$$

Comparison of Figure 1 with Figure 2a shows that the first Euler angle ϕ is the same as the driver band angle ϕ_3. Inspection of Figure 2b shows that the second Euler angle Θ is the same as the angle between the axes IS3 and OS3". The cosine of this angle is simply the direction cosine matrix element A_{33}, therefore $\Theta = \cos^{-1}(A_{33}) = \cos^{-1}(C_{33}/K)$. Finally, comparison of Figure 1 with Figure 2c shows that the third Euler angle ψ is equal to $180-\Theta_3$ in order to express the angle as a counterclockwise rotation. Note that Θ and $360-\Theta$ are solutions to the equation $\Theta = \cos^{-1}(A_{33})$ where $0 \leq \Theta \leq 180$. Inspection of Figure 2b shows that as soon as Θ exceeds 180 degrees, ϕ becomes $\phi+180$ since the appropriate intersection of driver band 3 with the receiver band lies in the hemisphere with OS3" normal at the top. Therefore, given a particular value for ϕ, Θ cannot exceed 180° and the $\cos^{-1}$ function will be single-valued under these circumstances. Furthermore, no possible attitudes will be excluded due to this restriction on Θ, since there are no restrictions on ϕ and ψ.

The quaternion elements in equation 7 can thus be expressed in terms of the band angles.

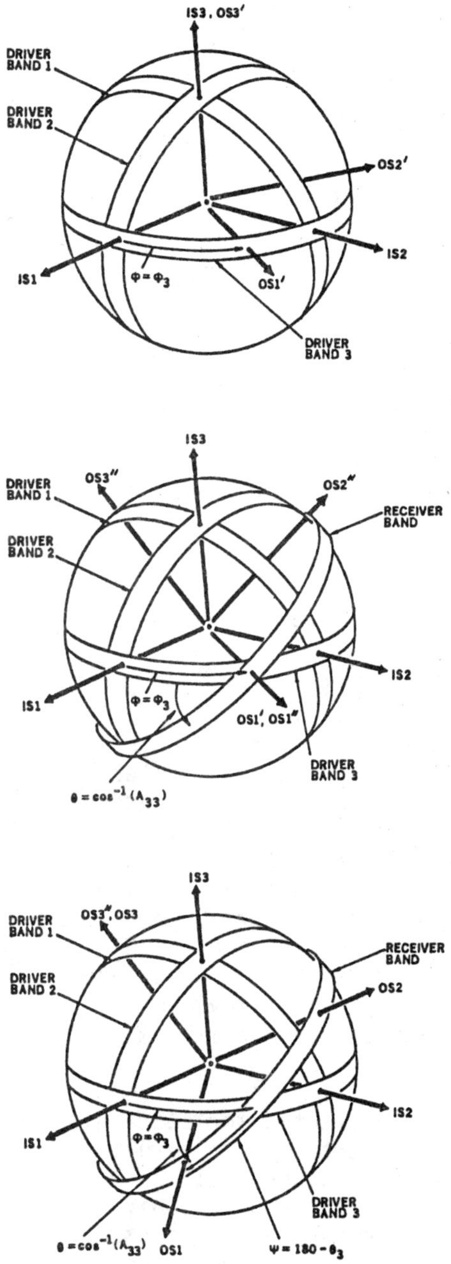

a) Counterclockwise rotation through the angle ϕ about axis IS3.

b) Counterclockwise rotation through the angle θ about axis OS1′.

c) Counterclockwise rotation through the angle Ψ about axis OS3″. In this case Ψ is depicted as a clockwise rotation and would have a negative value.

Figure 2. The 313 sequence of Euler rotations.

$$q_1 = \cos\frac{\phi_3+\Theta_3-180}{2}\ \sin\frac{\cos^{-1}A_{33}}{2}$$

$$q_2 = \sin\frac{\phi_3+\Theta_3-180}{2}\ \sin\frac{\cos^{-1}A_{33}}{2}$$

$$q_3 = \sin\frac{\phi_3-\Theta_3+180}{2}\ \cos\frac{\cos^{-1}A_{33}}{2} \qquad (8)$$

$$q_4 = \cos\frac{\phi_3-\Theta_3+180}{2}\ \cos\frac{\cos^{-1}A_{33}}{2} ,$$

where, if band 1 is rejected

$$A_{33} = \frac{\sin\phi_2\sin\phi_3}{\sin(\Theta_3-\Theta_2)} \qquad (9a)$$

and, if band 2 is rejected

$$A_{33} = \frac{\cos\phi_3\cos\phi_1}{\cos(\Theta_1-\Theta_3)} \ . \qquad (9b)$$

If band 3 is rejected, this algorithm is not applicable.

Table 1 displays the results for the 313, 323, 123, and 213 Euler sequences. Expressions for the three Euler angles ϕ, Θ, ψ in terms of the band angles are presented at the left and those for the quaternion elements in terms of the Euler angles are given at the right. Note that the analytical expression for the quaternion elements depends upon which Euler sequence is selected. Furthermore, it can be seen that the 123 sequence requires the direction cosine matrix element A_{31}, the 213 sequence requires the matrix element A_{32}, and the 313 and 323 sequences both require the matrix element A_{33}. However, the complete 3rd row of the direction cosine matrix will always be evaluated since these matrix elements are needed to determine which band is to be rejected.[6,8]

In practice, the 313 Euler sequence should be selected for quaternion extraction for all cases when band 3 is not rejected. The results using the 323 sequence under these conditions are identical, but Table 1 shows that the expression for the Euler angle ϕ is a bit simpler, using 313 sequence. For the case when band 3 is rejected, either the 123 or 213 sequences are applicable since, under these circumstances, bands 1 and 2 will be selected. Both algorithms involve the same amount of analytical complexity and since the results are identical, the choice of sequences is arbitrary.

Table 1

SUMMARY OF THE 313, 323, 123 AND 213 EULER SEQUENCES

313 Euler Sequence; Band 3 required	
$\phi = \phi_3$	$q_1 = \cos\frac{\phi-\psi}{2} \sin\frac{\Theta}{2}$
$\Theta = \cos^{-1}(A_{33})$	$q_2 = \sin\frac{\phi-\psi}{2} \sin\frac{\Theta}{2}$
$\psi = 180-\Theta_3$	$q_3 = \sin\frac{\phi+\psi}{2} \cos\frac{\Theta}{2}$
	$q_4 = \cos\frac{\phi+\psi}{2} \cos\frac{\Theta}{2}$
323 Euler sequence; Band 3 required	
$\phi = \phi_3 + 270$	$q_1 = \sin\frac{\psi-\phi}{2} \sin\frac{\Theta}{2}$
$\Theta = \cos^{-1}(A_{33})$	$q_2 = \cos\frac{\psi-\phi}{2} \sin\frac{\Theta}{2}$
$\psi = 270-\Theta_3$	$q_3 = \sin\frac{\psi+\phi}{2} \cos\frac{\Theta}{2}$
	$q_4 = \cos\frac{\psi+\phi}{2} \cos\frac{\Theta}{2}$

Table 1 (Continued)

SUMMARY OF THE 313, 323, 123 AND 213 EULER SEQUENCES

123 Euler sequence; Band 1 required

$\phi = \phi_1 - 180$	$q_1 = \cos\frac{\phi}{2}\sin\frac{\Theta}{2}\sin\frac{\psi}{2} + \sin\frac{\phi}{2}\cos\frac{\Theta}{2}\cos\frac{\psi}{2}$
$\Theta = \cos^{-1}(-A_{31}) - 90$	$q_2 = \cos\frac{\phi}{2}\sin\frac{\Theta}{2}\cos\frac{\psi}{2} - \sin\frac{\phi}{2}\cos\frac{\Theta}{2}\sin\frac{\psi}{2}$
$\psi = 90 - \Theta_1$	$q_3 = \cos\frac{\phi}{2}\cos\frac{\Theta}{2}\sin\frac{\psi}{2} + \sin\frac{\phi}{2}\sin\frac{\Theta}{2}\cos\frac{\psi}{2}$
	$q_4 = \cos\frac{\phi}{2}\cos\frac{\Theta}{2}\cos\frac{\psi}{2} - \sin\frac{\phi}{2}\sin\frac{\Theta}{2}\sin\frac{\psi}{2}$

213 Euler sequence; Band 2 required

$\phi = \phi_2 - 90$	$q_1 = \sin\frac{\phi}{2}\cos\frac{\Theta}{2}\sin\frac{\psi}{2} + \cos\frac{\phi}{2}\sin\frac{\Theta}{2}\cos\frac{\psi}{2}$
$\Theta = 90 - \cos^{-1}(-A_{32})$	$q_2 = \sin\frac{\phi}{2}\cos\frac{\Theta}{2}\cos\frac{\psi}{2} - \cos\frac{\phi}{2}\sin\frac{\Theta}{2}\sin\frac{\psi}{2}$
$\psi = 180 - \Theta_2$	$q_3 = \cos\frac{\phi}{2}\cos\frac{\Theta}{2}\sin\frac{\psi}{2} - \sin\frac{\phi}{2}\sin\frac{\Theta}{2}\cos\frac{\psi}{2}$
	$q_4 = \cos\frac{\phi}{2}\cos\frac{\Theta}{2}\cos\frac{\psi}{2} + \sin\frac{\phi}{2}\sin\frac{\Theta}{2}\sin\frac{\psi}{2}$

F. Results

Appendix II contains a sample FORTRAN listing which computes the third row of the direction cosine matrix and implements the 313 and 213 Euler sequences for direct quaternion extraction. This routine has proven to be quite space efficient compared to the indirect extraction procedure given in equations 4 and 5 (which require the complete direction cosine matrix). Table 2 displays the results of computations involving 4800 attitudes which systematically cover all orientations over an entire sphere. The calculations involve a simulator which outputs an attitude quaternion that was taken to represent a vehicle's exact attitude. Output from the simulator also included all band angles which were modified with an appropriate noise model to represent actual instrument readings. Attitude quaternions were then computed from the band angle outputs using the direct extraction procedure given in Appendix II and the indirect extraction procedure just described. The results are reported in terms of the quaternion angular error[10] which is defined as being the angle through which a *single axis* rotation would have to be performed to go from the attitude represented by the simulator's output quaternion to the attitude represented by the quaternion extracted from band angles. As can be seen, the results for the direct and indirect extraction are virtually identical, the quaternion angular error averaging roughly 0.04 degrees in both cases. Computations involving the 123 and 323 Euler sequences for the direct extraction routine yield similar results. It should be noted that when the band angle noise model is removed, the quaternion angular error is on the order of 0.001 arc second for both direct and indirect extraction. This deviation is so slight, it would indicate that it is due entirely to roundoff errors in the extraction routines as well as in the band angle generation scheme contained in the simulator.

Acknowledgements

The author would like to thank Jan Snapper, Bob Kramer, and Jon Kirby for allowing the time necessary to write this presentation. Special thanks goes to Judy Soman for her patient and excellent preparation of the manuscript.

Table 2

MAGNITUDE OF THE ERRORS BETWEEN SIMULATION ATTITUDE QUATERNION AND EXTRACTED ATTITUDE QUATERNIONS FOR 4800 TEST CASES SYSTEMATICALLY COVERING ALL ORIENTATIONS OVER AN ENTIRE SPHERE

	Simulation vs. Direct Extraction[a]	Simulation vs. Indirect Extraction[b]
Quaternion Angular Error[c]	0.0407 ± 0.0180 Degrees[d]	0.0401 ± 0.0183 Degrees[d]

a Direct extraction was performed utilizing the 213 and 313 Euler sequences described in the text and summarized in Table 1.

b Indirect extraction was performed by computing the entire direction cosine transformation matrix and extracting an attitude quaternion by means of equations 4 and 5.

c Quaternion angular error is the angle through which a *single axis* rotation would have to be performed to go from the attitude represented by the simulator's output quaternion to the attitude represented by the quaternion extracted from the band angles.

d Values are reported as: average error ± standard deviation.

Appendix I. Quaternion Mathematics and Rigid Body Rotations[2,7,9]

Let the real quantities (q_1, q_2, q_3, q_4) form the components of the quaternion $\tilde{q}$ as follows

$$\tilde{q} = \hat{q}_4 + \hat{i}q_1 + \hat{j}q_2 + \hat{k}q_3 \quad . \tag{AI.1}$$

Here q_4 is the quaternion scalar and $(\hat{i}q_1 + \hat{j}q_3 + \hat{k}q_3)$ is the vector portion of $\tilde{q}$. In the above expression, $\hat{i}$, $\hat{j}$, and $\hat{k}$ are complex numbers obeying the following commutation relations

$$\begin{aligned} \hat{i}\,\hat{j} &= -\hat{j}\,\hat{i} = \hat{k} \\ \hat{j}\,\hat{k} &= -\hat{k}\,\hat{j} = \hat{i} \\ \hat{k}\,\hat{i} &= -\hat{i}\,\hat{k} = \hat{j} \quad . \end{aligned} \tag{AI.2}$$

Certain complex 2 x 2 matrices (spin matrices) or real 4 x 4 matrices obey these commutation relationships. Along with the corresponding identity matrix (either 2 x 2 or 4 x 4) to represent the scalar portion, a matrix representation for $\tilde{q}$ can easily be developed.

The conjugate of $\tilde{q}$ is defined to be

$$\tilde{q}^* = q_4 - \hat{i}q_1 - \hat{j}q_2 - \hat{k}q_3 \quad . \tag{AI.3}$$

Quaternion multiplication is an operation which is frequently performed. However, it is not commutative. Consider the following quaternion product

$$\tilde{q}_A\,\tilde{q}_B = \left(q_{A_4} + \hat{i}q_{A_1} + \hat{j}q_{A_2} + \hat{k}q_{A_3}\right)\left(q_{B_4} + \hat{i}q_{B_1} + \hat{j}q_{B_2} + \hat{k}q_{B_3}\right). \tag{AI.4}$$

By use of the commutation relations given in equation AI.2, the product can be expressed as

$$\begin{aligned}\tilde{q}_A \tilde{q}_B = & \left(+q_{A_4} q_{B_4} - q_{A_1} q_{B_1} - q_{A_2} q_{B_2} - q_{A_3} q_{B_3}\right) \\ & + \hat{i}\left(q_{A_1} q_{B_4} + q_{A_4} q_{B_1} + q_{A_2} q_{B_3} - q_{A_3} q_{B_2}\right) \\ & + \hat{j}\left(-q_{A_1} q_{B_3} + q_{A_3} q_{B_1} + q_{A_2} q_{B_4} + q_{A_4} q_{B_2}\right) \\ & + \hat{k}\left(q_{A_1} q_{B_2} - q_{A_2} q_{B_1} + q_{A_3} q_{B_4} + q_{A_4} q_{B_3}\right) .\end{aligned} \tag{AI.5}$$

It should be noted that quaternion multiplication is associative when dealing with the product of three or more quaternions. The norm of $\tilde{q}$ is defined to be

$$|\tilde{q}| = \sqrt{\tilde{q}^*\tilde{q}} = \sqrt{\tilde{q}\tilde{q}^*} = \sqrt{q_1{}^2+q_2{}^2+q_3{}^2+q_4{}^2} \quad . \tag{AI.6}$$

In the consideration of the rotation of a rigid body, $\tilde{q}$ is always *normalized* to 1.

Euler's theorem states that any rigid body rotation can be expressed as a rotation about a single fixed axis of unit length, $\vec{e}$, through an angle, Φ. The quaternion elements q_1, q_2, q_3, q_4 are thus defined by

$$\begin{aligned} q_1 &= e_1 \sin\frac{\Phi}{2} \\ q_2 &= e_2 \sin\frac{\Phi}{2} \\ q_3 &= e_3 \sin\frac{\Phi}{2} \\ q_4 &= \cos\frac{\Phi}{2} \quad , \end{aligned} \tag{AI.7}$$

where e_1, e_2, e_3 are the components of the unit vector $\vec{e}$.

One can express a 3-dimensional vector $\vec{x}$, with scalar components x_1, x_2, x_3 along the basis set $\hat{i}$, $\hat{j}$, $\hat{k}$ respectively, as a quaternion with a scalar component of zero

$$\vec{x} = \hat{i}x_1 + \hat{j}x_2 + \hat{k}x_3 \quad . \qquad \text{(AI.8)}$$

A vector $\vec{x}$, described in a fixed system of coordinate axes can be expressed in a rotated system of coordinate axes as $\vec{x}'$ by

$$\vec{x}' = \mathbf{A}\vec{x} \quad , \qquad \text{(AI.9)}$$

where $\mathbf{A}$ is the direction cosine transformation matrix. The vector $\vec{x}'$ can be obtained as a similarity transformation of $\vec{x}$ using quaternion techniques. $\vec{x}'$ is given by

$$\vec{x}' = \tilde{q}^* \vec{x} \tilde{q} \quad , \qquad \text{(AI.10)}$$

where $\tilde{q}$ is the quaternion which corresponds to $\mathbf{A}$.[11] The correlation between $\mathbf{A}$ and $\tilde{q}$ is given in the text in equation 3. In the evaluation of $\tilde{q}$ by means of equation AI.7, the rotation angle Φ is defined as a counterclockwise rotation in going from the fixed (unprimed) system to the rotated (primed) coordinate system. It should also be noted that the rotation axis $\vec{e}$ is not affected by the operation and therefore its components e_1, e_2, e_3 are the same in both the fixed and rotated frames. The algebra of quaternions must be employed for the evaluation of $\vec{x}'$ in equation AI.10. It is also possible to express the vector $\vec{x}'$ in the fixed axis system by

$$\vec{x} = \tilde{q} \vec{x}' \tilde{q}^* \quad . \qquad \text{(AI.11)}$$

The application of two successive rotations, the first describable by $\tilde{q}_A$ and the second describable by $\tilde{q}_B$, has the following effect on the vector $\vec{x}$ in the fixed system

$$\vec{x}'' = \tilde{q}_B^* \tilde{q}_A^* \vec{x} \tilde{q}_A \tilde{q}_B \quad , \qquad \text{(AI.12)}$$

where $\vec{x}''$ is a representation of $\vec{x}$ in the doubly rotated coordinate axis system. Similarly,

$$\vec{x} = \tilde{q}_A \, \tilde{q}_B \, \vec{x}'' \, \tilde{q}_B^* \, \tilde{q}_A^* \quad . \tag{AI.13}$$

In general, the successive application of n rotations, describable by quaternions $\tilde{q}_{A_1}$, $\tilde{q}_{A_2}$,..., $\tilde{q}_{A_n}$ can be expressed as a single rotation, represented by quaternion $\tilde{q}_c$, where

$$\tilde{q}_c = \tilde{q}_{A_1} \tilde{q}_{A_2} \cdots\cdot \tilde{q}_{A_n} \quad . \tag{AI.14}$$

The inverse transformation can be shown to be

$$\tilde{q}_c^{-1} = \tilde{q}_c^* = \tilde{q}_{A_n} \, \tilde{q}_{A_{n-1}} \cdots\cdot q_{A_1} \quad . \tag{AI.15}$$

Appendix II. Example of a Direct Extraction Routine

Table AII.1 contains a sample FORTRAN listing of a direct quaternion extraction routine. The 313 and 213 Euler sequences have been employed in the procedure. Table AII.2 contains a list and description of all the data items contained in the FORTRAN listing. Note that the units in the text are degrees for conceptual simplicity. However, radians have been employed in this Appendix since the trigonometric functions in the FORTRAN programming language utilize them.

Table AII.1

SAMPLE FORTRAN LISTING OF A DIRECT QUATERNION EXTRACTION PROCEDURE USING THE 313 AND 213 EULER SEQUENCES

```
      DIMENSION PHIBAND(3),THETABAND(3),A(3),Q(4)
      PI     = 3.141592654
      SINPI  = SIN(PHIBAND(I))
      COSPI  = COS(PHIBAND(I))
      SINPJ  = SIN(PHIBAND(J))
      COSPJ  = COS(PHIBAND(J))
      FACTOR = 1.0/(SIN(THETABAND(I) - THETABAND(J)))
      A(I)   = FACTOR * (-COSPI * COSPJ)
      A(J)   = FACTOR * (-SINPI * SINPJ)
      A(K)   = FACTOR * ( COSPI * SINPJ)
      IF(I.NE.1)GO TO 10
      PHI    = PHIBAND(2) - (0.5 * PI)
      THETA  = (0.5 * PI) - ACOS(-A(2))
      PSI    = PI - THETABAND(2)
      SHALFPHI   = SIN((0.5)*PHI)
      CHALFPHI   = COS((0.5)*PHI)
      SHALFTHETA = SIN((0.5)*THETA)
      CHALFTHETA = COS((0.5)*THETA)
      SHALFPSI   = SIN((0.5)*PSI)
      CHALFPSI   = COS((0.5)*PSI)
      Q(1) = SHALFPHI*CHALFTHETA*SHALFPSI + CHALFPHI*SHALFTHETA*CHALFPSI
      Q(2) = SHALFPHI*CHALFTHETA*CHALFPSI - CHALFPHI*SHALFTHETA*SHALFPSI
      Q(3) = CHALFPHI*CHALFTHETA*SHALFPSI - SHALFPHI*SHALFTHETA*CHALFPSI
      Q(4) = CHALFPHI*CHALFTHETA*CHALFPSI + SHALFPHI*SHALFTHETA*SHALFPSI
      GO TO 20
10    CONTINUE
      PHI   = PHIBAND(3)
      THETA = ACOS(A(3))
      PSI   = PI - THETABAND(3)
      HALFDIFF   = (0.5)*(PHI - PSI)
      HALFSUM    = (0.5)*(PHI + PSI)
      SHALFTHETA = SIN((0.5)*THETA)
      CHALFTHETA = COS((0.5)*THETA)
      Q(1) = SHALFTHETA * (COS(HALFDIFF))
      Q(2) = SHALFTHETA * (SIN(HALFDIFF))
      Q(3) = CHALFTHETA * (SIN(HALFSUM))
      Q(4) = CHALFTHETA * (COS(HALFSUM))
20    CONTINUE
      IF(Q(4).GT.0.0)GO TO 40
      DO 30 INDEX=1,3
      Q(INDEX) = -Q(INDEX)
30    CONTINUE
40    END
```

Table AII.2

DESCRIPTION OF DATA ITEMS CONTAINED IN TABLE AII.1

Data Item	Definition	Units[a]
PHIBAND(3)	Band Angles ϕ_1,ϕ_2,ϕ_3	Radians
THETABAND(3)	Band Angles $\Theta_1,\Theta_2,\Theta_3$	Radians
A(3)	3rd Row of Direction Cosine Matrix	
PI	π	
SINPI, SINPJ COSPI, COSPJ, FACTOR	Trigonometric Functions of Band Angles	
I, J, K	Indices Related to Rejected Band	
PHI, THETA, PSI	Euler Angles ϕ,Θ,ψ	Radians
SHALFPHI, CHALFPHI, SHALFTHETA, CHALFTHETA, SHALFPSI, CHALFPSI	Trigonometric Functions of Half Euler Angles	
HALFDIFF	Half the Difference of Euler Angles ϕ,ψ	Radians
HALFSUM	Half the Sum of Euler Angles ϕ,ψ	Radians
INDEX	A Dummy Index	

a The units used in the text are degrees for conceptual simplicity. Since the trigonometric functions in the FORTRAN programming language use radians, they have been employed here.

REFERENCES AND FOOTNOTES

1. MX Reference Models, Prepared by TRW Defense and Space Systems Group, P.O. Box 1310, San Bernadino, CA 92402

2. James R. Wertz, Spacecraft Attitude Determination and Control, D. Reidel Publishing Company, Dordrecht, Holland, 1978

3. Allan R. Klumpp, "Singularity Free Extraction of a Quaternion from a Direction Cosine Matrix", Journal of Spacecraft and Rockets, VOL. 13, NO. 12, December 1976, pp. 754-755

4. Stanley W. Shepperd, "Quaternion from a Rotation Matrix", Journal of Guidance and Control, VOL. 1, NO. 3, May-June 1978, pp. 223-224

5. Carl Grubin, "Quaternion Singularity Revisited", Journal of Guidance and Control, VOL. 15, NO. 4, July-August 1978, p. 255

6. Peacekeeper MK21 Flight Program Design Analysis Algorithm and Equation Rederivation, Prepared by Logicon, Inc., 1983, 255 W. 5th St., San Pedro, CA 90733

7. Herbert Goldstein, Classical Mechanics, 2nd edition, Addison-Wesley Publishing Co., 1983

8. Gordon D. Niva, "Band Angle Selection Logic", Rockwell International Internal Letter, UCO/12240/278-032/25-01-23, October 1980

9. Jens Wittenburg, Dynamics of Systems of Rigid Bodies, B.G. Teubner, Stuttgart, W. Germany 1977

10. Mathematically, the quaternion angular error can be shown to be expressed by

$$\Delta\Theta = 2 \cos^{-1} \left[\left(\tilde{q}^*_{simulation} \; \tilde{q}_{extraction} \right)_4 \right],$$

where $\Delta\Theta$ is the quaternion angular error and the subscript 4 denotes the scalar portion of the indicated quaternion product.

11. Note that this convention for the transformation is not unique. If the transformation is expressed as

$$\vec{x}' = \tilde{q}\,\vec{x}\,\tilde{q}^*,$$

$\tilde{q}$ as defined in the text must be replaced with $\tilde{q}^*$ (as defined in the text). Furthermore, the direction cosine matrix $\mathbf{A}$ as defined in equation 3 must be replaced by its transpose $\mathbf{A}^t$. The effect of transposition is the same as changing the sign on quaternion elements q_1, q_2, q_3 in $\mathbf{A}$.

AAS 84-012

DIGITAL STANDARD STAR TRACKER

James P. McQuerry, Jr.*

The Digital Standard Star Tracker (DSST) is an electro-optical instrument which provides position data used for precise attitude determination.

The new DSST design uses flight-proven optical and sensor components from the BASD/NASA Standard Star Tracker (SST) programs while incorporating digital electronics techniques to improve producibility and reliability. This design approach has resulted in a new instrument capable of ≤ 10 arc second calibrated accuracy with 50 percent of the electrical components and only 10 percent of the electrical assemblies used in the SST.

* *Manager, Electrical Engineering Design, Electro-Optic Products, High Technology Products, Ball Aerospace Systems Division, P.O. Box 1062, Boulder, Colorado 80306*

INTRODUCTION

The Digital Standard Star Tracker (DSST) system meets performance requirements of Goddard Specification GSFC S-712-9 by using space-proven hardware with a minimum of modifications resulting in a low-risk, cost-effective design.

The DSST system is based upon BASD's NASA Standard Star Tracker (SST) systems produced for the Solar Maximum Mission (SMM), Landsat D, and Space Telescope spacecraft and similar to the Shuttle Star Tracker systems which have operated successfully on the Space Shuttles Columbia and Challenger. Subsystem components, designs, and technologies developed for other BASD trackers and star scanners were used where they were most beneficial to the DSST.

DIGITAL STAR TRACKER SYSTEM DESCRIPTION

The DSST system, shown in Fig. 1, is an electro-optical system which detects stars and generates position and intensity data over an 8° field of view. The main component of this system, the DSST star tracker assembly, uses accessory components including a lightshade, shutter assembly, bright object sensor, and protective window assembly. A brief description of each of these components follows.

Lightshade

The lightshade (LS) provides attenuation of stray light sources so that stars within a specified angle of the sun, moon, and the earth can be detected without degraded performance. The DSST shade is typically a two-stage design similar to shades used on previous BASD electro-optics programs. The shade may be designed by BASD to customer specifications as dictated by their system's requirements.

Shutter Assembly

The shutter assembly (SA) provides protection to the tracker's detector against potential damage from bright sources. The shutter contains a motor and a normally open clam shell shutter and is identical to the assembly on BASD's SST systems. The shutter is closed in front of the tracker lens upon receipt of a signal from the bright object sensor (BOS) which exceeds a preselected bright source protection level.

The LS/SA combination mounts on the opposite side of the spacecraft mounting base and is mechanically separate from other DSST components. Electrical connection to the SA is made via cables to a hard-mounted connector.

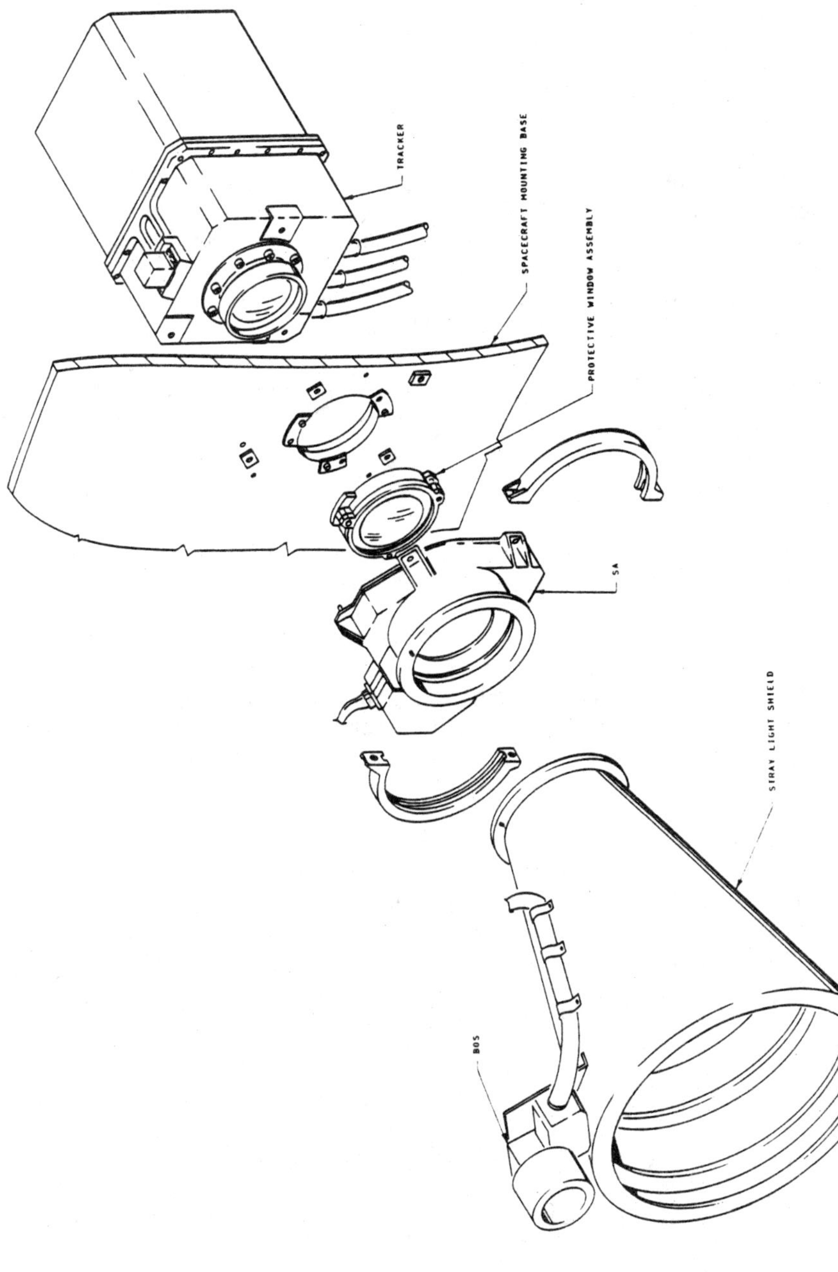

Fig. 1 Digital Standard Star Tracker System

Bright Object Sensor

The bright object sensor (BOS) consists of a shade, electronics, and a silicon detector. The BOS is mounted to the front outer diameter of the LS and aligned to its boresight axis.

The BOS silicon photodetector provides a signal which closes the shutter protecting the tracker detector. The shutter track angles may typically be 45° and 15° on the sun and earth respectively.

Protective Window Assembly

The protective window assembly (PWA) is a vital component of the DSST built-in test equipment (BITE) and provides protection for the ST tracker lens against a radiation environment. The PWA is 0.3 in. thick, flat identical to that of the SST systems. The flat also reflects a simulated star source into the tracker to effect a comprehensive, end-to-end self test of DSST performance.

The PWA mounts on the opposite side of the spacecraft mounting from other DSST components, thus providing reference to the spacecraft for update of tracker position after launch.

STAR TRACKER ASSEMBLY DESCRIPTION

The DSST assembly, the major component of the DSST system, provides the detection, signal processing, and star position and intensity data. Fig. 2 is a diagram of the tracker's internal configuration.

The tracker, whose dimensions are within an envelope of 6.5 in. x 9.4 in. x 12.25 in. maximum, weighs 18 pounds. The mechanical interface is identical to the SST, the tracker is mounted independently of the other components to one side of the spacecraft mounting base on precision-lapped pads which define its coordinate system.

The star tracker uses an image dissector tube to search and track stars in an 8° field of view of +5.7 visual magnitude or brighter. Following acquisition, a 0.4° track pattern follows the star throughout the field of view. The tracker provides precision two-coordinate position signals indicating angular line of sight to the star along with the star's intensity information. A block diagram of the DSST is presented in Fig. 3. The function of each block will be discussed briefly.

Sensor Head

The sensor head is the heart of the DSST. The lens collects and images radiant energy on the IDT photocathode surface. The sensor's field of view is a function of the lens focal length (70mm), and its sensitivity is a function of the lens aperture (58mm).

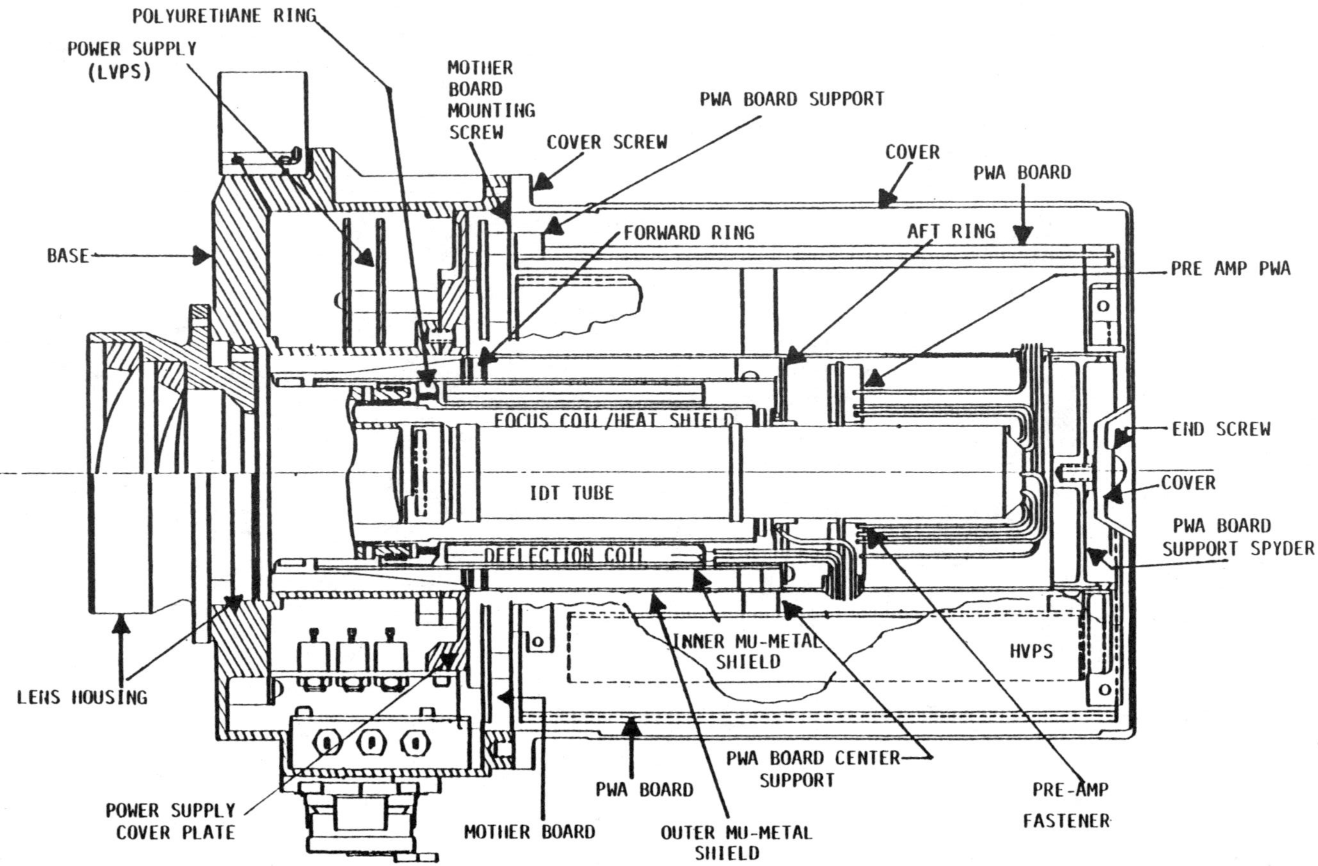

Fig. 2 Star Tracker

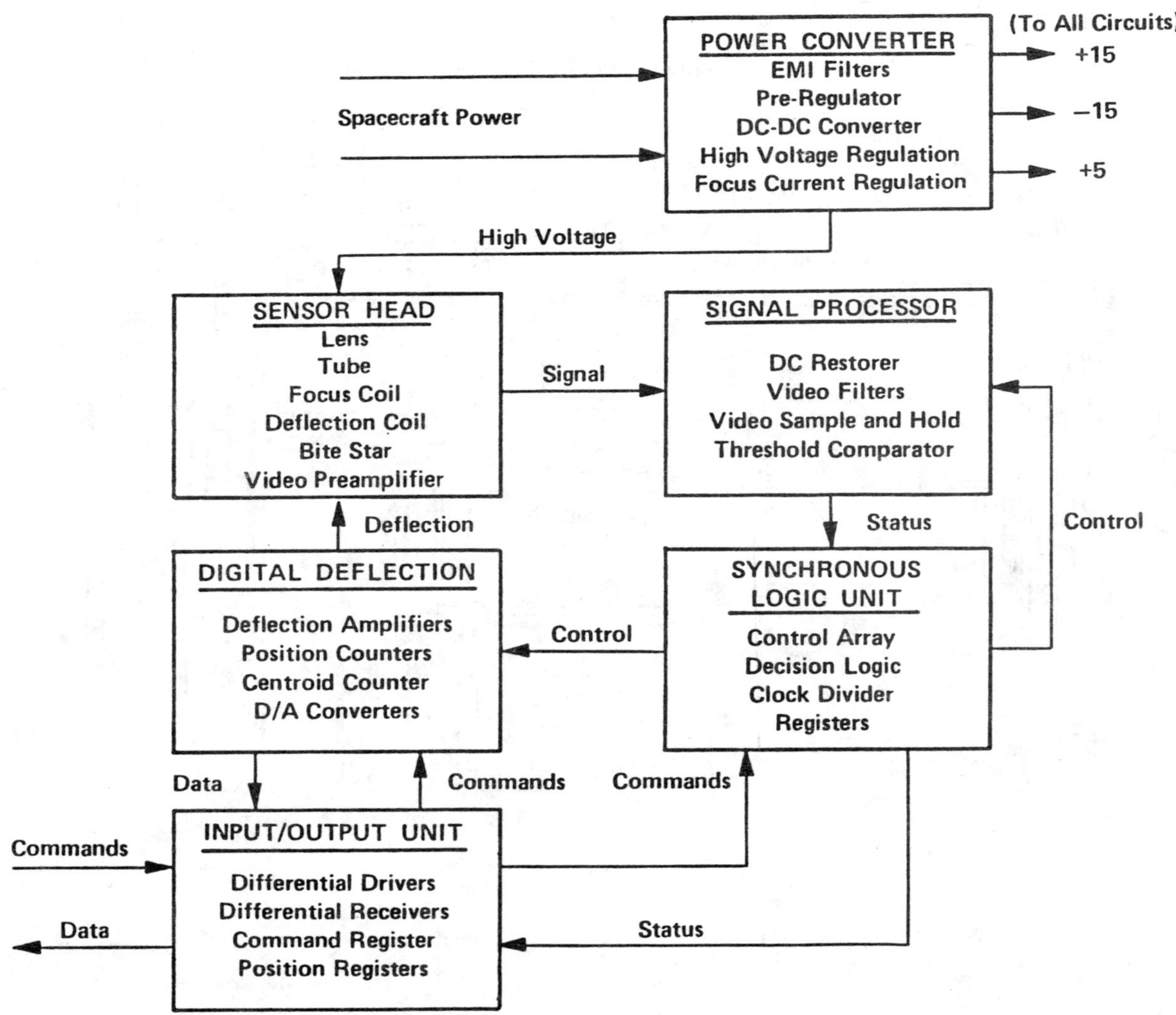

Fig. 3 DSST Block Diagram

The photosensor is an IDT whose photocathode produces an electron image corresponding to the optical image focused on photocathode surface. When an accelerating voltage is applied between the photocathode and a mechanical limiting aperture, only those electrons leaving a specific area on the photocathode will pass through the limiting aperture's instantaneous photocathode dimension (IPD). When a target image is positioned on the photocathode by the lens, a small stream of electrons is formed by the electron optic's electromagnetic focusing field. A deflection coil is positioned around the image section to provide magnetic deflection of the target image electron beam. By applying proper deflection coil currents, the electron beam is swept back and forth across the limiting aperture, causing a modulation of the electron beam.

Electrons going through the 8 mil aperture enter a secondary emission dynode string of 12 dynodes and multiply approximately 10^6 times within the IDT before entering the electronic signal processing circuits.

The IDT is a simple, rugged, reliable device requiring no cathode which limits life and consumes power.

Signal Processor

The signal processor conditions the video signal by DC restoring, filtering, and peak detecting. Sensitivity control based on user threshold commands is included as part of the signal processor.

Digital Deflection

The deflection circuits generate the full field search, offset search, and tracking waveforms for the sensor head deflection coils.

The digital track loop is made up of counters and position registers that maintain a track sweep on the star image. This modification from the SST has eliminated the need for approximately 40 select components used in the analog unit. Previously, the track loops were composed of analog sample and hold circuits, operational amplifiers, and over 50 precision select resistors.

Synchronous Logic Unit

The synchronous logic unit (SLU) is the operational control for the DSST. It monitors user commanded inputs and video to automatically set the mode of operation to search, track, or offset. In these modes, the SLU controls all internal functions in the tracker and generates signals which control the digital deflection loops. The logic is composed of state-of-the-art qualified memory components which allowed reduction of the DSST parts count to 50 percent of those required in the previous analog Standard Tracker design.

Spacecraft Interface (SCI)

The SCI unit, designed to Goddard Specification GSFC S-714-11, provides all buffering in the form of differential circuits for both command inputs and data outputs. User commands are received in the form of a 16-bit serial word and held in a register for the SLU.

Power Converter

The power subsystem consists of a power converter that takes the 28 ± 7 VDC input from the spacecraft and converts it into ±15 VDC and +5 VDC for tracker circuitry. A variable +1800 VDC supply and a fixed -400 VDC supply are used as driving voltages for the IDT. The focus current generator provides a regulated current to the IDT focus coil.

TRACKER SYSTEM OPERATIONAL DESCRIPTION

The DSST uses search, acquire, and track modes of operation identical to the SST.

The operation is best illustrated by describing the operation sequentially. When no power is applied, the shutter is open. Following "Power On," with no bright source in the BOS field of view (FOV), the tracker will automatically start in the search mode unless commanded otherwise. This mode is characterized by a scan pattern of an 8.0° square field of view as shown in Fig. 4. The H-coordinate pattern is a left to right scan with flyback to the left with a total cycle time per scan of approximately 128 milliseconds. There are 83 lines per frame which take 10.8 seconds to complete. During this mode, the tracker delivers no information on targets in the FOV. Rather, its objective is to locate a proper target about which information can be obtained. Four commandable threshold levels are available to set the minimum sensitivity from +5.7 to +2 visual magnitude. The search scan continues until the first star that exceeds the commanded threshold level is encountered. When this occurs, the star is acquired and the track scan begins.

When the tracker enters the track scan mode, it forms a smaller cross-pattern with a repetition rate of approximately 100 times per second, as shown in Fig. 5. The track scan pattern centers on the star image and a STAR PRESENT bit is generated in an output word, signifying the two-axis position and intensity analog data are available at the tracker output. As the star moves in the FOV because of vehicle attitude changes, the track scan pattern remains locked onto the star image and the position information is continuously updated. During this mode of operation, the tracker sees only the acquired star. It is not influenced by and provides no other information on other targets in the FOV. Tracking of the same star continues until either it leaves the FOV or a BREAK TRACK command is received from the user. When any of these events occur, the search mode is resumed at a point approximately 0.4° more positive in the V-coordinate from where the star was last tracked so that the same star will not be reacquired.

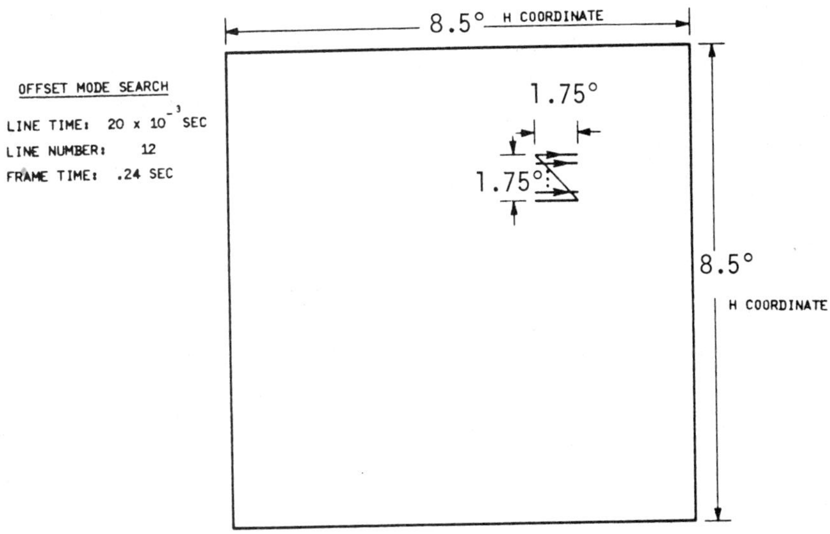

Fig. 4 Offset Mode Scan Geometry

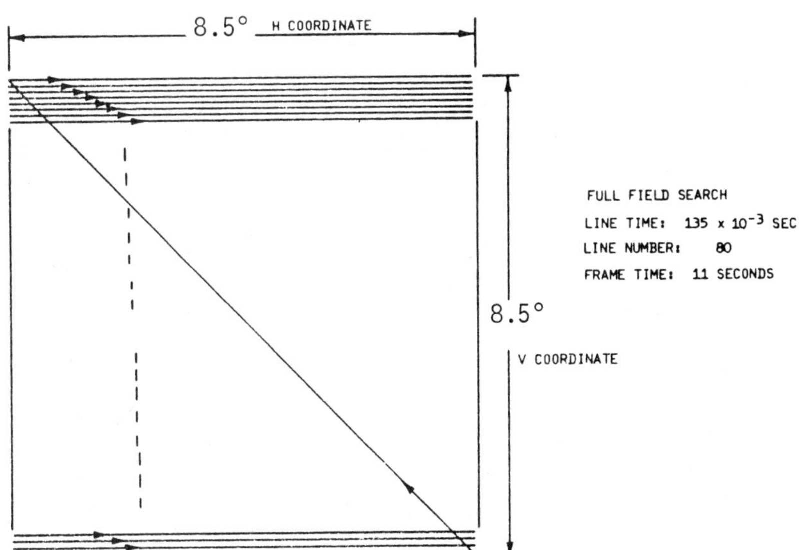

Fig. 5 Full Field Search Scan Geometry

The DSST has an offset mode capability identical to that of the SST. With this capability, the user can position, via the serial command word, a small search pattern roughly 1.75° square, as shown in Fig. 6, to any position in the 8.5° square FOV. The tracker searches this reduced area in .355 second with the same type pattern used to search the total FOV. If a star is acquired in the reduced field, it is tracked throughout the total FOV as described above. If the star is lost for any reason, the SST will return to the reduced scan at the position originally commanded.

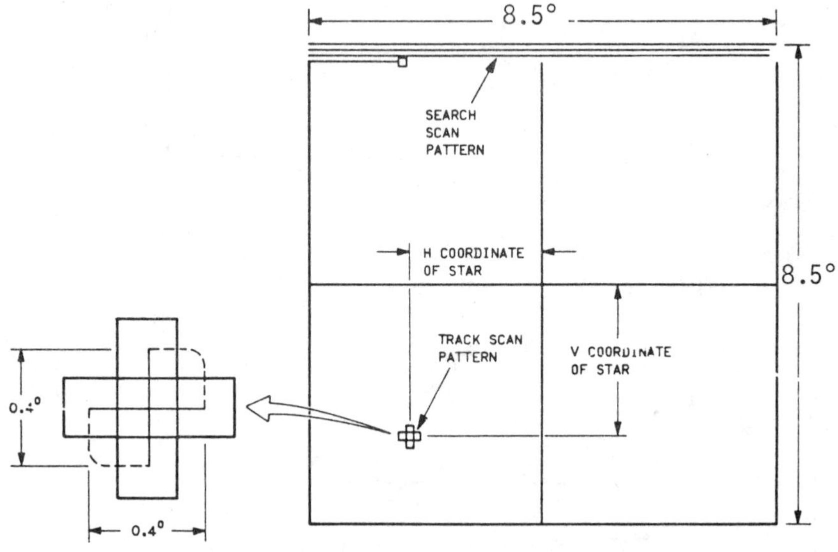

Fig. 6 Track Scan Geometry

The offset mode capability allows the user to reduce acquisition time and increase probability of acquisition on a particular star if coarse position information on that star is known prior to sending the command.

If at any time during any of the tracker modes of operation a bright object enters the BOS FOV, a bright object alert (BOA) digital signal is issued, causing the shutter to close in front of the PWA. The shutter automatically opens when the bright object leaves the BOS FOV. Shutter closure due to BOA can be overridden by a command bit on the serial command word.

The self-test capability permits a complete end-to-end check of DSST performance. A bright source is generated within the BOS causing a shutter closure, thus demonstrating proper operation of the protection system. An image, generated by an LED and prism assembly, is collimated by the tracker lens and reflected back into the FOV of the tracker from a silvered disc on the PWA quartz window. The return image is detected and tracked with its position and brightness compared to prior values by the spacecraft system to detect any shifts or changes. Because the window mounting is spacecraft and not tracker referenced, the tracker measurements of the artificial star are a true measure of its alignment to the spacecraft, thereby providing a means to determine if position shifts were induced during the launch environment.

CALIBRATION EQUATIONS

BASD's position calibration method for temperature, star intensity, and magnetic field has been successfully applied to five SST's and consistently provided 7 arc sec system accuracy.

DSST Coordinate System

The tracker output position data coordinate system axes are referenced in Fig. 7 and are defined as follows:

- o b is normal to the tracker mounting pads and positive toward the star field.
- o v is in the plane of the tracker mounting pads and normal to the roll pads line.
- o h is the plane of the tracker mounting pads and perpendicular to v.

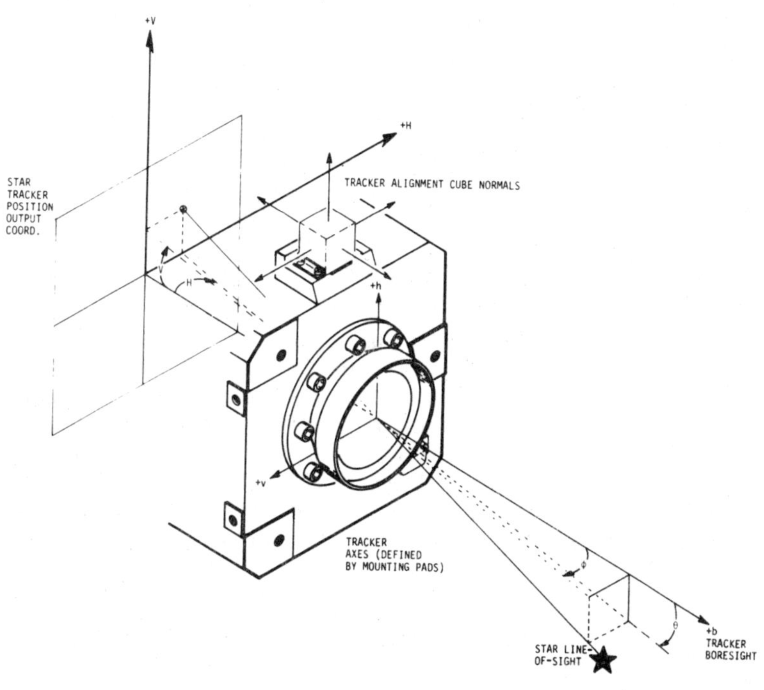

Fig. 7 ST Coordinate Definition

The position output coordinates are designated H and V. A positive-going H is generated by a positive rotation of the tracker about its H axis and a positive-going V by a positive rotation about its V axis. Each position output has a 12-bit resolution. Unlike a gimballed system, the tracker has deflection coils which measure horizontal and vertical angles independently; therefore, θ and ϕ, corresponding to H and V respectively, are star line-of-sight angles.

Systemmatic Error Calibration

Precision accuracy is only achievable in a wide field-of-view tracker like the DSST with calibration, thus requiring a detailed knowledge of the systemmatic errors over the total FOV for all of the environmental variables. Measurements based upon temperature, star intensity, and magnetic field are made on each tracker during formal acceptance test to obtain this knowledge; and the data are subsequently used to generate a unique set of calibration coefficients.

PACKAGING IMPROVED THROUGH DIGITAL DESIGN

Electronics design modifications permitted functional grouping of components for better producibility.

The tracker electronics is packaged on modular multilayer printed circuit boards with compatible electronics packaged together and all electronics requiring shielding, heat dissipation, or other special considerations packaged accordingly. All PWB's are designed for high density planar component mounting with a ground plane for signal return and shielding and a separate thermal plane for heat conduction to the chassis.

The high voltage power supply is packaged in a similar manner using planar circuit boards and encapsulated to eliminate possible corona problems.

CONCLUSION

The DSST system is a flight-proven, highly reliable, precision attitude determination instrument capable of providing position data to the spacecraft which may be calibration corrected to $\leq$ 10 arc second accuracy. The device incorporates state-of-the-art digital memory devices to replace 50 percent of the components previously required in the BASD/NASA Standard Star Tracker. With this reduction, the new design has eliminated the need for approximately 50 cordwood modules mounted upon four printed circuit boards. This has been replaced with five multilayer circuit boards. This reduction in circuit complexity and quantity has resulted in a reliable, producible navigation instrument.

DIGITAL STANDARD STAR TRACKER

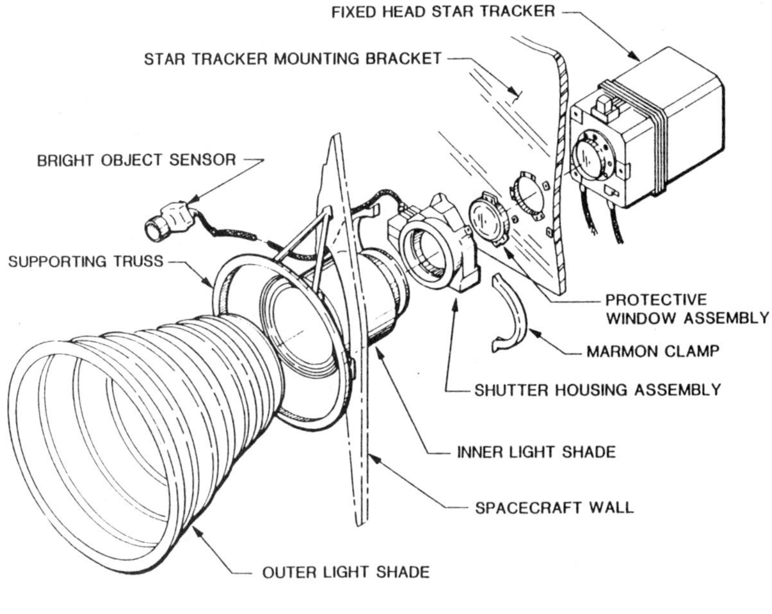

ADVANTAGES OF DSST

- IMPROVED PRODUCIBILITY
  - ELIMINATION OF CORDWOOD MODULES; CONVERSION TO MULTILAYER PC BOARDS
  - REDUCTION OF SELECTED TRIMMING RESISTORS FROM 45 TO 12
  - SIMPLIFIED TEST PROCEDURES
  - BETTER ACCESS DURING ASSEMBLY; FEWER HIDDEN COMPONENTS
- BETTER RELIABILITY
  - ELIMINATES 75% OF ANALOG CIRCUITS
  - 40% REDUCTION IN PARTS COUNT
- DESIGN IMPROVEMENTS
  - BETTER CIRCUIT STABILITY
  - IMPROVED DESIGN MARGIN

RECENT FLIGHT HISTORY

BASD has been a supplier of navigation determination devices since the 1960's. A compilation of recent star tracker flight history is as follows:

Spacecraft	Star Tracker Model Numbers	Launch Dates
SAS	CT-401, 2 ea.	1975
HEAO-1	CT-401, 2 ea.	1977
IUE	CT-501, 2 ea.	1978
HEAO-3	CT-401, 2 ea.	1979
MAGSAT	CT-401, 2 ea.	1979
SMM	SST, 2 ea.	1980
Shuttle (9 flights)	CT-411, 4 ea.	1981, 1982, 1983
LANDSAT	SST, 2 ea.	1982

DIGITAL STANDARD STAR TRACKER (DSST) SYSTEM DESCRIPTION

The Digital Standard Star Tracker (DSST) is designed to meet NASA standard fixed head star tracker GSFC Specification S-712-9:

1. The DSST provides information for user attitude determination calculations:
   - Analog star video intensity
   - Digital star line-of-sight angular position (two 12-bit words)

2. The DSST searches, acquires, and tracks star images to any position in the field-of-view.
   - +5.7 to +2.0 visual magnitude
   - 8.0 x 8.0 deg square field-of-view
   - 0.4 x 0.4 deg track field-of-view

3. The DSST is composed of flight-proven components from previous star tracker designs.
   - ITT S-20 response image dissector tube
   - 70 mm f/1.2 multi-element lens
   - Bright object sensor
   - Tube protection shutter assembly
   - Custom lightshade available

MODES OF OPERATION

The DSST may be commanded to any one of five unique modes of operation by the user's command interface:

1. OFF Mode:
   - No power applied to the tracker
2. AUTOMATIC SEARCH Mode:
   - Initial POWER ON Mode
   - Search Total Field-of-View (TFOV) 8 x 8 deg in a raster mode
   - Acquire first star encountered exceeding any one of the four commanded thresholds of detection
3. OFFSET SEARCH Mode:
   - This mode is the same as the Automatic Search Mode except a command must initiate the mode
   - The field-of-view is any one of 256 RFOVs (1.75 x 1.75 deg)
4. SELF-TEST Mode:
   - Closes Shutter
   - Turns on an internal LED in a fixed position in the TFOV
   - Mode is a complete systems test from optics to output data
5. TRACK Mode:
   - TRACK Mode may be entered from any of the above active modes when a star exceeds the detection threshold
   - This mode continues as long as the star is in the TFOV or upon termination by command

TRACK SCAN GEOMETRY

The tracker searches in a left to right raster scan and once a target is encountered which exceeds the threshold, a simple cruciform pattern is used to track the image.

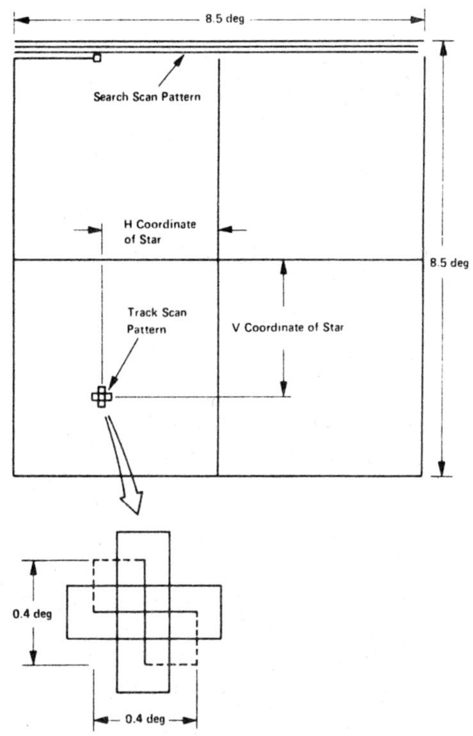

BLOCK DIAGRAM

The DSST is basically a digital implementation of the BASD analog SST. The basic functional blocks are unchanged in this design.

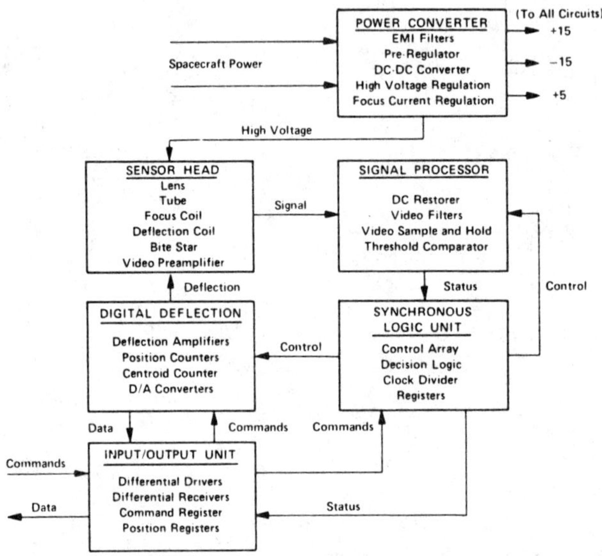

COMMANDS

There are five basic user commands which are sent via the DSST digital interface:

1. BREAK TRACK:
   - Return from track to either automatic or offset mode
   - Does not reacquire the same star
   - Blanks 0.45 deg of tracker field-of-view
   - Returns to the commanded RFOV if in offset mode

2. THRESHOLD SET:
   - Four selectable levels of detection thresholds
   - Does not change mode of operation
   - Acquires targets exceeding commanded level

3. OFFSET Mode:
   - Initiates/terminates the offset RFOV search mode

4. SELF-TEST Mode:
   - Initiates/terminates the self-test mode

5. SHUTTER OPEN:
   - Overrides internal logic inputs to engage the IDT protective shutter

INTERFACE DEFINITION

The DSST interface design is compatible to the NASA standard remote interface unit Specification GSFC S-714-11.

1. STANDBY II: A digital logic level which selects one of two redundant RIUs

2. COMMAND: Serial 16-bit digital word
   - Shutter open command enable
   - Threshold set enable
   - Threshold LSB
   - Threshold MSB
   - Shutter Open Command
   - Break track command
   - Self-test command
   - Offset mode command
   - Coordinates, 8 bits

3. DATA OUTPUT:
   - HMOD – A digital logic level which provides a scan synchronization output:

     Word One:
     - Horizontal star position (12 bits)
     - Self-test mode on
     - Optics shutter closed by BOA or self-test
     - Star present
     - Optics closed by TS

     Word Two:
     - Vertical star position (12 bits)
     - Four spare bits

STAR TRACKER PERFORMANCE

The DSST performance may be summarized into the following categories:

Field of View:
- 8 (−0, +0.5) deg square

Sensitivity:
- +5.7 visual magnitude, Class GOV

Vehicle Rate:
- ≤ 0.3 deg/sec

Search Mode:
- Scan Type: Raster scan (8 x 8 deg)
- Full Frame Scan Time: ≤ 11 sec

Offset Mode:
- Raster Scan: 1.75 x 1.75 deg
- Reduced Frame Scan Time: ≤ 0.25 sec
- Position Resolution: 0.5 deg increments over 8 x 8 deg

Track Mode:
- Scan Type: Unidirectional cross-scan
- Scan Time: 100 msec (50 msec/axis alternating)
- Position Output Data Rate: 10 updates/sec (each axis)

Output Signals:
- Analog: Star intensity, temperature
- Digital: Two 16-bit serial words consisting of 12 bits position information (each axis) and 8 bits status information

Position Output Accuracy:
- Noise Equivalent Angle: 10 arc sec/axis +5.7 M; 8.2 arc sec/axis +4.0 M
- Accuracy: 10 arc sec/axis

STAR TRACKER PARAMETERS

The DSST design parameters are reduced into its three major components: electro-optics, mechanical, and electrical.

ELECTRO-OPTICS

Photosensor:

- Generic: ITT Electron Tube Division F4012-RP Image Dissector
- Spectral Response: JEDEC S-20

Optics:

- Generic Type: Refractive, 4-element, low scatter, custom-designed lens
- Focal Length: 70 mm
- Aperture: 58.3 mm

Shutters:

- Type: Two overlapping blades, mechanically constrained open, driven closed by a motor
- Location: In front of optics

MECHANICAL

Size:

- Tracker: 12.24 in. long x 6.5 in. wide x 9.4 in. high
- Weight: 18 lb

ELECTRICAL

Input Voltage:

- 28 ± 7 Vdc

Input Power:

- 20 W (tracker); 5.9 W (shutter)

Interface:

- Power: 6-pin connector
- Command/Data: 40-pin connector

AAS 84-017

ROBOTICS IN ENGINEERING EDUCATION

Dr. Edward J. Bauman*
Donald W. Erickson*

This paper describes eight student projects that have been part of the construction of a "homemade" robot at the Electrical Engineering Department, University of Colorado, Colorado Springs.

INTRODUCTION

TO BE OR NOT TO BE is a Shakespearian character's dilemma; but that does not apply to robotics in Engineering Education. Robots present a fascinating and challenging adventure to almost everyone with engineering tendencies. Yet a robot can not be as presumptuous as Hamlet. A designer and creator will decide whether it will "to be or not to be" (isn't that true for us "Hamlets" as well)?

Almost everyone can find his niche in the design, for just as his counterpart, man, the intelligent robot must be "wonderfully and fearfully made". There is something for everyone and their talents. Robots may be designed to "have" or "have not" many human-like systems. Let me list the "be's" or "not to be's" of our homemade robot at the Department of Electrical Engineering at the University of Colorado, Colorado Springs, (UCCS).

To Think or not...	-	Microcomputer
To Interact or not...	-	Multiprocessor bus
To Move or not...	-	Wheel Drive system
To Grasp or not...	-	Manipulator Arm
To See or not...	-	Optical Sensor
To Speak/Hear or not	-	Voice Synthesizer/Recognizer
To Feel or not...	-	Pressure Gauge
To Communicate or not...	-	Radio Control Link

Because our system is "homemade" and an educational tool, whether a system "is" or "isn't" depends on how the latest innovative design projects are progressing.

Let's examine some past and present ones; some that worked and some that didn't!

* *Electrical Engineering Department, University of Colorado at Colorado Springs, Colorado 80907*

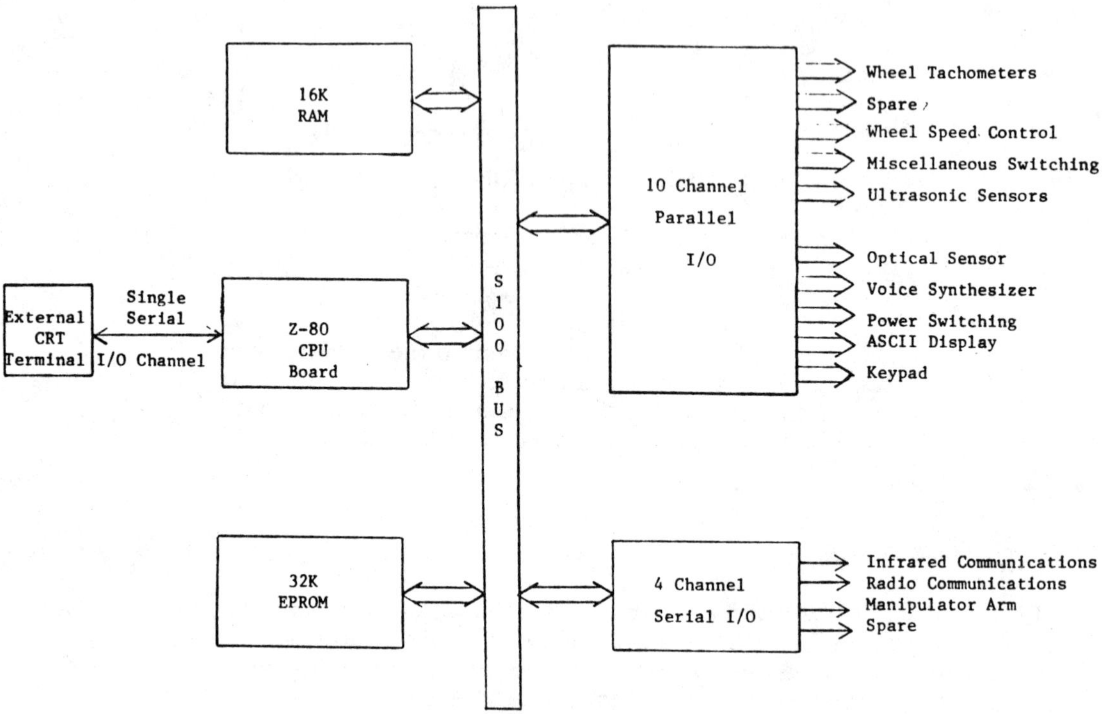

Fig. 1 The Microcomputer

Figure 1 shows the present microcomputer and S-100 bus structure with the present peripheral systems connected to the I/O channels.

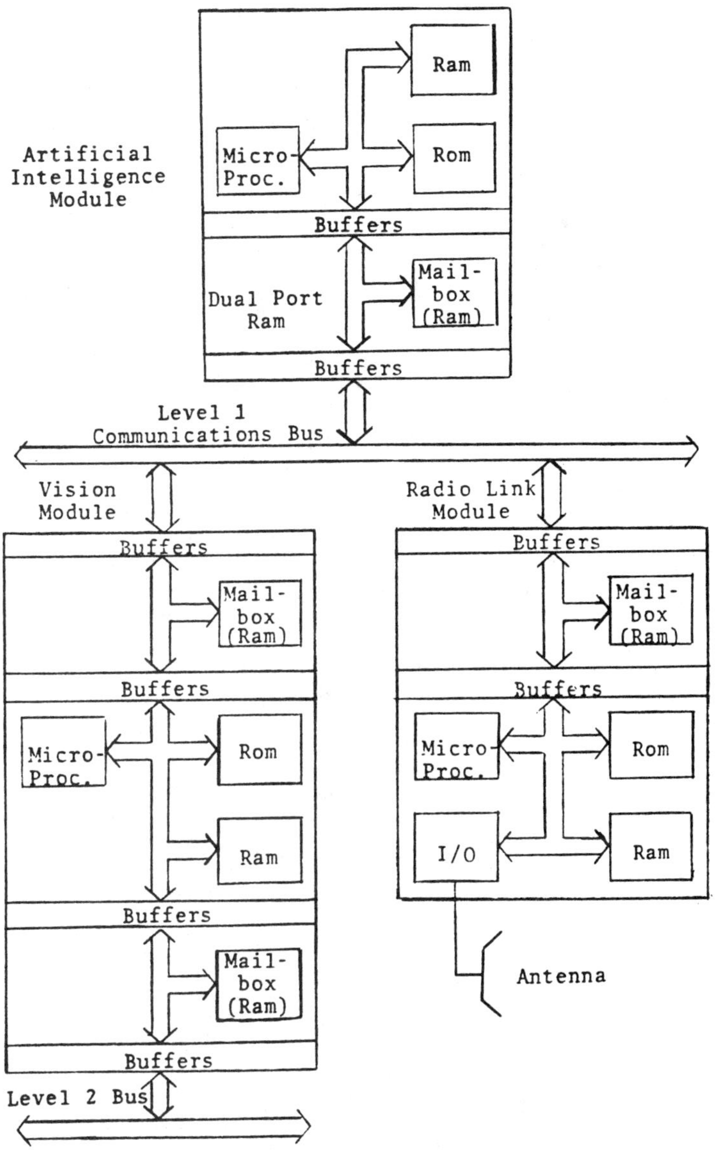

Fig. 2 Multiprocessor Bus

Figure 2 shows an example of three microprocessors in an architecture recently designed at UCCS (reference 1). It features a hierarchical bus system with dual port rams. In this example, commands from the radio link would be processed by the artificial intelligence module and sent via the level 1 bus to the vision module which would then communicate with lower processors over the level 2 bus.

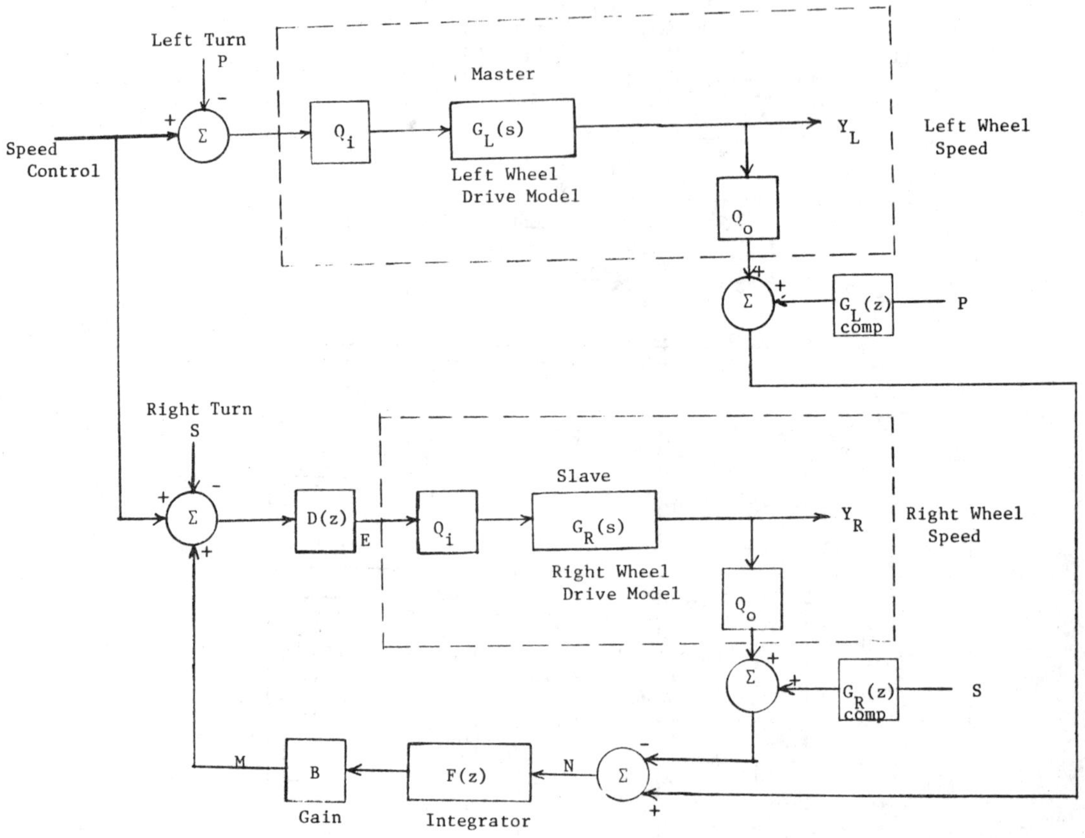

Fig. 3 Wheel Drive Control System Simulation

Figure 3 shows the wheel drive control system simulation. One feature is to consider the left wheel speed the master and use the error signal, M, to keep the two wheel speeds identical. D(z) is a digital filter which compensates for differences in the left and right drive systems. The P signal is added to reduce the voltage to the left wheel for a left turn. P is also added through a digital computer model ($G_L(z)$ comp) of the left wheel drive system. This maintains the right wheel at the same speed as before the turn signal was applied. In the same way the S signal is added for a right turn. The Q's are quantizers. Using this simulation, effects of both wheel system model errors and the quantizers were investigated.

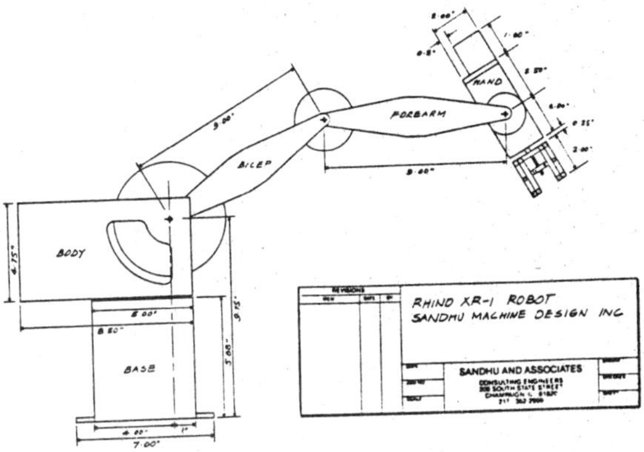

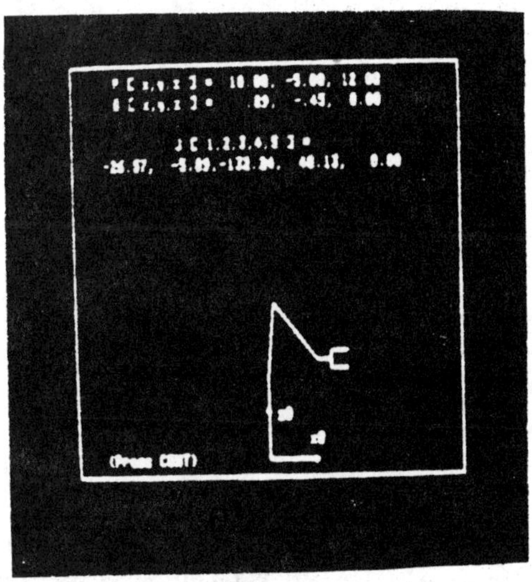

Manipulator Graphics Simulation of Arm

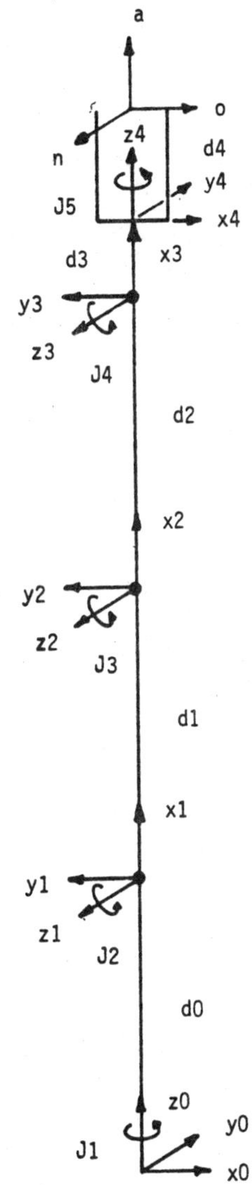

Manipulator Arm Coordinate System

Fig. 4 Manipulator Arm, Coordinate System and Computer Graphics Simulation

Figure 4 shows the Rhino arm built by Sandhu Machine Design, Inc. A coordinate system attached to the arm is also shown in Figure 4 along with a graphics display example of its simulation. The kinematic equations for the Rhino arm have been derived and are given in reference 2.

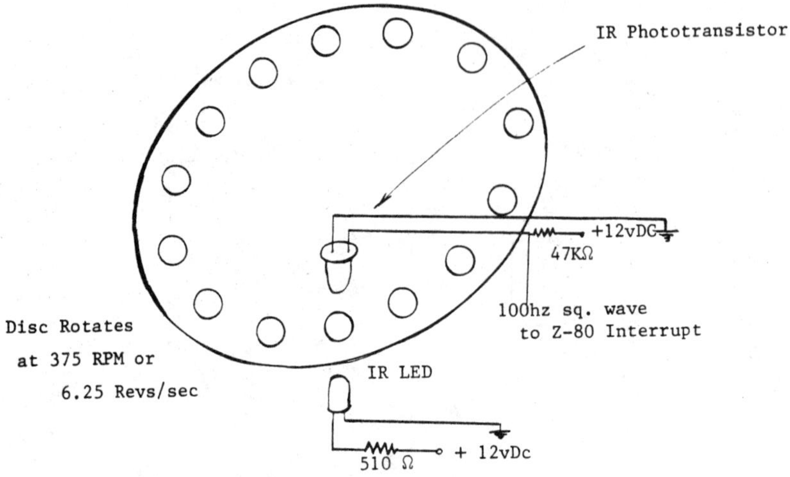

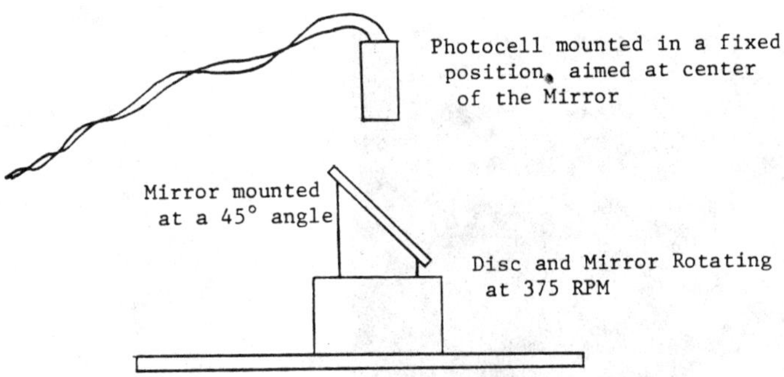

Fig. 5 Optical Surveillance Sensor

Figure 5 shows the optical surveillance sensor. A salvaged 8-track tape motor drives the disk which has a mirror mounted on it. The disk has holes spaced every 22.5 degrees which activate a phototransistor circuit which in turn interrupts the Z-80 microcomputer. When interrupted, the Z-80 reads the value of the output of a photocell. The mirror mounted on the rotating disk reflects light to the photocell. Because of this rotation, the light value in 16 directions is sampled every 0.16 second. By averaging four successive samples from the same hole, and differencing them from the last four averaged samples, the sensor can detect a moving object which has a different reflectivity than the background.

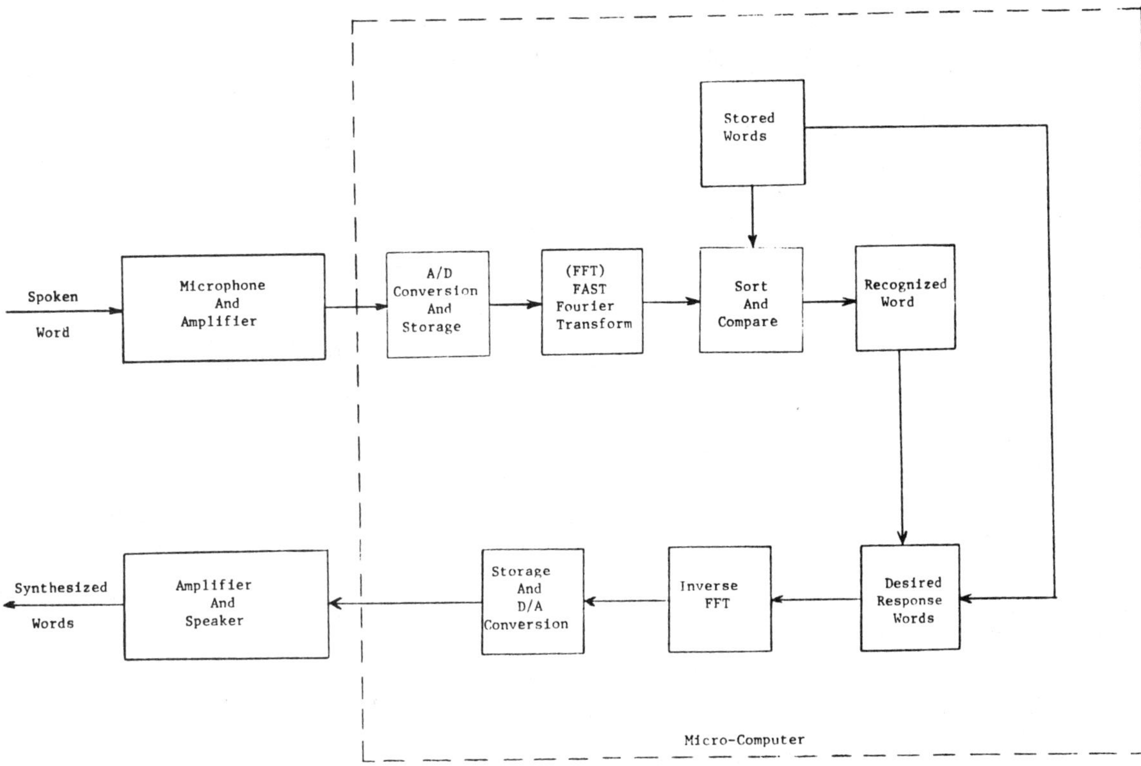

Fig. 6 Speech Synthesis and Recognition

At present, we have a standard VOTRAX voice synthesizer as part of our robot. Figure 6 shows a block diagram of a student project that was partially completed. A spoken word is recorded into the memory of the microcomputer. A Fast Fourier Transform (FFT) is taken of the word. The transform of the word is compared to previously stored transforms and "recognized". Response words are then chosen by the computer and an inverse FFT is generated to synthesize speech.

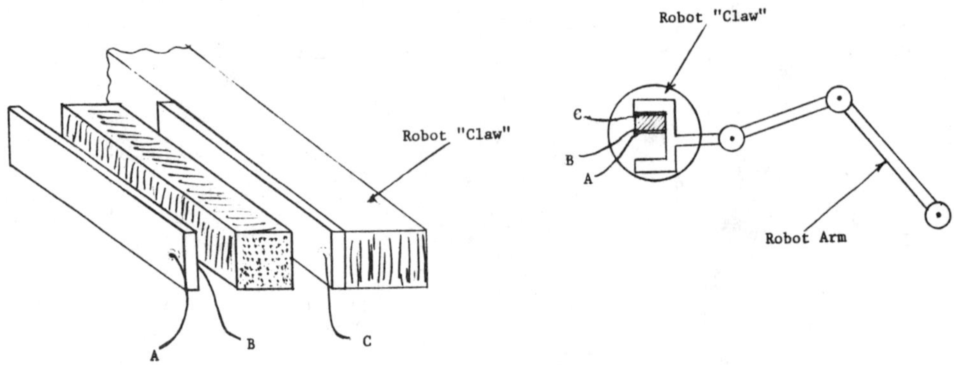

Fig. 7 Claw Pressure Sensor

Figure 7 shows an experimental pressure sensing device used on the Mechanical Claw of the Rhino arm. Using this device feedback determines when something has been grasped with sufficient pressure.

A sandwich is formed of two pieces of double sided printed circuit board material, each 2.5cm x 0.6 cm, and a similar size piece of carbon impregnated foam plastic.

The sandwich shown above is mounted on one side of the "claw". A short circuit between lead "a" and the frame of the claw indicates that the claw has been closed down enough to make contact, and that nothing which is an insulator is between the two sides of the claw.

The resistance between "b" and "c", varies from over 20,000 ohms with no pressure being applied to less than 200 ohms when a moderate amount of pressure is applied. The variable voltage from this resistance is applied to an analog-to-digital converter for interface to the computer.

The disadvantage of this sensor is that it will have a relatively short life time, since repeated cycles of crushing will cause the plastic foam to disintegrate.

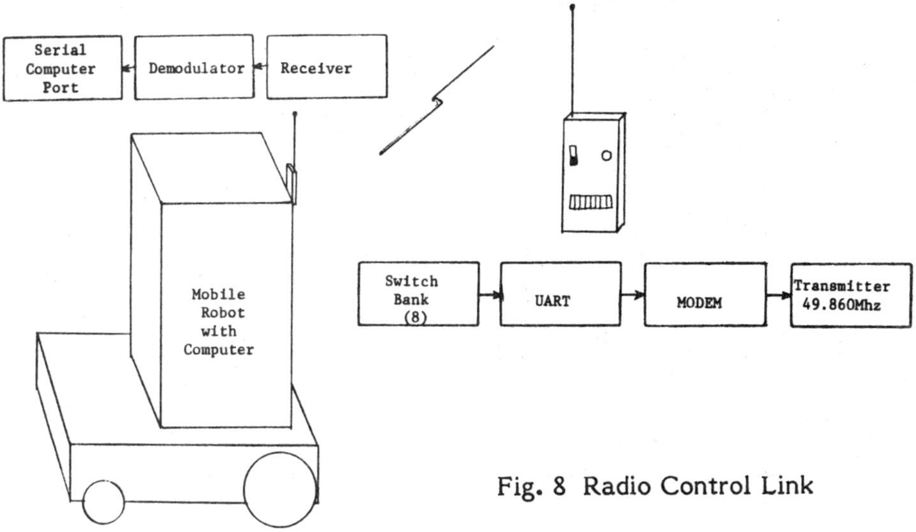

Fig. 8 Radio Control Link

Figure 8 shows the radio link that was developed as a means of exercising manual control without the inconvenience of cables. It was designed as a 300 baud serial link to the on-board computer. This design approach results in maximum versatility by enabling the robot software to determine what, if anything, is done as a result of inputs from the radio link.

The hand-held transmitter was built around an inexpensive walkie-talkie mounted in a permanent transmit mode. A UART and modulator were added to provide the 300 baud serial input to the transmitter.A matching walkie-talkie mounted on the robot is used as the receiver. The output from the receiver is fed to a demodulator and then to a 300 baud serial computer input.

A difficulty experienced with this arrangement is that the walkie-talkie used as the receiver element is so inexpensive, that it uses a super-regenerative receive circuit. This form of receiver tends to chop the 300 baud serial signal into something unrecognizable to the computer. Our upgrade plans are to obtain a super-heterodyne receiver for the robot.

CONCLUSION

On most research projects in the University, we must pay our students to do the work. When we mention robotics, the students will gladly work for free! Such is the mystic and adventure associated with almost any aspect of a robot.

REFERENCES

1. R.T. Schlater, A Hierarchical Multiprocessor Computer Architecture for a Robot, MSEE Thesis, University of Colorado, Colorado Springs, November, 1983.

2. A.I. Best, Computer Simulation of the Kinematics of the Rhino XR-1 Manipulator, MSEE Thesis, University of Colorado, Colorado Springs, November, 1983.

AAS 84-018

ARTIFICIAL INTELLIGENCE AND COMPUTER VISION FOR ADVANCED MANIPULATOR SYSTEMS

Dennis C. Haley*
David G. Morgenthaler*
James C. Sanborn*

This paper presents material displayed at the storyboard session of the AAS Keystone Conference. Included is the text of the narration of a video tape which presents applications of artificial intelligence and computer vision techniques to robotic systems. Also included are the display pages which provide a broad overview of the artificial intelligence and computer vision techniques and areas of research currently being pursued at Martin Marietta Denver Aerospace.

TEXT OF THE NARRATION FOR THE VIDEO TAPE

The Advanced Automation Technology Section at Martin Marietta is currently developing under various contracts and IRADs an offline programming capability employing artificial intelligence planning techniques which eliminate the conventional "teach mode" requirements and allow the working environment to become less constrained.

For the bulkhead inspection task shown, an operator specifies each measurement type and approximate location. This data is passed to an artificial intelligence planning system responsible for generating a sequence of operations to achieve the desired measurements.

Initially, the planner uses a geometric data base describing the bulkhead along with a knowledge base of information regarding measurement types to determine the precise location of all measurement points. Next, these points are sequenced according to their relative positions, as well as the type of tool required. This sequence is

* *Martin Marietta Denver Aerospace, P.O. Box 179, Mail Stop 0570, Denver, Colorado 80201*

then used to generate a full inspection plan. Plan execution occurs during real time in an online system complete with intelligent plan monitoring, capable of adapting the system behavior according to varying conditions.

Verification of the motion proposed by the offline programming system is provided by a robotic simulation program developed at Martin Marietta. The program executes the sequence of motions and produces a graphics representation such as that seen here.

The simulation program also provides a collection of analysis tools including a dynamic simulation capability which is driven by the same control algorithms used within the robotic system hardware. This allows analysis and verification of plans requiring force or torque control such as the ultrasonic web thickness measurements shown here. The ultrasonic sensor must be pressed against the bulkhead with a specified force to produce accurate readings. This sequence also shows a tool exchange which verifies that proper clearances are allowed during this operation.

A camera located near the wrist of the robot arm is used for guiding the arm and making inspection measurements. Automatic computer vision algorithms are used to identify inspection locations, provide accurate position feedback to the robot control system, and measure dimensions.

Shown here as the computer would see them are the intermediate steps of the process. Once an image is acquired, it is convolved with a blurring filter to reduce noise. Next, all the edges in the image are enhanced. Each image contains many edges so each must be examined for size and shape, and the principle edges identified and located. Insignificant edges are suppressed. The principle edge features are then used to guide the arm or to perform inspection measurements.

A high-speed CORDIC peripheral, developed by Martin Marietta, is being used to perform the coordinate conversions required for real-time control of the robotic arms. A two-dimensional coordinate transformation is shown here.

In conclusion, an intelligent robotic system which combines offline programming and artificial intelligence planning techniques with real-time, online intelligent plan monitoring will greatly increase the flexibility and productivity of robotic systems.

THE ROLE OF ARTIFICIAL INTELLIGENCE IN ROBOTICS

- ARTIFICIAL INTELLIGENCE (AI) RESEARCH AIMS AT PROGRAMMING MACHINES TO SOLVE PROBLEMS GENERALLY ASSUMED TO REQUIRE HUMAN REASONING CAPABILITIES.

- EXAMPLES OF EXISTING AI PROGRAMS INCLUDE:

  - R1 - A SYSTEM WHICH CONFIGURES DEC/VAX COMPUTERS
  - MOLGEN - A PROGRAM TO GENERATE AND SEQUENCE TASKS IN MOLECULAR GENETICS EXPERIMENTS
  - DEVISER - AN ON BOARD TASK SEQUENCER AND MONITOR OF SATELLITE FUNCTIONS.

- AI RESEARCH IN ROBOTICS APPLICATIONS CENTERS AROUND "PLANNING"; THE GENERATION AND SEQUENCING OF TASKS, WHICH, WHEN PERFORMED, WILL ACHIEVE SOME DESIRED STATE. THERE ARE SEVERAL EXAMPLES OF PLANNING APPLICATIONS IN ROBOTICS RESEARCH.

TASK PLANNING AND SEQUENCING

- o GIVEN A SET OF SYSTEM OPERATORS AND A SET OF TASKS TO BE PERFORMED, A PLANNER SEQUENCES THE TASKS ACCORDING TO CONSTRAINTS IMPOSED BY BOTH THE PLANNING ENVIRONMENT AND ROBOTIC SYSTEM.

- o RESEARCH IS UNDER WAY TO ENABLE PLANNERS TO "LEARN" THROUGH PAST EXPERIENCE, BETTER WAYS OF ACCOMPLISHING DIFFERENT TASKS.

- o THE REASONING DONE BY THESE PLANNERS OCCURS ON MANY LEVELS, ALLOWING THE PLANNER TO INITIALLY GENERATE PLANS IN ABSTRACT TERMS AND WORK TOWARDS OPERATOR SEQUENCES WHICH ARE EXECUTABLE BY THE ROBOTIC SYSTEM.

INTELLIGENT PLAN EXECUTION

- o DURING EXECUTION OF A PLAN BY A ROBOTIC SYSTEM, UNEXPECTED EVENTS MAY OCCUR. EXAMPLES ARE:
  - -- A BATTERY BECOMES DISCHARGED, CAUSING A FAILURE IN SOME SUB-SYSTEM
  - -- THE POSITION OF A MANIPULATOR IS MISCALCULATED, CAUSING DISCREPENCIES THROUGHOUT THE SYSTEM
  - -- A FILTERED VISION IMAGE CONTAINS UNEXPECTED NEW OBJECTS, OR DOES NOT CONTAIN EXPECTED OBJECTS.

- o RESOLVING UNEXPECTED CONDITIONS DURING REAL TIME PLAN EXECUTION IS A SIGNIFICANT AI RESEARCH GOAL. TOPICS INCLUDE:
  - -- FAULT DIAGNOSIS - USING SYSTEM KNOWLEDGE TO REASON ABOUT THE CAUSES OF FAULTS
  - -- AUTOMATIC PLANNING - GENERATION OF PLANS TO RESOLVE SPECIFIC FAULTS
  - -- CONDITIONAL PLANNING - PLANNING AHEAD OF TIME WITH CONTINGENCIES FOR UNEXPECTED FAILURES

PATH PLANNING

- PATH PLANNING IS THE SEQUENCING OF ROBOTIC MANIPULATOR MOTIONS BETWEEN TWO CONFIGURATIONS SO AS TO AVOID OBJECTS IN A WORKING ENVIRONMENT. THE TYPE OF PATH FOUND MAY BE DEPENDENT ON OPERATING CONSTRAINTS OF THE SYSTEM.
- CONCERNS IN PATH PLANNING ARE:
  - -- EFFICIENT REPRESENTATION OF OBJECTS, FACILITATING AN EFFICIENT, HIERARCHICAL REPRESENTATION OF THE WORK SPACE. THIS IS KNOWN AS GEOMETRIC MODELLING.
  - -- MODULAR OPERATING HEURISTICS, ALLOWING THE PLANNER TO FIND PATHS UNDER VARYING OPERATING CONSTRAINTS.
  - -- REAL TIME OPERATION

COMPUTER VISION -- WHAT IS IT?

COMPUTER VISION IS THE SCIENCE OF PROGRAMMING COMPUTERS WITH CAMERAS TO IDENTIFY, DESCRIBE, AND REASON ABOUT OBJECTS VIEWED BY THE CAMERA.

COMPUTER VISION CAN BE USED FOR:

- AERIAL PHOTOGRAPH INTERPRETATION (MAP MAKING)
- MANUFACTURING PIECEPART INSPECTION
- ROBOT GUIDANCE
- SPACECRAFT RENDEZVOUS & DOCKING
- SPACECRAFT NAVIGATION FROM GROUND CONTROL POINTS
- WEAPONS POINTING & AIMING
- MEDICAL DIAGNOSIS (E.G., X-RAYS)
- ANALYSIS OF SCIENTIFIC IMAGERY (E.G., ASTRO PHYSICS)

COMPUTER VISION -- HOW DOES IT WORK?

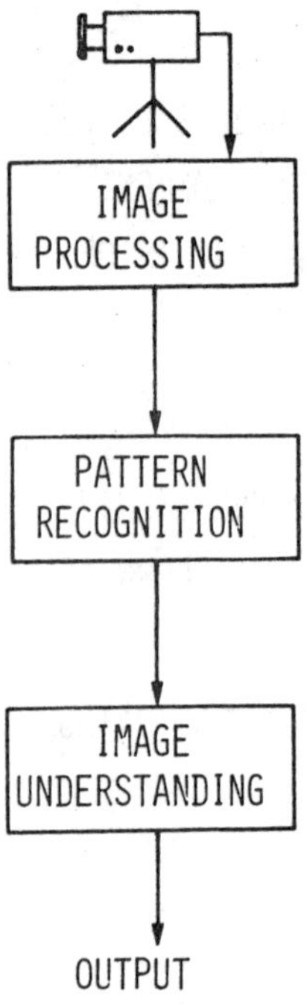

- DIGITAL CAMERA--DIVIDES AN IMAGE INTO OVER 64,000 SMALL AREAS, AND ASSIGNS EACH AREA A NUMBER ACCORDING TO HOW BRIGHT OR DARK THE IMAGE IS IN THAT AREA.
- IMAGE PROCESSING PROGRAMS CALLED FILTERS ARE USED TO IMPROVE THE IMAGE. THESE PROGRAMS CAN REMOVE NOISE, REMOVE BLUR, OR HIGHLIGHT EDGES FOR EXAMPLE.
- PATTERN RECOGNITION PROGRAMS ARE USED TO DETECT DISTINCTIVE IMAGE AREAS SUCH AS EDGES, LINES, CORNERS OR SURFACES OF OBJECTS.
- IMAGE UNDERSTANDING PROGRAMS ARE ARTIFICIAL INTELLIGENCE PROGRAMS THAT "REASON" ABOUT THE DISTINCTIVE IMAGE AREAS DETECTED IN ORDER TO PROVIDE A DESCRIPTION OF OBJECTS IN THE SCENE FOR OUTPUT.

COMPUTER VISION -- RESEARCH AREAS

- ROBOT GUIDANCE -- A DIGITAL CAMERA MOUNTED ON A ROBOT ARM IS USED TO GUIDE THE ARM TO VARIOUS LOCATIONS IN A WORK CELL. THIS CAN SAVE COSTS SINCE PARTS IN THE CELL NEED NOT BE PLACED ACCURATELY IN FIXED LOCATIONS.
- MANUFACTURING INSPECTION -- A SPECIAL DIGITAL LASER CAMERA, SIMILAR TO A "LASER RADAR", MEASURES RANGE TO ACCURACIES OF .001". DIGITAL IMAGES MADE WITH THIS CAMERA ARE USED TO FIND MANUFACTURING FLAWS IN AN F-15 BULKHEAD.
- STEREO VISION -- TWO CAMERAS, LIKE THE TWO EYES OF A HUMAN, ARE USED TO ENABLE THE COMPUTER TO PERCEIVE DEPTH. BEING ABLE TO PERCEIVE AND REASON IN THREE DIMENSIONS ARE CRITICAL FOR MOST GUIDANCE APPLICATIONS.

COMPUTER VISION -- RESEARCH AREAS

o SPACECRAFT RENDEZVOUS & DOCKING	--	A CAMERA ON ONE SPACECRAFT CAN TRACK A LIGHT PATTERN ON A SECOND SPACECRAFT, COMPUTING RANGE AND TRAJECTORY.
o SATELLITE IMAGERY ANALYSIS	--	A "SMART" SENSOR ONBOARD A SPACECRAFT CLASSIFIES IMAGE AREAS AS VEGETATED, BARE EARTH, SNOW, WATER, OR CLOUDS.
o COASTLINE TRACKING	--	A SPACECRAFT BORNE COMPUTER VISION SYSTEM CAN TRACK COASTLINES TO BE USED AS A NAVIGATION AIDE.
o WELDING FEED-BACK CONTROL	--	A COMPUTER VISION SYSTEM MONITORS A WELDING PROCESS, CONTROLLING WELD PARAMETERS SUCH AS VOLTAGE, CURRENT, WELD SPEED, FEED RATE, PENETRATION.

SOFTWARE SIMULATION OF MANIPULATOR SYSTEMS

PROVIDES KINEMATIC AND DYNAMIC MODELLING OF N-LINK SINGLE OR DUAL ARM MANIPULATOR SYSTEMS WITH THREE DIMENSIONAL VECTOR GRAPHICS DISPLAY.

APPLICATIONS:

- o OFFLINE PROGRAMMING FOR SCENARIO VALIDATION
- o SOFTWARE TESTBED FOR ARTIFICIAL INTELLIGENCE APPLICATIONS SUCH AS TRAJECTORY PLANNING AND COLLISION AVOIDANCE
- o SOFTWARE TESTBED FOR CONTROL ALGORITHM DEVELOPMENT AND TESTING
- o ANALYSIS TOOL FOR USE IN MANIPULATOR DESIGN WORK (CONFIGURATION, REACH ENVELOPMENT, ETC.)

- HIGH SPEED COORDINATE CONVERSIONS

- USED IN REAL TIME CONTROL OF SIX-DEGREE OF FREEDOM ROBOT

- APPLICATIONS IN

  - CARTOGRAPHY
  - VISION SYSTEMS
  - ROBOTICS
  - GIMBAL POSITIONING CONTROL

AAS 84-020

STRAPDOWN INERTIAL GUIDANCE PERFORMANCE IN SPACE

Michael J. Haley*
Harold B. Taylor*

Strapdown Inertial Guidance Systems have been incorporated on many U.S. space vehicles since the inception of the Space Program in the 1950's. The Hamilton Standard Division of United Technologies pioneered much of the early development of strapdown inertial systems beginning in the early 1960's, and is presently an acknowledged leader in the design, development and production of this technology. The purpose of this paper is to discuss the evolution of strapdown inertial system technology at Hamilton Standard with primary emphasis on demonstrated performance in space. Specific attention is directed toward the Redundant Inertial Measurement Unit presently in use on the Inertial Upper Stage. The extension of this current technology toward emerging space applications, requiring higher levels of autonomous fault tolerance, is discussed.

* *Hamilton Standard, 1690 New Britain Avenue, Farmington, Connecticut 06010*

INTRODUCTION

Hamilton Standard strapdown inertial systems incorporate a patented pulse width modulated technique which precisely controls rebalance torque to maintain the sensors in a nominal null position. In this mechanization, gyro digital rebalancing signals (pulses) provide a measure of the absolute rotation of the vehicle. Information concerning vehicle orientation obtained from gyro output pulses is used to transform similarly obtained accelerometer specific velocity output pulses to inertial coordinates for further navigational processing.

The pulse width modulated strapdown rebalance concept is unique in that it can be scaled to satisfy a broad spectrum of maximum angular rate, linear acceleration, resolution and dynamic environment requirements with a simply accomplished modification to the sensor loop electronics. This is a particularly important feature since it allows existing technology, and in many cases existing hardware, to be cost effectively adapted to satisfy the requirements of each particular application.

Past Accomplishments

Hamilton Standard developed the first man-rated strapdown Inertial Measurement Unit for the Lunar Module Abort Sensor Assembly (LM/ASA). From 1968 to 1972 the IMU performed successfully during one earth orbital and eight lunar flights. During Apollo XIII, the in flight explosion in the Command Service Module en route to the moon led to the termination of the scheduled lunar landing of Astronauts Lovell and Haise. In order to maximize limited resources, the Lunar Module was not jettisoned and was used as a lifeboat for the return to earth. As a power conservation measure, the LM/ASA was utilized for course corrections during the successful return to earth.

Hamilton Standard was the first to achieve boost vehicle production status for a strapdown Inertial Measurement Unit with the introduction of the Delta Inertial Guidance System (DIGS). From 1972 through 1980, the DIGS IMU completed 62 missions without a single guidance failure. This system was designed to maintain the high performance and reliability levels demonstrated by the LM/ASA using simplified packaging and manufacturing methods to improve reliability and reduce cost. Orthogonal triads of Hamilton Standard RI-1139E strapdown rate integrating gyros and Singer Kearfott 2401 accelerometers, with associated rebalance electronics, performed during flight to sense vehicle linear acceleration and angular rates. The sensed vehicle attitude and acceleration information was then provided to the Delco Electronics Guidance Computer for computation of Guidance and Steering functions.

The Viking Inertial Reference Unit (IRU) was designed to meet the stringent weight and power limitations associated with the Viking 1975 Mars lander mission. The IRU provided the inertial reference used during the Mars orbital, descent, landing and surface operating phases of the Viking mission. The two Viking space vehicles were launched in August and

September 1975, encountered Mars during the summer of 1976, and landed on the Martian surface on July 20, 1976 and September 3, 1976, respectively. This unit, which was designed to track oscillatory rates of ±100 degrees/second during the Martian parachute descent phase of the Viking mission, featured a fourth redundant gyro, fourth accelerometer and redundant electronics, thus increasing reliability during the year-long mission.

Present

Hamilton Standard currently manufactures the Redundant Inertial Measurement Unit (RIMU) for the Inertial Upper Stage (IUS). The RIMU contains five (5) Hamilton Standard RI-1010, pulse-width modulated, single degree of freedom, rate integrating, gas bearing gyros and five (5) Kearfott 2401, pulse-torqued, single axis, linear accelerometers, installed in a single cluster assembly. The vector directions of the sensor input axes are arranged asymmetrically about the surface of a cone whose half angle has its cosine equal to the square root of 1/3. This provides a globally uniform distribution of error, while maximizing the performance advantage afforded by the five like sensors. The cone axis is tilted from the X-RIMU (IUS thrust) axis by a rotation of approximately 41.2 degrees. This orientation places two of the gyro's output axes parallel to the gravity vector on the launch pad, thus optimizing prelaunch azimuth alignment accuracy. It also places the output axes of these gyros parallel to the IUS vehicle thrust axis. In addition, all gyro spin axes and accelerometer pendulous axes are perpendicular to the thrust axis which optimizes inertial navigation performance during boost and IUS burn phases of the mission.

Redundant Inertial Measurement Unit

Block Diagram

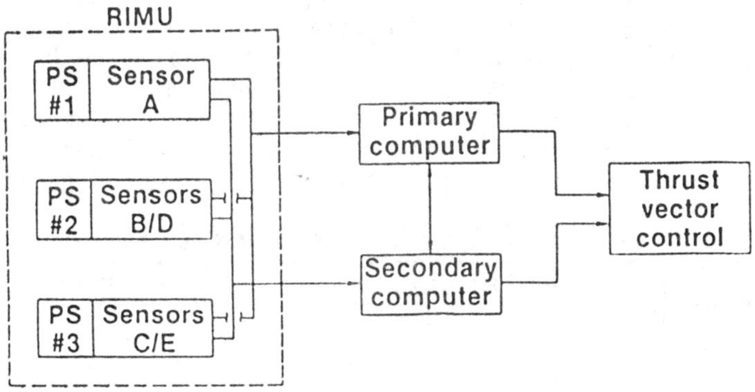

Figure 1

A "Kinematic Reversal" mechanism provides the capability of rotating the sensor cluster about a horizontal axis through an automatic sequence of four orientations, spaced 90 degrees apart. This is used during Earth pre-launch operation to perform a precise, self-contained alignment and calibration of the RIMU in order to provide a check of system readiness, and to increase mission accuracy.

The Inertial Upper Stage has completed its first two flights. On October 30, 1982, two Defense Satellite Communications System Vehicles, DSCS II and DSCS III were targeted for geosynchronous orbit at 22,973 and 22,817 nautical miles respectively. Launched aboard a Titan 34D/IUS, each DSCS Satellite was placed within 13 nautical miles of its target.

The second IUS mission was designed to place the first Tracking Data Relay System 1 (TDRS-A) spacecraft into geosynchronous orbit following deployment into a 153 nautical mile orbit by the Space Shuttle Challenger (STS-6) in April, 1983. Prior to completion of the SRM-2 burn, a motor problem occurred causing the vehicle to tumble. Subsequent events, including the eventual separation of the TDRS-A spacecraft and the use of its on-board fuel reserve to achieve geosynchronous orbit, are well documented. Indications are that the IUS guidance system would have met injection accuracy requirements had the motor problem not occurred.

Redundancy Management

The selected approach to the redundancy management/failure detection aspects of the RIMU design satisfies the IUS reliability level and single failure tolerant criteria. The choice of a computationally cross-strapped five sensor array provides the ability to detect at least one sensor channel or one power supply failure. Combined hardware BITE and computer software implemented diagnostics are used to detect RIMU subassembly failures. The key to successful RIMU operation is the ability to detect and isolate sensor performance degradation (i.e., "soft") failures. The magnitude of a soft failure is such that it tends to be masked by noise, but which ultimately degrades orbital injection accuracy if left to propagate throughout the mission. The FDI algorithms associated with the IUS RIMU solve this problem by means of a series of threshold checks upon the magnitude of "parity vectors" formed from the gyro and accelerometer data. The known relative placement of the inertial sensors permits the development of linear dependence equations for the sensor outputs. These linear dependence equations define the equivalance, or parity, among the sensor outputs and, as such, a failure in any sensor will cause a d viation from the expected parity. (Figure 2)

SENSOR FDI VIA PARITY VECTOR

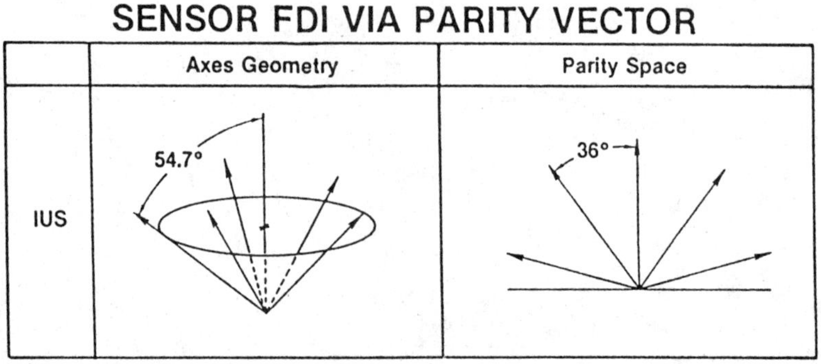

Figure 2

For the pentad array of sensors, the associated parity space is two dimensional, and the parity algorithm is relatively simple. The amplitude of the parity vector is a measure of the inconsistency in the redundant sensor data. The direction of the parity vector is toward the placement of the failed sensor input axis in parity space.

Future Directions

Emerging applications for strapdown inertial reference units in space demand extremely fine attitude resolution, low noise and a wide dynamic range and bandwidth capability. Methods to autonomously achieve several levels of fault tolerance must be exploited to achieve functional reliability over long time durations in the hostile space environment. Currently, Hamilton Standard is extending the IUS technology in developing an even more precise, inertial reference unit with additional levels of redundancy for such space applications. It features six gyros, with input axes arranged in a dodecahedron array. This provides a 3-fail operating gyro capability, and 42 viable combinations of the required three gyro set. The combination of hardware BITE and software FDI to accomplish redundancy management is a direct extension of IUS technology. The "parity vector" is still the basis for soft failure detection, but the "parity space" is now three-dimensional. Consistant with the reliability allocations, the power supplies and input/output interfaces are dual redundant. Standby redundancy of these elements provides a further extension of the reliability of the inertial reference function. Space vehicle form factor constraints are addressed by maximum use of proven advanced electronic packaging techniques. (Figure 3)

Strapdown Inertial Systems, with their inherent multiple channel fault tolerance, continue to be in the forefront as mission requirements become increasingly demanding. These systems will continue to provide the ultimate in performance and reliability for future Spacecraft and Upper Stage Boosters.

PRECISION ATTITUDE REFERENCE UNIT
SIMPLIFIED BLOCK DIAGRAM

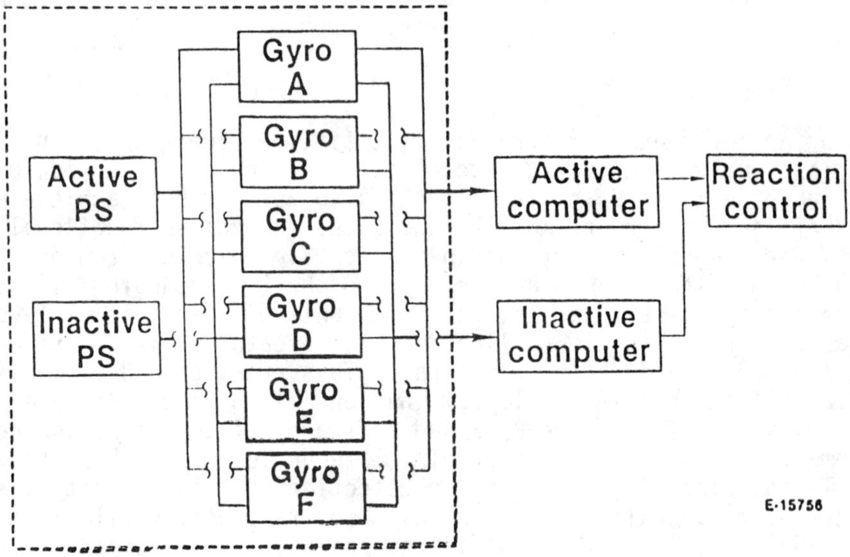

Figure 3

AAS 84-021

ISTP - NASA'S NEXT INTERNATIONAL COOPERATIVE PROGRAM*

Summary

Raymond S. Tatum+

The International Solar Terrestrial Physics (ISTP) Program is being planned as a major collaborative endeavor between the European Space Agency (ESA), the Institute of Space and Astronautical Science (ISAS) of Japan, and the National Aeronautics and Space Administration (NASA). The Global Geospace Study (GGS) will be a central element of the program.

Together with other programs now in preparation, it will be possible for the first time to look at the complete solar-terrestrial energy chain: first, the ejection and transport of energy and matter from the sun through interplanetary space; then, the coupling and transport of energy through the earth's magnetosphere; and finally, the deposition and dissipation of energy in the earth's upper atmosphere.

The chief complementary programs begin in the mid-1980's. Two International Solar Polar Mission (ISPM) spacecraft will be obtaining the first steroscopic view of the sun from latitudes high above the ecliptic plane and will be providing new insights into the three-dimensional structure of the interplanetary medium. Simultaneous measurements by the GGS WIND spacecraft will provide an in-ecliptic benchmark for out-of-ecliptic data from the ISPM. The latter can then be interpreted properly and related to the more familiar perspective of the sun and the solar wind.

During the same period, the Upper Atmosphere Research Satellite (UARS) Program is planned to conduct a global study of the chemistry and dynamics of the earth's mesosphere and stratosphere (the altitude range extending from about 15 to 85 kilometers). Simultaneous measurements by the four GGS spacecraft and by the UARS will result in the first coordinated investigation of the coupling and transport of solar wind plasma and energy transfer on the upper atmosphere.

The GGS spacecraft measurements of global behavior of geospace will also complement a number of investigations to be conducted from the Shuttle-Spacelab orbiting laboratory. Many active experiments, such as controlled

* *This paper appears in full in the microfiche supplement to this volume, AAS Microfiche Series, Volume 48.*

+ *Acting Project Manager, Code 407, NASA Goddard Space Flight Center, Greenbelt, Maryland 20771* chemical releases, electromagnetic wave injections, and particle beam accelerators, will benefit greatly from the diagnostic capability of the GGS Program and the real-time monitoring of magnetospheric activity, so that the Spacelab experiments can be conducted under known conditions.

The ISTP Program is the next logical thrust in solar-terrestrial research. Such an undertaking will be directly responsive to recent recommendations of the Space Science Board of the National Academy of Sciences.*

The ISTP Program will be a key element in the overall solar-terrestrial program in the 1980's. Together, the International Solar Mission, the Upper Atmosphere Research Satellite, the ISTP Program, and other complementary missions, such as Japan's EXOS D, will permit a coherent study of the flow of matter and energy from the sun through the heliosphere, the magnetosphere, and the upper atmosphere. The ISTP will also complement simultaneous investigations of space plasma near earth with the Shuttle Spacelab and of space plasmas near Jupiter with the Galileo orbiter.

Knowledge about the fundamental properties of geospace will have direct, significant applications to other studies in astrophysics and laboratory plasma physics. Plasma interactions are important throughout the solar system and the universe. Geospace uniquely provides an accessible example of a plasma environment in which significant processes can be studied directly.

A quantitative assessment of geospace inputs to the near-earth environment is feasible and necessary. The ISTP Program can develop the physical basis for understanding and monitoring geospace effects on satellite communications, space radiation environment, terrestrial power systems, exploration geophysics systems, and conceivably, some aspects of weather and climate.

The ISTP Program is defined scientifically and technically. The necessary technological capacity is available. Science definition studies were completed in 1983. Spacecraft definition studies are planned for 1984. Launches are feasible beginning in 1989. The ISTP Program is managed by the Goddard Space Flight Center for NASA's Office of Space Science and Applications, by the European Space Technology Center (ESTC) for ESA, and by the Institute of Space and Astronautical Science (ISAS) for Japan.

REFERENCES

1. *Space Plasma Physics: The Study of Solar System Plasmas*, Space Science Board, A.G.W. Cameron, Chairman (S.A. Colgate, Study Chairman), National Academy of Sciences, Washington, D.C., 1978.

2. *Solar System Space Physics in the 1980's: A Research Strategy*, Space Science Board, A.G.W. Cameron, Chairman (C.F. Kennel, Study Chairman), National Academy of Sciences, Washington, D.C., 1980.

3. *Solar-Terrestrial Research for the 1980's*, Committee on Solar-Terrestrial Research, J.V. Evans, Chairman (H. Friedman and D.S. Intriligator, Study Chairmen), National Academy of Sciences, Washington, D.C., 1981.

* *See References*

AAS 84-013

INTERFEROMETRIC ANGLE SENSOR*

Harrison B. Albert+
Lawrence M. Germann+

The following points are made about the Interferometric Angle Sensor: BASD's demonstration of an angle encoder with an automated calibration system, its sub-μradian accuracy, its use in demanding applications, the new technology involved (interferometry with wide angle ranges), its durability because of its design, information that calibration gives (three or more essential coefficients), how the calibration data are used (initialization at turn-on), which calibration to use for different applications, the effect of environmental changes on accuracy, environment vs. inter-target spacing, and which mathematical approach gives equations (iterative least squares).

This abstract was prepared by the Assistant Editor.

The paper consists essentially of presentation charts.

* *This paper appears in full in the microfiche supplement to this volume, AAS Microfiche Series, Volume 48.*

+ *Ball Aerospace Systems Division, P.O. Box 1062, Boulder, Colorado 80306*

AAS 84-015

RETROREFLECTOR FIELD TRACKER*

Frank E. Wargocki[+]
Susan E. Ross[+]

This paper discusses the Retroreflector Field Tracker's sensor characteristics, flight hardware, operation, detector/sensor head, processor overview, functional description and timing, and applications for its use.

This abstract was prepared by the Assistant Editor.

The paper consists essentially of presentation charts.

---

* *This paper appears in full in the microfiche supplement to this volume, AAS Microfiche Series, Volume 48.*

+ *Electro-Optic Products, High Technology Products, Ball Aerospace Systems Division, P.O. Box 1062, Boulder, Colorado 80306*

AAS 84-016

AIRBORNE LASER LAB: DESCRIPTION AND EXPERIMENTAL RESULTS*

Paul S. Shirley+

The Airborne Laser Lab is a flying laboratory designed to evaluate the technical feasibility of laser weapons. This paper provides a cursory look at the concepts employed in the development of a high energy laser system and in particular describes systems employed by the Airborne Laser Lab. Significant results obtained by the use of the lab are shown through the presentation of recently released information.

* *This paper appears in full in the microfiche supplement to this volume, AAS Microfiche Series, Volume 48.*

+ *Systems Engineer, Air Force Weapons Laboratory, ARWL/ARLB, Kirtland Air Force Base, New Mexico 87117*

Section III
RENDEZVOUS DOCKING AND ORBIT SERVICING

SESSION III

Chairman:	H. Buchanan NASA Marshall Space Flight Center
Co-Chairman:	Stephen Z. Szirmay NASA Jet Propulsion Laboratory
Local Chairman:	Roy M. Scott Ball Aerospace Systems Division

The following paper was not available for publication:

AAS 84-031 An Overview of Space Station Rendezvous and Berthing Operations, L. Powell, NASA Headquarters

The following paper numbers were not assigned:

AAS 84-037 to -040

AAS 84-032

THE FIRST ROLE OF THE MANNED MANEUVERING UNIT IN SPACE

J.C. Harcinske, R.A. Schein,
A.M. Ray, D.J. Cwynar*

The Manned Maneuvering Unit (MMU) is a propulsive backput that provides precision maneuvering capabilities to the shuttle astronaut. An astronaut equipped with a space suit, portable life support system, and MMU can operate independently of the orbiter systems up to 500 feet from the orbiter. Future enhancements would allow the MMU to travel from 2000 to 3000 feet or more in support of operations that cannot be performed in closer priximity to the orbiter.

HISTORICAL DEVELOPMENT OF THE MMU

The MMU technology dates back to NASA's Gemini program. This program defined two maneuvering units: an Astronaut Maneuvering Unit developed by the Air Force that was never tested in space, and a Hand Held Maneuvering Unit that was sucessfully tested in space.

The Skylab experiment, M509, was an Astronaut Maneuvering Research Vehicle that tested a hand held maneuvering unit and three attitude control modes for backpack maneuvering unit. Skylab also flew experiment T020, a Foot Controlled Maneuvering Unit.

After Skylab, simulation evaluations of seventeen different control modes and five different hand controller configuations resulted in the control mode and hand controller configurations used in the present MMU.

The Extra-Vehicular Mobility Unit (EMU) concept was developed for the lunar surface activities, and was refined for shuttle applications. The MMU was designed as a separate attachment to the EMU to provide a simple, reliable, and precise propulsive maneuvering capability for the astronaut. The EMU makes the MMU design much simpler by freeing the astronaut from the encumbrances of the life support umbilical used on the Gemini and Skylab missions.

* *Martin Marietta Aerospace, P.O. Box 179, Denver, Colorado 80201*

MANNED MANEUVERING UNIT PRESENT CONFIGURATION

The MMU (figure 1) is a self-contained, six degree of freedom mobility system. It uses compressed gaseous nitrogen expelled from small thrusters to provide forces and moments for translational and rotational motions. Motion commands are input by the pilot through two hand controllers. One hand controller is used for rotational commands, one for translation commands. The hand controllers generate electrical signals that are processed by two independent, redundant control electronics modules. The control electronics then sends thruster firing commands to the appropriate thrusters. Table 1 contains system weight limits for the light specification case and for the

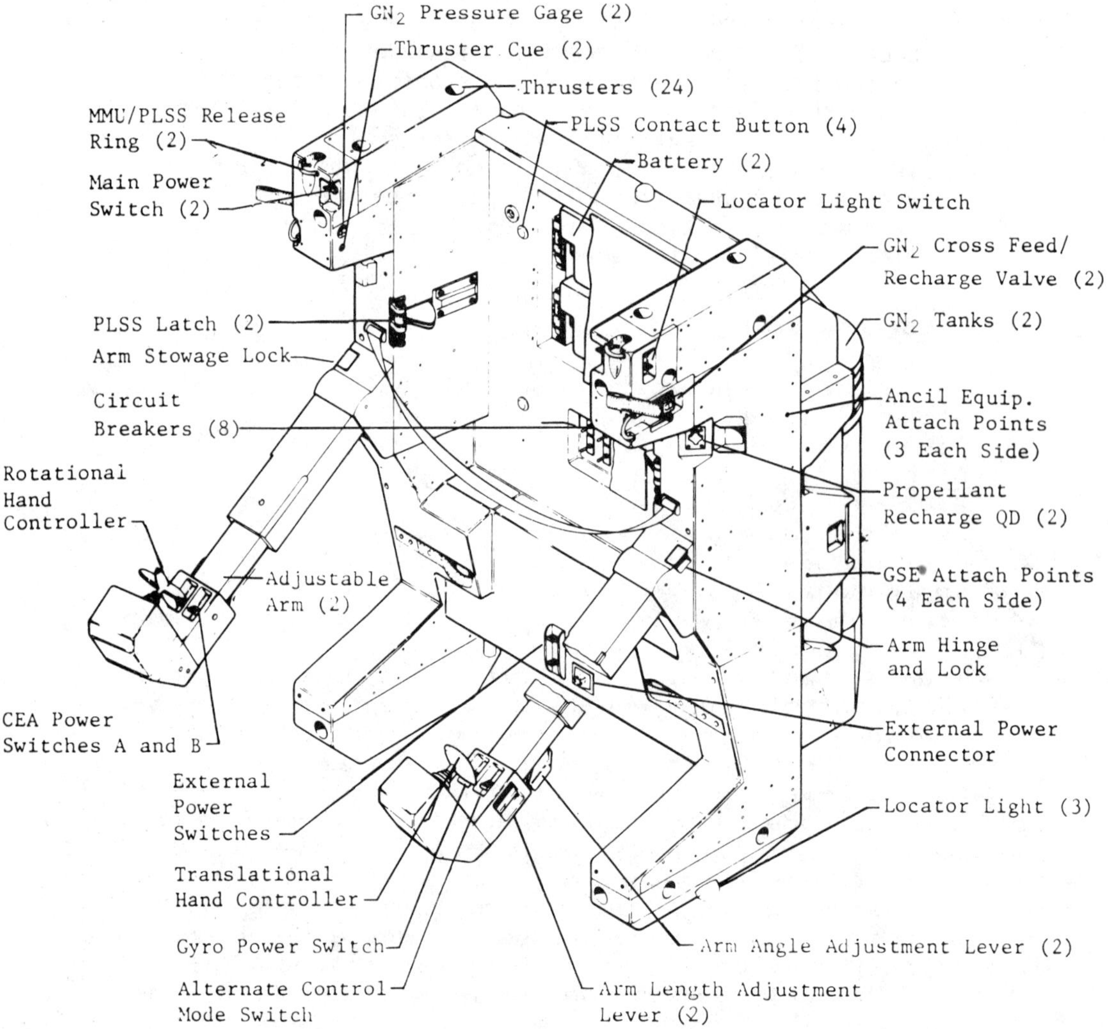

Figure 1
Manned Maneuvering Unit (MMU)

Table 1

MANEUVERING PERFORMANCE

Spec Light Limit			Spec Heavy Limit		
		Pounds			Pounds
5% Female		103	95% Male		215
Spec EMU (10 lb H_2O Used)		199	EMU (CEI Spec)		210
Spec MMU (23.2 lb GN_2 Used)		314.8	MMU (0.0 lb GN_2 Used)		338
TOTAL		616.8	TOTAL		763
Nominal Thrust	= 1.70 lb		Nominal Thrust	= 1.70 lb	
Translational Acceleration	= 0.356	FPS^2	Translational Acceleration	= 0.290	FPS^2
Delta Velocity*	= 81.7	FPS	Delta Velocity*	= 66.3	FPS
Pitch Acceleration	= 9.7	DPS^2	Pitch Acceleration	= 6.8	DPS^2
Yaw Acceleration	= 10.4	DPS^2	Yaw Acceleration	= 7.9	DPS^2
Roll Acceleration	= 9.7	DPS^2	Roll Acceleration	= 7.1	DPS^2
*SPECIFIC IMPULSE = 66 seconds.					

heavy case human, EMU, and MMU combinations. The performance values for these cases are included. These values do not include any additional equipment that may be carried by the MMU to support a particular mission.

When the MMU is not in use, it is stored in a flight support station (figure 2) located in the forward end of the orbiter payload bay. The flight support station is a don and doff station, and a service recharge station for the MMU.

The MMU is comprised of four subsystems: electrical, propulsion, thermal control, and structures and mechanisms. A brief description of each follows.

The electrical subsystem is illustrated in figure 3. This system includes switches to enable different control modes, propellent isolation valves, locator lights, power conditioning circuits, gyros, and the control electronics.

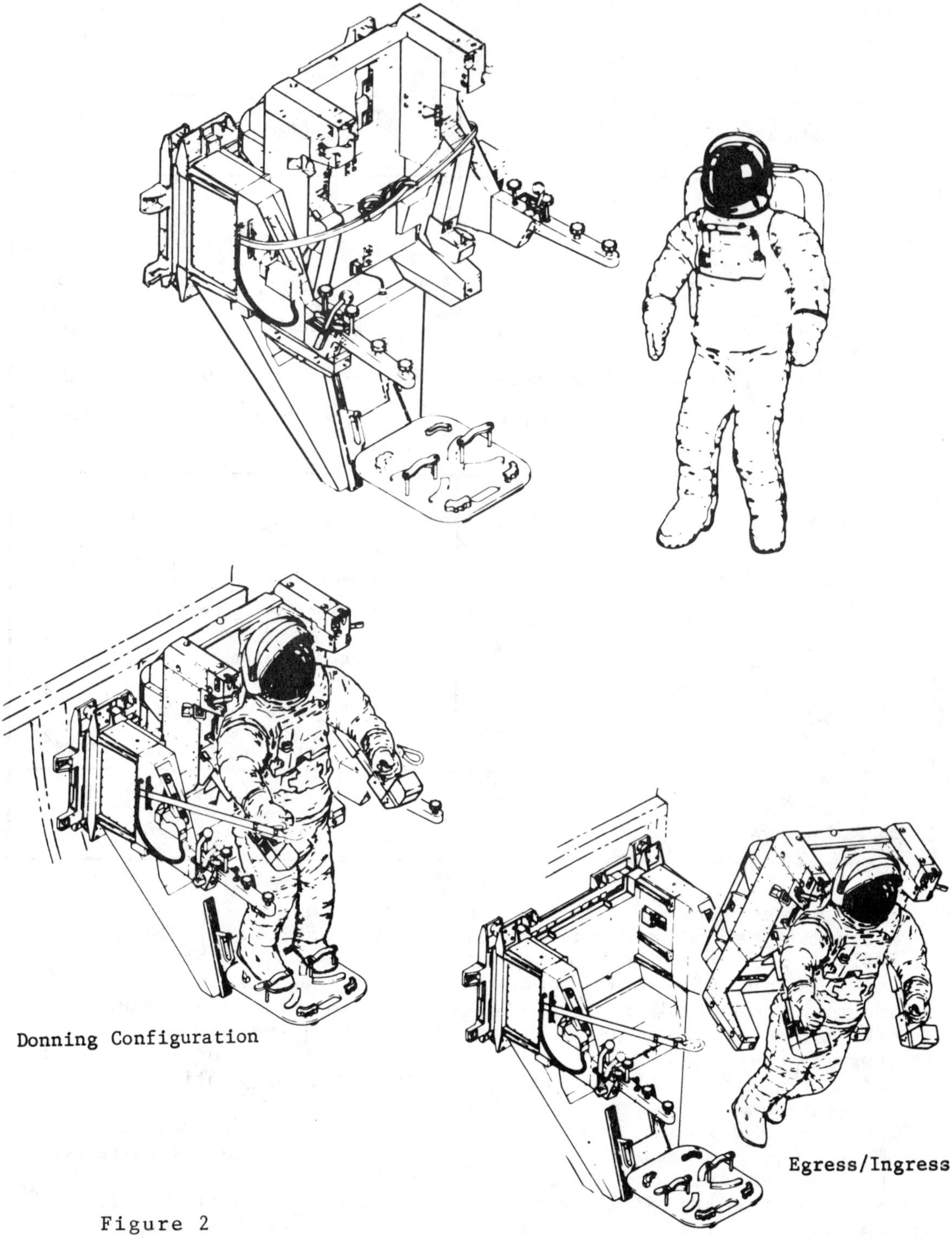

Figure 2
MMU Flight Support Station

Block Diagram of One of Two Redundant Systems, A&B

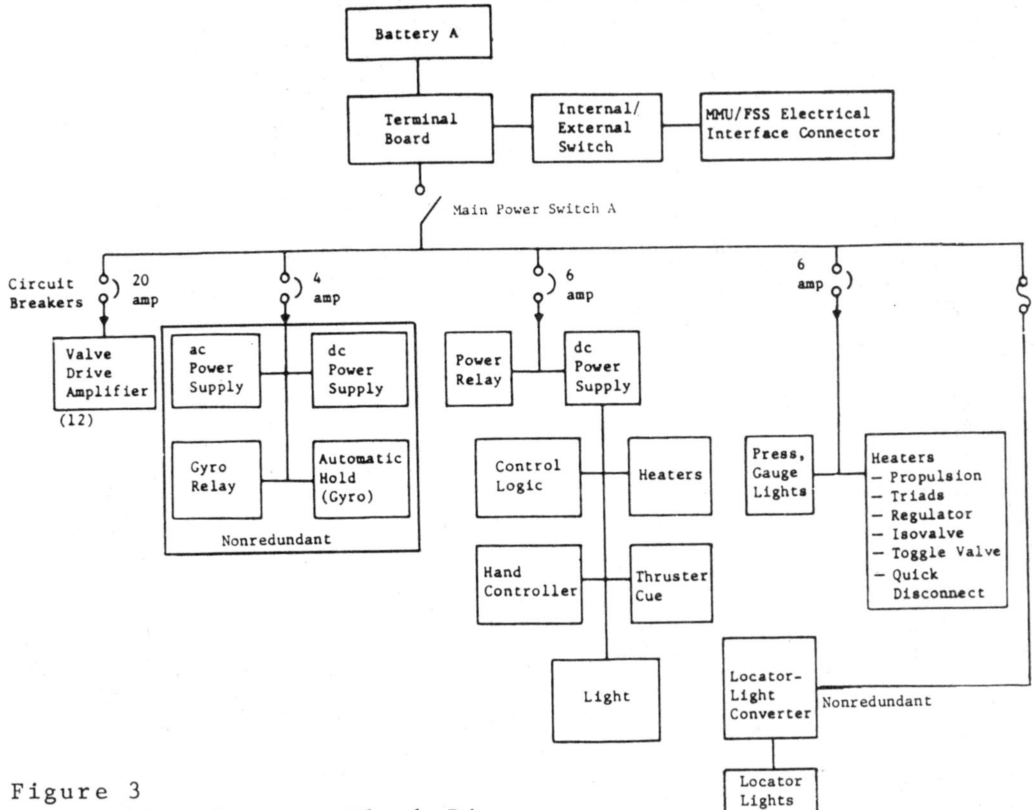

Figure 3
Electical Subsystem Block Diagram

Each of the MMUs redundant electronic systems is powered internally by a dedicated silver zinc battery. Each battery power is switched by a main power switch. When in the on position, power is applied to the appropriate control electronics power switch, gyro power switch, alternate control mode switch, locator light switch, propellent gauge light, and heaters.

The control electronics assemblies shown in figure 4 process hand controller inputs and gyro inputs to provide six degree of freedom control authority. The control electronics module contains gyros, thruster select logic, valve and thruster drive amplifiers, and power supplies.

The control electronics has two redundant circuits that are normally operated simultaneously. MMU flight with only one of the redundant circuits operational is possible. In this mode, the translational control authority is halved in all three directions, and rotational control authority is unaffected.

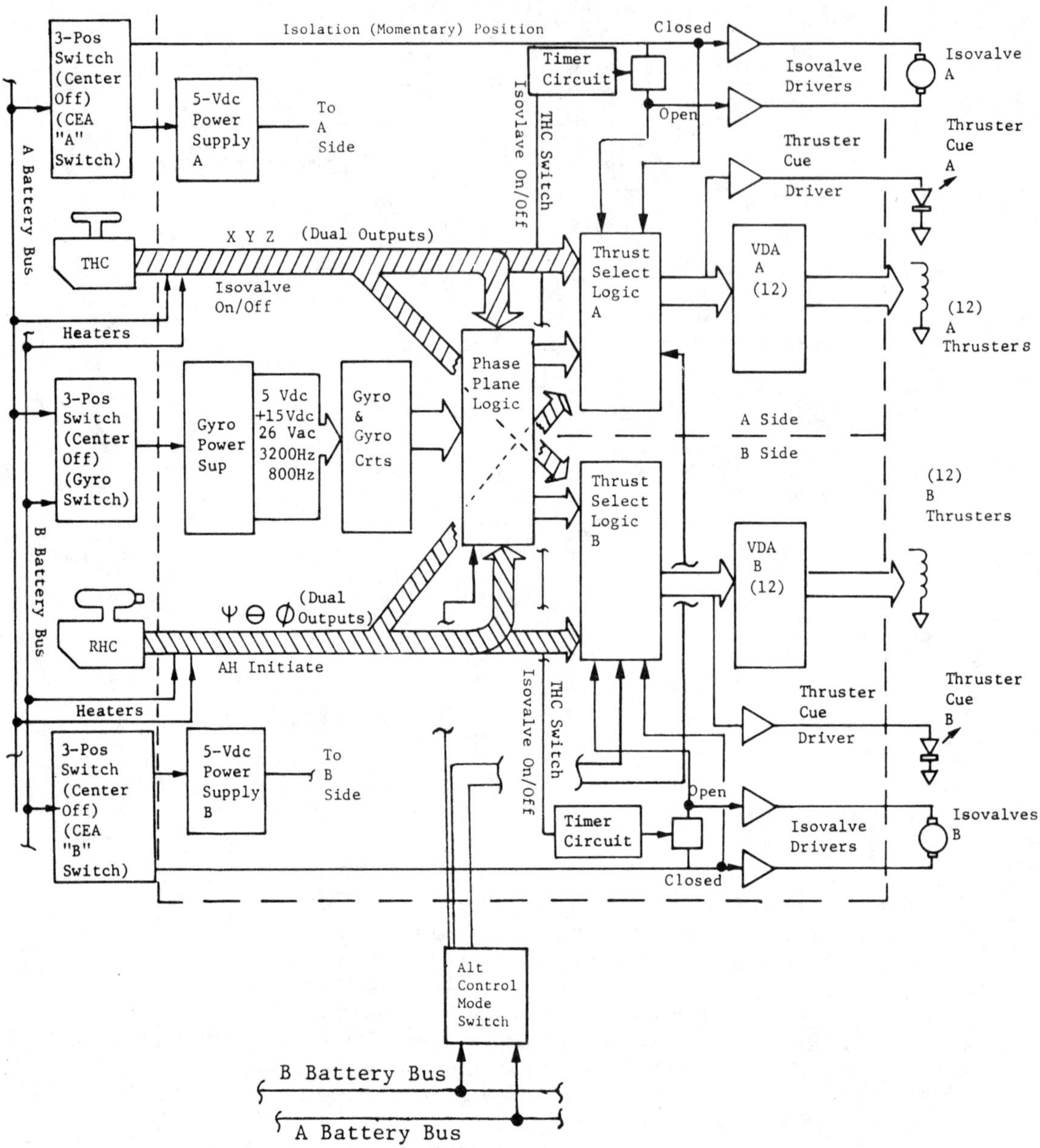

Figure 4
CEA Block Diagram

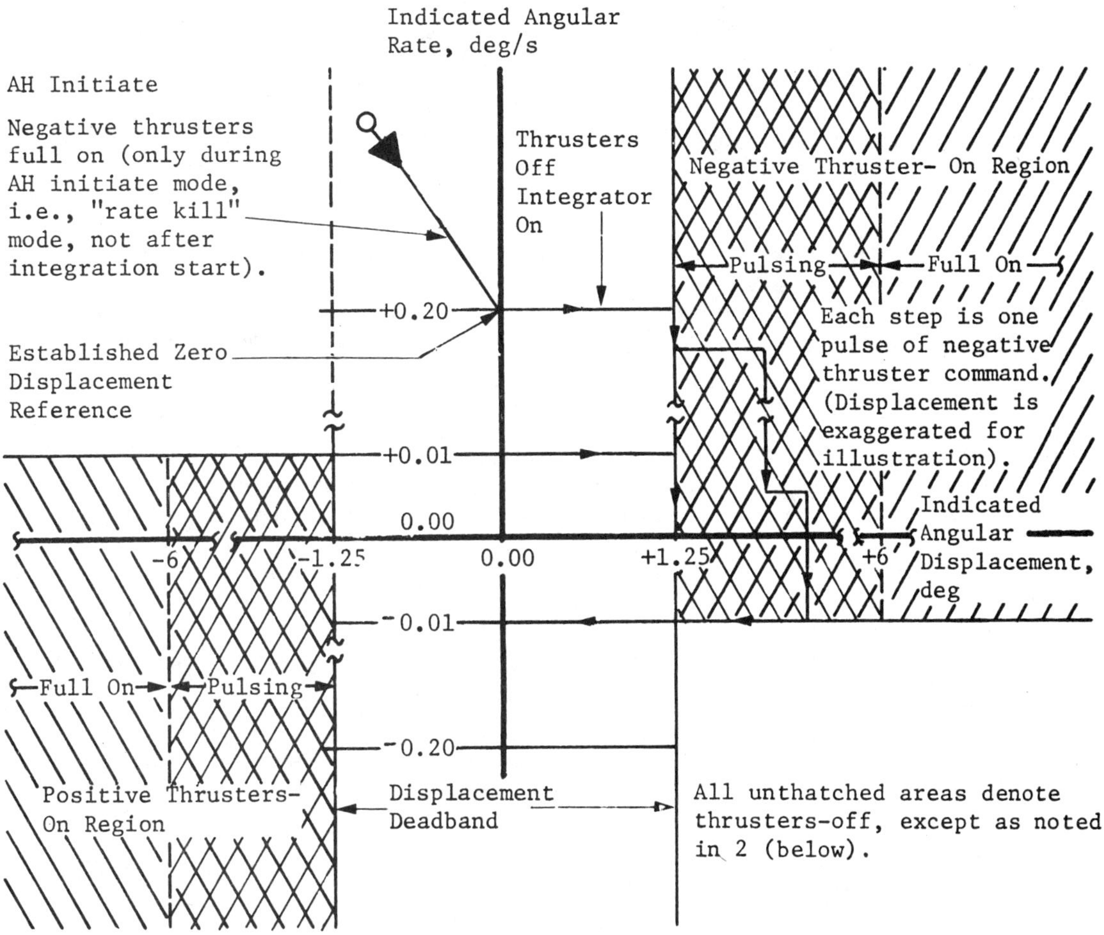

Note:

1. Typical convergence limits cycle when AH is initiated while MMU rate is greater than +0.2 deg/s. All values are nominal (for illustration only). Effect of system delays is not shown. Sketch represents each axis.
2. Also, THC commands may be in progress during AH, causing thrusters to fire regardless of the AH displacement deadband. The CEA imposes a 0.5-deg/s indicated angular rate limit when THC commands are present during AH, which may cause thruster firings within the displacement deadband.

Figure 5
Limit Cycle

Command inputs from the two hand controllers are seen as switch closures by the control electronics. When one or more hand controller switch closures are detected, the control electronics refers to a table containing thruster select logic for that combination of switch closures. The control electronics subsequently activates the proper electrical thruster drive signals to effect the motion commanded.

The MMU is also equipped with an inertial attitude hold function that is activated by a push button on the rotational hand controller. When activated, the attitude hold establishes and maintains all rotational MMU rates at a near zero value. When the rates have been nulled, the attitude hold system enters a phase plane, illustrated in figure 5.

There are two ways to exit the attitude hold mode. A rotational command input from the hand controller will disengage attitude hold only in the axis commanded. When

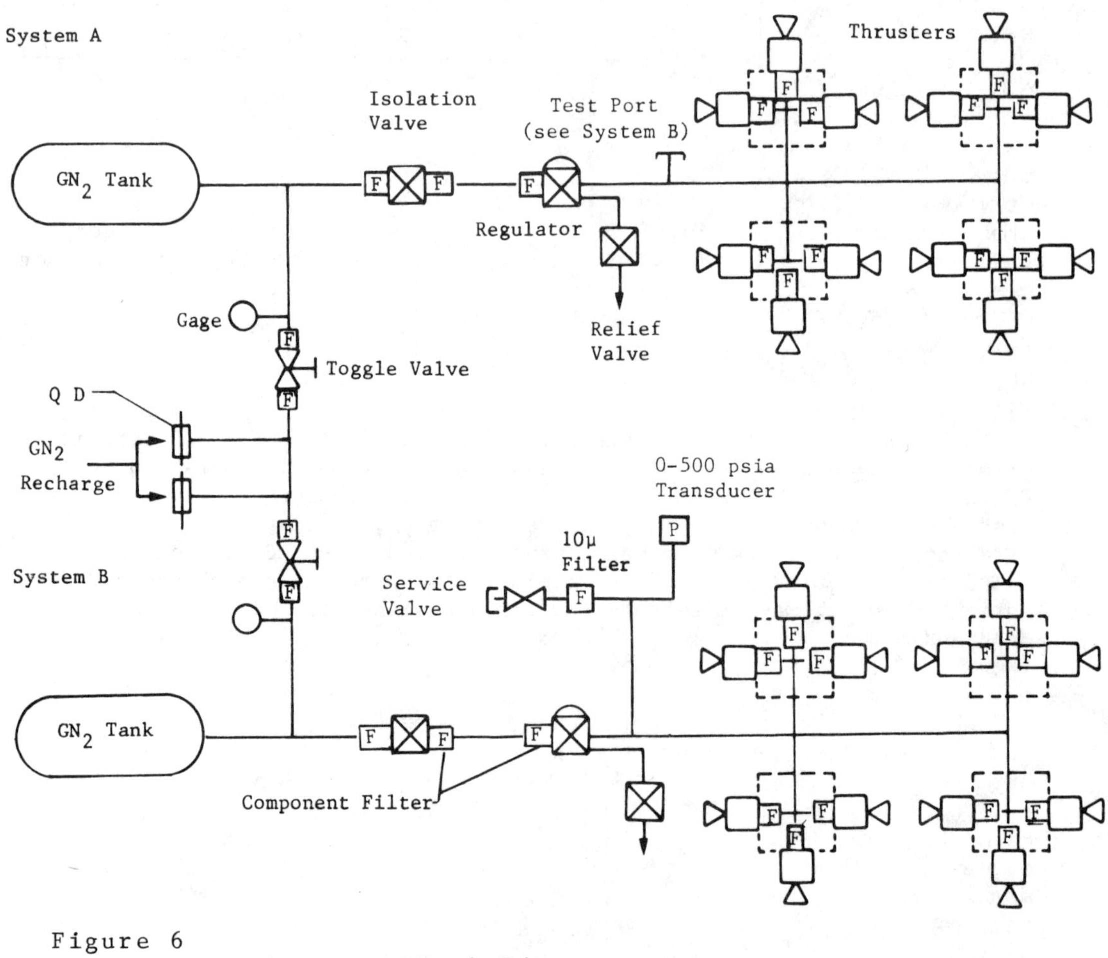

Figure 6
Propulsion Subsystem Block Diagram

the attitude hold switch is pressed again, only the axis or axes not already in attitude hold will be initialized. The second method of disengaging attitude hold is to remove power from the gyro. The latter method disables attitude hold in all three rotational axes.

Two additional control modes have recently been added to the MMU control system to support the Solar Maximum repair mission. Each mode is pilot selectable via an alternate control mode switch located on the left control arm. The axis inhibit control mode, when selected, inhibits the attitude hold function in one ground selectable axis. This allows the MMU pilot to use attitude hold in two rotational axes without disturbing a previously set rotation rate in the third axis. For the Solar Maximum repair mission, the yaw axis is selected as the attitude hold inhibited axis.

The satellite stabilization control mode uses an alternate set of thruster select logic tables. This mode is designed to fire additional thrusters to compensate for a large mass attached to the front of the MMU, i.e., a satellite.

The MMU propulsion subsystem is illustrated in figure 6. It is divided into two sides; each redundant, each controlling twelve thrusters. Each side is fuelled by a separate propellent tank capable of holding 3000 psi of gaseous nitrogen. The two tanks are connected by pilot operated crossfeed valves, used to equalize the pressure in the two tanks. Downstream from each tank are isolation valves. The isolation valves may be closed by the pilot to conserve propellent in the event of a leak in one system.

A pressure regulator reduces the pressure to the thrusters to approximately 200 psi, and a relief valve protects the thrusters from damage in the event of a regulator failure. The low pressure nitrogen is manifolded to four sets of three orthogonal thrusters for each redundant side. The twelve thrusters per side are physically placed such that thrust is balanced even if only one system is being used. The thrusters are solenoid operated, the drive signals coming from the control electronics.

The propulsion system is also equipped with two quick disconnect fittings for recharging the propellent tanks from the orbiter systems, and with illuminated mechanical tank pressure gauges.

The thermal control subsystem uses thermal coatings, multilayer insulation, thermal isolators, thermostats and heaters for temperature control within the MMU.

The MMU structure is made entirely of aluminum to minimize weight. The MMU alone weighs approximately 300 pounds. The control arms are adjustable in length. The arms can also be rotated down from the flight position to a stowed position to provide a larger work envelope for the pilot at a work station.

SOLAR MAXIMUM REPAIR MISSION OVERVIEW

The first real use of the MMU will be on the Solar Maximum satellite repair mission, presently scheduled for April 1984.

The Solar Maximum satellite, figure 7, is a scientific mission designed to gather more data about the sun. Shortly after launch, the satellite attitude control system failed. This rendered much of the scientific instrumentation useless. The MMU is to play a key part in the on-orbit repair of the Solar Maximum.

At 265 nautical miles above the earth, the MMU pilot will don the MMU, and perform a short checkout flight in the orbiter payload bay. The checkout will ensure that the MMU is fully operational before beginning the mission.

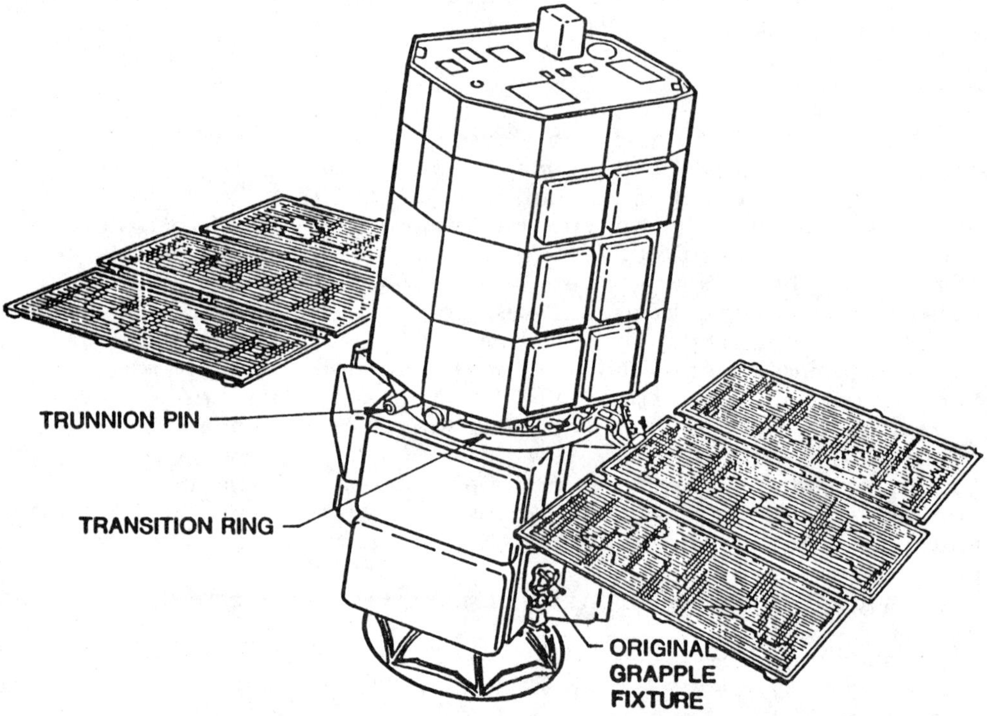

Figure 7
Solar Maximum Mission Satellite

Then, the MMU pilot will fly over to a storage locker. With the help of another suited crewman, a trunnion pin attachment device will be installed on the control arms of the MMU (figure 8).

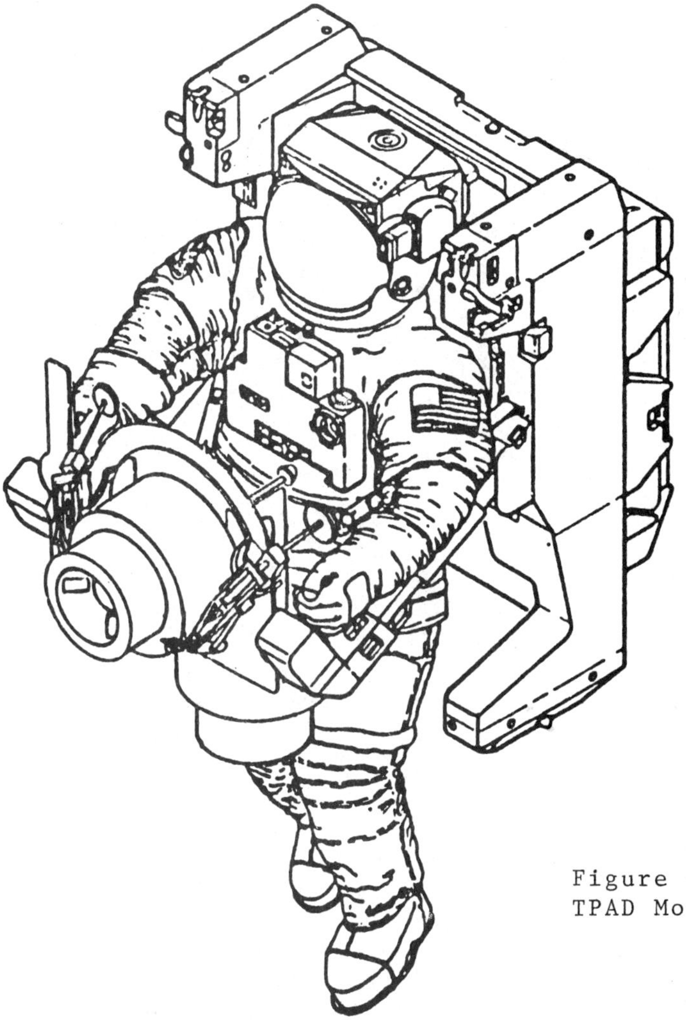

Figure 8
TPAD Mounted on the MMU

The trunnion pin attachment device, or TPAD, is illustrated in figure 9. The primary purpose of the attachment device is to allow the astronaut flying the MMU to attach to the Solar Maximum satellite at one of the hollow, threaded metal tubes (trunnion pins) on the satellite. Once firmly attached to the trunnion pin, the MMU and Solar Maximum are essentially one body, and the MMU control systems can effectively stop the rotational motion of the Solar Maximum.

Before attachment can occur, however, the MMU must fly the 200 feet to the satellite and match the one degree per second rotation rate of the satellite. During the rate matching operation, the "axis inhibit" attitude hold mode of the MMU may be engaged.

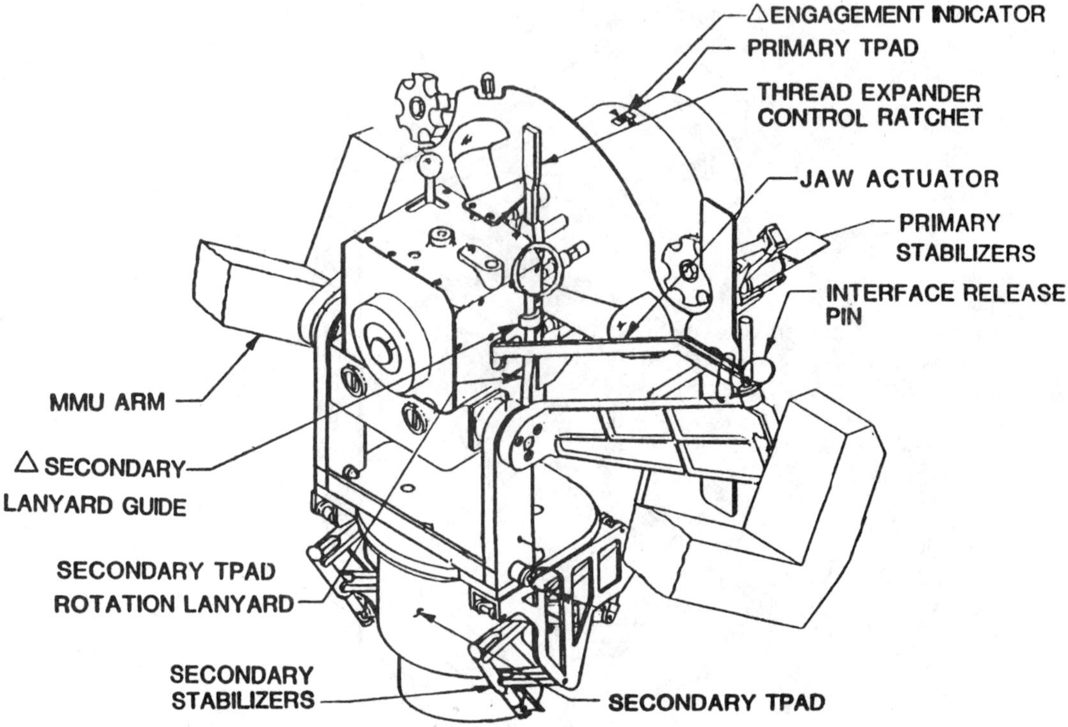

Figure 9
TPAD Assembly

Matching the rotation rate of the satellite is one of the key features of the MMU. If the MMU were not able to accomplish this feat, the rescue and on-orbit repair of the Solar Maximum might not be a feasible operation. The orbiter maneuvering systems do not allow the orbiter to accomplish a rate match at the angular velocity and with the precision required. If the orbiter is stationary and the manipulator arm is used to grapple the 5000 pound rotating body, the possibility of damage to the satellite and/or to the orbiter would be increased, and the time required to effect attachment could be prohibitive.

Once attached to the satellite, the MMU pilot will use the switch-selectable MMU attitude hold mode specifically designed to stabilize spinning satellites by firing additional MMU thrusters to compensate for the large inertia and mass offset in the center of gravity of the combined MMU and Solar Maximum. When the satellite rotation has stopped, the MMU will remain attached to the satellite. The MMU will align the Solar Maximum with the orbiter, and provide any additional stabilization required while the Solar Maximum is being attached to the manipulator arm. After attached to the manipulator arm, the satellite will be placed into a special repair fixture in the orbiter payload bay. The repair fixture, known as a Flight Support Station, is illustrated in figure 10. When the satellite is securely fastened into the repair fixture, the MMU pilot will disengage the satellite and return to the MMMU flight support station for doffing.

Dynamic MMU flight training for this mission is being done at the Space Operations Simulator (SOS) at Martin Marietta in Denver, Colorado.

The Space Operations Simulator allows high fidelity simulatons of gravity-free MMU flight. The SOS allows simulation of long range translations from the orbiter to the Solar Maximum satellite, final approaches (from approximately thirty five feet), rotational rate matching with target rotational rates from zero to five degrees per second, MMU dock with the satellite, payload bay MMU checkout flights in a full-scale payload bay mockup, and MMU malfunction training.

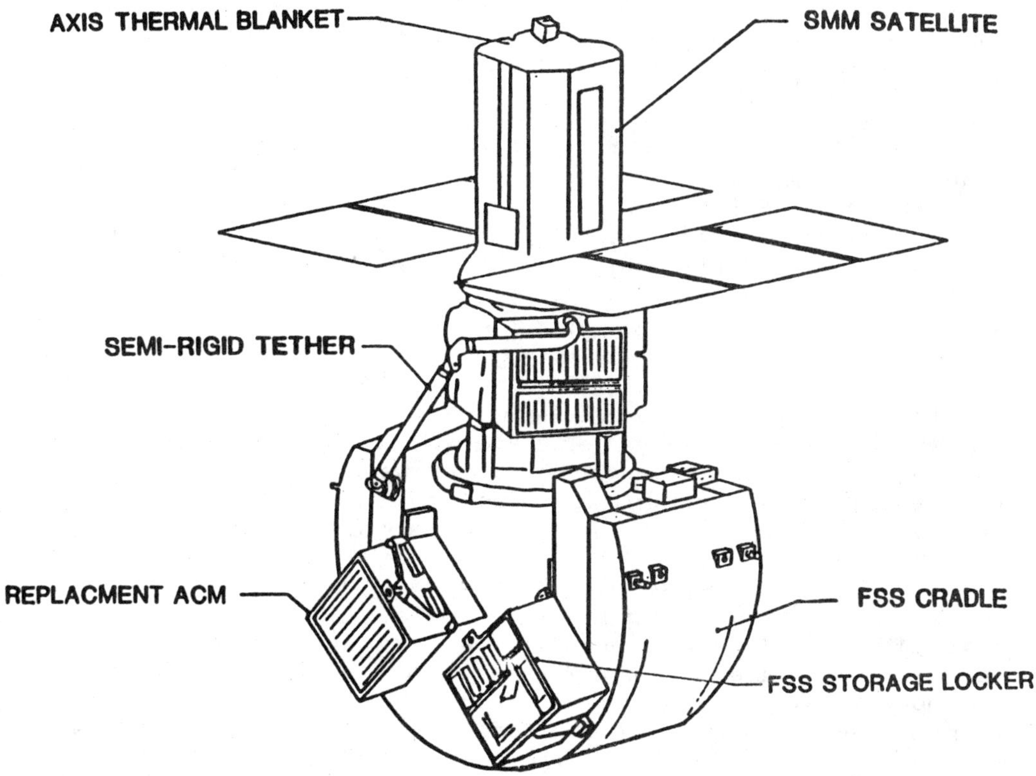

Figure 10

Solar Maximum Berthed in Flight 16 Support Station

The final approach, rate match, and dock simulations are accomplished using a man-rated six degree-of- freedom carriage with a high fidelity MMU trainer mockup, and a high fidelity rotating full-scale mockup of the Solar Maximum satellite.

The long range translation simulations are accomplished with a high- fidelity scale model of the Solar Maximum mounted on a three degree-of-freedom target gimbal, a video camera mounted on the six degree-of-freedom carriage, and a six-foot by nine-foot video screen with the pilot positioned in front of the screen. The pilot is placed in a mockup of the MMU. The relative motion seen by the MMU pilot on the display screen is of maximum fidelity.

The carriage, the three degree-of-freedom target gimbal, and the full scale satellite mockup are driven by computer-controlled servo systems. The computer is programmed to respond to MMU/pilot interfaces (switches, lights, gauge readouts), and to model orbital mechanics, centers of gravity, motion dynamics, and more than 200 MMU malfunctions.

Tests conducted at the Martin Marietta SOS have indicated that the translation from the orbiter to the Solar Maximum, dock, and stabilization will consume less than twenty percent of the available MMU fuel, and approximately thirty minutes of time.

Simulations have also been done to determine the feasibility of using the MMU to maneuver the Solar Maximum into the repair fixture in the payload bay. Preliminary results of these simulations indicated that the MMU is capable of accomplishing this task as well.

MMU FUTURE CAPABILITIES AND ROLES

The current model of the MMU required a sixteen month development and delivery schedule. To meet the schedule, current space-qualified hardware and a conservative design were used to minimize qualification testing requirements. The result was a larger, heavier unit with relatively simplistic pilot displays; mechanical pressure gauges and two LED thruster cues.

An MMU having several enhancements over the current version has been on the drawing boards at Martin Marietta since 1976. The enhanced MMU features are integral to any near term upgrades to the current MMU, in particular the microprocessor based control electronics assembly. The microprocessor would reduce the pilot workload and allow more flexibility in task performance. The thruster selection logic would be automatically altered for peak performance by the microprocessor when the MMU is transporting payloads, such as beams for a large space structure.

Through an interface with the EMU, audible thruster cues could be linked into the astronaut communications system. A caution and warning system would provide MMU health and status information on the twelve character display on the EMU.

Physically, the MMU can be made smaller, lighter, and easier to maintain. Increased capacity in the nitrogen tanks, 4000 psi versus the present 3000 psi, would provide a larger delta velocity capability. This would enable longer and more comprehensive missions.

Long term improvements include a range and range rate display, selectable proportional rate control, sun sensor, and a horizon scanner. All of these additions contribute to the unique abilities of the current MMU in mission duration, efficiency of operation, and versatility.

The future roles currently envisioned for the MMU are numerous. Satellite servicing will become a reality in April, 1984. Routine and contingency servicing of the Space Telescope would be greatly simplified by the MMU. The Space Station program is dependent on the MMU. Structure assembly in space is another proposed use.

CONCLUSION

With the flight of STS-11, the MMU will evolve from an experiment to a useful, dynamic tool for orbital operations. Scientific investigations, observations, and rescue operations conducted with the MMU will increase mans potential in space. In fact, the usefullness of the MMU is limited only by the imagination.

BIBLIOGRAPHY

1. The MMU Systems Data Book, Martin Marietta Corporation, Denver Aerospace Division, August, 1983

AAS 84-033

LASER DOCKING SYSTEM

Harry O. Erwin *

The Laser Docking System (LDS) consists of passive docking aids (reflectors) placed on the target vehicle in a known location and orientation. These reflectors are acquired and tracked by means of a modulated laser beam located on the interceptor vehicle. The LDS enables the interceptor vehicle to analyze the return (reflected) signal in order to determine both relative position and relative attitude of the target vehicle during stationkeeping and docking.

Laser ranging experiments have been accomplished at NASA, and from these experiments have evolved laser docking concepts. The concepts include angle and attitude measurement which are capable of providing all of the information needed for automatic docking control by the interceptor vehicle. Several designs are being compared. Plans include the development and testing of: first a breadboard model, then an engineering model, and finally qualification and flight systems.

* *Head of Microwave and Laser Section, NASA Johnson Space Center, Houston, Texas 77058*

INTRODUCTION

Future Earth-orbital operations will require sensors capable of supporting rendezvous, stationkeeping, and docking between various vehicles, satellites and space stations.

There are several terms used when discussing rendezvous, stationkeeping, and docking which require definition. Generally, these operations involve two vehicles: a passive vehicle which does nothing other than maintain its present state, and an active vehicle which moves to effect the operation. The passive vehicle is called the target and the active vehicle is called the interceptor.

The target vehicle may be cooperative or noncooperative. A cooperative target assists the interceptor by helping it to obtain the information needed for the intended operation. The target can help by carrying aids such as reflectors, or transponders, or by making measurements with its own sensors and transmitting the information to the interceptor. Passive aids are those that require no power from the target. Passive aids are preferable but may not always be feasible.

Rendezvous is the maneuvering of the interceptor into the same orbit and phase as the target. Rendezvous requires that the interceptor match both the target's position and velocity as opposed to interception, in the militiary sense, which merely requires that the positions be identical.

For our purpose, a rendezvous maneuver can be considered to have two steps. Initially, the target is beyond the range of the interceptor's sensors. The interceptor obtains the target's state either from a third party (such as a ground station) or from the target itself. The interceptor then calculates a trajectory designed to affect a rendezvous, based upon knowledge of both its own and the target's states. Due to various errors, this rendezvous maneuver will not deliver the interceptor to the precise position, relative to the target, that is desired.

Once the target comes within range of the interceptor's sensors, the second phase of the rendezvous begins. The interceptor now conducts maneuvers based upon direct measurements of the relative position of the target. The errors in the original trajectory are corrected and the rendezvous is completed when the two vehicles are within about 100 meters of one another.

Rendezvous requires that the relative range and bearing (and rates) of the two vehicles be known, but does not require knowledge of their relative attitude.

Docking is the physical joining of the interceptor and the target. Since each vehicle must contain a docking mechanism of some sort (even if it's just a grappling fixture), docking requires a cooperative target. A cooperative target is also required because successful docking requires data accuracy beyond that achievable without aids.

There are two types of docking mechanisms: hard (impact) and soft (non-impact). Hard-docking mechanisms can tolerate nonzero impact velocities. In fact, hard-docking mechanisms are designed to use the energy of impact to latch up. Soft docking mechanisms are designed to operate with a zero-impact velocity.

In the past, hard-docking mechanisms have been employed. These are suitable for the docking of two small - or medium-size vehicles. In the future, we can expect docking between large, medium, and small vehicles in all possible combinations and, in many cases, a soft-docking mechanism will be required.

Docking requires that the target's range, bearing, and attitude (and rates) be known. This is in contrast to rendezvous which does not require attitude. The maximum range for the docking data is 100 meters, which is the range at which the docking maneuver is expected to begin. The docking maneuver continues until physical contact is made between the two vehicles.

Stationkeeping is the maintenance of constant relative positions, and possibly attitudes, between the interceptor and the target. Stationkeeping would normally take place within 10-1000 meters of the target. It is assumed that any sensor that provides data adequate to support rendezvous and docking will also support stationkeeping

Table 1 shows the types of space activities which require sensors such as the Laser Docking Sensor for measurement of position or attitude.

TABLE 1

CAPABILITIES REQUIRED FOR FUTURE MISSIONS WHICH CAN BE MET BY LASER DOCKING SENSOR

	RENDEZVOUS	STATIONKEEPING	DOCKING
Deployment		X	
Construction	X	X	X
Operation	X	X	X
Inspection	X	X	
Repair	X	X	X

Retrieval	X	X	X
Cargo Transfer	X	X	X

"X" DENOTES NEED FOR SENSOR MEASUREMENT OF POSITION OR ATTITUDE

ATTITUDE DETERMINATION

Since docking requires that the interceptor be able to determine the relative attitude of the target, some consideration must be given to ways to accomplish this.

There are only two ways to determine attitude within the passive-aid constraint. The first method determines attitude from range and bearing measurements to three (or more) reflectors in a known configuration on the target. The second method determines attitude from bearing-only measurements of four (or more) reflectors in a known configuration on the target.

The first method (range and bearing measurements) is best because range measurement capability is needed anyway (to support rendezvous), the required bearing accuracy is less than required by the second method, and fewer reflectors are needed. The coplanar docking aids (reflectors) are assumed to be placed in a known arrangement around the circumference of a 1-m-diameter, or smaller, circle, as shown in Figure 1.

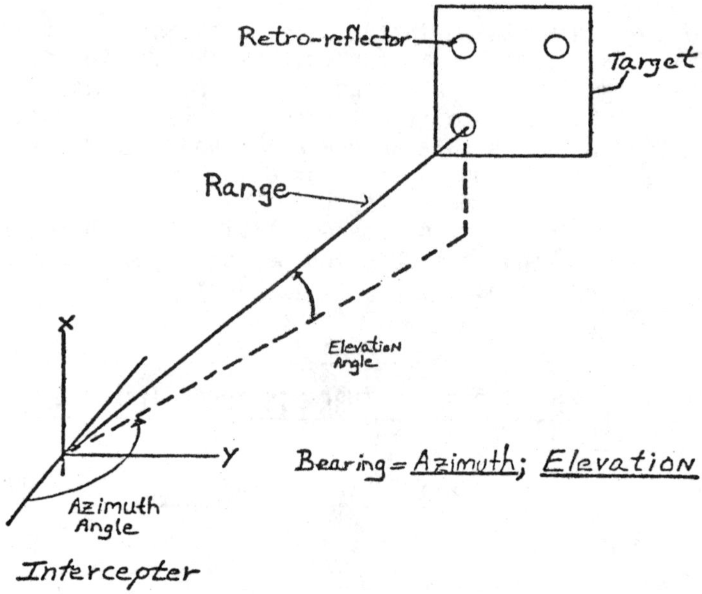

FIGURE 1

TARGET GEOMETRY

Calculating Attitude From Range And Bearing

The range, R_i, and bearing, θ_{ei} and θ_{ai}, of each reflector in the array are measured and then the attitude of the target is calculated from this data. There are several ways to describe the relative attitude of two vehicles. Euler's angles are frequently used in analytical mechanics but for our purposes, pitch, θ_p, yaw, θ_y, and roll, θ_r, as shown in Figure 2, are employed.

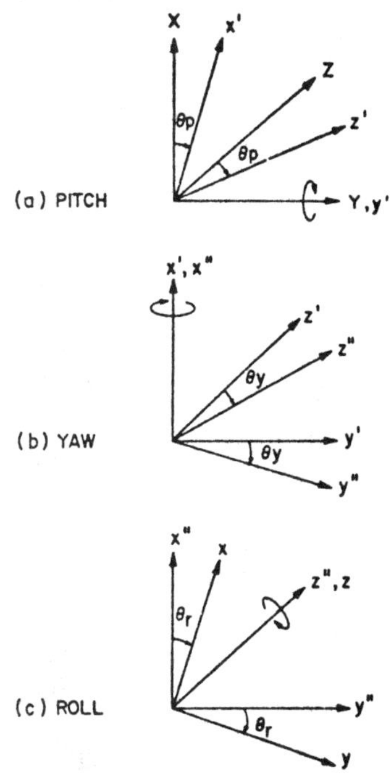

Figure 2

Attitude Angles

The relationship between the coordinates of a point in the x, y, z reference frame and in the X,Y,Z frame are:

where
$$A_1 = \begin{bmatrix} \text{Cos}\ \theta_p & 0 & -\sin\theta_p \\ 0 & 1 & 0 \\ \sin\theta_p & 0 & \cos\theta_p \end{bmatrix}$$

$$A_2 = \begin{bmatrix} 1 & 0 & 0 \\ 0 & \cos\theta_y & -\sin\theta_y \\ 0 & \sin\theta_y & \cos\theta_y \end{bmatrix}$$

$$A_3 = \begin{bmatrix} \cos\theta_r & -\sin\theta_r & 0 \\ \sin\theta_r & \cos\theta_r & 0 \\ 0 & 0 & 1 \end{bmatrix}$$

$$\begin{bmatrix} X \\ Y \\ Z \end{bmatrix} = A_1 \begin{bmatrix} x' \\ y' \\ z' \end{bmatrix}$$

$$\begin{bmatrix} x' \\ y' \\ z' \end{bmatrix} = A_2 \begin{bmatrix} x'' \\ y'' \\ z'' \end{bmatrix}$$

$$\begin{bmatrix} x'' \\ y'' \\ z'' \end{bmatrix} = A_3 \begin{bmatrix} x \\ y \\ z \end{bmatrix}$$

and
$$\begin{bmatrix} X \\ Y \\ Z \end{bmatrix} = A \begin{bmatrix} x \\ y \\ z \end{bmatrix}$$

$$A = A_1 A_2 A_3$$

$$= \begin{bmatrix} \cos\theta_p \cos\theta_r - \sin\theta_p \sin\theta_y \sin\theta_r & -\cos\theta_p \sin\theta_r - \sin\theta_p \sin\theta_y \cos\theta_r & -\sin\theta_p \cos\theta_y \\ \cos\theta_y & \cos\theta_y \cos\theta_r & -\sin\theta_y \\ \sin\theta_p \cos\theta_r + \cos\theta_p \sin\theta_y \sin\theta_r & -\sin\theta_p \sin\theta_r + \cos\theta_p \sin\theta_y \cos\theta_r & \cos\theta_p \cos\theta_y \end{bmatrix}$$

If there are two vectors, P_0P_1 and P_0P_2, having a known orientation in the X, Y, Z frame, as shown in Figure 3 then we can determine A as

$$A = \begin{bmatrix} (X_1 - X_0) & (X_2 - X_0) & (X_3 - X_0) \\ (Y_1 - Y_0) & (Y_2 - Y_0) & (Y_3 - Y_0) \\ (Z_1 - Z_0) & (Z_2 - Z_0) & (Z_3 - Z_0) \end{bmatrix} \times \begin{bmatrix} x_1 & x_2 & x_3 \\ y_1 & y_2 & y_3 \\ z_1 & z_2 & z_3 \end{bmatrix}^{-1}$$

where

$$\begin{bmatrix} x_3 \\ y_3 \\ z_3 \end{bmatrix} = \begin{bmatrix} x_1 \\ y_1 \\ z_1 \end{bmatrix} X \begin{bmatrix} x_2 \\ y_2 \\ z_2 \end{bmatrix}$$

The values of the elements of A may be all we need. If the actual angles θ_p, θ_y, θ_r, are needed they can be determined numerically.

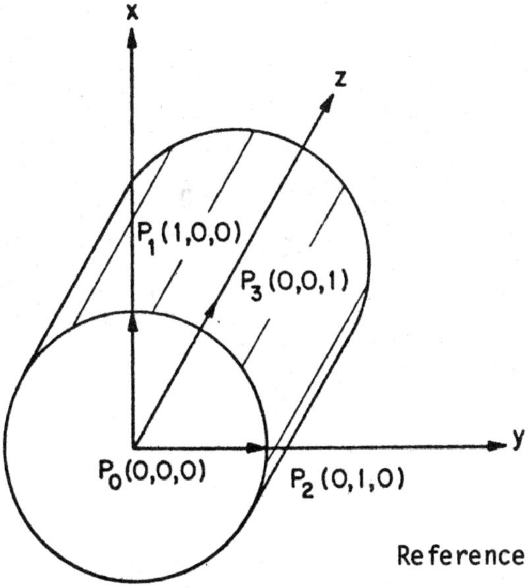

Figure 3

Reference vectors for attitude determination.

DOCKING SENSOR DESIGN

The preceeding section defined the measurements which must be taken in order to determine the position and attitude of a target vehicle. To turn these requirements into hardware is the next logical step. There are a variety of sensors that can be used to measure range and bearing. The field can be narrowed however by examining the performance requirements, the conditions of use, and the available technology.

Illumination Source

An active source is needed to transmit energy to the target. This transmitter has a beamwidth requirement that is determined by Figure 4.

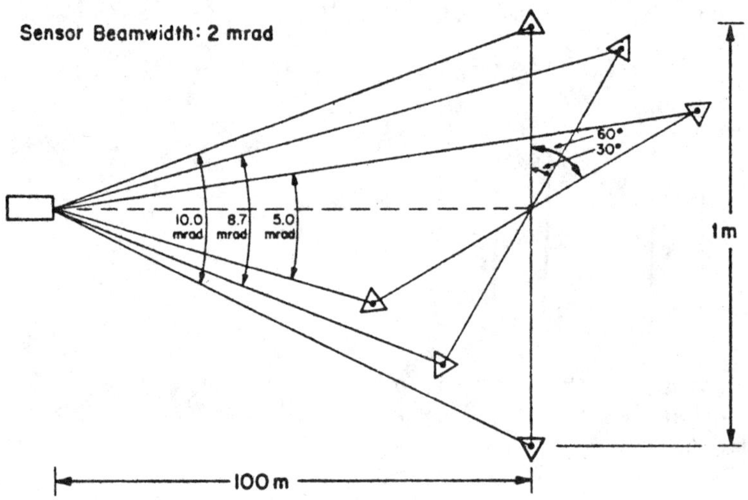

Figure 4

Angular separation at 100 m of reflectors spaced 1 m apart.

It can be seen that to resolve the three targets at 100 meters with tilt angles up to 60 degrees requires a beamwidth of smaller than 2 mrad. Figure 5 shows that for a reasonable sized transmitter aperature, the transmitter wavelength must be in the laser region of the electromagnetic spectrum to have a beamwidth of 2 mrad.

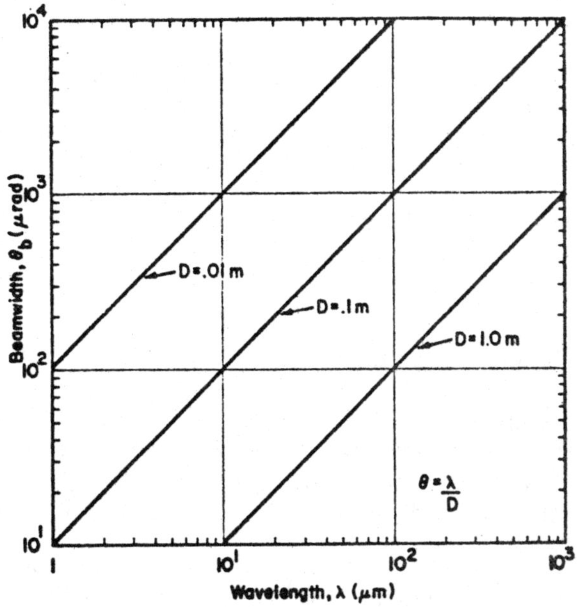

Figure 5

The beamwidth of the radiation from a uniformly illuminated circular aperture.

There are only three types of lasers that have the output power level, efficiency, and lifetime required for an optical radar: the CO_2 laser, and Nd:YAG laser, and the (GaAlAs) semiconductor laser.

The characteristics of these lasers are compared in Table 2. The semiconductor laser is considered superior to the other two because it does not require active cooling (like the Nd:YAG laser does for all but the lowest power levels), its associated detector does not require cooling (as do those associated with the CO_2 laser), it has high efficiency (unlike the Nd:YAG laser), and it does not require an external modulator (as do both the CO_2 and Nd:YAG lasers). The one disadvantage of the semiconductor laser is the present maximum output power limit of about 50 mW. However, this level will undoubtedly be exceeded in the future and, if necessary, the outputs of several lasers can be combined to reach higher power levels.

Table 2

Candidate Lasers for the R&D Sensors

Laser	Wavelength	P^*_{max}	Mode	Efficiency	Lifetime	Detection	Detector	Modulation
CO_2	10,600 nm	100 mW	cw, Pulse	20 %	10^4 h	Heterodyne	Cooled	External
Nd:YAG	1,060 nm	10 mW	cw, Pulse	1 %	10^4 h	Direct	Uncooled	External
GaAlAs	830 nm	50 mW	cw, Pulse	10 %	10^5 h	Direct	Uncooled	Direct

*Without active cooling

Ranging Modulation

The transmitted signal must be modulated so that the range to the target can be determined. Since we have selected a directly modulated semiconductor laser as the trasmimitter source, only intensity modulation (IM) can be employed.

Tone ranging has been selected for the docking sensor because it is practical, relatively simple, has no minimum range requirement, and has the necessary precision.

Tone Ranging is described in Figure 6. It is necessary to modulate the outgoing laser energy with more than one tone to avoid ambiguous range readings. The phase comparison of the long wavelength modulation gives approximate range while the short wavelength modulation gives accuracy.

Figure 6

TONE RANGING

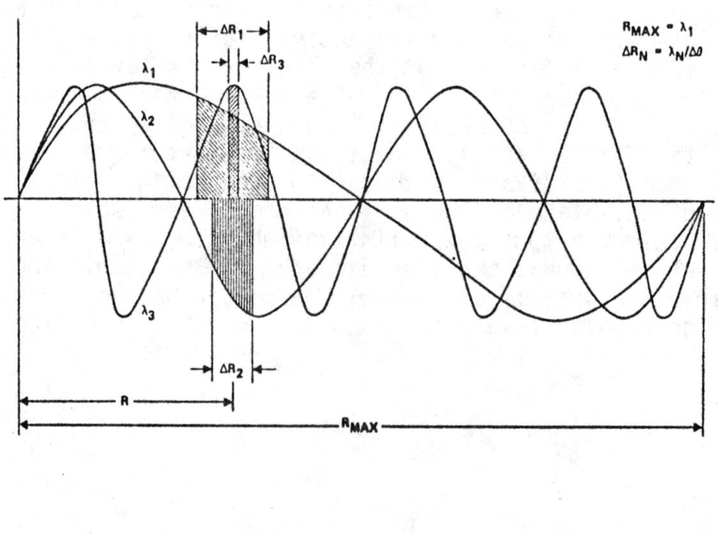

Transmitter Beam Steering

To point a narrow transmitter beam anywhere in a target field of view requires a device similar to the ones shown in Figures 7 and 8.

Figure 7

Galvanometer Deflector

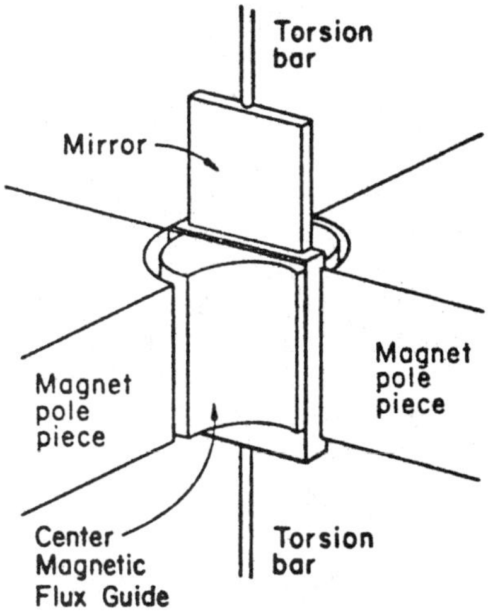

Figure 8

Piezoelectric Deflector

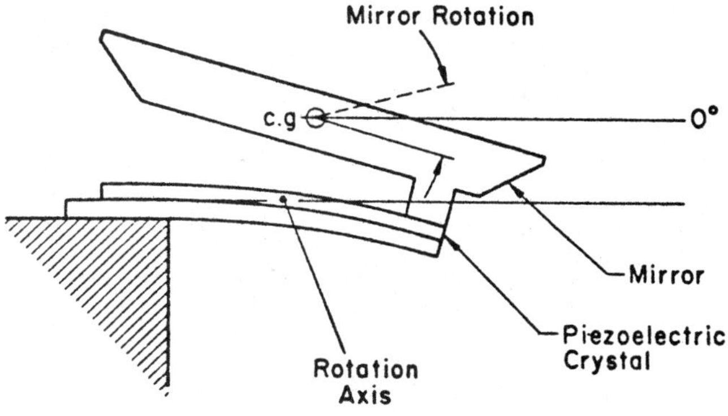

The device used in the Laser Docking System breadboard is a pair of digital to analog driven galvanometers with mirrors on the orthogonal axes.

Receiver Image Steering

A receiver scanning mechanism must be able to:

- Scan a narrow (2 mrad) beam almost instantaneously, at random over a 0.5-rad fov. (At close ranges the individual reflectors will lie at the extremes of the fov.)
- Have a large enough aperture to gather a useable signal at the maximum range.
- Operate in the portion of the spectrum around 830 nm.

Within the selected spectral range there are only two scanning mechanisms that can move from point to point in the fov quickly enough. The first method, mechanical beamsteering, uses mirrors mounted on piezoelectric deflectors or galvonometers to steer the optical beam; the sensor itself does not move. The second method, electron beamsteering, uses magnetic coils to steer the photoelectrons emitted by a photocathode which has the sensor fov imaged upon it. This is the method used in the image-dissector receiver.

Depending upon the requirement for aperature and field of view either the mechanical scanning of a receiver mirror or the electronic scanning of an image dissector field of view is necessary. The deciding factor on which of these types of receiver scanning to use is max range to the target. The light gathering ability of the small mechanically scanned mirrors limits this usage to short ranges. For docking purposes either type of receiver scanning is acceptable, although sharing a scanned mirror with the transmitter has obvious advantages.

Reflectors

Reflectors are required on the target vehicle both to increase the strength of the returned signal and to give the sensor distinct points of known positions to track. Cube-corner (or corner-cube) reflectors, as shown in Figure 9, are used.

Figure 9

Cube-Corner Reflector

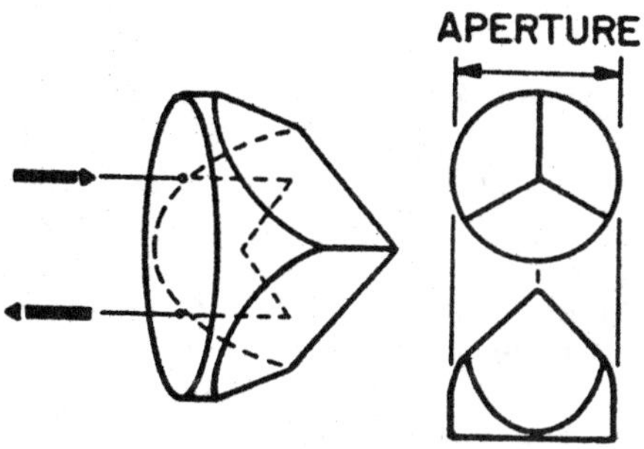

Total System

A potential system block diagram is shown in Figure 10. There are several variations possible in this implementation but the basic concept of using a narrow beam laser to determine angle and a tone ranging circuit to determine range still apply.

Figure 10

A Possible Laser System Block Diagram

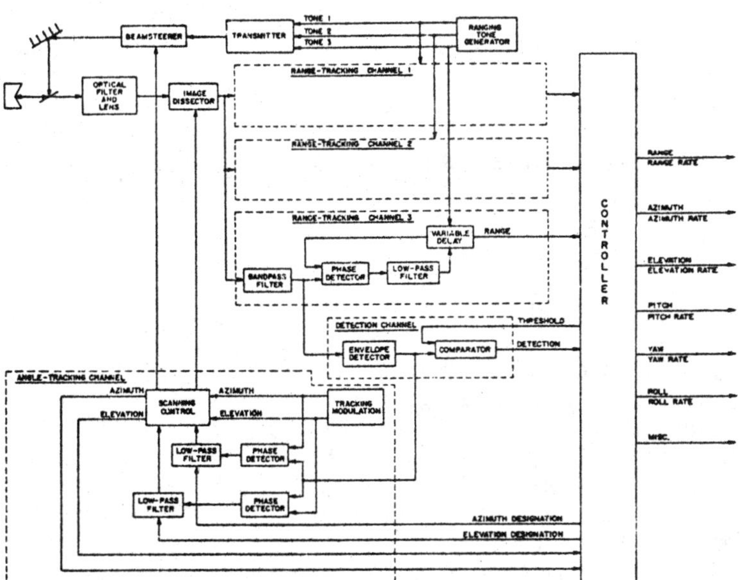

This system has been breadboarded in the Laser Labs at the Johnson space Center and has been tested and improved several times.

FLIGHT EXPERIMENT

The purpose of a planned flight experiment of the Laser Docking System is to demonstrate a laser system capable of measuring position and attitude between two stationkeeping or docking vehicles. The laboratory developed system will be upgraded to flight demonstration quality and attached to the Orbiter at a convenient location in the payload bay. Small passive retro-reflectors will be placed on the targets to be docked with. The Laser Docking System will then track these reflectors in angle and in range and this information will be used to calculate the complete position and attitude needed to perform an automatic docking or stationkeeping. Figure 11 shows a typical configuration for an automatic stationkeeping formation.

Figure 11

Automatic Stationkeeping With A Target

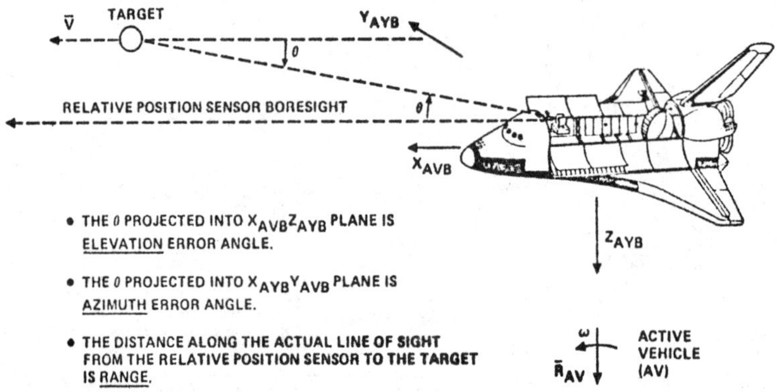

WHY A LASER DOCKING SYSTEM?

The precision of measurements required for automatic dockings is possible only with a laser type of system. Automatic stationkeeping and docking capability needs to be developed because it will save fuel and crew time and will improve the safety of these maneuvers. Using a sensor such as the Laser Docking System to stationkeep frees the Orbiter Ku-band Rendezvous Radar/Communications System for data transmission. If the Ku-band system is being tied up tracking, the communications channels through TDRS at S band have very limited data capability.

Another benefit of a laser based docking sensor is that the small size, weight, and power of such a sensor will allow it to be used on smaller vehicles such as the Orbital Maneuvering Vehicle (OMV). A microwave system even if it could solve the target attitude measurement requirement would impose a greater impact on a small vehicle than an optical system. The optical wavelength also makes possible smaller retro-reflectors to be placed on the targets. A standard configuration for an array of these reflectors to allow attitude determination needs to be defined and applied to all future retrievable targets.

AAS 84-034

RENDEZVOUS AND DOCKING WITH REMOTE PILOTED VEHICLES

James D. Micheal*

The first demonstration, by NASA, of a successful rendezvous and docking of two vehicles in space was achieved by two Gemini spacecraft in July 1966. This paved the way for the successful rendezvous and docking missions of both the Apollo and Skylab programs. All of these docking missions have involved piloted spacecraft, however, and the capability to dock a remote piloted vehicle to another unmanned spacecraft has not been established by flight test, thus far, by the U.S. The first occasion for entertaining such a docking requirement was in 1977 when a Teleoperator Retrieval Vehicle (later named the TRS) was considered for rescue of the Skylab. The TRS program was well into the development phase when it was canceled because the Skylab orbit decay advanced more rapidly than originally predicted, and it became axiomatic that the Shuttle would not be ready in time. Now, with the advent of the Orbital Maneuvering Vehicle (OMV), rendezvous and docking with remote piloted vehicles resurfaces as a major technology item. The man-in-the-loop control system requirements for such a system is the subject of this paper. Since so many similarities exist between the TRS and the unfolding OMV concept, a review of the TRS control system baseline along with selected design trades which led to that baseline will be discussed. The degree to which the TRS control system requirements and baseline apply to the OMV will be addressed followed by a concluding discussion of the major issues currently facing the control system designer on OMV.

* *NASA Marshall Space Flight Center, Huntsville, Alabama 35812*

INTRODUCTION

The Teleoperator Retrieval System (TRS) program which began in 1977 was to provide a remote piloted vehicle capable of rendezvousing and docking with the Skylab. Subsequent to the dock, the TRS was to then either reboost the Skylab to a higher orbit or deboost it to a desirable reentry position. The TRS vehicle was a 10,000 pound class vehicle equipped with a compliment of 24 RCS engines for manual control, 32 kit engines off modulated for automatic control, strap down gyros, and a TV camera to provide target viewing. The TRS was to be flown from a control station situated in the Orbiter aft flight deck and was to be the first U.S. sponsored docking of two vehicles in space involving a remote piloted vehicle. Canceled in January 1979, due to a slippage in the Shuttle schedule, the TRS was well into the development phase with some hardware already fabricated. For example, the man-in-the-loop control system design was essentially complete, supported by extensive simulation with actual crew member participation at both MSFC and the prime contractor site. This work remains cogent today because there has been continued interest in the TRS functional concept since its cancellation. Over the past four years, studies at MSFC have evolved an Orbital Maneuvering Vehicle (OMV) concept. The OMV, though dissimilar in configuration and appearance (disc shaped as opposed to the rectangular solid TRS) to the TRS, is functionally similar in many respects and acutely congruent with regard to man-in-the-loop control. Many of the same issues affecting the man-in-the-loop control system design on TRS are being raised in connection with the OMV. It therefore behooves the control system designer on OMV to familiarize himself with control system baselined for the TRS,[2] as well as the issues, and considerations which led to that baseline.

TRS PROGRAM ISSUES RELEVANT TO THE MAN-IN-THE-LOOP CONTROL SYSTEM DESIGN

- Thruster Size.
- Communication Time Delays and TV Bandwidth Compression.
- Range/Range Rate Radar.
- Tumbling Targets.
- Shimmed Docking Interface.
- Control System Definition.

Thruster Sizing

Sizing the RCS thruster for the TRS vehicle as well as an OMV establishes the acceleration capability for the vehicle which in turn dictates to a large extent the vehicle handling qualities. Determining the acceleration capability of a vehicle is an iterative process involving trades of vehicle thruster size, location and population, fuel requirements, etc. The resulting vehicle handling qualities must also be acceptable to the pilot or payload specialist who will be flying the mission. Pilot evaluation of the handling qualities of many different kinds of vehicles over the years have yielded an acceleration capability envelope which is used as a design guideline. As the figure shows, vehicles with RCS control, such as the Shuttle (large thrusters), Apollo, Gemini, MTV, and TRS have exhibited acceleration capabilities which fall for the most part within this envelope. Current MSFC OMV baseline calls for 24 RCS thrusters (8 orthogonal triads) each with 15 pounds thrust. These thrust levels produce acceleration capabilities and projected manual mode handling qualities well within the acceptable range. However, as competition heightens and the OMV program progresses, the tendency will be to lower these thrust levels, and in fact some consideration is presently being given to 5 pound RCS engines. These thrust levels produce an acceleration capability which falls outside the recommended envelope (see Fig. 1) and is not far removed from that which was initially set for TRS. The initial TRS RCS engines were sized at 2.5 pounds. Simulation results revealed a vehicle with a very "sluggish" response making emergency and/or evasive maneuvers difficult to perform. From the pilot's standpoint, a vehicle this slow

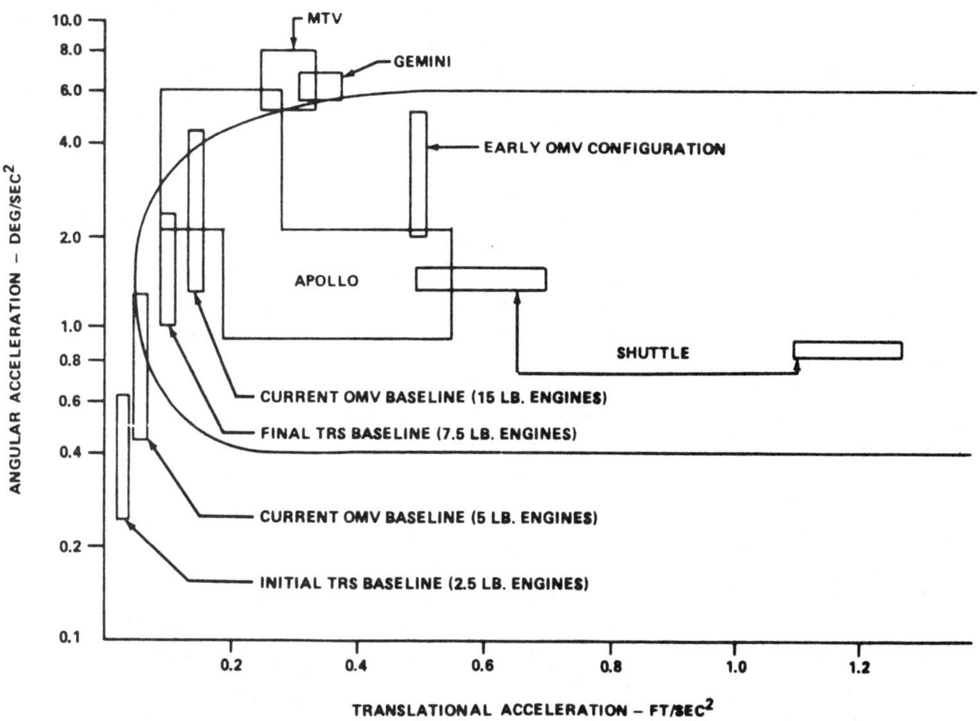

Fig. 1 Acceleration Capability

to respond means there will be a significant time lag between when the pilot inputs a command to the vehicle to the time he detects the resulting vehicle motion on the TV monitor. As a result, even the experienced pilot would often over command the vehicle, an especially critical error if done in the last few seconds of the docking sequence. Since further time lags will be introduced through communication delays for OMV, careful examination of the thruster sizing requirement becomes essential.

Communication Time Delays and TV Bandwidth Compression

While low acceleration levels produce apparent time lags for the pilot as he views the TV image of the target, communication time delays and lower TV frame rates represent actual time lags. The actual communication delay time is estimated at between 1.5 to 2 seconds and occurs primarily from accumulated delays endemic to the ground network. Control of the TRS from the ground was considered early in the TRS program and simulation results indicate that a pilot could accommodate this kind of delay while flying a docking mission with a three-axis stabilized target. Assuming a reasonable acceleration capability for the OMV, these results suggest that an OMV could also accept these ground network delays. Verification of OMV performance in this regard is needed and should further take into account the variability of the delay and its affect on OMV applications involving disabled or tumbling spacecraft.

Another communication requirement which could influence the man-in-the-loop control system design on OMV is that of TV bandwidth compression. This was not an issue during the course of the TRS program. But, following the reentry of the Skylab and prior to cancellation of the TRS project, TV bandwidth compression was considered as means of increasing the TRS to target vehicle acquisition range without accompanying increases in either power or antenna requirements. Simulation results showed that a pilot could fly a docking mission with a baselined TRS vehicle and a target vehicle undergoing moderate tumble (less than 1 deg/sec) with TV frame rates down to 5 frames/sec. Indications are that a suitably equipped OMV could accept similar frame rate reductions. What needs to be confirmed, however, through simulation, is that the combination of communication time delays and TV bandwidth compression still comprise a tractable problem for the OMV.

Range/Range Rate Radar

The most controversial issue raised during the TRS program that affected man-in-the-loop control appertained to the radar requirement. This radar was to provide chase (TRS) to target (Skylab) vehicle range and range rate information for control during the flyover from approximately 500 ft through the closure and dock phase. The focus of the controversy centered around the closure and dock phase, for there was more of a general agreement that the flyover phase could be successfully completed without the assistance of radar. Extensive simulation studies were commissioned at MSFC and at the prime contractor site. Astronauts Jack Lousma, Vance Brand, and Karol Bobko participated in these studies which sought to define the pilot's performance in flying the final closure phase with and without radar. These studies conclusively showed that pilot performance was enhanced when provided with range and range rate information (range rate data being the most critical). Improved docking accuracies and increased probability of a successful dock; higher pilot confidence level; and the ability to accommodate targets undergoing tumbling rates in excess of 1 deg/sec were achieved when radar was added. However, these studies also convincingly showed that there was a high probability of executing the final closure phase without radar augmentation. For this reason and the fact that radar implementation would be a long lead-high cost item, it was not baselined for TRS. Though not approaching the controversial stage as yet, disparate views prevail with regard to this issue on OMV. Much can be gained by reviewing the TRS studies, though a decision on such a critical issue as the radar requirement could be premature and impolitic at this time in the program. Considering the envisioned multimission role of the OMV, an expanded inquiry as to the projected need for radar will become necessary.

Tumbling Targets

Following the final Skylab mission in 1974, Skylab was placed in a gravity gradient stabilized configuration. Early in the TRS program (1977), Skylab was thought to be still g-g configured. As a consequence, initial TRS concepts and studies reflected a requirement for docking with three-axis stabilized, nonspinning spacecraft. Although these early studies showed that increased acceleration capability, decreased communication delay times, added control augmentation (to be discussed below) and added instrumentation such as radar, all increased performance; with enough training and time (mission time), even an under powered and relatively ill-equipped chase vehicle could be maneuvered to a dock with a stable target. Renewed observations of the Skylab during the first quarter of 1978, however, revealing a coning Skylab out of its g-g orientation[3] led to a requirement on the TRS for accommodating target vehicle spin rates of 1 deg/sec in roll and .5 deg for in pitch and yaw, respectively. The impact of this new requirement was felt immediately: low thrust levels which at first were manageable through training became unmanageable, a control system designed for stabilized target situations became inadequate; time delays became more critical; etc. In general, the closure and dock phase of the mission became more problematical for both the designer and the pilot. The problem, however, was not an insoluble one for TRS, a satisfactory design was developed, and neither should it be for OMV. Though the initial OMV concepts and proposed first missions do not involve dynamic targets, eventual use of the OMV for capture and retrieval of disabled spacecraft is planned. Considering the potential impact of the tumbling target requirement on the man-in-the-loop control system as well as other system elements, factoring in as much capability to the design of the initial concept as possible will serve to produce a more cost effective and a more versatile OMV vehicle.

Shimmed Docking Interface

Another issue, holding relevance to future OMV development, was that of thrust vector alignment for the mated TRS/Skylab. The mated TRS Skylab principal X axis was offset from the docking axis (which was along the geometric center line) by some 10° and the combined center of gravity (C.G.) was some .8 meters above the center line.[4] The automatic control system required that the TRS, main engine thrust be directed at points on a circular path about the combined C.G. To satisfy this requirement the TRS docking ring was shimmed 4.3 degrees producing the alignment at dock illustrated in Fig. 2 below. Many other system elements were involved, but the problem faced by the man-in-the-loop control system was that the control axes and camera viewing axis were no longer aligned with the docking probe. To facilitate probe to drogue alignment during final docking maneuvers, two alternatives were considered: 1) To shim the camera into alignment with the probe; or 2) to displace the camera by the distance subtended at the probe end by the 4.3° shim. Parameterizations of this problem were simulated with astronaut participation resulting in selection of option 2. This insured placement of the end of the probe at the drogue opening when the target sighting aide TBar was aligned with the TV monitor reticle. In addition, orthogonality would be maintained between the viewing axis and the control axis. OMV will likely be faced with very much the same problem, but the solution must be adaptable to a variety of different configurations and circumstances. Knowledge of the ramifications of this issue will be important to the selection of a docking interface, propulsion system, control system (both automatic and man-in-the-loop) and other related OMV elements.

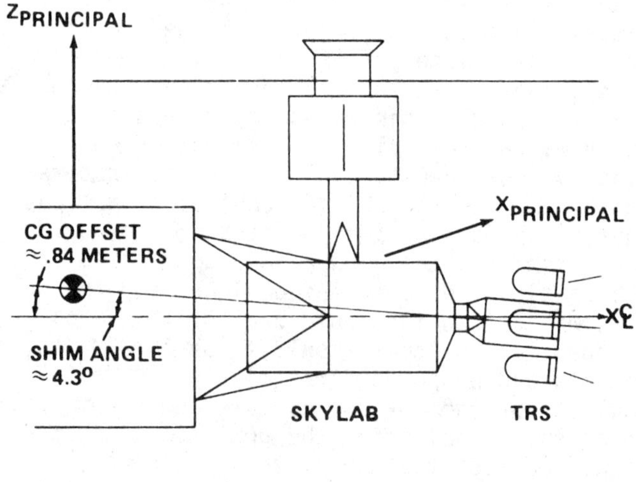

Fig. 2 TRS/Skylab Mated Alignment

Control System Design Issues/Baseline

Resolution of the thrust vector alignment problem, as well as other issues, spawned derived requirements for the control system that led to a continually increasing level of augmentation and complexity. The TRS was to have been flown from a control station residing in the Orbiter aft flight deck. The initial control system concept was a rudimentary, open loop acceleration commanded system for both translation and rotation with few pilot options or modes. As the TRS concept matured and as the derived requirements together with the self-imposed requirements became more demanding, the discrete system became inadequate. The closed loop rate proportional attitude system was incorporated to enable the pilot to better cope with tumbling targets. A rate commanded attitude with "rate hold" was found to be much more desirable in dealing with dynamic targets. In addition, operating the attitude system closed loop, allowing for automatic compensation of disturbances added a degree of automation which removed a large measure of the workload from the pilot. A pulse mode was added to provide higher resolution of control during final maneuvers just prior to dock by offering switch-selectable, variable pulse width, translational acceleration commands. To assist in station keeping operations. LVLH and inertial hold modes were added allowing for TRS maintenance of either local vertical or inertial attitude reference. The TRS man-in-the-loop control system baseline[2] that evolved illustrated is the functional block diagram of Fig. 3.

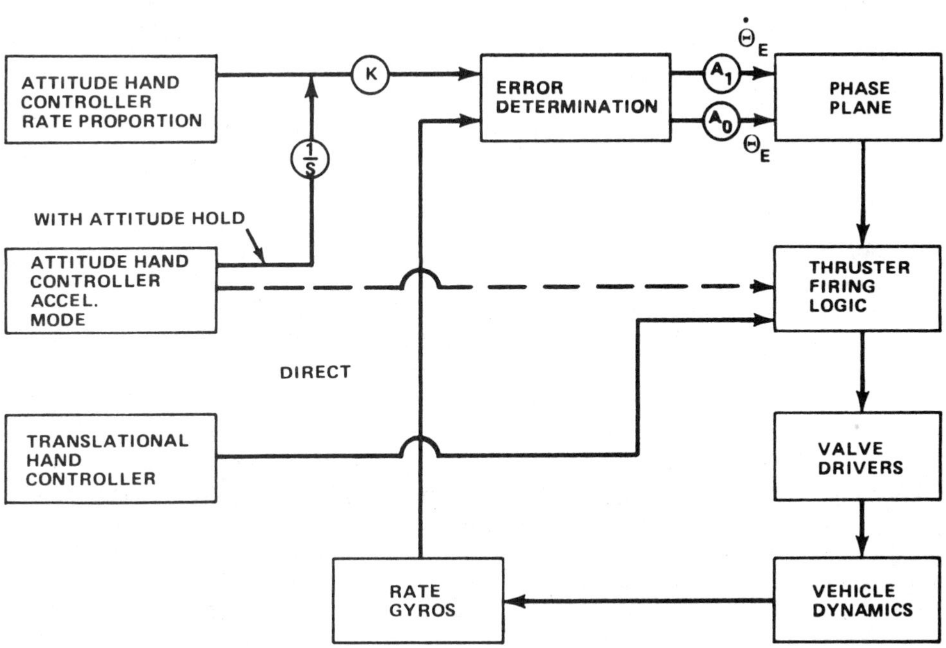

Fig. 3 TRS Control System Block Diagram

CSD Baseline Description

Attitude Hand Controller - In the rate proportional mode, the commanded angular rates are proportional to the control stick deflections related by a second order function. Rate hold capability was implemented by way of a switch on the hand controller. When the switch is engaged the commanded rates on all axes will be sampled and held until removed by pilot command. Except during attitude hold the acceleration mode operates as an open loop system. Three possible states exist for each axis (positive, negative, or zero) and are input directly to the thruster firing logic. When the attitude hold mode is selected, by actuating a switch on the controller, the command signal is integrated and fed through the phase plane for attitude hold control.

Translational Hand Controller - The translational hand controller was configured to operate in the acceleration mode. As above, there exist three possible states for each command axis. Outputs from the control stick are fed directly to the thruster firing logic to be processed simultaneously with the attitude command signals.

Attitude and Attitude Rate Error Determination - The TRS vehicle attitude errors were computed by differencing the commanded rates (desired) from the measured (actual) attitude rates.

$$\dot{\underline{\Theta}}_E = \dot{\underline{\Theta}}_A - \dot{\underline{\Theta}}_D \tag{1}$$

The attitude error signal is derived by integrating the time rate of change of the Quaternion[5] Q_{AD}. Quaternion Q_{AD} is a four element array relating the desired reference frame to the actual.

$$\dot{Q}_{AD} = \frac{1}{2}\,\overset{\approx}{Q}_{AD}\begin{pmatrix}\dot{\underline{\Theta}}_A\\ 0\end{pmatrix} - \frac{1}{2}\,\overset{\equiv}{Q}_{AD}\begin{pmatrix}\dot{\underline{\Theta}}_D\\ 0\end{pmatrix}^{\dagger} \tag{2}$$

$$Q_{AD} = \int \dot{Q}_{AD} \tag{3}$$

Assuming small angle approximations the first three elements of the Q_{AD} Quaternion are related to the attitude errors, Θ_E by"

$$\Theta_E(i) = 2 * Q_{AD(i)} \tag{4}$$

Phase Plane - The control system attitude phase plane comprises a functional logic methodology for the evaluation of vehicle attitude and attitude rate errors. The decision to issue a command and the nature of the command is determined by comparisons of the error signals to the gated logic deadband, see Fig. 4 below. The quadratic switching lines together with the phase plane constants $R_{(i)}$ (attitude rate limits) and $E_{(i)}$ (attitude error limits) establish the deadband boundaries. These

---

† Footnote: See notation table on page 13 for definitions of $\overset{\approx}{Q}_{AD}$ and $\overset{\equiv}{Q}_{AD}$.

boundaries are a function of vehicle acceleration capability and will, in general, be different for each control axis and flight regime. There were a total of 12 different sets of phase plane parameters for TRS including the automatic and storage mission phases.

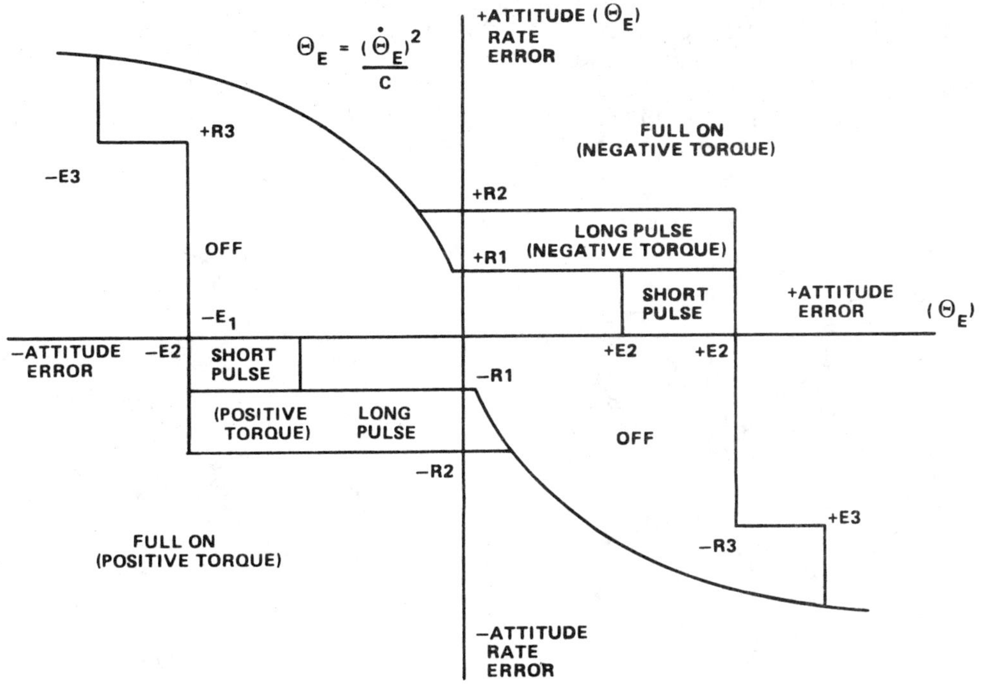

Fig. 4 Phase Plane Logic Diagram

Thruster Firing Logic - The Thruster Firing Logic (TFL) was designed to accept inputs from the control system phase plane, translational hand controller, and when operating in the discrete mode, the rotational hand controller. With six degrees of freedom each comprising three states (+1, -1, 0), the TFL algorithm must be prepared to process a total of 3^6 or 729 different command combinations. For purposes of providing constant major thrust vector acceleration and deceleration together with avoiding target thrust impingement, in close, X axis thrusters were reserved for X axis translation only. This effectively reduced the total number of remaining command combinations to 3^5 or 243-engine firings which populated a matrix, table look-up. With 24 thrusters to choose from, there was more than one possible solution for each command. The TRS TFL objectives were to provide maximum acceleration, whenever possible, by selecting four thrusters per command axis and choosing the axis which provided maximum torque. Provisions for inhibiting redundant firings and dealing with engine-out contingences were also incorporated.

TRS vs OMV Simulation Comparative Study

In an attempt to quantify comparisons between the TRS and the OMV with respect to the man-in-the-loop control problem. The TRS control system baseline was incorporated into a simulation at MSFC where direct comparisons of performance could be made. A general description of the vehicle models[6,7] used are given below in Fig. 5 (more recent OMV concepts are heavier with larger engines, but the geometry remains essentially the same) followed by a discussion of the study objectives, approach, and results.

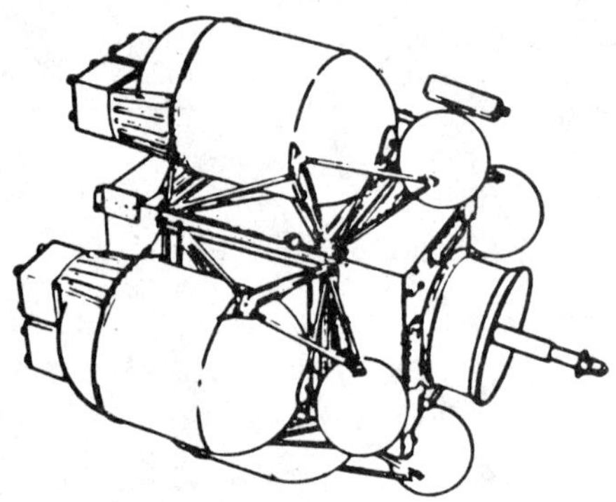

TRS CONCEPT

LENGTH = 79.7 INCHES
WIDTH = 47.0 INCHES
HEIGHT = 47.0 INCHES

WEIGHT (LBS)	INERTIA (SLUG–FT^2)		
	IXX	IYY	IZZ
9020.	3440.	2250.	2571.

THRUSTER COMPLEMENT
8 ORTHOGONAL TRIADS

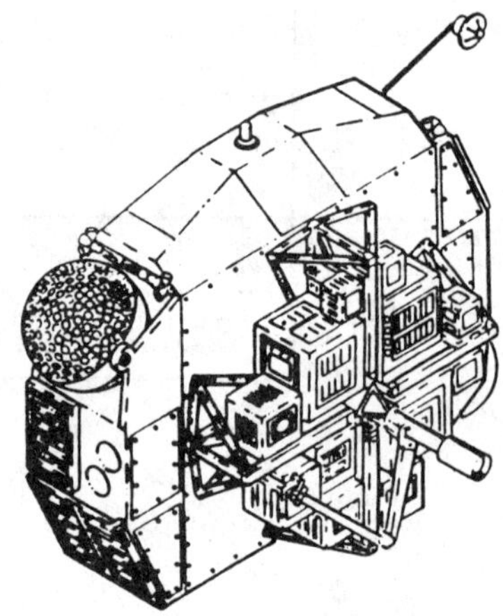

EARLY OMV CONCEPT

LENGTH = 32 INCHES
DIAMETER = 156 INCHES

WEIGHT (LBS)	INERTIA (SLUG–FT^2)		
	IXX	IYY	IZZ
3000.	1345.	732	739.

THRUSTER COMPLEMENT
8 ORTHOGONAL TRIADS

Fig. 5 Configuration Descriptions

Objective - The unique geometry and resulting mass characteristics of the disc configured OMV represents a contrast to the "box" configured TRS. The objective of this study was to examine the manual mode handling qualities of this new configuration using the TRS as a benchmark for comparison. In addition, this comparison would enable an assessment of the relevance of previous TRS studies to today's OMV control system issues.

Approach - MSFC's Target Motion Simulator was used for the study. Its composed of a linear track equipped with two appositioned three degree of freedom gimbals, one for a TV camera and the other housing a scale model of the target vehicle; a control station with hand controllers and TV monitor; and a host computer with the resident math models. Except where configurational differences dictated changes, the TRS control system baseline was used for both the TRS and the OMV. Each vehicle was flown against both three axis stabilized, nonspinning targets and dynamic targets undergoing complex tumble from a range of about 80 ft with initial azimuth and elevation offsets of ±10 degrees.

Results - Close examination of the results, summarized below reveal few marked differences between the performance of the two configurations and tend to support earlier estimations of the comparative manual mode handling qualities. However, one trend does seem apparent; that is, the superior ability of the OMV configuration in handling the higher target vehicle tumble rates. Except for elapsed time, in which there is little difference, the OMV registers superior performance parameters across the board as compared to the TRS for tumbling rates above .7 deg/sec. (For example, note the divergence in performance evidenced in plots of the docking accuracies.) The higher thrust-to-

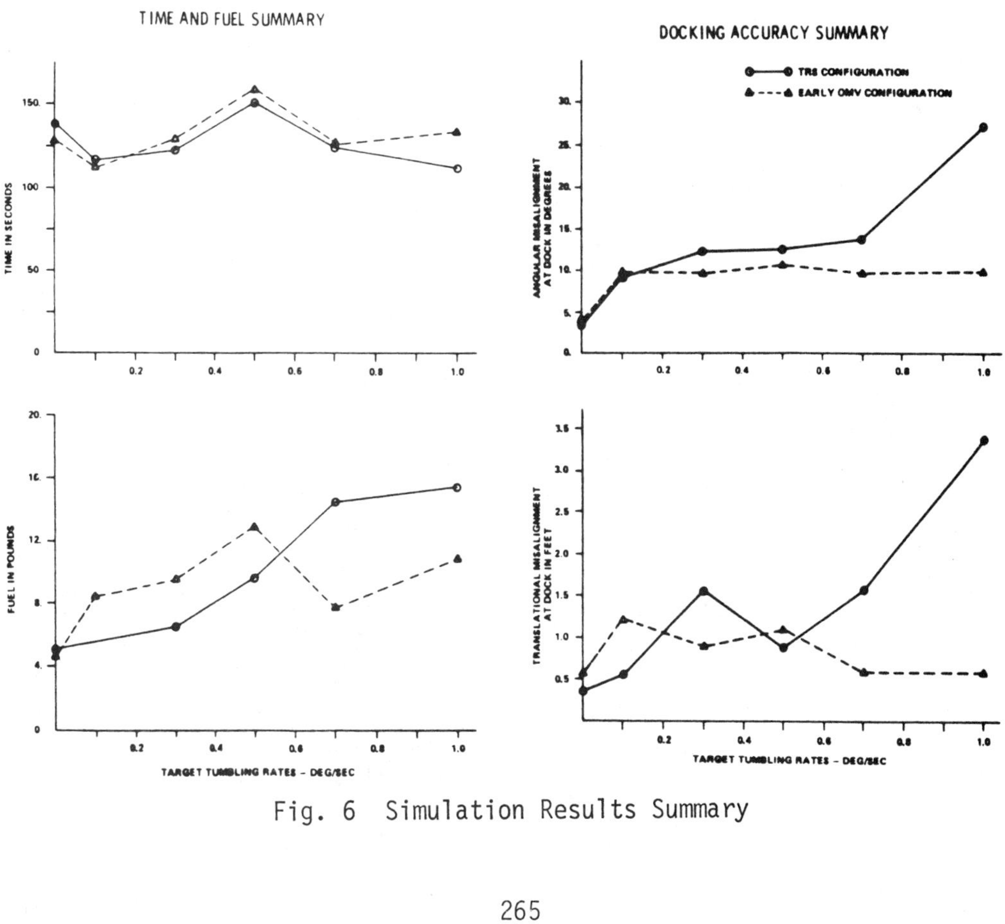

Fig. 6 Simulation Results Summary

weight ratio of the OMV (early concept) results in a higher translational acceleration capability as noted in Fig. 6. This improvement in performance with measured translation acceleration is expectable. Increased translational acceleration (especially X axis) enables the pilot to begin tracking the target vehicle drogue at an increased range, a feature from which our previous man-in-the-loop simulations has shown to be important in achieving a successful docking with a tumbling target vehicle.

Study Results Postscript

The author served as the only test subject (pilot) for the simulation study. Lack of recent experience in flying the Target Motion Simulator resulted in a noticeable learning curve bias on the early simulation results. Fig. 7, below, shows a plot of the RSS of the chase vehicle probe to target vehicle drogue, Y- and Z-axis translational offsets just prior to dock plotted for the chronology of simulation runs. This is a representative performance parameter and the wide dispersions in the data for the first 20 to 30 simulation runs illustrates this learning curve effect. The fact that these first 20 to 30 runs were made with the TRS configuration will tend to moderate some of the differences in the results obtained. Another factor which would

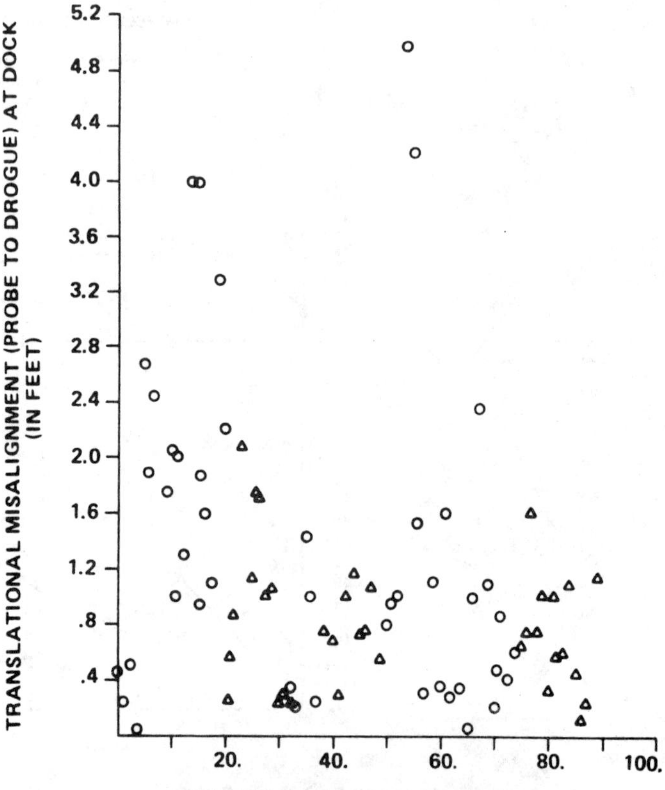

Fig. 7 Docking Accuracies Per Case

tend to normalize comparative performance, if included, is that more recent OMV configurations are heavier and would not share the thrust to weight ratio advantage of the earlier baseline. In summary, study results show that from the standpoint of acceptable manual mode handling qualities, a suitably equipped disc configuration OMV should be capable of performing a docking mission with a stabilized target as well as a target vehicle undergoing relatively high tumbling rates. The study results further indicate that previous TRS man-in-the-loop control system studies and trades remain relevant to the current and projected OMV control system issues.

CONCLUSION

Designing a man-in-the-loop control system for a remote piloted vehicle is a problem with many solutions. The control system baselined for the TRS represents one solution based on one set of requirements. Other comparable solutions reflecting different combinations of vehicle instrumentation, handling qualities, and control augmentation are possible, if not likely. The control system design process entails parameterization of system variables through simulation with the human pilot in the loop. The ensuing iterative process begins to bring shape to the control system design. But since the affect of every combination of system variable could not possibly be determined, the order in which they are taken will, perforce, strongly influence the form of the final solution. For example, a decision to exclude radar as a requirement for OMV early in the program, though supportable by simulation, would according to the TRS experience diminish the man-in-the-loop control system performance. This reduction in performance will impinge to some degree upon the future accommodations of other derived requirement for capture of disabled spacecraft. In other words, to fully assess the impact of requirements which affect pilot performance, and in turn, the control system design, requires that they be considered collectively and subjected to extensive simulation. Review of the evaluation of the TRS control system baseline should be helpful in this regard, offering a sound initial control system concept for OMV preliminary design studies and needed insight as to the relative importance of the issues of communication delays, radar, tumbling targets, etc., to the design of a man-in-the-loop control system for the remote piloted vehicle.

NOTATION

$$\tilde{\tilde{Q}} = \begin{bmatrix} +Q4 & -Q3 & +Q2 & +Q1 \\ +Q3 & +Q4 & -Q1 & +Q2 \\ -Q2 & +Q1 & +Q4 & +Q3 \\ -Q1 & -Q2 & -Q3 & +Q4 \end{bmatrix}$$

$$
\bar{\bar{Q}} = \begin{bmatrix} +Q4 & +Q3 & -Q2 & +Q1 \\ -Q3 & +Q4 & +Q1 & +Q2 \\ +Q2 & -Q1 & +Q4 & +Q3 \\ -Q1 & -Q2 & -Q3 & +Q4 \end{bmatrix}
$$

REFERENCES

1. Design Criteria for Control of Space Vehicles, Vol III. Part 5, Rendezvous and Docking, Contract NAS8-11494, August 1, 1966.

2. Control System Stability and Dynamic Response, Vol I, TRS-GNS Definition Document, Contract NAS8-32821, October 1978.

3. H. J. Buchanan, M. S. Hopkins, and Z. J. Galaboff, "Uncontrolled Dynamics of the Skylab Vehicle," Journal of Guidance and Control, Vol 4, No. 3, May-June 1981.

4. A. C. Cobb, "Skylab Mission Final Mass Properties Report," NASA Memo S&E-ASTN-SA, March 5, 1974.

5. S. N. Carroll and H. J. Buchanan, "System Equations for Automatic Docking," NASA Memo ED15-79-8, August 20, 1979.

6. Teleoperator Retrieval System Mass Properties Status Report, Contract NAS8-32821, Report No. TRS-SE-10, Issue 5, September 1978.

7. W. E. Agan, A Letter to NASA/MSFC, "TMS Docking and Simulation Data," January 15, 1981.

AAS 84-035

CONTROL OF A FLEXIBLE ROBOT FOR SPACE APPLICATIONS

Richard Gran*

The use of a robot device for such applications as berthing, satellite retrieval, and space activities necessitates a careful understanding of the problems caused by the interactions of the control system and the structural dynamics of the robot. This paper explores those interactions and shows that they are created by the following:

- The speed with which the robot must move
- The fact that the flexible dynamics change with the orientation of the robot
- The fact that the critical objective is the control of the location of the "end effector"
- The inherent non-colocation of the actuators for the robot, which are usually joint controllers, with the end effector
- That the measurement of the end effector location and velocity are critical for proper control of the motion of the "rigid body" while internal state measurements such as strain are critical for the flexible motion
- That the changing geometry necessitates some form of adaptation

* *Dr. Richard Gran, Research and Development Center, Grumman Aerospace Corporation, Bethpage, New York 11714*

INTRODUCTION

Figure 1 is a schematic of a typical robot that shows the nomenclature used for describing the interaction of the flexible dynamics with the control of the rigid body motion. To understand structural motion we must describe the mathematical modeling of the structural dynamics. The three aspects of this modeling question are: the modeling of the dynamics, the modeling of the actuators and sensors, and the modeling of the disturbances. Without a thorough and careful understanding of each of these, the control system will be doomed to failure. Even if these are handled in a substantially correct way there are still issues that must be understood and accounted for that could cause the control system to operate unstably or with extremely poor characteristics. Among the issues that must be understood are:

- o The bandwidth of the controller is determined by the speed of response of the robot, and by the disturbances that are induced

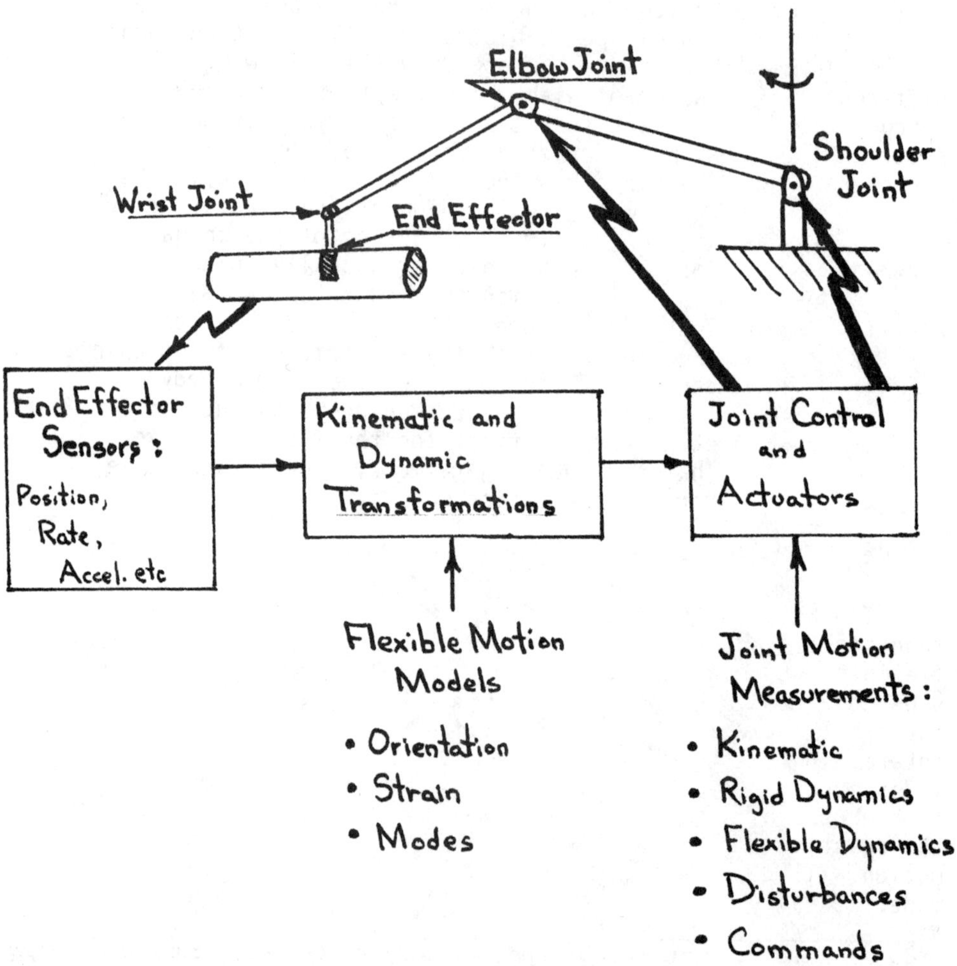

Figure 1 Typical Flexible Robot for Space Applications

by the robot motion - as a consequence of the flexibility of the robot, either robot commands must be made so slow that the motion does not interact with the structure, or the robot must explicitly control the structural dynamics.

- There is a set of vibration modes that will unstably interact with the control system when sensors and actuators are not colocated. Therefore, if the robot must use end effector position as a feedback variable, the inherent non-colocation of the actuators with this sensor will result in problems.
- When the robot's geometry changes, the mode shapes and mode vibration frequencies will change. Therefore, any control system that contains filters that are tuned to specific modes could be driven unstable by the filters. This means that a robot controller that contains structural motion compensation must adapt to the changing positions of the robot arm.

DISCUSSION

If we start the modeling part of the control design with a partial differential equation description, then the concepts discussed in the introduction can be described in terms of the transfer functions for the system. Thus if the transfer function of the combined rigid body and structural dynamics model that results from the mode description is:

$$H_{jk}(s) = \sum_{i=0}^{\infty} \frac{A^{jk}}{(s^2 + a_i s + b_i)}$$

where:

A_{jk} is the coefficient that determines the amount of control or disturbance that enters mode i from the location j and the amount of motion from mode i that is measured by the sensor at location k.

a_i and b_i are the coefficients of the second order linear differential equations that describe the motion of each of the vibration modes.

The block diagram that results from the use of these transfer functions is shown in Fig. 2. The first characteristic of robot flexible motion and control interaction that can be discerned from this block diagram is the fact that the control bandwidth and the speed of the robot motion determines the level of control/structure interaction. Thus as the control system gains and bandwidth increase, it is more likely that the structural frequencies (which are the square root of the b_i) will fall within the bandwidth. If this happens, then the relative amplitudes of the structural motion and the rigid body motion will determine the amount of energy that appears in the vibration mode. Clearly this is a function of the magnitude of the coefficient A_{jk}.

A far more important characteristic of structural interactions is

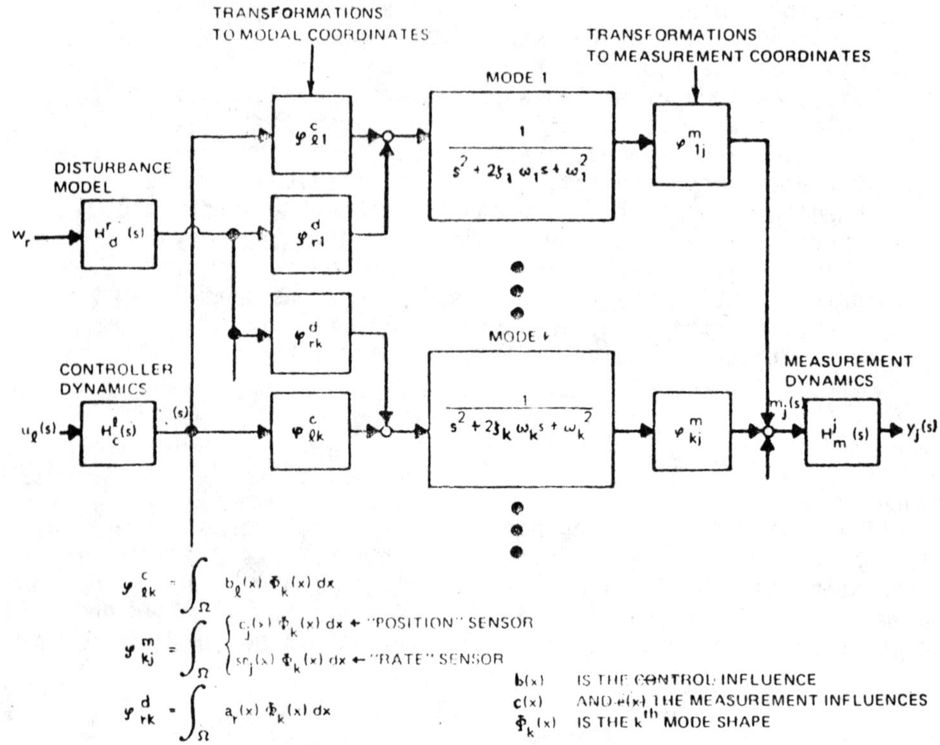

Figure 2 Block Diagram of a Robot's Interactive Structure, Measurement and Actuator Dynamics

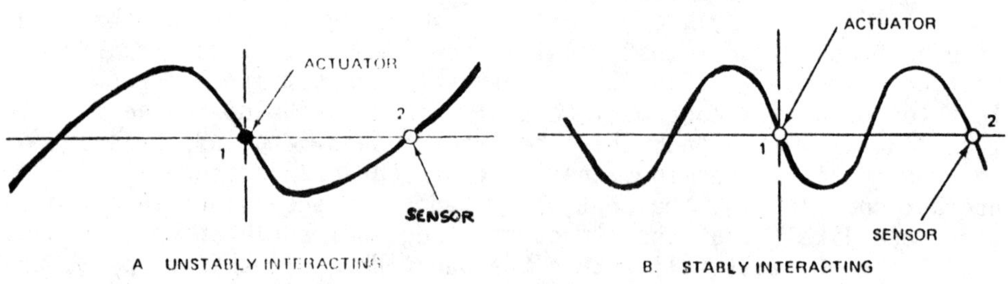

Figure 3 Stable and Unstable Mode Interactions

the sign of the A_{jk}. From the structural analysis, the value of A_{jk} is the product of the mode matrix elements (from a NASTRAN analysis) corresponding to the location j and the location k. There is a matrix element for every position on the structure that is modeled in the NASTRAN analysis for every mode that is determined. Thus if the actuator and sensor are in the same location, the product of these mode matrix elements is always positive, and there can be no sign change in the A_{jk}. If, however, the actuator and sensor are not at the same location then the possibility exists that some of the A_{jk} will be negative. In fact it can be easily shown that there always is an infinite number of negative A_{jk} when there is not exact co-location of the actuators and sensors. The saving fact is that the frequencies at which these first appear are usually so high that there is no control energy in that range. The problem is illustrated by Fig. 3. Here we have defined two kinds of structural/control interaction - "stable" and "unstable". In the stable interaction, the sign of the A_{jk} is positive, and in the unstable case the sign is negative. Because of the 180° phase shift introduced by the negative sign, when actuators and sensors are not co-located and the control system gain at the structural frequency is greater than unity, there will be an instability at the structural frequency.

Let us now conceptualize a robot's motion in terms of these interactions. First, the most important feature of a robot in space applications (for berthing, capture, work etc.) is the location of the end effector. Secondly, control is almost always at the joints. Thus there is an inherent lack of co-location in a robot. This means that any control system with a bandwidth that overlaps the structural frequencies will very likely be unstable. To control the stability, a 180° phase shift must be introduced at the frequency of the unstably interacting modes. This is accomplished in the compensator (the Kalman filter in an LQG design or with notch filters etc . in classical designs). There is a significant problem with this design approach.

As the robot moves, and its geometry changes, the mode frequencies are changing. A robot with a wide bandwidth is designed using a set of structural frequencies based on one geometry. In this design the control compensation that provides the phase shift needed to compensate the unstable interactions will be "tuned" to the assumed mode frequency that is unstable. As the robot geometry changes, this mode frequency does not remain invariant. Thus the filter that "stabilized" the original design model could actually cause the modified geometry to be unstable. This, then is the fundamental issue in robot control design.

There is a large body of control theory that has been developed for large space structures control that can be applied to the robot application. References 1 and 2 describe in more detail the theory of unstably interacting modes, and Ref. 3 describes a technique for adaptive control that can be used to track mode frequencies as the geometry changes. The use of an adaptive approach permits the control system to account for the unstably interacting modes in the presence of geometry changes. The alternative to this approach is a very stiff robot with structural frequencies that are well removed from the control system frequencies, or a drastic reduction of the bandwidth of the controllers so that structural interactions do not occur. Note that the consequence of the bandwidth limitation is the time required to perform maneuvers. In the space environment, where time is at a premium, this is not a viable alternative. Thus the control system design issues being discussed here are important from all aspects of robot design, planning and scheduling.

REFERENCES

1. Gran, R., "Finite Dimensional Controllers for Hyperbolic Systems", Proceedings of the Third VPI&SU/AIAA Symposium on Dynamics and Control of Large Flexible Spacecraft, June, 1981.

2. Gran, R., "Control of Flexible Structures: A Systematic Overview of the Problem", Proceedings of the Workshop on Applications of Distributed System Theory to the Control of Large Space Structures", NASA JPL Publication No. 83-46, July, 1983.

3. Gran, R., "On-Orbit Testing for Large Space Structures", AIAA 17th Aerosciences Meeting, New Orleans LA, Jan. 1979.

AAS 84-036

DOCKING OF A SPACECRAFT WITH AN UNRESTRAINED ORBITING STRUCTURE*

David A. Levinson[+]
Thomas R. Kane[++]

Simulation results are presented for a series of docking maneuvers involving a rigid body and an unrestrained, nonuniform beam in orbit, the underlying analysis is set forth, and extension of the methodology to the simulation of docking of a rigid body with more complex orbiting structures is discussed.

The subject of spacecraft docking, of interest since the days of the Gemini program, has acquired new importance with the advent of the Space Shuttle. In this context, what merits particular attention is the docking of a rigid body with an unrestrained, flexible structure.

Early analytical treatments of the subject of docking were confined, for the most part, to the interaction of two rigid bodies. Some attention was directed toward the docking of a rigid body with a rigid body carrying flexible appendages, and, in 1970, an attempt was made to develop a general purpose computer program for simulating the docking of two flexible bodies. The present paper has the twofold purpose of providing a detailed algorithm for the simulation of the docking of a rigid body with a nonuniform, unrestrained beam in orbit, and of bringing to light a number of issues that must be faced in the course of producing any simulation of the docking of a rigid body with an unrestrained, flexible structure.

---

* *This paper appears in full in the Journal of the Astronautical Sciences, Vol. xxxi, No. 1, pp23-48, January-March, 1983. It was presented at the AIAA/AAS Astrodynamics Conference, August 9-11, 1982. This work was supported in part by the Lockheed Missiles and Space Company Independent Research Program, and in part by the National Aeronautics and Space Administration under Grant NAG 1-97 to Stanford University.*

\+ *Senior Scientist, Lockheed Palo Alto Research Laboratory, Palo Alto, California 94304*

++ *Professor of Applied Mechanics, Department of Mechanical Engineering, Stanford University, Stanford, California 94305*

Section IV
VEHICLE GUIDANCE, CONTROL, AND PAYLOAD EFFECTS

SESSION IV

Chairman:	Lt. Col. David L. Younkin Air Force Space Division
Co-Chairman:	B. Drake The Aerospace Corporation
Local Chairman:	Richard T. Yoshida Martin Marietta Denver Aerospace

The following paper numbers were not assigned:

AAS 84-047 to -050

AAS 84-041

IUS PROPULSION, GUIDANCE AND CONTROL - AN INTEGRATED DESIGN

H. Sokoloff*

The integrated design of the IUS propulsion, guidance and control systems is discussed. The utilization of two solid rocket motors (SRMs) and a liquid hydrazine reaction control system (RCS) requires bold concepts in guidance and control system design and consumables management. The effect of payload properties and requirements on total system design is considered together with experiences from two recent flights.

INTRODUCTION

The Inertial Upper Stage (IUS) is the newest and most sophisticated space launch vehicle upper stage in the United States inventory. Designed for launch from the Space Shuttle, it has been made compatible with the Titan 34D expendable booster system. Expanding on concepts developed during the Transtage and Centaur programs, the IUS is unique in its own right, employing innovations which have presented challenging problems to be solved by the aerospace scientist.

Based on considerations of Shuttle crew safety, Orbiter bay size and system reliability, the selection of a solid rocket motor propulsion concept is a departure from convention and is the source of the more challenging problems. The IUS utilizes the most up-to-date avionics and mechanical systems and has been designed to be at least single fault tolerant.

The brains of the IUS consists of a Redundant Inertial Measurement Unit (RIMU) and two parallel avionics strings. Each string is under the control of its own digital computer which processes its own data, continually checks itself and communicates with the other. One string is in control of the mission until a fault is detected within it, whereupon the other takes control. No single avionics fault can cause a mission failure.

The two SRMs provide the energy for mission orbit transfers. The larger SRM1 propels the IUS from low earth orbit into an highly elliptical transfer orbit with apogee at synchronous altitude whereupon the smaller SRM2 circularizes the final orbit. Potentially large SRM impulse dispersions are minimized by applying velocity vernier corrections after each SRM firing with the liquid hydrazine propellant RCS.

*Mr. Sokoloff is a Project Engineer at The Aerospace Corporation, P.O. Box 92957, Los Angeles, California 90009

The SRM and RCS propulsion systems also provide the muscle for the Flight Control system (FCS) to orient the IUS during SRM and RCS burns and maintain attitude control throughout the mission. For SRM burns, pitch and yaw attitude control are achieved by thrust vector control (TVC) of the SRM nozzle which rotates about the "tech-roll joint", an IUS design innovation. During the vernier and coast flight phases, attitude control is effected by the RCS through software-computed moment applications about the vehicle center of mass. Roll control during SRM burns is also performed with the RCS.

IUS mission success is dependent on the operation of many interrelated functional elements. The Propulsion systems provide orbital transfer energy and effect attitude control. The Navigation system, through the body-mounted RIMU gyros and accelerometers, generates inputs for the software computed IUS state vector - inertial position, velocity and body attitude. The Guidance system determines how best to achieve mission end conditions from the current IUS state by optimizing the energy remaining while keeping within mission constraints. The Flight Control system performs the logic to orient the IUS and maintain attitude control as required by the Guidance system and/or the Mission Data Load (MDL). The Mission Sequencing function ties the mission operations together by interfacing with all other software functions and maintaining proper timing of operations. Redundancy Management monitors the state of the RIMU, avionics and software systems and provides fault detection and switchover. The Communication system reports the status of all systems by way of the Telemetry function and is capable of receiving uplink commands through the Command function.

IUS PROPULSION SYSTEMS

The basic function of the SRM primary propulsion system is to provide energy for the IUS mission orbit transfers (Fig. 1). SRM1 fires in the park orbit to effect the transfer orbit, and SRM2 circularizes the final orbit at synchronous altitude.

The SRM propulsion system is also used to perform the pitch and yaw TVC function during the burns. By rotating the thrust vector about the tech-roll joint, powered flight attitude control is achieved using linear control concepts. Sophisticated digital logic transforms the linear concepts into a form which maintains attitude control in response to Guidance commands and in the face of dynamical disturbances, parasitic oscillations, and system nonlinearities.

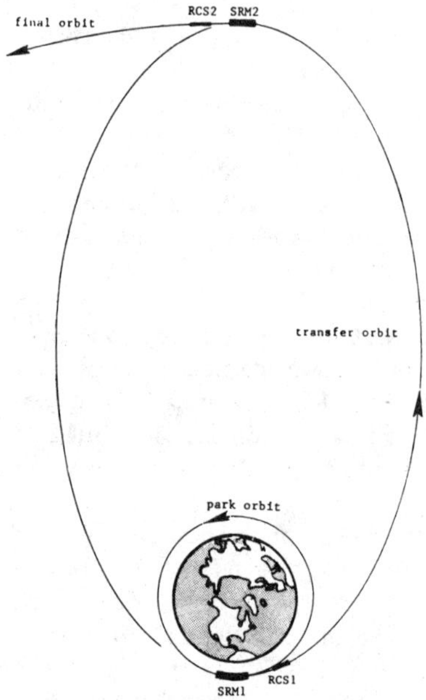

Fig. 1 IUS Orbit Transfers

The RCS secondary propulsion system also performs several functions. It applies velocity vernier corrections to the SRM impulse dispersions and provides translational velocity when required, such as for collision avoidance between the IUS and its separated payload. In addition, the RCS is used to maintain three-axis attitude control during all coast phases of flight, together with roll control for SRM burns.

The IUS RCS is comprised of the twelve thruster set (Fig. 2) and its propellant feed system. Control moments are achieved by operating the thrusters in pairs*: +pitch → N1N2, -pitch → N7N8, +yaw → N4N5, -yaw → N10N11, +roll → N3N9, -roll → N6N12. During all coast flight phases and for roll control during RCS burns, thrusters are normally off and pairs are turned on for attitude control. However, in the velocity vernier mode, all eight rear-facing thrusters are turned on for velocity addition and appropriate thruster pairs are turned off for attitude control.

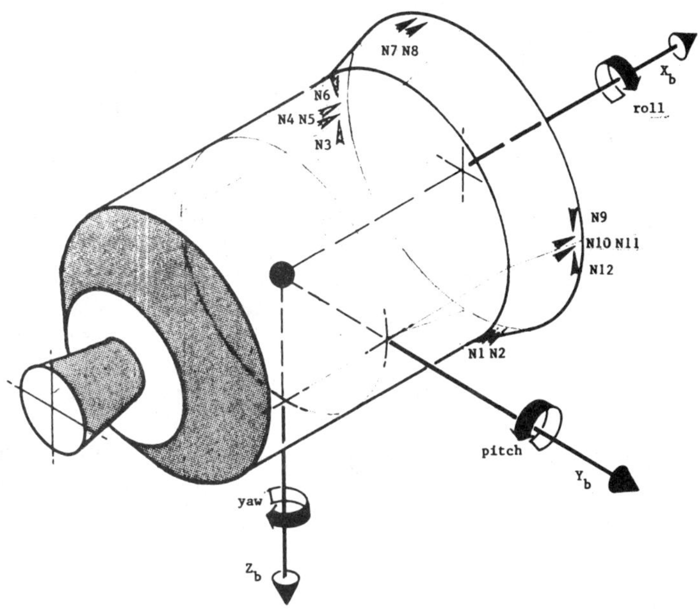

Fig. 2 RCS Thruster Configuration

The RCS thrust level is determined by the blowdown type propellant feed system (Fig. 3). The spherical propellant tanks (either one, two or three depending on mission design) each contain two compartments separated by a flexible diaphragm. About 122.5 lb of hydrazine propellant (N_2H_4) is loaded into one compartment, while the other is pressurized to about 380 psi with gaseous nitrogen (GN_2). Initially, the thrust level is about 30 lbf per thruster; but as propellant is expended during the mission, the GN_2 compartment enlarges and the feed system pressure decreases, yielding a concomitant reduction in thrust level. Typically, there is about a 4:1 blowdown ratio from a full tank to an empty tank, and the thrust level at propellant depletion is about 8-10 lbf.

*Thrusters generally operate in pairs, but are commanded individually. When minimum impulse is desired, single thruster commands are issued.

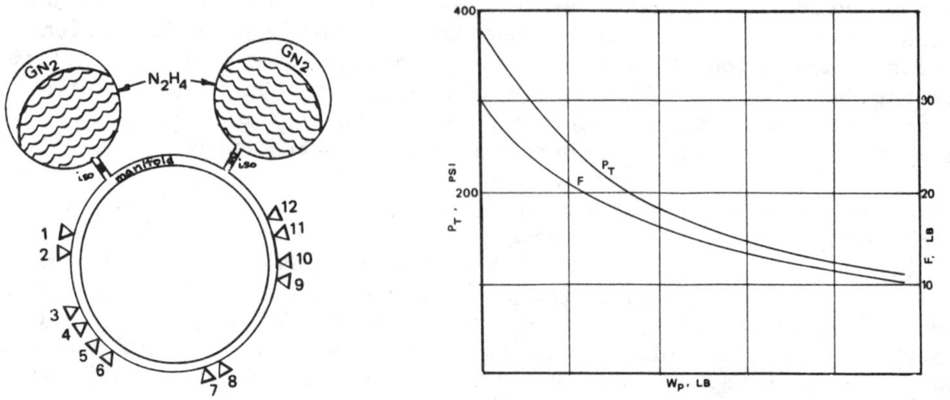

Tank Pressure, Thrust vs Weight of Propellant Consumed

Fig. 3 RCS Propellant Feed System and Parameters

Redundancy management provisions have been included for the flight control aspects of the propulsion systems. Independent dual string avionics are utilized in the TVC system to the extent that either of two drive motors can control the actuator shaft. Any fault detected in the string controlling the mission will result in a switchover to the other string.

RCS redundancy management is considered in two parts. Valve driver commands can be issued by each computer and the dual string concept holds to that point; however, RCS configuration redundancy management employs the concept of masking redundancy - that is, redundancy invisible to the operating system. To protect against valve <u>open</u> failures, each thruster has two valves in series. If one fails <u>open</u>, the other maintains thruster integrity. If a valve fails <u>closed</u>, its thruster becomes inoperative; but its parallel partner continues to apply commanded thrust at half the level of the pair. During the IUS mission design cycle, analysis and simulation testing ensure that no single RCS valve <u>closed</u> failure will compromise the mission.

MISSION DESIGN

IUS mission design begins with a definition of the payload and its requirements, followed by a performance study to determine how best to achieve mission objectives utilizing two SRMs and a fixed RCS propellant load. Optimal targeting is an iterative process that considers the IUS/payload mass properties, mission requirements, SRM1 and SRM2 energies, and total RCS propellant usage - statistically minimizing the latter. While payload weight and orbital objectives define the gross demands on the IUS, study trade-offs may include SRM offloads, the number of RCS tanks, use of an extendable exit cone (EEC) to enhance SRM2 performance, and the integration of the RCS performance potential with that of the SRMs. A fully loaded IUS can place about 5000 lb of payload into synchronous orbit when launched from the Space Shuttle. Lighter payloads can be accommodated by either off-loading the SRMs or by applying sub-optimal orbit transfers. The number of RCS tanks affects system weight as well as performance, while the EEC can enhance SRM2 performance by about five percent.

Where maximum performance is not required, the mission design targets SRM1 to achieve the desired transfer orbit energy, and the RCS1 vernier corrects SRM1 dispersions only; but the RCS2 vernier is selected to enhance SRM2 performance in addition to correcting its impulse dispersions. SRM2 nominal performance is affected by the RCS1 vernier. Greater than nominal RCS1 propellant expenditure causes the IUS to be lighter for the SRM2 burn, increasing its nominal performance. As a result, the expected RCS2 vernier is reduced, yielding a partial recovery in total RCS propellant expended.

The RCS propellant budget is an important IUS consumables concern. The major contribution to it is from the correction of SRM dispersions as described above. The allocation for coast attitude control is essentially independent of SRM dispersions and is based on vehicle mass properties, the thruster configuration, and attitude control operations. Requirements for incremental velocity are based on mass properties alone. Dispersions associated with coast control and incremental velocity addition are small compared to those of the SRMs.

The RCS propellant budget allocation for the IUS mission is separated into two categories - nominal and dispersed. For vernier burns associated with SRM performance, nominal entries are based on the current mission design, which may change somewhat from the time the SRM sizes are frozen. Dispersed entries are the result of the iterative design process, taking into account the SRM statistics and the correlation between the RCS1 and RCS2 consumptions. Coast control consumption is calculated for nominal conditions and parameters, and dispersions are taken about those nominals. A typical RCS propellant budget is shown in Table 1 for the two tank configuration.

Table 1 TYPICAL RCS PROPELLANT BUDGET

		NOMINAL	DISPERSED
COAST CONTROL	PARK ORBIT	9.4	
	TRANSFER ORBIT	33.7	15.3
	FINAL ORBIT	3.6	
		46.7	15.3
VELOCITY	RCS1	23.9	
	RCS2	76.1	56.0
	CAM	4.0	
		104.0	56.0
	TOTAL	150.7	58.1

USABLE PROPELLANT	242.0
NOMINAL + DISPERSED	208.8
EXCESS MARGIN	33.2

As a precaution against uncertainties in the knowledge of SRM dispersions and to protect against worse than "3-sigma" performance, backup thresholds are provided for each RCS vernier burn so that enough propellant is available to complete the mission, even at the expense of reduced orbital objectives. A measure of the propellant consumed in flight is provided by the on-board software; and if the on-board calculation exceeds a threshold, the velocity vernier is terminated. For RCS1, the 175 lb threshold value is remote, being well beyond "3-sigma" dispersions. The 220 lb RCS2 threshold could be reached for unfavorable combinations of SRM1 and SRM2 performance, and would result in a slightly degraded final orbit; but statistical analyses limit this occurrence to a "3-sigma" probability.

GUIDANCE

The IUS Guidance system operates on the current state of the mission together with nominal performance expectations for the remainder of the flight in order to achieve final orbit parameters. The guidance philosophy (termed Gamma Guidance) has been developed to be compatible with the performance dispersions of the SRMs. Tested for a wide range of conditions in pre-flight simulation, Gamma Guidance adapts to existing state conditions in the mission and modifies the remaining flight segments (ARCs) to keep within allowable constraints.

The IUS Guidance mission consists of eight ARCs (Table 2). Guidance calculations, or call-ups occur at different intervals in each ARC. SRM and RCS burn ARCs (2, 4, 6, 8) are the most dynamic and call-ups occur every 10 sec. The coast ARCs between SRM and RCS burns are relatively short, with call-ups every 50 sec; but during the long coast periods (ARCs 1, 5), they occur at intervals of 200 to 500 sec.

Table 2 GUIDANCE MISSION ARCs

ARC NO.	MISSION SEGMENT
1	Park Orbit Coast
2	SRM1 Burn
3	Coast: SRM1 to RCS1
4	RCS1 Burn
5	Transfer Orbit Coast
6	SRM2 Burn
7	Coast: SRM2 to RCS2
8	RCS2 Burn

In ARC 1, the Guidance system uses updated park orbit state vector information to determine the SRM1 burn ignition time and its inertial orientation to achieve mission objectives, assuming nominal system characteristics thereafter. In ARC 2, state vector dispersions from the burn are calculated and steering commands are issued to fulfill end conditions. SRM impulse dispersions are the basis for the ARC 3 calculations which determine the ignition time and orientation for the RCS1 vernier burn (ARC 4). During ARC 4, attitude orientation and cut-off time are calculated at each call-up. In ARC 5, the long transfer orbit coast, call-ups every 500 sec refine the ignition time and orientation for SRM2. ARCs 6, 7, and 8 are similar to ARCs 2, 3, and 4 except that some mission constraints have been eliminated with flight progress.

The Gamma Guidance concept is described with the aid of Table 3. The control vector $\bar{\lambda}$ provides the options for effecting the mission, while the constraint vector $\bar{Y}$ determines the mission conditions to be achieved. $\bar{X}_F$ is the predicted final state vector based on the applied control and the remaining IUS energy. Finally, the Γ matrix gives the incremental relationship between the control and constraint vectors.

At each call-up, the Guidance system seeks a solution iteratively for the parameters to be used in the current or next ARC based on predicted conditions. Generally, a solution converges in one or two iterations.

The Gamma Guidance logic is shown in Fig. 4. The final state vector is predicted from the initial control vector and the remaining energy. The constraint vector and control matrix are computed. If the computed and desired constraint vectors are within ϵ_1, the control vector is acceptable and the process stops. If not, linearized system equations are

Table 3 GAMMA GUIDANCE DEFINITIONS

SYMBOL	VARIABLE	EXAMPLES
$\bar{\lambda}$	CONTROL VECTOR	SRM IGNITION TIME; POINTING ANGLES; RCS VELOCITY
$\bar{Y}$	CONSTRAINT VECTOR	ECCENTRICITY; ARGUMENT OF PERIGEE, INCLINATION; RAAN
$\bar{X}_F$	PREDICTED FINAL STATE VECTOR	POSITION (X,Y,Z); VELOCITY ($\dot{X},\dot{Y},\dot{Z}$)
Γ	CONTROL MATRIX	$\Gamma = \frac{\partial \bar{Y}}{\partial \bar{\lambda}}$

solved for an incremental correction $\delta\bar{\lambda}$ to the control vector. If $\delta\bar{\lambda}$ is less than ϵ_2 or if eight iterations have occurred, the process is halted. Otherwise, $\delta\bar{\lambda}$ is added to the previous control vector and the process is repeated. If a solution is not achieved in eight iterations, a fallback position to the last converged solution is used and a new solution is attempted at the next call-up.

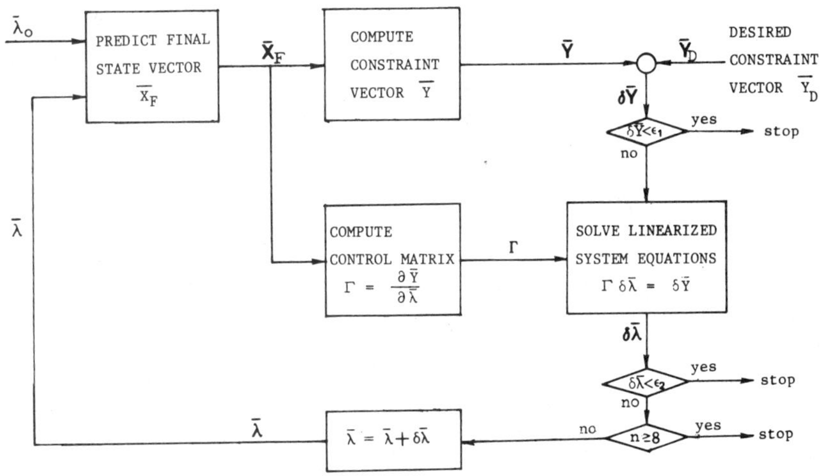

Fig. 4 Gamma Guidance Functional Flow Diagram

FLIGHT CONTROL

The Flight Control system (FCS) is responsible for maneuvering the IUS to the attitude command quaternion and maintaining that orientation to within prescribed limits, even in the face of disturbance torques. Just before and during SRM burns, the command quaternion is determined by the Guidance function. At all other times, command quaternions are either preset values or are derived from the state vector.

There are two aspects to the FCS (Fig. 5) - powered flight and coast control. Powered flight uses the TVC for attitude control about the pitch and yaw axes during SRM burns, while coast control applies RCS thruster moments at all other times, including roll control for SRM burns. Both systems interface with input functions via the Steering Mode Selection logic, yielding error quaternions for pitch, yaw and roll. From there, functional paths separate, each system processing data differently and interfacing with different control effectors; however, both utilize the Signal Conditioning Unit (SCU) which converts digital outputs to analog signals. The effectors cause motions about the IUS body axes which are sensed by the RIMU and are fed back as inputs to the attitude control software.

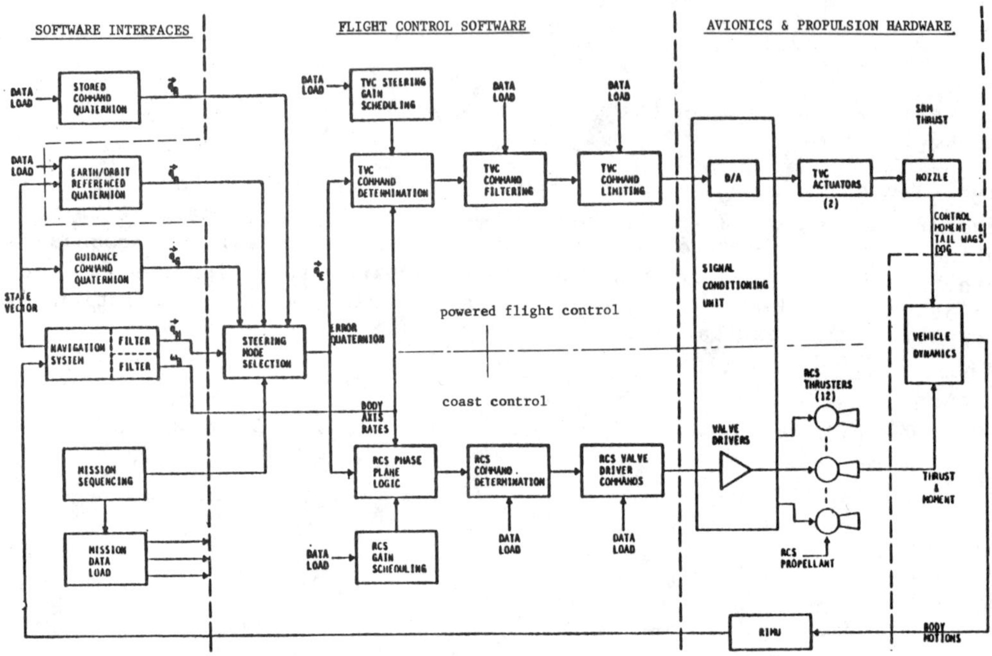

Fig. 5 Flight Control System Functional Block Diagram

The transitions from coast to powered flight control (ARCs 1→2, 5→6) are set by the Guidance system through Mission Sequencing (MSEQ) logic, while the transitions back are based on sensed accelerations and velocities calculated by the on-board software.

Powered Flight Attitude Control

Thrust vector control of the IUS SRM nozzle produces angular accelerations about the pitch and yaw axes which are proportional to the nozzle deflections. Therefore, the initial design of the powered flight control system for a particular payload is based on linear control techniques. In the simplified powered flight control system block diagram of Fig. 6, the flight control system operates on the Guidance-generated input command and on the RIMU-sensed vehicle attitude and attitude rate. Appropriate weighting functions, or gains, are applied to each signal in accordance with linear control laws and IUS system parameters. Unwanted high frequency signals sensed by the RIMU are filtered and the system is controlled as if it were a rigid body. The deflection command to the actuator is proportional to the sum of the attitude error and attitude rate; but because of actuator and nozzle dynamics, the TVC response is similar to a second order system.

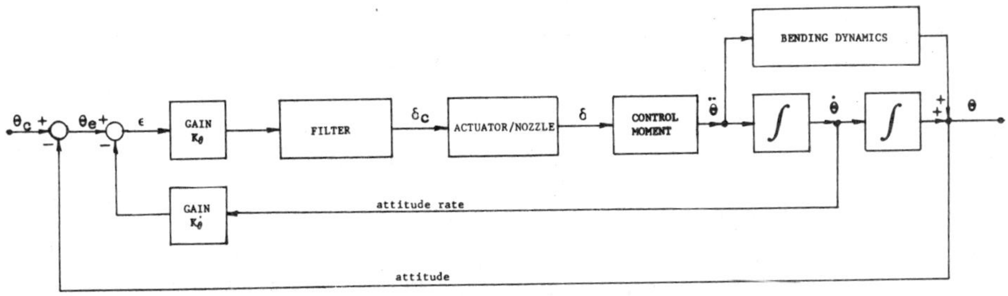

Fig. 6 Powered Flight Control Simplified Block Diagram

Final selection of control system parameters, both gains and filter coefficients, is determined with nonlinear digital simulations which include the representative characteristics of the IUS system. Examples of nonlinearities encountered are related to the TVC deflection process. The finite nature of digital-to-analog (D/A) converters yields signals which quantize the actuator commands. If the desired command lies between two quanta, a limit cycle results. Another nonlinearity lies in a characteristic of the tech-roll joint itself - in particular, coulomb friction. It may take a command of two or three quanta to overcome coulomb friction in the nozzle, causing discontinuous deflections of the TVC and applying structural loads to the IUS which must be carefully analyzed and verified to be within IUS and/or payload specifications.

Coast Attitude Control

The coast attitude control system utilizes the RCS thrusters as control effectors, but only in two states - <u>on</u> and <u>off</u>. Linear techniques do not apply to this system and the principal design tool is the phase plane. A plot of attitude rate vs. attitude error (Fig. 7), the phase plane lends itself to <u>on-off</u> control system analysis since <u>on</u> trajectories are represented as parabolas while <u>off</u> trajectories are horizontal lines.

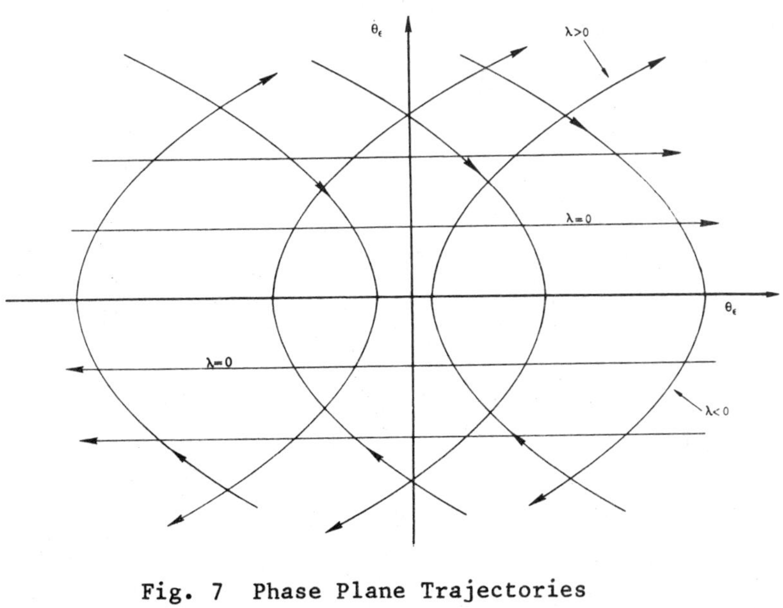

Fig. 7 Phase Plane Trajectories

Parabolic trajectories are shown for negative angular acceleration ($\lambda<0$; e.g., -pitch→N7N8) for positive angular acceleration ($\lambda>0$; e.g., +pitch→ N1N2), with the shape of the parabola dependent on the magnitude of λ. Coasting trajectories ($\lambda = 0$) are represented by horizontal lines, their directions determined by the residual attitude rate.

IUS coast attitude control operation is based on the logical application of thruster firings - for both on and off states. The switching logic is shown in Fig. 8, with the three zones indicated - negative angular acceleration, positive angular acceleration and coasting, or no angular acceleration. Every 40 ms, the coast control logic determines its state within the phase plane, compares it to the switching boundaries, and calculates the amount of thrusting (if any) needed to achieve the desired condition. If less than 40 ms is required, then the calculated on time is issued. If more than 40 ms is required, then an on time of 40 ms is issued. In the next computation cycle, the process is repeated. A series of concatenated 40 ms pulse commands is seen by the thrusters as one long command.

The coast control system phase plane switching boundaries are derived from several sources. Attitude (ADB) and rate (RDB) deadbands are functions of the limit cycle requirements. Parabolic switching lines are functions of the mass properties and the RCS thrust level. Mass properties are pre- calculated, but the thrust level is greatly affected by mission operations (particularly RCS1) and an in-flight on-board software calculation is made. Finally, attitude rate limit lines (RTLIM) define the attitude rates at which orientation maneuvers are performed, and are the eigenvectors of the attitude error quaternion. During IUS design phases, thruster valve closed failures are evaluated, ensuring that the phase plane switching boundaries are valid with only one thruster firing.

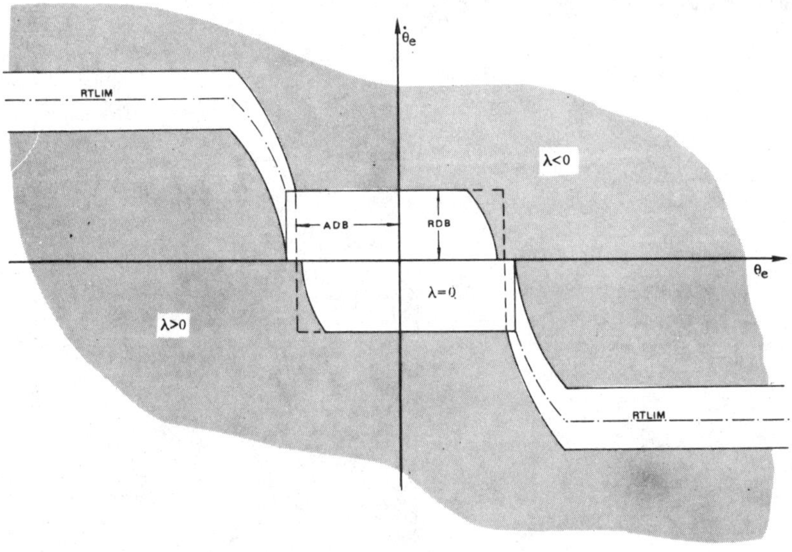

Fig. 8 Coast Control System Switching Logic

IUS CAPABILITIES/PAYLOAD REQUIREMENTS

The IUS has been designed to perform as a standard space launch vehicle upper stage vehicle, and has built-in generic capabilities which cover a wide range of requirements for potential payload users. The on-board software is flexible and can easily be adapted to new payload requirements. The generic IUS performance and design requirements are defined in a system specification, while the requirements for a particular IUS/payload mission are defined in an interface control document (ICD). An iterative process matches the IUS capabilities to the payload requirements until a compatible arrangement is reached. Studies are conducted to determine if the IUS can meet requirements not contained in the system specification; and if so, they are agreed to in the ICD. Requirements which cannot be met are either withdrawn or waived. In some cases, the on-board software is modified to accommodate specific requirements.

IUS Capabilities

The basic IUS design capabilities are found in the IUS System Specification (Ref. 1) for all functional subsystems, including Guidance and Control. For synchronous missions, Guidance/Navigation system 3-sigma accuracies are defined in terms of position and velocity after final orbit injection. The geosynchronous design reference mission capability applies to a range of payloads from 1950 lb to 5000 lb.

The 5000 lb design reference payload is 40 ft long and 10 ft in diameter with very large moments of inertia. The payload is restricted in its lateral center of mass offset, depending on its position relative to the IUS/payload interface. This constraint limits static offsets of the TVC. In the design process, worst case TVC deflections are limited to 90% of the maximum capability (SRM1 $\rightarrow \pm 4$ deg; SRM2 $\rightarrow \pm 7$ deg).

Powered flight linear stability margin specifications for nominal system parameters include conventional gain and phase margin criteria for both rigid body design and bending mode characteristics. For the IUS, propellant slosh is a secondary consideration since it is limited to the RCS tanks and payload propellant. When the combined parameter variations and analysis uncertainties represent 3-sigma conditions, the stability margin requirements are one-half the nominal requirements.

The coast control system specifications on attitude error and rate represent guaranteed minimum values for the design reference payloads. Attitude deadbands are selectable and range from a minimum ± 0.5 deg for fine mode control to ± 5.0 deg (and more) for coarse mode control. The transition from coarse mode to fine mode control is accomplished within 100 sec.

The IUS design has been validated for the maximum and minimum payloads in the design reference mission.

Payload Requirements

Payload and mission specific requirements are contained in an ICD in which IUS and payload contractors specify the unique requirements of the mission design. For the Tracking Data Relay Satellite (TDRS), the following are pertinent requirements found in the IUS/STS/TDRS ICD (Ref. 2).

The park and final orbits required by the TDRS are compatible with those of the design reference mission. TDRS mass properties fall within the design reference payload limits and its stowed cantilevered first bending mode frequency is greater than 5.0 hz, minimizing IUS powered flight control system interference. The following are areas requiring further analysis prior to ICD agreement.

TDRS Aft Tank Configuration - The TDRS propulsion system configuration is unique and presented an area of concern for powered flight control stability. Two hydrazine tanks, each with propellant and pressurant separated by a flexible diaphragm are located along the thrust axis of the IUS/TDRS (Fig. 9). The forward tank, with 950 lb propellant is configured conventionally with the propellant aft of the diaphragm. The aft tank, containing 350 to 550 lb propellant, is inverted with the propellant forward. The propellant response to thrust acceleration and its effect on mass properties and powered flight control stability was uncertain.

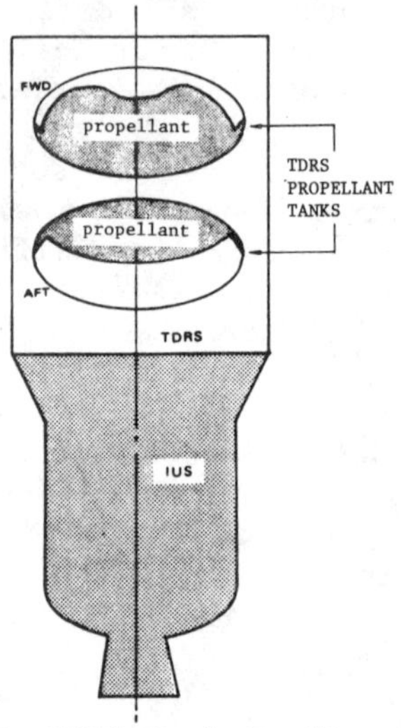

Fig. 9 TDRS Tank Configuration

3 deg/sec Thermal Roll - The TDRS thermal roll rate requirement about an axis of minimum inertia coupled with potential energy dissipation from the on-board propellants presented an RCS propellant consumption concern.

Low Frequency Dynamics - After the basic mission has been completed and just prior to payload separation, the IUS RCS is disabled and the TDRS appendages are deployed. When the RCS is subsequently reactivated, the deployed configuration is highly flexible with a first bending mode potentially as low as 0.5 hz, with maximum allowable loads specified at some of the hinge points. The coast attitude control system is required to maneuver the IUS/TDRS to the payload release orientation and achieve fine mode attitude control for the spacecraft separation, while keeping within the allowable loads.

PAYLOAD EFFECT ON DESIGN

In general, the TDRS properties and requirements fit within the IUS capabilities. Although weighing 5000 lb, the TDRS is relatively compact and its moments of inertia are easily accommodated; but the aforementioned areas were analyzed to determine if they fit within IUS capabilities or if design modifications to the IUS software were to be made.

TDRS Aft Tank Configuration - The uncertainty of the TDRS aft tank propellant stability and its effect on powered flight control required special analysis. A plexiglass mockup tank was outfitted with a real tank diaphragm. Tests were conducted for both the conventional and inverted configurations. As suspected, symmetry in the inverted tank represented an unstable equilibrium and the propellant always flopped to one side (Fig. 10). Further extensive testing led to the conclusions that 1) the propellant always flopped, 2) the location of the offset around the longitudinal axis was random and 3) once offset, the propellant tended to remain stable under all expected dynamical conditions.

The powered flight control system was analyzed over the range of mass properties associated with the random nature of the offset, and was found to meet stability requirements under all conditions. For mass properties predictions, the propellant was assumed to be centered, since the direction was unpredictable; but the extremes of the SRM ignition transient were easily accommodated by the powered flight control system.

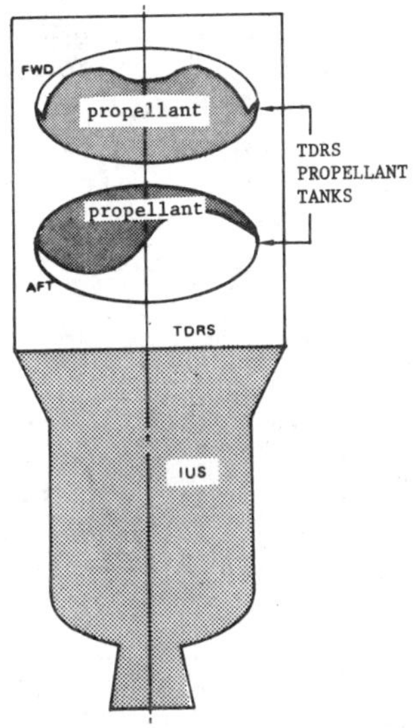

Fig. 10 TDRS Aft Tank Offset

3 deg/sec Thermal Roll - Concern arose relative to the 3 deg/sec roll rate about an axis of minimum inertia, affected by non-zero cross products of inertia and energy dissipation in the on-board propellants. If the tendency to increase the coning angle beyond the attitude deadbands was great, excessive RCS propellant usage would result. The problem was resolved by choosing appropriate dead bands. Prior to each thermal roll period, the IUS was stabilized in the fine control mode with minimum attitude rates about each axis. Simultaneous with the 3 deg/sec roll rate command, attitude deadbands were increased to 5 deg. Thus, when coning increased, its effect on propellant consumption was delayed until late in each thermal roll period and RCS usage was not excessive.

Low Frequency Oscillations - When the low frequency dynamical effects of the deployed appendages were included in the IUS simulation, it was found that the generic coast control logic was not capable of handling the situation and a software redesign was effected for the deployed condition.

The introduction of a finite impulse response (FIR) filter in the attitude rate computation solved the problem. Basically a moving window averaging filter, the FIR filter produces average rates over many sample intervals; however, the coast control computations could no longer fit into the 40 ms minor cycle and the logic had to be modified further. Each 40 ms, only one control channel was addressed - first roll, then pitch, and finally yaw. This effectively increased the minor cycle time from 40 ms to 120 ms, but no problems were encountered because of the benign environment. Simulation studies indicated that acceptable control was achievable without exceeding the loads specifications on the deployed appendages.

FLIGHT EXPERIENCE

The IUS has flown two synchronous missions with the ultimate result that all payloads achieved their planned orbits. In the first flight, a tandem pair of DSCS II/III communication satellites was placed into perfect orbits by the IUS. In the second flight, a TDRS payload was placed in a sub-synchronous orbit by the IUS, which failed in the final seconds of its SRM2 burn, but after the AFSCF* commanded IUS/TDRS separation, the TDRS spacecraft saved the mission through its capability to stabilize from a tumble condition and apply the final orbit corrections itself.

IUS/DSCS II/III

The maiden IUS flight was atop a Titan 34D expendable booster, and its DSCS II and III payloads were placed into perfect orbits. Two incidents marred the result. First, the impulse delivered by SRM1 was about 0.4% low, causing the RCS1 vernier to be much larger than anticipated, although within allowable dispersions. Second, about 30 minutes into the transfer orbit, a telemetry switch failed and no further telemetry was received. For the rest of the mission, knowledge of IUS performance was not known; however, the IUS is an autonomous upper stage, and should not have been affected by the telemetry system anomaly. From the end results as determined from tracking data and payload monitoring, all other functions performed very well. The RCS1 vernier burn was so large that only proper Guidance and Control System operation could have produced the outcome. All observed data prior to the switch failure indicated excellent Guidance and Control system operation.

IUS/TDRS

The first IUS flight from the Space Shuttle carried a TDRS spacecraft and experienced a number of unexpected occurrences. Most were minor but one was significant enough to cause a mission failure. Post-flight analysis provided the IUS program an opportunity to thoroughly evaluate the design and make adjustments to prevent future recurrences.

As in the first flight, SRM1 performance was about 0.4% low, and the resulting RCS1 vernier burn was correspondingly long. In addition, a number of relatively minor anomalies occurred, including the following:

1. During the in-bay star scan, light flashes from the Orbiter RCS jet firings were viewed as stars by the scanner. Erroneous attitude data was passed to the A computer after the star scan, which propagated into a transfer orbit apogee error of 220 n.mi. The B computer received correct data.

2. The RIMU reconfigured from 5 to 3 gyros three times in flight during attitude maneuvers due to improper fault detection thresholds. Each time, time, the 5-gyro configuration was reset by ground command.

3. Prior to SRM2 ignition, an illegal address fault was detected in the A computer and reconfiguration was effected to the B string.

---

* U. S. Air Force Satellite Control Facility, Sunnyvale, California

None of the above occurrences was sufficient to cause a flight failure although the final orbit would have been in error. The major anomaly of the IUS/TDRS flight was the collapse of the tech-roll seal with about 21 sec remaining in the SRM2 burn. The resulting nozzle offset caused the vehicle to spin up to approximately 180 deg/sec transversely. Normally, such an occurrence would be catastrophic, as for a time it appeared to be. The high spin rate interfered with the acquisition of real time data by the AFSCF and no effective telemetry was seen. Attempts were made to separate the spacecraft by ground command, at first unsuccessfully; but about one hour after the anomaly occurred, a short burst of data was received, indicating a hangup in the mission sequence. A real-time fix was generated to bypass the hangup, it was uplinked to the IUS, and separation was again attempted - this time with only minutes of battery power remaining. The fix was apparently successful. About one hour after separation from the IUS, the TDRS monitors reported acquiring the tumbling spacecraft, with its control system attempting to stabilize from the high rate. Once the TDRS was stabilized in the deficient orbit, an assessment was made of its health. The TDRS was thoroughly checked out and preparations were made to use some of its excess on-board propellant to effect many small orbit transfers and nudge itself into the proper orbit.

The spinup anomaly afforded the investigators an opportunity to return to basic engineering principles to verify that the sporadic and seemingly unintelligible signals telemetered during the spinup and beyond had been adulterated by several effects. First, most of the five rate gyro outputs were saturated during the spinup, and were transformed into erroneous body rate signals. Second, angular acceleration and rate effects on the RIMU instruments which were far from the center of mass caused erroneous linear accelerometer readings. Through a combined effort of the Boeing Aerospace Company and The Aerospace Corporation, a spinup scenario was defined which was compatible with a tech-roll seal collapse, the telemetered data and the oscillating signal strength traces. A six degree-of-freedom spinup simulation produced telemetry signals which were comparable to the flight data, satisfying the Government investigating committee.

The results of the two IUS flights were very satisfying from the standpoint of the Guidance and Control functions. In both flights, the Guidance system accepted the deficiencies in SRM1 performance and did its job well. Powered flight control operation was nominal during the SRM burn periods, as was coast control operation during both missions.

The IUS/TDRS flight has provided a great opportunity to thoroughly evaluate IUS performance, which has been taken to ensure a more reliable product for future flights.

REFERENCES

1. System Specification, Performance and Design Requirements for the DOD STS, Vol. 3, Basic Requirements for the Inertial Upper Stage System Segment, U.S.A.F., 1982.

2. IUS/STS/TDRS Interface Control Document, The Boeing Company, 1982.

AAS 84-042

GAMMA GUIDANCE – DESIGN AND RESULTS FOR TWO FLIGHTS

RANDALL H. KUHNS* AND DR. GRANT C. COOMER**

The primary purpose of both IUS-1 and IUS-2 missions was to place communications satellites (TDRS-1, DSCS II, DSCS III) in geosynchronous orbit. The guidance scheme implemented in the Inertial Upper Stage (IUS) to perform this task is an iterative technique known as Gamma Guidance. This paper presents the IUS-1 and IUS-2 unique mission designs for controlling the Gamma Guidance algorithm for these two flights. The results of post-flight analysis are presented for each mission. The results show that Gamma Guidance was accurately directing IUS to the proper orbit in each case.

INTRODUCTION

Two Inertial Upper Stage (IUS) flights have occurred to date. The first, a Titan T34D/IUS-2 launch vehicle combination, placed a dual payload consisting of a DSCS II and a DSCS III communications spacecraft into geosynchronous orbit. The second flight was the STS-6/IUS-1 mission which attempted to place TDRS-1 into geosynchronous orbit. The IUS vehicles were guided via an onboard explicit guidance algorithm known as Gamma Guidance. This paper will present the particular guidance strategy implemented to control the algorithm for these two flights and present data on algorithm performance.

GUIDANCE OR NAVIGATION

The term "guidance" is often used to discuss both guidance and navigation effects. Such will not be the case in this paper. Guidance in this context will consider only the control of burn times and directions to make the final orbit achieved, as indicated by the navigation system, equal to the desired orbit. The navigation system will be considered as "truth," and navigation error will not be discussed except where needed for completeness. Fig. 1 depicts the flow of information among the various systems which cause the vehicle (IUS) to attain the proper orbit.

* *Senior Specialist Engineer, Boeing Aerospace Company, Seattle, Washington*

** *Specialist Engineer, Boeing Aerospace Company, Seattle, Washington*

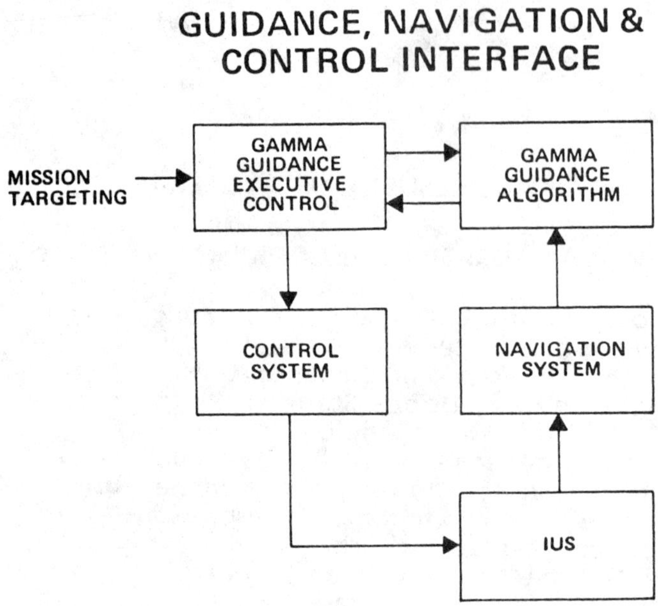

Figure 1

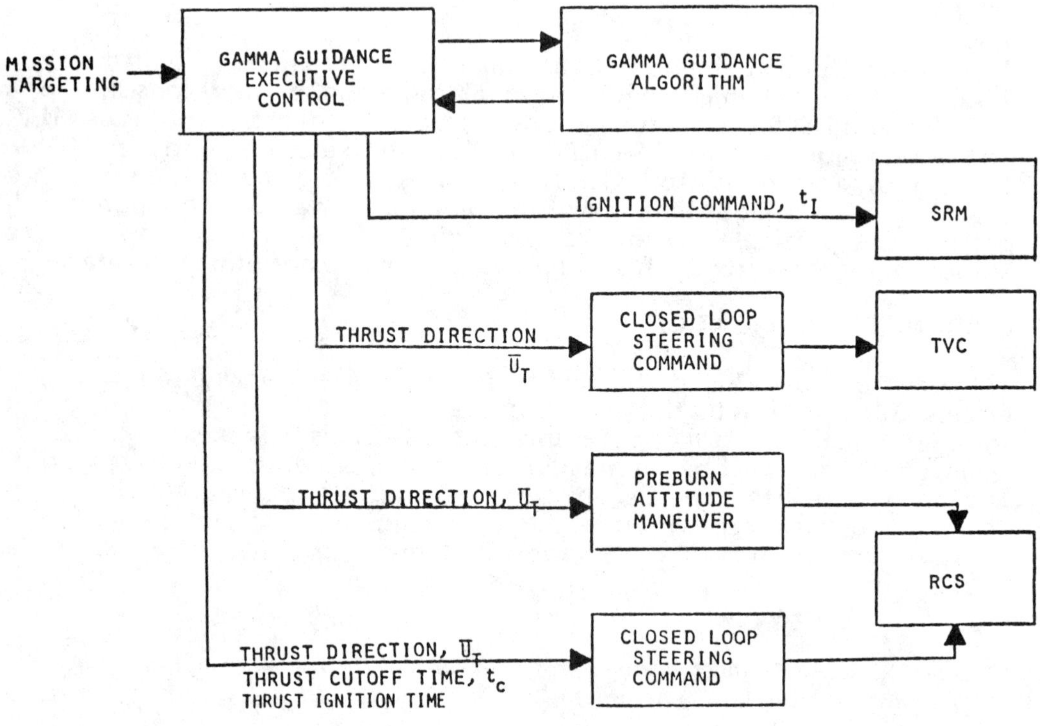

Figure 2 Gamma Guidance Interface with Control System

GAMMA GUIDANCE

The IUS guidance problem consists of determining burn times and directions for combinations of Solid Rocket Motor (SRM) and Reaction Control System (RCS) burns to place the IUS payload in the required mission orbit. The RCS burns, called vernier burns, are performed with small liquid rocket motors to compensate for variations in the total impulse provided by the SRMs. The commands generated by Gamma Guidance for the control system are presented in Fig. 2. A solution to this problem was presented in Reference 1. In particular, Gamma Guidance solves a two-point boundary value problem in ordinary differential equations where one boundary is the current IUS state vector and the second is the required mission orbit.

IUS guidance is an explicit guidance scheme mechanized as an onboard, autonomous real-time algorithm. Once the algorithm is initiated in flight, it requires no input or support from the ground. The algorithm performs all guidance functions including onboard targeting, midcourse updates, and closed loop guidance for both SRM and RCS burns. The algorithm is an iterative technique whose major functions can be seen in Fig. 3. δY represents the error in the mission orbit elements. $\delta\lambda$ represents the change in control elements, burn times and directions, needed to drive δY toward 0. At any given update, iterations continue to be performed until each element of δYor $\delta\lambda$ becomes arbitrarily small.

ALGORITHM CONTROL

The Gamma Guidance algorithm is a general algorithm capable of guiding IUS to many different mission orbits dependent only on the amount of energy available. This adaptability was achieved by implementing a series of menus that allow the mission planner to select menu options that will satisfy requirements for a particular mission. The menu options available were described in detail in reference 2, and the options used will be shown.

IUS-2

The first flight of an IUS, IUS-2, occurred October 30 and 31, 1982. IUS-2, carrying its DSCS II and DSCS III satellite payloads, was placed in an approximate 80x95 nautical mile park orbit by a Titan T34D launch vehicle. After coasting for 61 minutes in park orbit, the SRM-1 and RCS-1 burns were performed over a 14-minute period to place the IUS on a transfer orbit to geosynchronous altitude. During the 5-hour, 2-minute transfer coast, IUS-2 assumed a particular inertial attitude and performed continuous roll maneuvers to maintain the payloads within thermal requirements. SRM-2 and RCS-2 burns were performed over a 6-minute period to achieve geosynchronous orbit. DSCS II was released 90 seconds after reaching orbit. A 5-feet-per-second separation maneuver was performed prior to release of DSCS III. After DSCS III release the IUS performed a collision avoidance maneuver and shut down.

Fig. 4 presents a pictorial view of the IUS-2 flight depicting the key events. The IUS mission is divided into phases and arcs as shown in Fig. 5. Each phase consists of two arcs, a coast arc and a burn arc. Tables 1 and 2 indicate

GUIDANCE UPDATE: SEQUENCING THROUGH THE ALGORITHM UNTIL $\delta\overline{Y}$ IS ZEROED

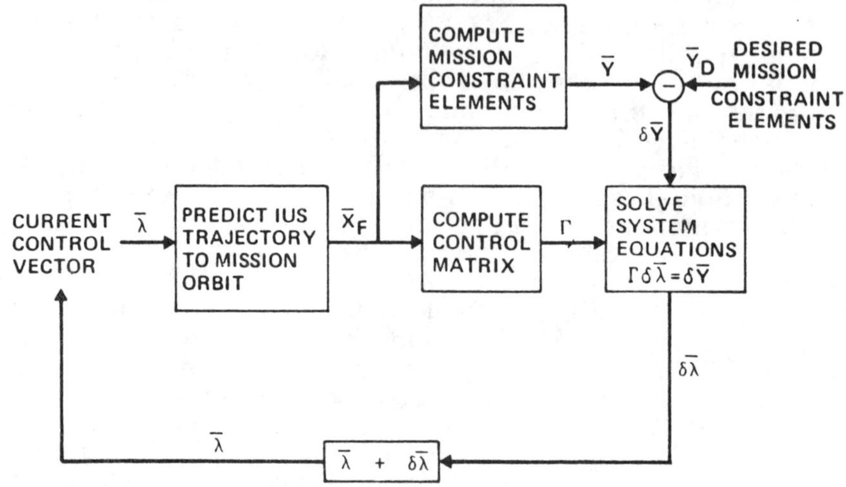

Figure 3

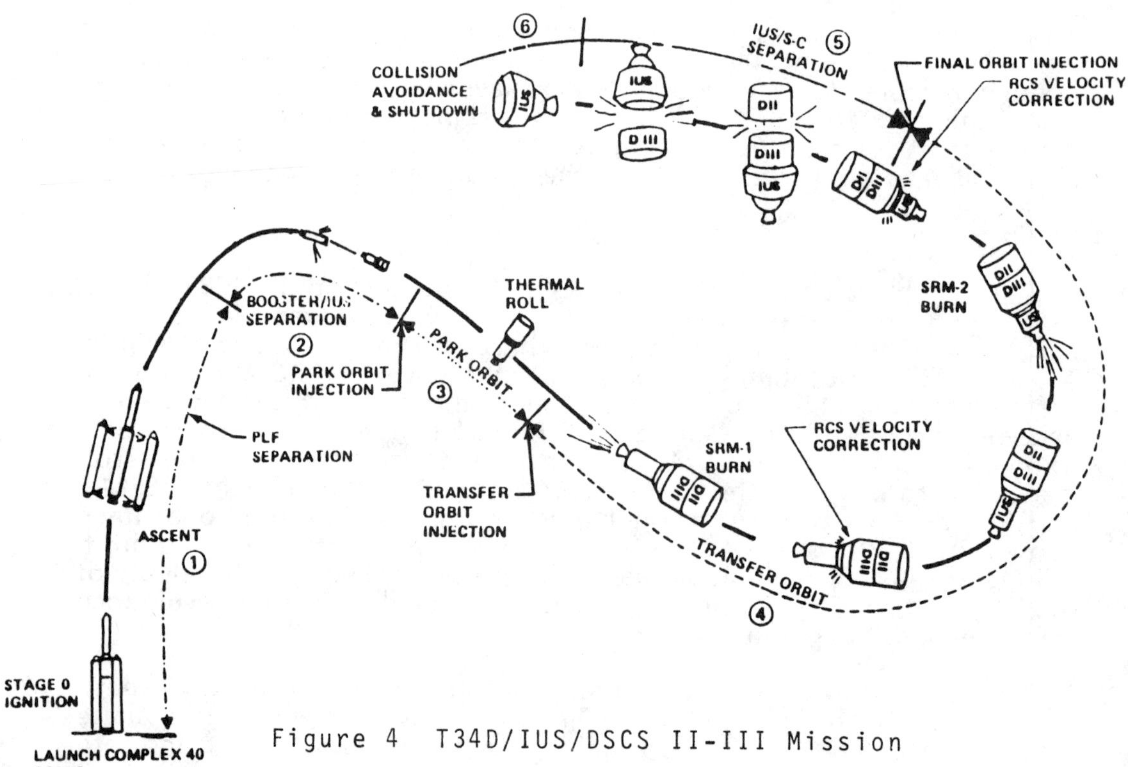

Figure 4 T34D/IUS/DSCS II-III Mission

PRECISE TRAJECTORY CONTROL ACHIEVED BY SRM & RCS BURN SEQUENCE

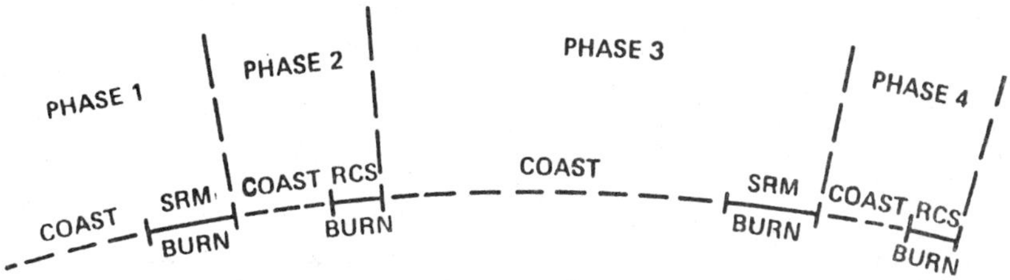

Figure 5

TABLE 1

IUS-2

ITEM	CONSTRAINT VARIABLE MENU (MC)		CONSTRAINT VARIABLES SELECTED **MISSION ARC NUMBER**								
			1	2	3	4	5	6	7	8	9
1	RP(1)	PREDICTED POSITION (1)									
2	RP (2)	PREDICTED-POSITION (2)									
3	RP (3)	PREDICTED-POSITION (3)									
4	VP (1)	PREDICTED-VELOCITY (1)									
5	VP (2)	PREDICTED-VELOCITY (2)									
6	VP (3)	PREDICTED-VELOCITY (3)									
7	RM	POSITION-MAGNITUDE	■	■	■	■					
8	VM	VELOCITY-MAGNITUDE									
9	SFPA	SINE-FLIGHT-PATH-ANGLE	■	■	■	■	■	■	■	■	
10	TEL	TANGENT-EARTH-LONGITUDE									
11	HV (1)	ANGULAR-MOMENTUM-VECTOR (1)	■	■	■	■	■	■			
12	HV (2)	ANGULAR MOMENTUM-VECTOR (2)	■	■	■	■	■	■			
13	HV (3)	ANGULAR-MOMENTUM VECTOR (3)	■	■	■	■	■	■	■	■	
14	HM	ANGULAR-MOMENTUM-MAGNITUDE							■	■	
15	EV (1)	ECCENTRICITY-VECTOR (1)									
16	EV (2)	ECCENTRICITY-VECTOR (2)									
17	EV (3)	ECCENTRICITY-VECTOR (3)									
18	C_3	TWICE-ORBITAL-ENERGY									
19	SC (1)	HYPERBOLIC-ASYMPTOTE-SCALED (1)									
20	SC (2)	HYPERBOLIC-ASYMPTOTE-SCALED (2)									
21	SC (3)	HYPERBOLIC-ASYMPTOTE-SCALED (3)									
22	DVSRM (1)	SRM1-VELOCITY-IMPULSE-MAGNITUDE	■	■							
23	DVSRM (2)	SRM2-VELOCITY-IMPULSE-MAGNITUDE	■	■	■	■	■	■			
24	DVSRM (3)	SRM3-VELOCITY-IMPULSE-MAGNITUDE									

■ USED □ NOT USED

TABLE 2

IUS-2

SELECTABLE IMPULSE ARC	GUIDANCE PARAMETER	CONTROL VARIABLE MENU (MCV)	CONTROL VARIABLES SELECTED — MISSION ARC NUMBER 1	2	3	4	5	6	7	8	9
SRM_1	ΔV_{SRM1}	λ_1	■	■							
	ΔV_{SRM2}	λ_2	■	■							
	ΔV_{SRM3}	λ_3	■	■							
	t_i	λ_4	■								
RCS_1	ΔV_{RCS1}	λ_5			■	■					
	ΔV_{RCS2}	λ_6			■	■					
	ΔV_{RCS3}	λ_7			■	■					
	t_i	λ_8									
SRM_2	ΔV_{SRM1}	λ_9	■	■	■	■	■	■			
	ΔV_{SRM2}	λ_{10}	■	■	■	■	■	■			
	ΔV_{SRM3}	λ_{11}	■	■	■	■	■	■			
	t_i	λ_{12}	■	■	■	■	■				
RCS_2	ΔV_{RCS1}	λ_{13}					■	■	■	■	
	ΔV_{RCS2}	λ_{14}					■	■	■	■	
	ΔV_{RCS3}	λ_{15}					■	■	■	■	
	t_i	λ_{16}									
SRM_3	ΔV_{SRM1}	λ_{17}									
	ΔV_{SRM2}	λ_{18}									
	ΔV_{SRM3}	λ_{19}									
	t_i	λ_{20}									
RCS_3	ΔV_{RCS1}	λ_{21}									
	ΔV_{RCS2}	λ_{22}									
	ΔV_{RCS3}	λ_{23}									
	t_i	λ_{24}									
TIME OF EARTH-RELATIVE-EVENT		λ_{25}									

■ USED   □ NOT USED

the constraints and control variables used for the IUS-2 flight (Y's and λ's, respectively in Fig. 3). Constraining the angular momentum vector, the radius magnitude and the sine of flight path angle at mission orbit injection was equivalent to constraining semi-major axis, eccentricity, inclination and right ascension of the ascending node of the mission orbit. The selected control variables in Table 2 were the elements Gamma Guidance was free to vary in order to achieve the mission orbit (i.e., satisfy the constraints).

Approximately 30 minutes after RCS-1 burnout, telemetry from IUS-2 was lost. No further telemetry was received during the remainder of the flight, due to a failure in an antenna switching device. DSCS II signal strength was monitored during the flight. From this and other sources, the times of certain salient events such as SRM-2 and RCS-2 burns could be deduced. The injection errors of the spacecraft were later observed to be less than 10% of requirements.

IUS-1

The second flight of an IUS occurred April 4 and 5, 1983. IUS-1 carrying the TDRS-1 tracking and data relay satellite was deployed from the Space Shuttle Challenger on the STS-6 mission. Fig. 6 depicts the key events for the planned IUS-1 flight.

IUS-1 with the TDRS payload was deployed from the shuttle in a 153 NM altitude parking orbit approximately 10 hours after launch. The vehicle coasted in park orbit for 55 minutes after deployment until SRM-1 ignition. The SRM-1/RCS-1 burn sequence took place over an 11-minute period to inject the vehicle into transfer orbit. The transfer orbit coast lasted slightly more than 5 hours. SRM-2 ignited and burned normally for 83 seconds, at which point a failure in the nozzle suspension system caused the nozzle to lock in a single position. With the nozzle unmoveable, the vehicle tumbled and normal flight operations ceased. The last guidance update planned during the SRM-2 burn occurred prior to the tumble. No further guidance was performed during the flight. The TDRS spacecraft was separated from IUS-1 approximately 2 hours after the tumble began. The TDRS orbit at separation had radius of apogee 200 nautical miles below geosynchronous radius, radius of perigee 7400 nautical miles low. The orbit inclination was 2.4° as opposed to the desired 0° inclined orbit.

Tables 3 and 4 indicate the constraint and control variable strategy for this mission. The semi-major axis, eccentricity and inclination were the orbital elements required to be constrained. The desired final orbit was to have been geosynchronous, circular with zero inclination.

Features of the IUS-1 flight compared to IUS-2 that caused different constraint menu selections were the fact that IUS-1 was near maximum payload and had a zero inclined desired mission orbit. The first was the reason for not constraining SRM-2 delta velocity magnitude in arcs 1 and 2; the second was the cause for different constraint selection in arcs 7 and 8.

The control variables selected for IUS-1 were the same as for IUS-2 with the exception of mission arc 1, the park orbit coast. Gamma Guidance was not

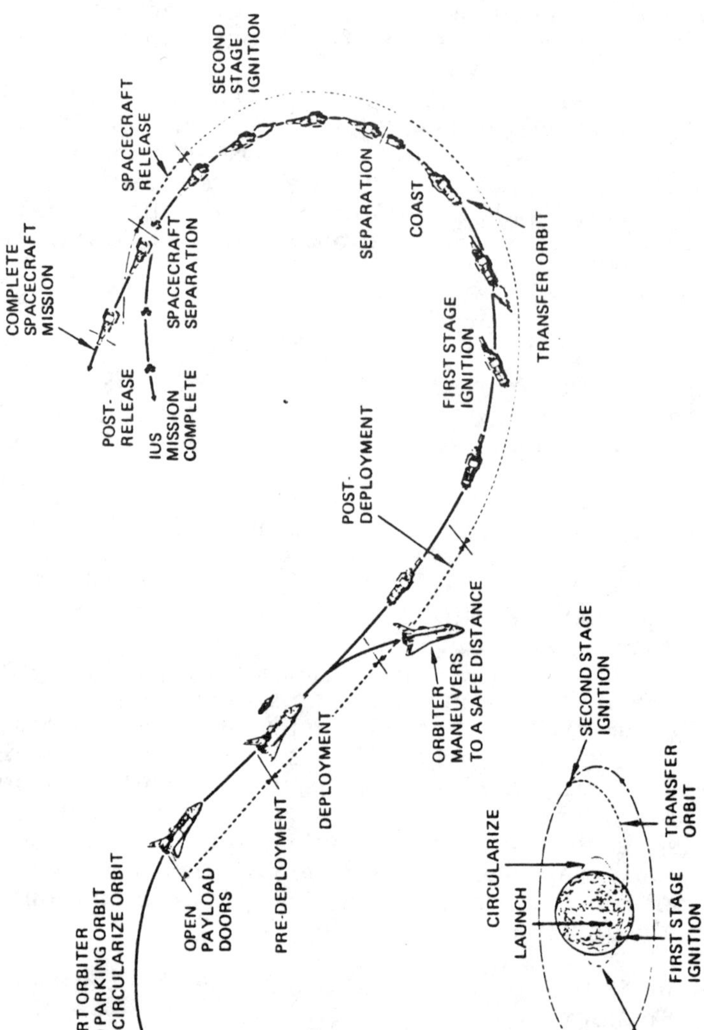

Figure 6 Planned IUS-1/TDRS-1 Mission

TABLE 3

IUS-1

ITEM	CONSTRAINT VARIABLE MENU (MC)		CONSTRAINT VARIABLES SELECTED MISSION ARC NUMBER								
			1	2	3	4	5	6	7	8	9
1	RP(1)	PREDICTED POSITION (1)									
2	RP (2)	PREDICTED-POSITION (2)									
3	RP (3)	PREDICTED-POSITION (3)									
4	VP (1)	PREDICTED-VELOCITY (1)									
5	VP (2)	PREDICTED-VELOCITY (2)									
6	VP (3)	PREDICTED-VELOCITY (3)							■	■	
7	RM	POSITION-MAGNITUDE	■	■	■	■					
8	VM	VELOCITY-MAGNITUDE									
9	SFPA	SINE-FLIGHT-PATH-ANGLE	■	■	■	■	■	■	■	■	
10	TEL	TANGENT-EARTH-LONGITUDE									
11	HV (1)	ANGULAR-MOMENTUM-VECTOR (1)	■	■	■	■	■	■			
12	HV (2)	ANGULAR MOMENTUM-VECTOR (2)	■	■	■	■	■	■			
13	HV (3)	ANGULAR-MOMENTUM VECTOR (3)	■	■	■	■	■	■			
14	HM	ANGULAR-MOMENTUM-MAGNITUDE							■	■	
15	EV (1)	ECCENTRICITY-VECTOR (1)									
16	EV (2)	ECCENTRICITY-VECTOR (2)									
17	EV (3)	ECCENTRICITY-VECTOR (3)									
18	C_3	TWICE-ORBITAL-ENERGY									
19	SC (1)	HYPERBOLIC-ASYMPTOTE-SCALED (1)									
20	SC (2)	HYPERBOLIC-ASYMPTOTE-SCALED (2)									
21	SC (3)	HYPERBOLIC-ASYMPTOTE-SCALED (3)									
22	DVSRM (1)	SRM1-VELOCITY-IMPULSE-MAGNITUDE	■	■							
23	DVSRM (2)	SRM2-VELOCITY-IMPULSE-MAGNITUDE			■	■	■	■			
24	DVSRM (3)	SRM3-VELOCITY-IMPULSE-MAGNITUDE									

■ USED □ NOT USED

TABLE 4

IUS-1

SELECTABLE IMPULSE ARC	GUIDANCE PARAMETER	CONTROL VARIABLE MENU (MCV)	CONTROL VARIABLES SELECTED — MISSION ARC NUMBER 1	2	3	4	5	6	7	8	9
SRM_1	ΔV_{SRM1}	λ_1	■	■							
	ΔV_{SRM2}	λ_2	■	■							
	ΔV_{SRM3}	λ_3	■	■							
	t_i	λ_4									
RCS_1	ΔV_{RCS1}	λ_5			■	■					
	ΔV_{RCS2}	λ_6			■	■					
	ΔV_{RCS3}	λ_7			■	■					
	t_i	λ_8									
SRM_2	ΔV_{SRM1}	λ_9	■	■	■	■	■	■			
	ΔV_{SRM2}	λ_{10}	■	■	■	■	■	■			
	ΔV_{SRM3}	λ_{11}	■	■	■	■	■	■			
	t_i	λ_{12}	■	■	■	■	■				
RCS_2	ΔV_{RCS1}	λ_{13}					■	■	■	■	
	ΔV_{RCS2}	λ_{14}					■	■	■	■	
	ΔV_{RCS3}	λ_{15}					■	■	■	■	
	t_i	λ_{16}									
SRM_3	ΔV_{SRM1}	λ_{17}									
	ΔV_{SRM2}	λ_{18}									
	ΔV_{SRM3}	λ_{19}									
	t_i	λ_{20}									
RCS_3	ΔV_{RCS1}	λ_{21}									
	ΔV_{RCS2}	λ_{22}									
	ΔV_{RCS3}	λ_{23}									
	t_i	λ_{24}									
TIME OF EARTH-RELATIVE-EVENT		λ_{25}									

■ USED □ NOT USED

allowed to vary the impulse time (and therefore the ignition time) of the first SRM burn. The impulse time was kept at the park orbit node to ensure energy was not wasted.

Two events, one prior to deployment from the Shuttle, the other 19.5 minutes prior to SRM-2 ignition combined to have an effect on guidance during the IUS-1 flight. Prior to deployment, a star scan maneuver was performed to correct IUS attitude errors that had accumulated over the 5 hours since launch. Light flashes associated with the orbiter attitude thrusters are believed to have caused abnormal processing of the star sighting data, with the result being different IUS attitude references in the two redundant IUS computers. From that point on, accelerometer data were resolved differently in the two machines and navigation diverged between the two computers. The second event was a switch from the prime computer (computer A) to the redundant machine (computer B) as the computer in control. This second event was the result of an attempt to write in protected memory by computer A. The combined effect of these two events was to switch between computers with significantly different perceptions of the transfer orbit for the SRM-2 burn. The ramifications of this on the flight of IUS-1 and Gamma Guidance evaluation are described in the following sections.

RESULTS

There are several ways to measure the performance of the Gamma Guidance algorithm. Among these are number of iterations, amount of time per update, and the ability to handle off-nominal conditions. The predictability of inflight performance with ground-based simulation prior to flight is important to ensure reliable and adequate mission planning. The final test is the ability of the guidance system to accurately guide the vehicle to the final desired mission orbit in the presence of dispersions from subsystem weights, performance and models.

Table 5 summarizes data from IUS-1 and IUS-2 on guidance algorithm processing time and convergence history. No more than 4 iterations were required for any guidance update in both missions. In Fig. 3, this corresponds to 4 repetitions of the constraint error δY evaluation. Only 3 repetitions of the solution loop $\Gamma \delta\lambda = \delta Y$ are required since the algorithm is exited once the constraint error tolerances on δY are satisfied. Background processing time can only be measured to the nearest half-second from telemetry. In general, it can be seen that each iteration required a half-second or less of real time to perform. The CPU time required was considerably less because the Guidance algorithm is background processing. In no case did an update require more than two seconds to complete. Data were not available for IUS-2 after the antenna-switching failure or from either flight during periods out of contact with ground stations.

Both flights exhibited lower than expected SRM1 added velocity. Gamma Guidance compensated by requiring larger RCS1 burns. As shown in Table 6, the total actual added velocity delivered in each flight corresponded to the total expected. The actual numbers are slightly larger due to the extra velocity required to overcome the gravity loss effects of obtaining more delta

TABLE 5

GAMMA GUIDANCE CONVERGENCE RESULTS

TYPE	ARC	UPDATE RATE	IUS-1		IUS-2	
			MAX NUMBER ITERATIONS PER ARC	PROCESSING TIME	MAX NUMBER ITERATIONS PER ARC	PROCESSING TIME
Park orbit coast	1	every 200 (sec.)	NA*	NA	4	3 (sec.)
SRM1 burn	2	every 10	4	2 (sec.)	3	1.5
Pre RCS Coast	3	every 50	4	1.5	4	2.0
RCS-1 burn	4	every 10	3	1.5	3	1.5
Transfer orbit coast	5	every 500	2	.5	3	2.0
SRM-2 burn	6	every 10	3	.5	NA	NA
Pre-RCS coast	7	every 50	NA	NA	NA	NA
RCS-2 burn	8	every 10	NA	NA	NA	NA

* *NA = Not available*

TABLE 6

GUIDANCE CORRECTION FOR OFF-NOMINAL SRM-1 BURN

	IUS-1		IUS-2	
	Expected ΔV (ft./sec.)	Actual ΔV (ft./sec.)	Expected ΔV (ft./sec.)	Actual ΔV (ft./sec.)
SRM-1*	8030.36	8015.30	8183.74	8145.70
RCS-1	30.79	47.45	8.46	49.85
TOTAL	8061.15	8062.75	8192.20	8195.55

* *Includes impulse obtained from the outgassing of motor case liner material after fuel burn completion*

velocity with the later-occurring RCS burn. Small control system pointing errors during the burns may also have contributed to RCS requirements.

These two flights provided the first opportunity to correlate flight results with ground-based simulation data. The data in Table 7 demonstrate close agreement. The simulation data were obtained by supplying a navigation state vector telemetered early in the transfer orbit coast. These results were compared with telemetered arc end times and burn directions from a guidance update near the time of the telemetered state vector used in the simulation. The burn directions represent the unit pointing vector in inertial coordinates for the remaining burns needed to achieve mission orbit.

The widely divergent navigation states between computer A and the backup, computer B, on the IUS-1 flight demonstrated the robustness of the Gamma Guidance Algorithm for handling unusual situations. Prior to the SRM-1 burn, both computers indicated approximately the same navigation state and the respective guidance solutions were nearly the same. During the burn, as a result of the differing attitude references between the two machines, the burn was not performed in the proper direction according to the computer B. Since computer B was not in control, its commands were not followed for controlling the vehicle. Guidance in this machine reacted to what appeared to be a deteriorating situation by calling for a very large RCS1 burn to correct the increasingly large errors. These commands were also ignored since computer A was in control.

The guidance in computer A functioned in a normal manner since it was in control and the burns were in the proper direction according to its attitude reference. During the transfer coast Gamma Guidance in computer B kept converging, finding a solution of SRM2 and RCS2 burn times and directions that would solve the desired mission orbit constraints of angular momentum and sine of flight path angle, as shown in Table 3. Table 8 presents data of the guidance situation that existed in computer A just prior to the switch to computer B in control and in computer B just after the switch. The switch occurred approximately 5 minutes prior to SRM2 ignition. As shown, computer B navigation indicated a transfer orbit that would not achieve geosynchronous altitude for radius of apogee. The guidance-predicted mission orbit elements shown in Table 8 reveal how well Gamma Guidance in each computer would attempt to achieve the desired mission orbit, based on its perception of the transfer orbit. Based on tracking data, computer B had the best indication of the actual transfer orbit.

Table 9 presents the accuracies with which the DSCS II and DSCS III spacecraft were injected into orbit. The errors represent the combined navigation, guidance and control effects. Guidance error could not be separated from the total for this mission since telemetry was not available. Normally, guidance errors would have been evaluated by computing mission orbital elements from a navigation state vector telemetered immediately after completion of the RCS2 burn. Comparing these orbital elements with those desired would indicate combined guidance and control error without the inclusion of navigation error. Simulations indicate a combined guidance and control error of less than .5 nautical mile in semi-major axis, for example.

TABLE 7
SIMULATION CAPABILITY OF GAMMA GUIDANCE

	IUS-1 ARC END TIME (1)		IUS-2 ARC END TIME (1)	
ARC	GROUND(2) (Sec.)	FLIGHT (3) (Sec.)	GROUND (2) (Sec.)	FLIGHT (3) (Sec.)
5	22386.24	22386.31	22440.80	22440.89
6	22490.92	22490.97	22528.84	22528.94
7	22711.04	22711.11	22749.40	22748.95
8	22799.96	22798.65	22817.08	22810.059

(1) Guidance planned/predicted arc end times from umbilical disconnect
(2) Ground simulation values based on arc 5 telemetered state vector
(3) Telemetered values from an update near the state vector in (2)

	IUS-1 BURN DIRECTION (4)		IUS-2 BURN DIRECTION (4)	
ARC	GROUND	FLIGHT	GROUND	FLIGHT
SRM-2 X	.79963	.79963	-.91430	-.91397
Y	.44961	.49961	-.20738	-.21002
Z	-.39804	-0.39804	-.34792	.34722
RCS-2 X	.78830	.78831	-.76283	-.90674
Y	.46864	.46863	-.60936	-.21689
Z	-.39870	-0.39869	.21625	.36163

(4) SRM/RCS burn inertial unit pointing vectors

TABLE 8
IUS-1 GUIDANCE SOLUTIONS

ARC	COMPUTER A				COMPUTER B			
	GUIDANCE ARC END TIMES (Sec.)	BURN UNIT POINTING DIRECTIONS X	Y	Z	GUIDANCE ARC END TIMES (Sec.)	BURN UNIT POINTING DIRECTIONS X	Y	Z
5	22386.31				22503.00			
6	22490.97	.8001	.4489	-.3979	22607.66	.8148	.4241	-.3953
7	22711.11				22828.66			
8	22798.65	.7890	.4675	-.3986	23001.418	.8002	.4466	-.4002

ORBITAL ELEMENTS	TRANSFER ORBIT NAVIGATION	GUIDANCE-PREDICTED FINAL ORBIT	TRANSFER ORBIT NAVIGATION	GUIDANCE-PREDICTED FINAL ORBIT	DESIRED MISSION ORBIT
SMA	13183.27 (NM)	22766.84 (NM)	13094.48 (NM)	22768.35 (NM)	22766.84 (NM)
ECC	.72695	.0000386	.725195	.0081	0
INC	26.1685°	.00000171°	25.9528°	.00013	0°
RAAN	299.755°	UNDEFINED	299.778°	UNDEFINED	UNDEFINED
AP	179.8438°	121.04°	179.1138°	-58.5727	UNDEFINED
RA	22766.85 (NM)	22767.72 (NM)	22590.53 (NM)	RP = 22583.91 (NM)	22766.84 (NM)

TABLE 9
IUS-2 INJECTION ACCURACY*

DSCS II INSERTION ERROR (EPOCH: SPACECRAFT SEPARATION)

Parameter	Target Value	Actual**	Error	Allowable Error
Semi-Major Axis (n.mi.)	22793.	22780.	-13	+ 120 † -102
Eccentricity (n.d.)	0.0011	0.0013	+ 0.0002	+ .008 † -.001
Inclination (deg.)	2.484	2.493	+ 0.009	+ .085 † -.074
RAAN (deg.)	277.612	277.016	-0.596	+ 2.3
Drift Rate (deg/day west)	0.5	0.567	+ .067	± 2.91
Injection Longitude (deg)	93	92.53	-.47	± 5°

DSCS III INSERTION ERROR (EPOCH: SPACECRAFT SEPARATION)

Parameter	Target Value	Actual**	Error	Allowable Error
Semi-Major Axis (n.mi.)	22817.	22804.	- 13.	+ 120 † -102
Eccentricity (n.d.)	0.0000	0.0006	+ 0.0006	+ .008 † -.001
Inclination (deg.)	2.483	2.502	+ 0.019	+ .085 † -.074
RAAN (deg.)	277.610	276.995	-0.615	+ 2.3

* Boeing 6D guided nominal at separation event (includes 1 ft/sec separation impulse)

** Orbit determination from SCF

† Orbit errors equivalent to STS-100 specification errors (in RTN Coordinates)

This table provided by the Air Force

* *Combined navigation and guidance results*

IUS is capable of delivering payloads to orbit with great accuracy, as the small injection errors in Table 9 indicate.

CONCLUSIONS – LESSONS LEARNED

Based on data from the first two IUS flights, several conclusions can be drawn regarding Gamma Guidance Algorithm performance.

First, the algorithm converges quickly and reliably using little computer background processing time. The algorithm, in flight, performs nearly as predicted in ground-based simulation, thereby increasing confidence that the preflight-chosen values for algorithm control will have the desired effects and accuracy. Flight data show that the algorithm performs well in compensating for vehicle dispersions such as the noted SRM delta velocity dispersions. The algorithm has been shown to be robust in the face of unusually large off-nominal conditions such as those present in computer B of the IUS-1 flight. IUS-2 demonstrated that the algorithm can guide payloads to orbit with great accuracy.

Based on the experience garnered on these first two flights, there are no indications that the algorithm need be modified or that the design philosophy and technique for choosing flight unique algorithm control parameters should be altered.

ACKNOWLEDGMENT

Developed under Air Force contract F04701-78-C-0040 (IUS). Cognizant Technical Managers, Cpt. Charles Vono and Lt. Cindy Halliday of the IUS Program Office. The authors wish to thank Carol Fisher for word processing and preparing the tables.

REFERENCES

1) J. W. Hardtla, "Gamma Guidance for the Inertial Upper Stage (IUS)," *AIAA Guidance and Control Conference Proceedings*, Palo Alto, California, August 7-9, 1978.

2) J. W. Hardtla and Loren A. Schultz, "Inertial Upper Stage (IUS) Multi-Mission Adaptability via Gamma Guidance Software Menu Tables," *AAS Annual Rocky Mountain Guidance and Control Conference*, Keystone, Colorado, January 30 – February 3, 1982

AAS 84-043

CENTAUR D-1A GUIDANCE/SOFTWARE SYSTEM

Andrew L. Gordan *

The main body of this paper describes the evolution of the Centaur D-1A Guidance and Software System. Specifically, the performance of the explicit guidance equations, using a linear tangent steering law. Inherent flexibility exists in the equations in that they have multimission capability. They can accommodate both earth-orbital and earth-escape missions with either one or two Centaur burns. They can also guide for multi-burn earth orbital missions. The Centaur performance is indicated in terms of optimality (propellant usage), accuracy, flexibility and computer requirements. In the course of the Centaur Guidance development substantial changes and improvements have been made and more improvements are on the way for the Shuttle/Centaur Guidance. It is the intent of this paper to describe, provide insight into, and identify certain unique aspects of the individual Centaur flight profiles. Mission profile(s) are described narratively with some numerical data given in cases where it may be useful.

* *NASA Lewis Research Center, Cleveland, Ohio 44135*

INTRODUCTION

The purpose of this paper is to describe the basic philosophy of the Centaur D-1 Guidance modules and its application to powered flight. It is a guidance scheme which can compute an optimum thrust attitude as an explicit solution to a two-point boundary-value problem. That is, the commanded thrust vector is found by a direct solution of the appropriate equations of motion subject to the initial boundary condition of the vehicles instantaneous state and final boundary condition at thrust termination. The guidance scheme is truly an explicit guidance scheme in that it will retain its optimization properties under vehicle perturbations without any loss in accuracy at cutoff. The instantaneous state of the vehicle-velocity, position, longitudinal acceleration, and gravitational acceleration are available from the on-board Centaur navigation system.

The objective of the Centaur guidance system is to provide the maximum amount of explicit guidance for a set of space mission objectives such as:

-- Deliver a payload into a specified elliptical orbit about the earth for a wide range of apogees, perigees, inclination with respect to the equatorial plane and/or longitude of the ascending node. Argument of perigee and true anomaly at injection may or may not be constrained.

-- Deliver a payload into a pre-specified transfer conic (ellipse, parabola, or hyperbola) which will intercept a "target" point in N-body inertial space at a fixed time.

-- Deliver a payload into a minimum delta V transfer conic (conic parameters determined in flight) which will intercept a "target" point in N-body inertial space. Time-of-arrival may or may not be specified.

For reasons of launch window, tracking coverage, fuel economy, and vehicle design limitations, it may be desirable to attain the final desired orbit through a series of alternate thrusting periods and free-flight coasts. The Centaur Guidance System handles these multiple-burn type missions in a series of "stages," steering to appropriate interim conics for each free-flight coast with the final burn guided to achieve the desired terminal conditions.

INERTIAL MEASUREMENT GROUP (IMG)

If a perfectly accurate and flexible hardware system(s) could be built, the rocket thrust vector could be pre-programmed as a function of time, and the mission objectives could be satisfied without closed-loop guidance. In reality, the rocket engine and other vehicle hardware will not perform ideally; the thrust vector will differ in magnitude and direction from the model. This difference can be sensed (directly or indirectly) by the guidance sensors (gyros and accelerometers). For Centaur these sensors are provided by the Honeywell Inertial Measurement Group

(IMG). These sensors will in themselves have errors, but the errors are much smaller than the deviations in the engine system. The purpose of guidance is then to command the rocket thrust vector attitude and engine on and off commands based on the sensed position, velocity, and acceleration to meet the mission objectives.

Gimbal Platform - Approximately 13 years ago, a decision on a gimbal platform for the Centaur D1 Inertial Reference Unit (IRU) was made on the data base established through the 29 flights of the Centaur D. The decision was made because of the knowledge developed on the set of problems peculiar to gimbal platforms against the uncertainty of new state of the art strapdown systems and the reluctance to go with a yet-to-be-discovered set of strapdown problems. Figure 1 below is a simplified block diagram depicting the Centaur D-1A Guidance System.

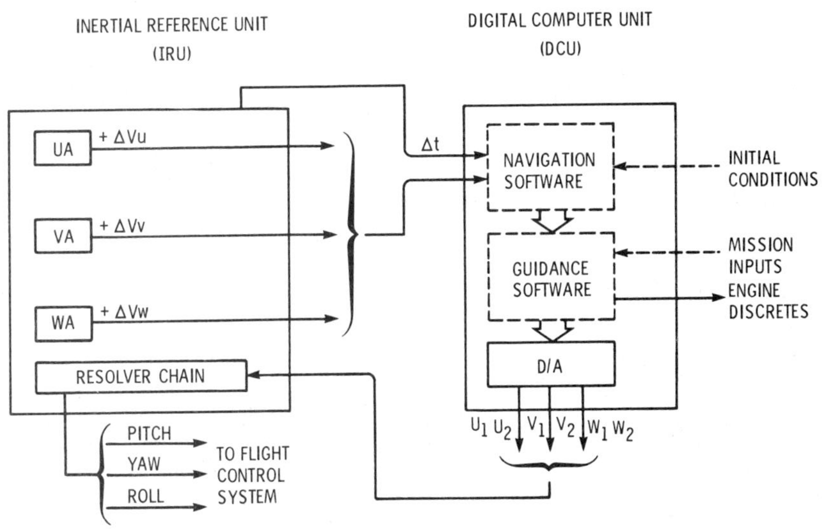

FIGURE 1. Centaur D-1 Guidance System

IMPLICIT VERSUS EXPLICIT GUIDANCE EQUATIONS

A terse statement of the guidance problem as given above is the process of determining steering commands, and engine on-off commands, using data obtained from the guidance sensors, which will lead to mission success. For the Centaur inertial guidance system described in this paper, this process is oriented toward a self-sufficient set of equations that are programmed in the Centaur Digital Computer Unit (DCU).

Guidance equations are usually termed "explicit" or "implicit." Explicit guidance equations solve the problem of how to get from an instantaneous position and velocity vector to the desired cutoff conditions according to an optimal principle (i.e., minimum propellant utilization).

Implicit guidance equations basically rely heavily on data derived from preflight nominal trajectories. The guidance philosophy may be to either try to steer back to a nominal trajectory, or to steer optimally to the target. If the latter approach is chosen, a much greater number of pseudo-nominal trajectories are required to account for possible vehicle or state perturbations. Within the range of perturbations considered, the guidance equations will work satisfactorily. However, for vehicle conditions outside this range, the scheme does not adapt as accurately as explicit equations will.

Implicit Targeting - The general approach to targeting an implicit set of guidance equations is based upon determining the coefficients of a polynominal expansion about a nominal trajectory. These expansions provide an implicit guidance scheme with the ability to continuously steer to a nominal flight path. Guidance targeting is contained within the guidance polynominals obtained for the implicit scheme. The preflight guidance targeting effort involves the determination of the coefficients for these polynominals. The optimality of this approach depends on the validity of the nominal trajectories used to generate the polynominals. For the implicit type which returns to a nominal trajectory, a greater amount of propellant is usually required, since the original path may no longer be the optimum one if perturbations, in state or the vehicle, have occurred.

Explicit Targeting - As discussed in implicit guidance targeting, the number of mission targeting trajectories required are roughly the same for both implicit and explicit schemes and are usually obtained in the same manner. Other than these, nearly all of the remaining complex preflight guidance coefficient calculations required for implicit guidance are eliminated with explicit schemes.

EVOLUTION OF CENTAUR D-1A GUIDANCE; RATIONALE FOR

Prior to Atlas/Centaur-30, all Centaur D guidance was based on implicit guidance. After an exhaustive analysis and simulation of features of the guidance equations, i.e., implicit versus explicit, the advantage of explicit guidance was clearly recognized. A set of explicit guidance equations was then developed that incorporated some of the best features of each guidance equation set provided by NASA, GDC, and TRW. The criteria for selection of the equations were accuracy, performance (optimality in the sense of minimum propellant expenditure) and flexibility. Flexibility in a set of guidance equations implies the ability to guide a wide spectrum of missions with a minimum of equation changes. The Centaur D-1 equations are extremely flexible and reliable as indicated by the results shown in the subsequent discussions.

Mission Success Record - The first flight of the D-1 Centaur occurred on April 5, 1973, preceded by 29 flights of the first generation Centaur in which 4 out of 7 R and D flights were successful and 19 out of 21 operational flights were successful. Of the 34 second generation D-1 Centaur flights through August 6, 1981, all operational missions were successful except two "no-trials" where booster problems precluded Centaur from those two mission attempts, and one R and D flight failure. In tabular form, Centaur's flight record is summarized below:

Centaur	R and D Flights	Operational Flights
D	4 successes out of 7 (1 no-trial not included)	19 successes out of 21
D-1	1 failure (out of 1)	33 successes out of 33 (2 no-trials not included)

Of the 33 successful D1 missions, 20 involved placement of communication satellites into earth orbit and 13 involved injection of planetary spacecraft into escape orbit.

Accuracy - Payload placement accuracy is an important part of the mission success record. The combination of the highly accurate inertial guidance system together with flexible guidance software provides the Atlas/Centaur with the capability to achieve precise payload injection conditions for many different types of missions. The injection precision has been demonstrated on many interplanetary, near-earth orbital, and synchronous orbit missions.

On Pioneer missions to Jupiter, the guidance requirement included precision orientation of a solid-rocket kick stage to achieve the proper final planetary intersect conditions. The major error source on these missions was the solid rocket impulse uncertainty.

A summary of Atlas/Centaur mission accuracies for various types of missions together with flight results and mission requirements are tabulated as follows:

	Transfer Orbit Error (Centaur 2nd Burn)	
Typical Synchronous Orbit Missions - INTELSATS	Flight Data	Guidance System 3 σ
Perigee (N.Mi.)	0.03-1.2	2.5
Apogee (N.Mi.)	3.7-26.3	77
Inclination (Deg.)	Not Available	0.024

Lunar and Interplanetary Mission Accuracy	Midcourse Correction Requirement (MCR) m/sec.		
	Mission Requirement	Flight Data	Guidance Sys. 3 σ
Surveyor	<50	1-6	17
Mariner Mars	<13.5	1-2	9
Mariner Venus Mercury	<13.5	7	7
Pioneer 10	<117	14	117
Pioneer 11	<108	40	108

Examples of Titan/Centaur accuracies based upon flight results and guidance accuracy analysis studies.

Typical Heliocentric Mission Helios A	Mission Requirement	Flight Data	Guidance System 3 σ
Perihelion (A.U.)	+0.01	0.0004	0.0015
Eclipic Inclination (Deg.)	+0.60	0.02	0.32

Titan/Centaur Interplanetary Missions	Midcourse Correction Requirement (MCR) m/sec.		
	Mission Requirement	Flight Result	Guidance System 3 σ
Viking A	<15	3.9	11.1
Viking B	<15	4.7	11.1

The flight record has demonstrated that the D1 Centaur has a mission success probability of ~98 percent. Continuation of this perfect operational flight record, the mission success probability will reach > 0.99 in time for its use in the Shuttle program.

UNIQUENESS OF CENTAUR D-1 EQUATIONS

The explicit nature of the D-1 Centaur equations means that targeting is much simplified with a minimum number of mission dependant constants, in contrast to D Centaur. Also, the incorporation of the linear tangent steering law allows optimum trajectory profiles which were not available under the D Centaur program. The technique used is to integrate numerically the equations of motion to the predicted cutoff time to generate injection parameter error signals. The integration is then repeated to produce partial derivatives of the injection parameters with respect to the steering coefficients. Finally, a linear system of equations is solved for corrections to the steering coefficients. It would be prudent at this point to mention the use of Calculus of Variations (COV). The D-1 yaw steering laws are designed to minimize performance loss due to out-of-plane steering. This is accomplished by the use of COV to compute the nominal components of the commanded vector, i.e., yaw steering nodes; target vector, orbital inclination, or orbital inclination and node. The COV technique reaches the same end conditions as the corresponding guided trajectory, but in addition establishes a lower bound for fuel requirements.

CENTAUR D-1 DIGITAL COMPUTER UNIT (DCU)

A major characteristic of the D-1 Centaur is the integration of several functions through the use of a powerful airborne Digital Computer Unit (DCU). The DCU plays a role in several functions; navigation, guidance, control, sequencing, propellant utilization, propellant tank pressurization, and instrumentation and telemetry. Its speed, storage capacity, and input/output also provides additional capacity for growth. A variety of DCU I/O are shown in Figure 2 below.

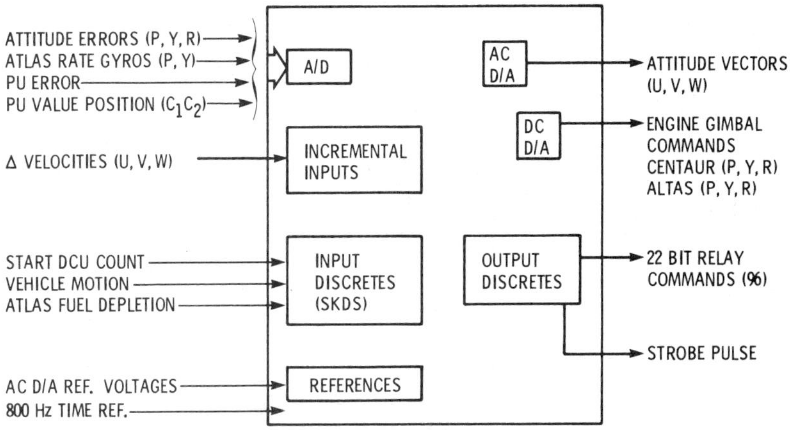

FIGURE 2. DCU Input/Output

The use of a powerful digital computer has permitted many functions to be done by software, which are normally done by other hardware. In particular, mission-peculiar requirements are handled by software where practical, permitting changes to be expressed in mission constants.

The DCU is a stored program, random access core machine. Memory is composed of 16,384 words of 24 bits each. Hardware interlocks prevent changing contents of 12,280 of these words, however, when special laboratory equipment is connected the contents can be changed. The flight program and telemetry formats are loaded into this area, and cannot be altered on the vehicle. Memory cycle time is nominally three microseconds.

For the purpose of this paper the remaining discussion will address the DCU as part of the guidance system. It is assumed that the reader is well versed in the details of digital computers and as such these details will not be considered in this paper.

DCU SOFTWARE - To provide insight into the functional aspects of the guidance/navigational system, a brief description of the DCU software is provided.

As discussed above, the DCU plays a significant role in the Centaur D-1A astrionics system. The software is an integral part of the DCU. As pointed out previously, the D-1A software incorporates many functions done by hardware on Centaur D. This reduction in hardware allows the vehicle configuration to remain static while adding in the software the design flexibility for mission peculiar requirements.

The D-1A software was designed to satisfy specific objectives in the areas of cost, reliability, launch simplicity, response and resiliency. For example, to provide resiliency, the software is designed to remain intact and functioning in the unforeseen event of failures in the extended system hardware. Its task is to achieve maximum flight success in spite of system failures.

SOFTWARE DESIGN CONCEPTS - Within the D-1A software specific concepts have been developed to achieve the following objectives:

Modularity: A modular software concept fulfills the requirements for a cost effective and flexible software system. The concept classifies software into two categories: an executive software system that remains unchanged through all missions, and a set of mission-or-vehicle-peculiar task modules that can be selected from a library and adopted for the current mission. The task modules can be scheduled by the executive at different frequencies during the flight. They can be turned off or reactivated for different phases of flight, and interrupted at any time during their operation. Since the modules do not communicate with each other, but only through the Executive, they are assured of consistent sets of data. Program changes are inserted at the module level, and checked at the module and integrated program level (all task modules operating together as a system). Figure 3 below illustrate the flexible modularity.

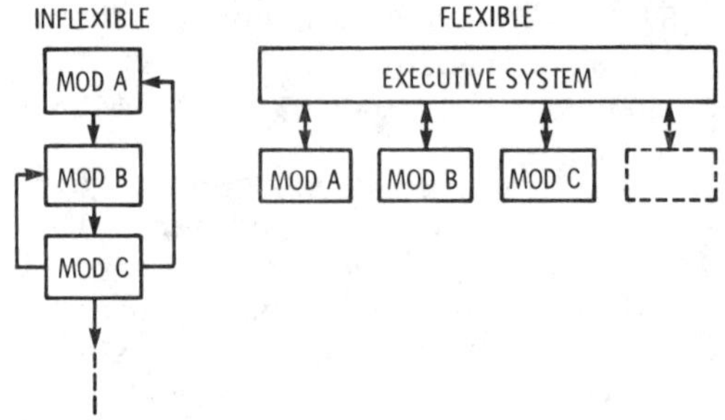

FIGURE 3
A Flexible Kind of Modularity

Independence from Interrupts: The design of the D-1A software is such that any combination of two or more modules can be in a simultaneous state of interrupt; furthermore, the interrupt is allowed at any location in the modular program. Thus the modular task programs are completely independent of the interrupts, and each one can be coded as if it were the only module in the computer. This "independence" concept is implemented by task scheduling and real-time interrupt system service techniques, at the DCU level, and by appropriate design of all modules only at the final program level.

SOFTWARE DEVELOPMENT

Functional tasks are designed as separate software modules, and are developed and checked out in a parallel process. The checkout of the software is subdivided into the task level and integrated level. The software system structure is designed general enough so that after the first integrated checkout has occurred, revised or new modules can be incorporated with minimum effort. Once a revised module is completely checked out, it is added to the modular library. Figure 4 depicts the flow of software development.

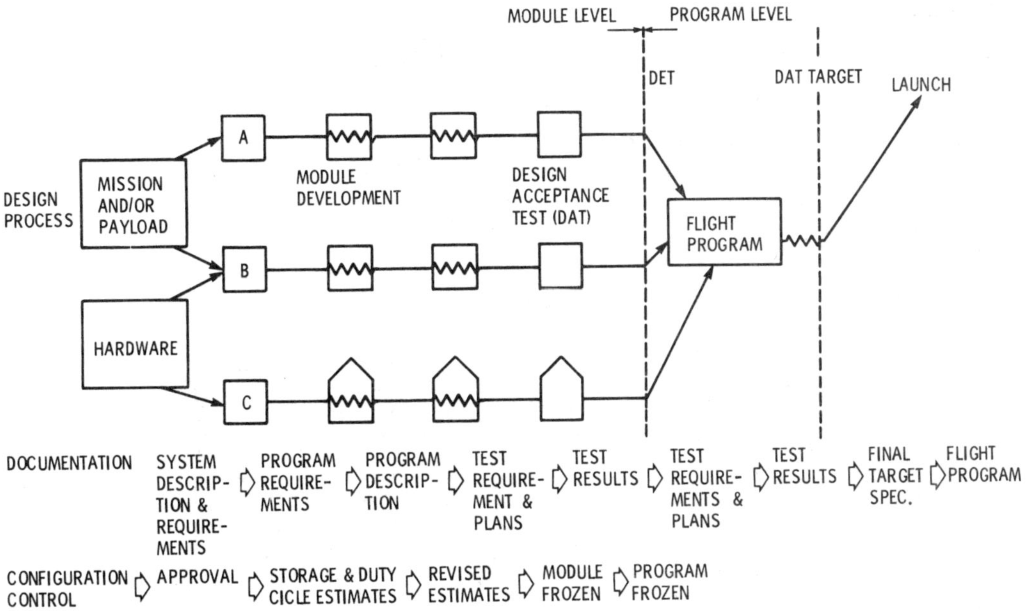

FIGURE 4. Software Development - Parallel Module Design and Checkout

Contingency Software. - A prime uniqueness of the DCU/Flight software is that it is designed failure tolerant. In the event of certain external equipment failures (such as, unscheduled thrust termination, or failure to start a stage) the software provides the capability of selecting reasonable alternative strategies.

For nonstandard environments (such as large steady-state drop in thrust level) the software senses the environment and makes appropriate adjustments to the trajectory. The recovery techniques are designed so that the mission is achieved within the performance capability of the launch vehicle. Every conceivable extreme nonstandard environment cannot be protected; however, a reasonable balance is achieved between software complexity and protection attained.

DCU/GUIDANCE FUNCTIONAL TASKS

The functional tasks performed by the DCU are coded in program modules. These modules are listed in the task table from which they are called by the executive for operations at the proper time to ensure correct module frequency. Some of the modules, e.g., navigation, operate throughout flight, while others are scheduled for only certain phases, e.g., Powered Autopilot. The modules do not interface with each other. Data flow is controlled by the data management portion of the executive. The following is a brief description and pictural representation of these modules pertinent to the Guidance function.

Navigation

Function. Furnish position, velocity, and acceleration data to guidance.

Method. Integrate in true inertial coordinate system, having previously converted for known platform drifts.

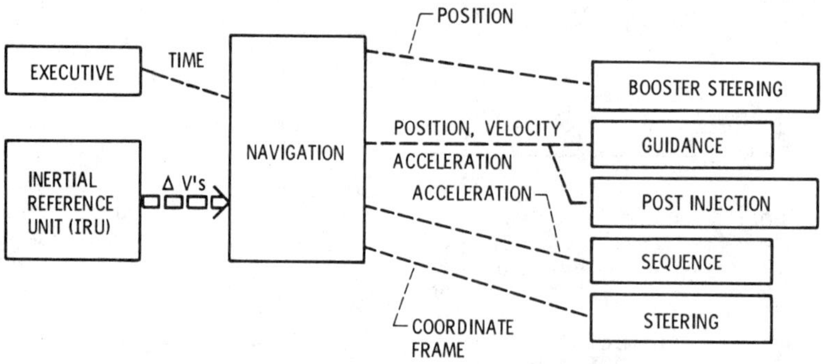

Guidance

Function. Determine steering coefficient data for optimizing the trajectory and furnish engine cutoff time to the sequencer.

Method. Assumes a near-optimum linear tangent steering law in pitch and a calculus of variation (COV) steering law in yaw.

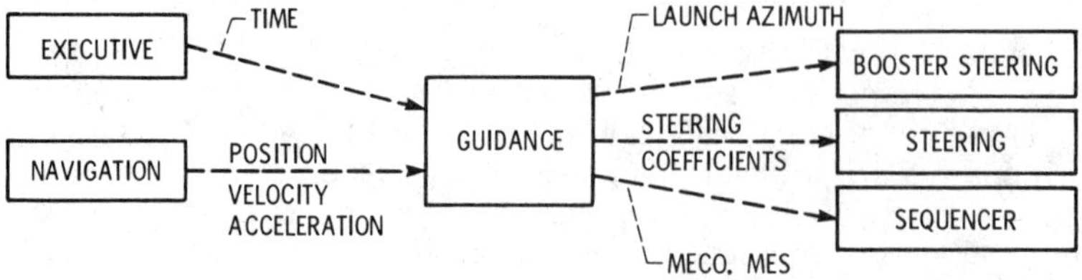

Steering

Function. Furnish the desired vehicle attitude to the platform resolver chain.

Method. Compute the desired vehicle attitude from the guidance-supplied steering coefficients.

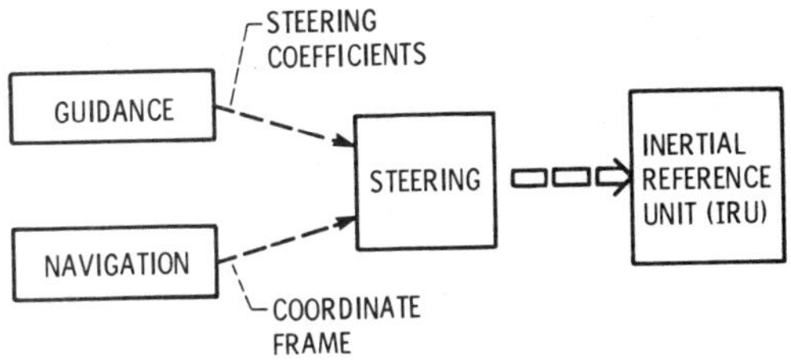

Attitude Rate (ATRAT)

Function. Furnish rate information to powered and coast phase autopilots.

Method. Computes the time derivative of the attitude rate signal.

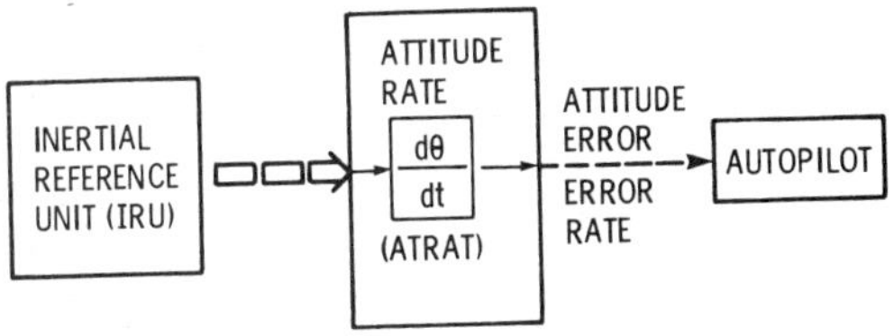

Powered Phase Autopilot

Function. Maintain control stability during main engine firings and control the vehicle axes to the desired attitude.

Method. Engine gimbal angle output commands are computed using attitude errors and error rates as inputs to control laws.

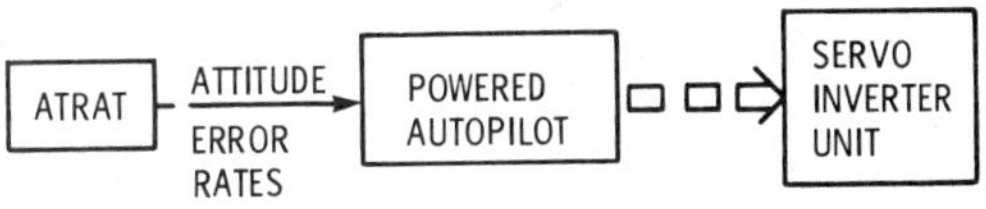

Coast Phase Autopilot

Function. Control the vehicle attitude during coast phase maneuvers.

Method. Command N_2H_4 attitude control engines in on/off mode.

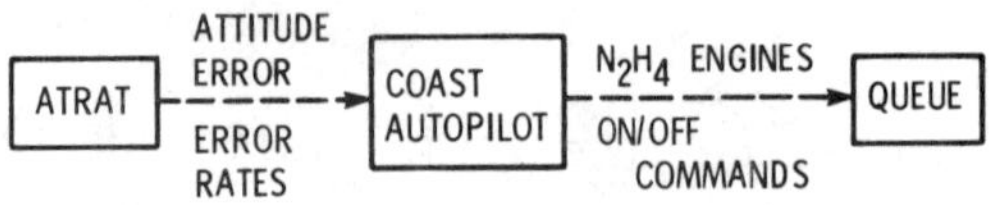

Booster Steering

Function. Steer the booster in pitch, yaw, and roll in an open-loop manner during ascent through the atmosphere.

Method. Using polynomials in altitude, generate attitude as a function of altitude.

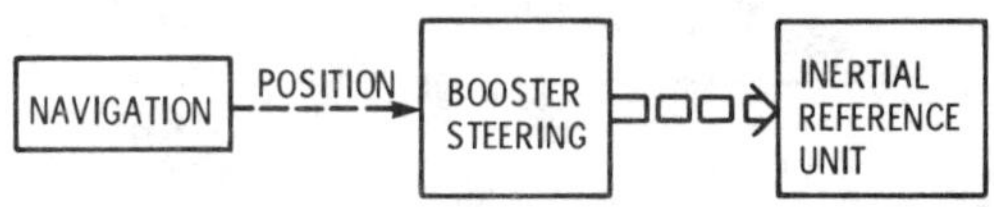

Post Injection

Function. Provide steering coefficients to point the vehicle for separation and retromaneuver.

Method. Output to the resolve chain roll and pitch axes pointing vectors.

Sequencer

Function. Generate discretes for sequencing of all events during flight.

Method. Perform various tests to determine time to issue event discretes or to accept from selected modules for module dependent discretes.

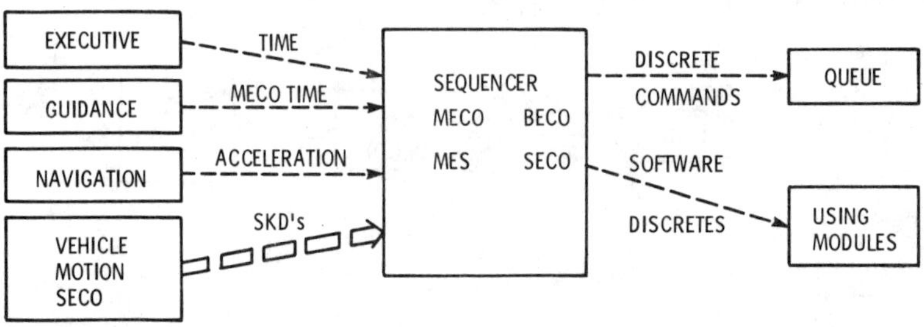

Propellant Utilization (PU)

Function. Maintain a proper ratio of LH_2 and LO_2 in the tanks to preclude a premature depletion of one or the other.

Method. Monitor the propellant ratio error signal and command the PU values to a position which will null the error.

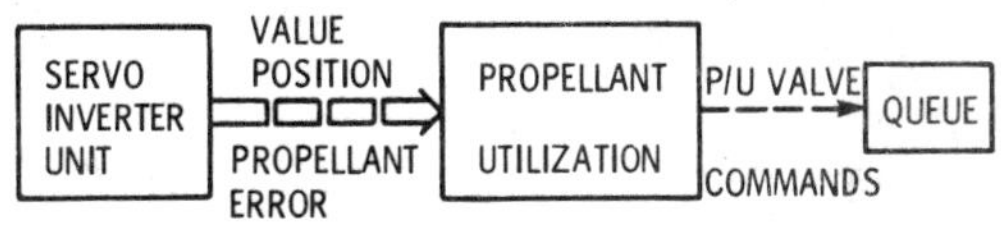

QUEUE

Function. Resolve any potential conflict of simultaneous requests for switch action.

Method. Assign priorities to switch requests and command switches to the priorities.

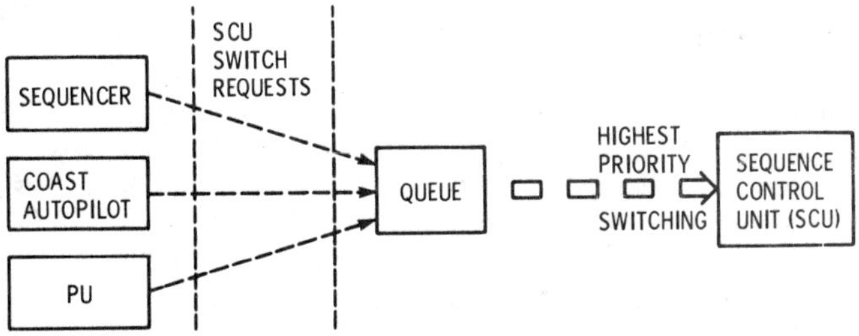

An example of the software modules linked together to perform a specific function, in this case Centaur powered phase steering, is depicted below in Figure 5.

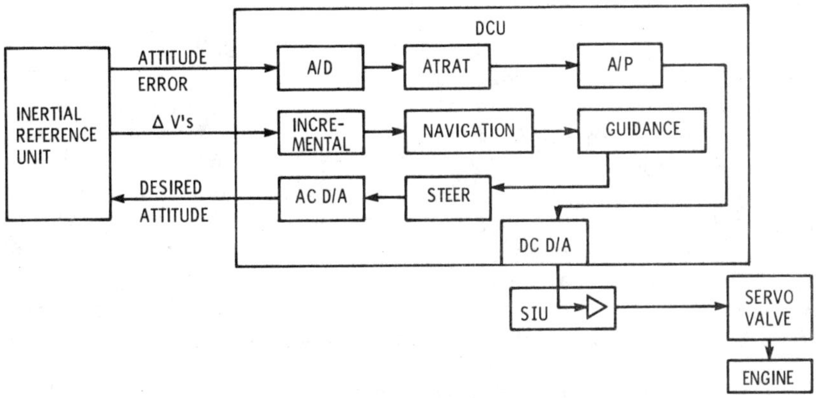

FIGURE 5. Module Combination Example

SHUTTLE/CENTAUR

The adaptation of Centaur as an upper stage in the Space Transportation System is a key element in NASA's space transportation plans. The primary missions for the STS/Centaur combination is Galileo and ISPM in 1986.

Software Flexibility. The versatility of Centaur software and the excess payload capability for existing missions provide mission flexibility by allowing many options for contingency planning. For example, Centaur software capability increases mission flexibility by allowing Centaur deployment and/or mission initiation on any revolution in the parking orbit. Orbit parameters can be selected from previously validated multiple-targeting sets as a function of revolution to account for mission initiation delays. The software contingency options have been flight-proven; typical examples are automated in-flight re-targeting capability for HEAO launches (Atlas/Centaur) and provision for contingency parking orbit revolution for Voyager launches (Titan/Centaur).

Guidance and Navigation. The guidance and navigation functions will be implemented using the D1 Centaur inertial measurement group (IMG) for measurement of vehicle accelerations and the digital computer unit (DCU) for computation of vehicle position and velocity and generation of the required steering signals. The D1 Centaur IMG will be slightly modified for DOD missions to improve gyro torquing command accuracy to accomplish the attitude update and azimuth alignment. A gyrocompassing mode will be used for inertial azimuth alignment of the inertial element for Shuttle missions.

Although IRU accuracy is sufficient to meet the mission requirements for deployment eight hours after liftoff, by navigating from the ground up without external navigation or attitude assist, extended period in the Orbiter payload bay may necessitate navigation and attitude updates. Navigation update can be provided, if necessary, from the Orbiter via the PSP interface. Attitude update can or will be accomplished by a Ball Aerospace Systems star scanner.

Deployment Delays. A few words concerning the DCU real time. The DCU is synchronized to real time (GMT) through the GSE before liftoff. Since the system is navigating from the ground up, the Centaur is insensitive to delays in deployment on the Centaur from the Orbiter. This permits the mission to be accomplished on any deployment opportunity. For those missions that have a time-dependent target condition (such as ascending node), the appropriate target-orbit data can be selected from prestored options within the DCU once the deployment time is established. The synchronization of DCU time and GMT thus makes the Centaur system autonomous, and obviates the need for Orbiter crew-initiated functions to initialize the software or communicate the deployment time.

Adaptability. Within the framework of system maturity, changes are continually evaluated to improve Centaur capability and performance. As an example, several block changes have been incorporated into the IMG that have resulted in an order of magnitude improvement in predicted reliability from the 1250 hour MTFB initial specification requirement to an actual MTBF of 12,500 hours. Two factors drive the continual upgrading of Centaur's hardware and software:

- o New technology accounts for improved electronic parts, components, and techniques.
- o New mission requirements demand increased capability; from more computational capacity to increased accuracy.

As an added note, one of the most important pre-launch tests conducted just prior to launch is the recalibration of the inertial reference platform. Values of gyro drift (as well as other error source parameters) are updated from this test and are stored for use in the navigation computations to compensate for errors caused by gyro drifts, etc. The new calibration values are compared by the Computer Controlled Launch Set (CCLS) to the previous values for significant changes. This technique will detect "soft" failures in gyros from out of tolerance drift values.

CONCLUSION

The Centaur D-1A guidance equations sets and software are extremely flexible and provide accuracy and performance. A two fold advantage emerges by the use of explicit over that of the Centaur D implicit guidance sets. First, the preflight targeting requirements have been shown to be much less for the explicit scheme than for an implicit scheme. Second, the mission targeting subphase can itself be performed by the partial, or full use of an explicit guidance scheme: optimum capability demonstrated by past Centaur D-1A missions.

The flight software is kept simple to reduce checkout costs and minimize storage required. Maximum use of prime sequencing software (back-up software included) is made, and the software is mission and vehicle independent to the maximum extent possible.

The reliability of the D-1A software is enhanced through controls from module inception through flight. These controls ensure that the management of the software will be thorough and complete. A Change Requirement System is a formalized procedure for initiating, approving, and recording changes to modules or programs.

In summary, it has been the purpose of this paper to demonstrate to the reader the highlights of the Centaur D-1A guidance scheme and to provide a qualitative feeling for the different levels of effort involved. Levels included the desirability of using explicit guidance, and the uniqueness of the software.

REFERENCES

1. Shuttle/Centaur Airborne Component Procurement Specification Star Scanner. Specification No. 65-04001, General Dynamics Convair Division, Oct. 23, 1981.

2. Birnbaum, M. M.; Bunker, R. L.; and Tavolacci, J. T.: A Radiation-Hardened Star Scanner for Spacecraft Guidance and Control. Journal of Guidance, Control, and Dynamics, vol. 6, no. 1, Jan.-Feb. 1983, pp. 39-46.

3. Bliss, G. A.: Lectures on the Calculus of Variation. University of Chicago Press, 1946.

4. Brush, Richard G.: Bilinear Tangent Yaw Guidance. Guidance and Control Conference, Boulder, CO, Aug. 1979. American Institute of Aeronautics and Astronautics, 1979, pp. 250-264.

5. Courant, R.; and Hillert, D.: Methods of Mathematical Physics. Vol. 1. Interscience Publishers, Inc., 1953.

6. Roberts, R. E.: Engineering Discussion of Centaur D-1 Guidance Equations. General Dynamics Convair Division, GDCA-BKM71-007, Apr. 1977.

7. Roberts, R. E.: Guidance Equations and Preliminary Results Using Linear Tangent Steering Laws for Titan/Centaur Missions. San Diego, California; General Dynamics Convair, GA-6923, July 1979.

AAS 84-044

NAVIGATION & CONTROL CONSIDERATIONS FOR SPACE BASED ORBITAL MANEUVERING SYSTEMS

Larry Brandon *

The concept of space basing a vehicle, although not new, is still a relatively open design area. There are still numerous trades and analyses that must be performed to determine what sensors are needed, what level of autonomy is justified, what parameters should be optimized and what level of man intervention, if any, is needed if the vehicle's mission includes docking with another vehicle.

The space based vehicle must determine its own state and that of its target. It can do this with no outside aid or it can depend almost totally on outside support. Weight, power, volume and software are some of the design parameters of interest. Then, depending on the mission, it may be desired to optimize time, propellant, accuracy or some other performance index.

This paper will highlight some of the navigation and control considerations in the design of a space based vehicle and design elements will be discussed where appropriate. No particular system design will be recommended, but rather it is hoped that additional thought will be provoked in the area of space basing design.

* *NASA Marshall Space Flight Center, Huntsville, Alabama 35812*

INTRODUCTION

Space basing is not a new concept. It is the standard operational mode of a large percentage of satellites that have been launched. Historically, this space basing concept consists of the initial placement of a satellite in a particular orbit with fixed altitude, inclination and eccentricity. Additionally, the spacecraft is normally kept in a single attitude best suited for its mission. That attitude may be inertially fixed, local vertical, spin stabilized, scanning or other, but it is generally maintained throughout the life of the spacecraft.

Some spacecraft, in particular geosynchronous, do move their orbital position during their life or they control their orbital position to maintain a particular orbital station. This maneuvering or stationkeeping is quite limited, is always commanded through a ground control station, onboard computation is kept to a minimum, and the process is done very slowly.

The Space Shuttle has an orbital maneuvering capability, but its capability is limited to small changes in altitude and, although it has a substantial computing capability onboard, its orbital maneuvering is generally done in close concert with ground support. Additionally, the Shuttle Orbiter is unique in that it is manned and man has control over many aspects of its operation. Therefore, unique requirements in the Orbiter operations philosophy and implementation are encountered, leading its design away from that of a space based vehicle. Indeed, numerous add-on systems must be considered to extend the Orbiter's operation in space even out to thirty days.

Thus, although the concept is not new, the particular requirements encountered in a space based, unmanned, orbital maneuvering system presents an opportunity to develop new designs to meet those requirements. Some of those requirements which will be discussed are storage mode(s), sensing/ methods of activation, navigation schemes, target/mission determination, rendezvous and docking schemes, reliability considerations and commonality between low energy and high energy maneuvering vehicles.

ORBITAL STORAGE MODE

A key design consideration that may drive the overall configuration of a vehicle is the required electrical power level and the method of power generation while in storage on orbit. The required power is in turn driven by whether the vehicle is stored in an all up mode with essentially all systems on or whether the vehicle is stored with many of the electrical subsystems off.

Since batteries as a primary power source are not a viable option for a space based vehicle, power consumption is not as critical a driver as it could be in the selection of the storage mode. However, power is still of concern since the size of arrays, fuel cells or whatever other primary power source is selected is determined by the power requirements. Therefore, for the space basing concept, the selection of navigation and control hardware should be made with a concern for power requirements.

Another consideration in storage mode selection is component lifetime, both active life for the all up storage mode and shelf life for the powered down storage. Also, for the powered down mode, where periodic activation of equipment is required, the failure rate associated with cycling components on and off would need to be considered.

When considering navigation and control, the key benefit of the full up storage mode is the essentially instant reaction time in starting a mission since no warm up or alignment period would be required. However, at least one IMU does not need warm up or calibration: it is the ring laser gyro IMU. In all cases of powered down storage, though, an attitude update would be required after powering up.

The storage mode selection, then, will probably depend heavily on the mission requirements since speed of response may be the most critical aspect of the storage mode. The other considerations are important, but are of a secondary nature.

ACTIVATION

No matter what storage mode is selected, the space based vehicle will have criteria for mission activation. The two basic modes by which activation can be initiated are either from the ground or from onboard.

The ground based method may be best for servicing missions or payload orbit to orbit transfer missions where key elements of the space based vehicle's mission must be uplinked anyway.

For some other types of missions, however, such as interception, observation, or periodic movement, the ground based activation sequence may not be as attractive. In particular, the ground may not be able to observe as well, there may be a delay in mission initiation if relay satellites must be used and, finally, man would be involved, decreasing the autonomy of the mission. If the vehicle is in a low earth orbit, however, the ground, by collecting data from many sources, could provide a much larger area of observation.

If the activation is based on onboard sensing, a family of sensors may be required if the vehicle has multimission applications. The sensing could be based on target velocity, target illumination, target proximity, target beacon, target anomalies, target trajectory/orbit, or perhaps even target behavior over a period of time. The target in the case of an observation mission may even be a sensed activity on the surface of the earth.

In some cases, the sensor may be a radar, either RF or laser depending on requirements. If the target is cooperative, the ranges achievable for moderate power levels may be fairly large. However, if the target is uncooperative, requiring skin tracking, the range may be quite limited and useful only for proximity operations or close in rendezvous data.

In other cases, a telescope with or without filters may be used to sense the target. The sensed parameter could be reflected light, engine plumes, or obscuration of the background by the target. Other sensors could be sensing infrared sources, energetic particles, RF emissions, etc.

Finally, for the onboard sensing and activation mode, it must be determined if the sensing will be full time or periodic. This determination will depend on power requirements to some degree as previously discussed, but probably will depend on the mission or mission types for which the space based vehicle is intended. If the mission(s) are to develop long term data, then a periodic sensing scenario is appropriate; but, if the mission requires fast response to unanticipated events, then a full time sensing/monitoring concept is needed.

NAVIGATION

Navigation, when discussing a space based vehicle, has numerous aspects, such as autonomy versus nonautonomy, degree of autonomy, ground based versus onboard versus tracking satellites, target state determination and more.

One of the primary functions of the space based vehicle's navigation system is the determination of its own position and velocity. This determination can be with ground support or without. If the ground is used, there are several tracking systems that could be used. These are TDRSS, STDN/DSN and NORAD. The TDRSS is generally implemented such that the navigation solution is derived through ground based data reduction. The data used by the ground is doppler shifted due to the relative motion between the spacecraft and the TDRS. After several data takes followed by the navigation solution, the ground uplinks the state vector to the spacecraft: the solution can take several hours. If enough onboard computational capability is dedicated to the problem, the TDRSS equations could be implemented onboard and, thus, be more autonomous and perhaps a little faster. The TDRS ephemeris would have to be maintained onboard and would have to be updated periodically. A serious limitation to TDRSS is its limited range. For operational altitudes near or above geosync, other navigation schemes must be used.

The STDN/DSN and NORAD sytems are ground based radar systems and, thus, must always have the ground involved in any navigation determination. For some classes of missions, this poses no real handicap, such as for payload placement or for periodic shifts of a payload from one orbit to another. These missions do not necessarily require autonomous state determination so systems cost, complexity, or other considerations will drive the subsystem selections.

For missions where autonomy, speed of operation, or other considerations demand onboard processing, there are several viable options for the space based vehicle's navigation determination. One of these options is the Global Positioning System (GPS), a navigation system based on data received from an array (18 ultimately) of satellites orbiting at an altitude of about 10,300 n.m. at an inclination of 63 degrees. The space based vehicle uses an onboard receiver/processor to receive data from typically four of the GPS satellites, then determines three position components and three velocity components of the state vector and, additionally, a time correction to an onboard clock. The user does not transmit to the GPS satellite and no wait for access is required since the GPS satellites transmit continuously. Typically, convergence even from an unknown initial state only requires a few minutes and convergence from a known state can be somewhat less than a minute. State vector accuracies obtainable range down to 10 meters and 0.01 meters per second to, perhaps more typically, 50 meters and .05 meters per second. Although not originally intended for use by space vehicles, the GPS is usable out to at least geosync altitudes and perhaps further depending in part on the receiver and antenna on the user vehicle. All in all, the GPS appears to be a very useful navigation aid providing high accuracy, fast convergence and autonomy.

Another autonomous navigation system is based on the use of the space sextant (or similar systems). Such a system has a reference gyro assembly, a computer and two telescopes to track the moon's limb and certain bright reference stars. By including the moon's epheremis, the state vector of the space vehicle can be determined as well as attitude. Convergence times from an unknown state can be long and accuracy will probably be in the one hundred to several hundred meters range. The system is still in a development state and its ultimate usefulness and accuracy are still to be demonstrated.

A final navigation technique that provides a degree of autonomy has been mentioned earlier. It is the onboard processing of TDRS data to yield the state vector. Processing takes several hours (up to ten hours), requires TDRSS epheremis updates, including time, and requires five minutes of data each orbit. Accuracy is good to about 100 meters, but because of the other considerations, it is not as viable an autonomous system as the previously described techniques. Overall, the GPS will probably be desired in most applications.

Another area of consideration for a space based vehicle is the determination of the target state vector. The target may only be a new location in orbit or it may be the position of another spacecraft, but in either case the target state vector can be determined onboard or it can be uplinked from an external source such as a ground station. Using the ground control concept definitely reduces the complexity of required spacecraft subsystems, but at the expense of autonomy and speed of response in some applications. The ground link is appropriate for some applications involving a space based vehicle such as transferring a payload from one orbit to another or for moving to a new orbit position in support of earth resources and so forth.

Where autonomy or other considerations require it, the target state vector may be determined onboard the space based vehicle. There are then two primary categories of targets, cooperative and uncooperative. If the target is cooperative, then it could have an RF transponder allowing tracking of the target by an RF radar system on the space based vehicle. The tracking range would be limited by power considerations and antenna size, but for reasonably sized RF radars, the range could be expected to be several hundred to perhaps a thousand nautical miles.

Another type of cooperative target could have optical reflectors, such as corner cube reflectors so that a laser radar could be used by the space based vehicle for tracking.

Again, depending on reflector size and laser power, ranges of several hundred nautical miles could be expected. Finally, in the extreme of cooperative targets, a communication link could be established with the target vehicle and the state vector of the target could be transferred to the chase vehicle.

In all these cases, except perhaps the last, range is limited because of power, physical size, and line of sight constraints. Where range limitations are unacceptable, the ground must become involved or other schemes for tracking devised.

A second category of targets are those that can be considered uncooperative. The RF and laser radars are again candidates, but in this case their operation will depend on surface refelctions from the target, thus, the range can be quite limited. Additional range can be obtained by increasing power and/or antenna size, but ranges even out to 100 nautical miles are probably impractical to implement on a space based vehicle. Also in this category is a telescope based tracking system. The sensor at the focal point of the telescope could be sensitive to visible, ultraviolet, infrared or some other wavelength. Detection of the target would depend on light reflected off the target or on emitted radiation from the target. The emitted radiation's source could be internal heat, exhaust plumes, experiments or other. A beneficial combination of functions might be had if the telescopes on the space sextant (mentioned earlier) could also be used in a target sensing mode as just described.

It is seen, then, that the navigation function for a space based vehicle can vary from the very simple to the very complex. A key driver in choosing the type navigation is the mission, which can vary from a simple repositioning of the space based vehicle to autonomous acquisition and docking with an uncooperative target.

RENDEZVOUS AND DOCKING

The ability of one spacecraft to rendezvous and dock with another spacecraft is often a key requirement of mission planners and, for that reason, a great deal of work has been done in that area and is continuing to be done. Almost all aspects of the problem have been worked to some degree and here we will try to bring out some of those aspects of rendezvous and docking that should be considered in a space based vehicle's design. Here again, the basic issues of autonomy, level of ground interaction, and target cooperativeness must be addressed.

For those missions where it is acceptable, the use of a man-in-the-loop with a television camera as one of the primary onboard sensors is certainly one of the least risky docking schemes. Additional docking aids, such as a radar for range and range rate data, may be required, but television overlays and/or scene processing software may be sufficient and may be preferred. The man-in-the-loop scheme is dependent on a communications link to the ground and, since a significant amount of time is required for docking, a direct link to the ground for low earth orbit operations is not viable due to very limited overflight times. Thus, a link through a communications relay satellite such as the TDRS is required.

For missions where greater autonomy is desired, the docking function must be implemented onboard the space based vehicle. Assuming that the target vehicle is stable, there are basically two categories of docking that must be considered. They are docking with a known cooperative target and docking with an unknown target. At least two schemes for autonomously docking with a cooperative target seem viable; a scene recognition scheme using a television camera as a sensor and, secondly, a pattern recognition scheme using a laser and retroreflectors. The scene recognition method employs arrays of data describing the target vehicle's shape stored in onboard memory. The target image generated by the onboard television camera is processed to yield a similar data set and additional onboard processing compares the data sets to determine from what direction the target is being viewed. This technique can require huge amounts of memory and very long times for processing, but requires very modest hardware.

The pattern recognition scheme depends on the target having a set of corner reflectors mounted in specific patterns. A laser on the space based vehicle is used to illuminate the reflectors and the coordinates of each return reflection are stored. Through appropriate processing, the relative attitude of the target vehicle can be determined allowing the space based vehicle to perform the appropriate maneuvers for docking. This scheme does not require as large a data base as the scene recognition scheme and processing should be somewhat faster, but it may be more susceptible to false signals and it does require prior planning to fit the target vehicle with retroreflectors.

The second category, docking with an unknown target, can generate the most severe requirements for the space based vehicle. The first requirement would be to recognize the shape of the target vehicle. A television camera would be a likely sensor and with appropriate imaging a set of data could be generated outlining the target vehicle's shape. Many sets of such data would have to be generated requiring appropriate maneuvers of the space based vehicle. After a sufficient set of data has been created, the onboard processor must apply an algorithm to 'decide' what the target vehicle's shape is and what would be the best direction from which to approach the target. After these decisions have been made, a type of universal grappling device mounted on the space based vehicle would be used to grapple the target vehicle. This scheme would require extended operation time, a large amount of computer memory for data storage, and sophisticated algorithms for data reduction and decision making. If autonomy is not necessary or desired, then a communications link to the ground could be established with man-in-the-loop control which would greatly simplify the onboard requirements.

A final class of target vehicle that is gaining more attention is the uncooperative tumbling spacecraft. This class includes disabled spacecraft, expended spacecraft, debris and other. A space based vehicle could be used extensively in the retrieval of this class of targets and, since these targets may have low to high attitude rates, a whole new set of requirements are generated. These requirements and possible solutions are being assessed by MSFC and other groups.

RELIABILITY

For the space based vehicle, reliability is a key concern since maintenance is limited or nonexistent. The reliability must be high to reduce life cycle costs by limiting failures, to meet safety requirements if man is involved and to minimize risk to highly critical missions. When reliability is increased, the initial cost of a subsystem is higher, but when life cycle costs are considered the total costs can be less. Of course, when the mission is highly critical, the cost is secondary to mission success. Reliability must be a key concern when designing any subsystem for a space based vehicle and the navigation and control subsystems are especially so since they are so critical to the success of any mission.

SPACE BASING APPLICATIONS

Two key applications of the space basing concept could be the Orbital Maneuvering Vehicle (OMV, formerly TMS) and the Orbit Transfer Vehicle (OTV). The OMV is designed for payload placement, retrieval and servicing at altitudes of several hundred nautical miles from its reference orbit. It can be based in low earth orbit or in a geosynchronous orbit or in almost any other orbit including high inclination orbits if needed. Since it will be used for placement as well as retrieval of payloads, all the topics covered in this paper are appropriate for the OMV and should be considered in its design.

The OTV is a high energy stage whose primary mission is the placement and retrieval of payloads in geosynchronous orbits. The OTV would be based in a low earth orbit and would receive payloads from the Orbiter or other earth launch vehicles or would be bringing payloads back to the Orbiter for return to the ground.

The OMV and the OTV would have differences in the range required for sensors and in control implementation, but the space basing aspects as a whole are similar for both vehicles.

SUMMARY

Several topics in the area of navigation and control for space based vehicles have been touched on in this paper. Much of the design of a space based vehicle is dependent on the mission definition, the storage mode selected and the degree of autonomy desired for the vehicle. Navigation and docking concepts have been presented that are appropriate for a space based vehicle. These concepts are by no means the only viable concepts, but they do reflect some of the available techniques and they cover the ranges of autonomy and target cooperativeness. Also briefly discussed was the necessity of addressing the reliability of a space based vehicle's navigation and control subsystem. Reliability can be a key driver both in mission success and cost. It is hoped that the presentation of this paper will provoke some additional thought and discussion on navigation and control considerations for space based orbital maneuvering systems.

AAS 84-045

GUIDANCE and CONTROL FOR THE TRANSFER ORBIT STAGE (TOS)

R.B. Schroer*

The Transfer Orbit Stage (TOS) is a new shuttle compatible, three (3) axis stabilized perigee stage designed to transfer payloads to geostationary orbit. Avionics sophistication has been minimized and extensive use has been made of existing hardware and software. Guidance requirements have been satisfied by innovative use of Shuttle services. The attitude control system is designed around an off-the-shelf gyro, a straight-forward microprocessor-based controller, and flight proven TVC actuators. Launch site operations have also been simplified.

INTRODUCTION

The Transfer Orbit Stage (TOS) is a low cost, solid propellant, three axis stabilized, perigee stage being developed by Martin Marietta under contract to Orbital Sciences Corporation. TOS will be operational in late 1986 to transfer spacecraft from low earth orbit to geosynchronous and other high energy (including planetary) orbits (Figure 1).

Figure 1. Transfer Orbit Stage (TOS)

* *Martin Marietta Denver Aerospace, P.O. Box 179, Denver, Colorado 80201*

High reliability and low cost have been achieved through use of flight proven designs. Both the main propulsion solid rocket motor and reaction control system (RCS) thrusters were derived from Inertial Upper Stage (IUS). Avionics hardware and software from other programs have been used to minimize overall costs.

TOS avionics is made up of three subsystems: 1) Attitude Control, 2) Event Sequencing, and 3) Electrical Power. The Attitude Control Subsystem (ACS) provides TOS pointing and stabilization in all phases of flight from deployment through spacecraft separation. TOS derives its attitude from the Shuttle guidance and navigation (G&N) system by having the Shuttle point to the desired TOS attitude. The attitude control system combines a flight proven gyro and thrust vector control system with a simple microprocessor-based control computer.

Mission timing is initiated upon physical separation from the Shuttle. The sequencer has been designed to be fault tolerant to satisfy Shuttle safety requirements. To enhance reliability, redundancy has also been incorporated into the battery power system. Figure 2 is a functional block diagram of the TOS avionics system.

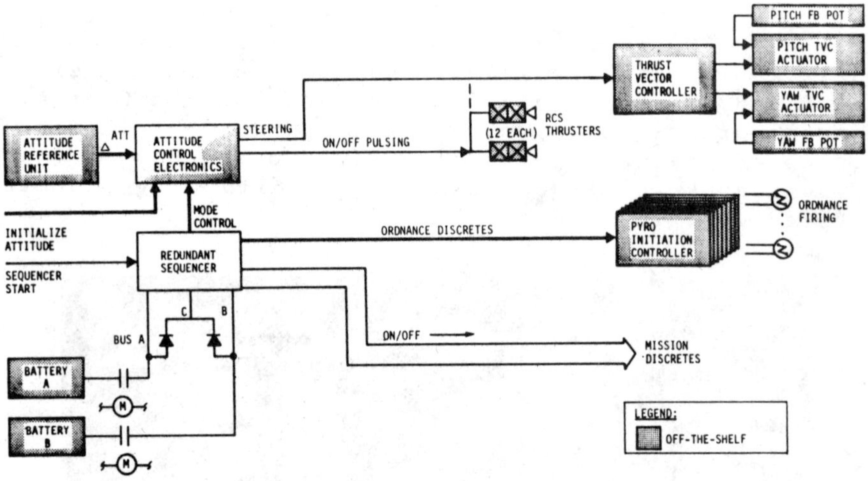

Figure 2. TOS Avionics Block Diagram

MISSION SCENARIO

TOS is transported into low earth orbit by the Shuttle in the unpowered (passive) state. About one orbit before deployment it is turned on and checked out. Prior to deployment TOS attitude will be established by pointing the Shuttle bay to the desired TOS perigee burn attitude. Once the perigee burn heading has been established, TOS gyros are initialized and deployment sequencing is initiated from the aft flight deck. Mission timing starts at physical separation. To satisfy Shuttle safety requirements the RCS will be inhibited from deployment until a safe separation distance has been achieved. Upon enabling the RCS, TOS will recover from its uncontrolled state and reacquire the original burn attitude.

The stage then coasts for half an orbit until it reaches the nodal crossing. Prior to ignition the solid motor will be armed and the attitude deadband tightened. During the perigee burn, attitude is controlled by gimballing the solid rocket motor (SRM) engine nozzle in pitch and yaw. Roll is controlled with the RCS thrusters. At SRM burnout, pitch and yaw control reverts to RCS operation in the wide deadband. Pointing maneuvers prior to separation or thermal maneuvers during the coast can be provided. After separation TOS performs a collision avoidance maneuver to prevent subsequent recontact with the spacecraft.

GUIDANCE APPROACH

Guidance simplicity has been achieved by using the inherent accuracy of the Shuttle guidance, navigation and control (GN&C) system. TOS pointing is established by maneuvering the Shuttle to the desired TOS burn attitude and initializing the TOS attitude. Using this technique, the basic TOS accuracy is comparable to the PAM stage. Initializing the TOS on orbit in this manner eliminates ground support equipment (GSE) for guidance alignment, simplifies mission targeting, and minimizes avionics software.

TOS error sources include shuttle contributions, TOS gyro errors, and TOS CG uncertainty. Shuttle error sources are G&N accuracy, and mechanical alignment errors between its navigation base and the payload bay attach points. The major TOS errors are gyro bias and scale factor. These heading errors, when combined with the CG uncertainty, are:

Heading Errors	
Orbiter Induced	0.872 deg
TOS Gyro	1.0 deg
CG Uncertainty	0.5 deg
RSS	1.417 deg

Combining these heading errors with the 0.532% SRM delta V uncertainty yields:

Heading Errors	Apogee Altitude Error (nmi)	Inclination Error (deg)
0	431.2	0.011
1	466.2	0.240
1.4	493.0	0.340
2	529.5	0.479

This satisfies the TOS accuracy requirement of less than 600 nautical miles apogee error and 0.7° inclination.

CONTROL SYSTEM IMPLEMENTATION

The Attitude Control Subsystem (ACS) provides three-axis control of TOS in both powered and coast flight. The vehicle is controlled in powered flight by pitch and yaw gimballing of the SRM nozzle. Roll, pitch and yaw attitude during coast is controlled with reaction control system (RCS) thrusters. Powered flight roll control is also provided by the RCS. The control system is made up of 1) Reference Gyros, 2) Attitude Control Electronics, and 3) Thrust Vector Control System

Reference Gyros

The TOS attitude reference is a three axis, Singer-Kearfott SKIRU, strapdown gyro. The package contains two Gyroflex, dynamically tuned, two degree of freedom gyros with associated electronics. The Gyroflex gyro is the basis of the Shuttle IMU and has a MTBF of 20,000 hours. The gyros operate in the analog rebalance mode with rebalance current metered through a precision resistor. Incremental angle pulses are generated by an accurate voltage-to-frequency converter that monitors the resistor voltage. The basic gyro rebalance rate was increased by rewinding the torquer coils, and replacing the Alnico magnets with high torque platinum cobalt magnets. When the Shuttle is pointed to the desired heading TOS attitude is initialized by enabling pulse accumulation in the control electronics. This establishes the inertial attitude for perigee burn as well as providing a reference for subsequent maneuvers.

To simplify launch operations the gyro is installed without special alignment or calibration. Guidance coefficients are determined at acceptance and burned into a PROM that is installed in an external connector. This eliminates guidance alignment equipment and minimizes software and procedures.

Attitude Control Electronics

Hardware - The attitude control electronics is a microprocessor based digital controller. It accepts incremental gyro inputs and generates analog and discrete outputs to the main engine actuators and RCS thrusters. The microcomputer has been designed around a 16-bit microprocessor and supporting circuits. Part qualification and susceptibility to single event upset were primary selection criteria. Comprehensive trades were conducted that considered design maturity, performance, and the availability of support chips and support software.

The 16,000 16-bit word memory is a combination of programmable read only memory (PROM) and random access memory (RAM). Fixed control system parameters are stored within the internal memory of the unit. Mission dependent parameters are burned into PROM that is packaged in externally removeable connectors. This approach to the mission data load (MDL) simplifies launch operations and minimizes GSE by eliminating memory loading and and verification on the pad. Fault tolerant features include a 6-bit error detection and correction code, watch-dog timer, background self-test, and selected reasonableness tests. The design also includes built-in-test and monitoring for ease of checkout.

Input circuits include up/down counters that accumulate gyro pulses and input gates that accept mode control discretes from the sequencer. Digital/analog converters generate steering signals to the TVC controller to gimbal the SRM nozzle actuators. Twelve (12) discrete output drivers control pulsing of the RCS thrusters. A functional block diagram of the Attitude Control Electronics is shown in Figure 3

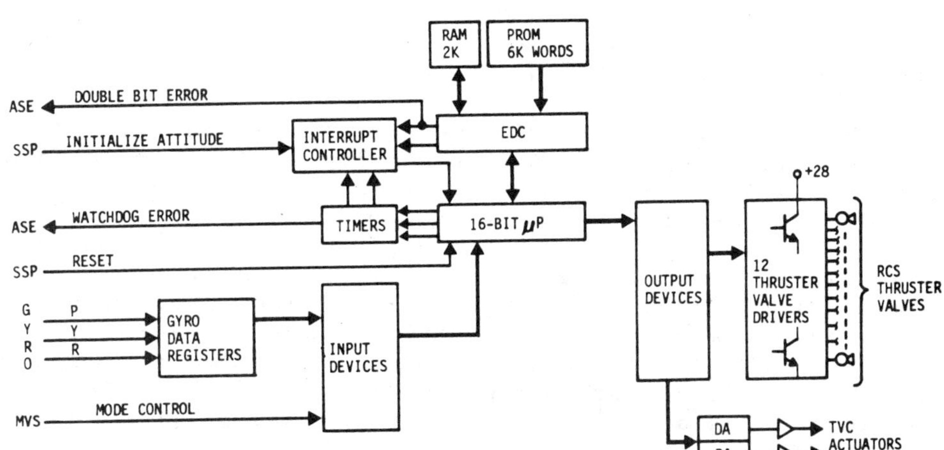

Figure 3. Attitude Control Electronics Block Diagram

Software - The gyro processing software accumulates gyro data and develops analog and discrete outputs to maintain stable control during flight. Attitude errors are developed as the difference between the attitude command and the attitude reference. Maneuvers for thermal control and spacecraft separation pointing are selectable as part of the mission data load (MDL).

In the powered flight (autopilot) mode, the computer performs gain incrementing and digital filtering. Gain is incremented to follow a decaying exponential that traces the inverse of the SRM thrust/time history. This approach maintains nearly constant stability margins except near burnout, where thrust magnitude is unpredictable. Near burnout gain is held constant to ensure stability thru thrust tail-off. Cascaded, second order, recursive filters are used to compensate for flexible body dynamics. Conditioned signals are combined with the nozzle pre-aiming command before being applied to the D/A converter. Mission unique powered flight attitude gain, rate gain, and filter coefficients are selectable with an external PROM as part of the MDL.

In coast control operation, gyro outputs are processed through a software representation of the non-linear switching phase plane. On/off pulses to the RCS thrusters are generated as a function of attitude and attitude rate. A phase plane has been implemented with near optimum convergence and propellant consumption characteristics. It is redefined in-flight to compensate for changing vehicle inertia as well as accommodate spacecraft unique performance requirements. Coast control parameters, fixed in PROM, include attitude deadband, pulse width and related switching characteristics.

Thrust Vector Control (TVC) System

The TVC system, along with the SRM it controls, was derived from the IUS program . It consists of a TVC controller, two internally redundant electromechanical actuators, and two potentiometers.

The controller provides both the servo electronics and power drive for actuator control. A servo amplifier compares actuator position with the controller input command. The design of power drive is based on a variable frequency, pulse width modulation (PWM) system. It is a hybrid between a true PWM system and a pulse frequency modulation system that offers overall lower power dissipation.

The electromechanical actuator is designed around a rare-earth dc torque motor with solenoid-operated gears for locking. Installation of the actuators is unusual in that the bracket and actuator form a ball-and-socket at the output shaft end, as opposed to a rod end bearing. This arrangement reduces weight, but increases nozzle cross-coupling.

"Mirror-image" potentiometers on the nozzle, 180° opposite the actuators, are electronically summed within the controller to accurately measure nozzle deflection and minimize cross coupling errors. A pair of normally meshed solenoid-operated gears, which must be retracted before operation, provide fail-safe position locking.

CONCLUSION

To produce a simple, three axis stabilized, upper stage, TOS has been designed to a very basic set of requirements. Flight proven hardware and software have been used to maximize reliability and minimize costs. Avionics operations have also been streamlined to eliminate gyro alignment and computer loading, thereby minimizing in-line launch pad processing.

Studies are under way to develop an Apogee and Maneuver Stage, compatible with TOS, that provides cost effective delivery into the final geostationary orbit. This apogee stage design will make maximum use of TOS avionics and be modular so that it can be tailored to spacecraft unique requirements.

AAS 84-046

INTEGRATED LAUNCH VEHICLE AND SPACECRAFT AVIONICS - A SEARCH FOR A FEASIBLE OPTION

J.D. Gilchrist*

A recurring theme in the development of upper stage vehicles is the concept of reducing costs by sharing a portion of the avionics equipment with the spacecraft. With spacecraft becoming more sophisticated and the increasing requirement of autonomous operations for DoD spacecraft, the concept of using spacecraft avionics to help guide, navigate and control the upper stage and possibly the booster stage of a launch vehicle becomes increasingly appealing from a cost viewpoint. This paper reviews the two basic guidance systems in use today for launch vehicles - inertial guidance and those systems relying on ground control - and addresses the applicability of using spacecraft avionics for the upper stage GN&C function. The avionics functional requirements and typical implementations are given for upper stage vehicles and also for spacecraft. An identification is made of the spacecraft avionics subsystems that are appropriate for upper stage vehicles. This commonality of avionics requirements between the upper stage vehicle and spacecraft is given for inertial-guided as well as those vehicles requiring ground-based commands. Finally, the potential advantages of sharing avionics equipment are summarized.

INTRODUCTION

Launch vehicle costs are rising at an alarming rate and it is of interest to examine alternatives which may provide lower-cost options to the current IUS or future Centaur inertially-guided upper stages for use with the STS. The cost of the avionics subsystems is a

* *Principal Director, Guidance and Control Division, The Aerospace Corporation, P.O. Box 92957, Los Angeles, California 90009*

significant portion of the total upper stage cost. This is especially true when the system has redundancy to meet increasing reliability requirements. Such is the case with IUS. This paper addresses the recurring theme in the development of upper stage vehicles of reducing costs by sharing a portion of the avionics equipment with the spacecraft. With spacecraft becoming more sophisticated and the increasing requirement of autonomous operations for DoD spacecraft, the concept of using spacecraft avionics to help guide, navigate and control the upper stage and possibly the booster stage of a launch vehicle becomes increasingly appealing from a cost viewpoint. This paper reviews the two basic guidance systems in use today for launch vehicles - inertial guidance and those systems relying on ground control - and addresses the applicability of using payload avionics for the upper stage GN&C function. The avionics functional requirements and typical implementations are given for upper stage vehicles and also for spacecraft. An identification is made of the spacecraft avionics subsystems that are appropriate for upper stage vehicles. This commonality of avionics requirements between the upper stage and spacecraft is given for inertial-guided as well as those vehicles requiring ground-based commands. Finally, the potential advantages of sharing avionics equipment are summarized.

When considering the concept of sharing spacecraft avionics with the upper stage, two possible configurations are possible. One could be called "integrated", meaning that the upper stage is mated with or bolted on to the spacecraft structure externally. A second configuration called "integral" has the upper stage propulsion system built into the spacecraft structure. This latter configuration consists of a single vehicle that performs the function of an upper stage as well as the on-orbit function of the spacecraft. The "integral" configuration seems especially meaningful for future DoD missions where on-orbit repositioning or maneuver requirements exist.

GUIDANCE SYSTEM ALTERNATIVES

Guidance and navigation systems for boosters and upper stages can be divided into two broad categories, ground-command guidance and inertial guidance systems. Inertial guidance systems are self-contained and require no ground-based support after lift-off. On-board inertial sensors supply navigation and attitude data to the on-board computer that generates steering commands and discrete commands to the attitude control system. Radio guidance systems for boosters are included in the ground-command category. In this case, the application is limited to a time interval where the booster vehicle is within the line-of-sight to the ground radar tracking system. The tracking data (range, elevation and azimuth) are processed in the ground-based guidance computer. The resulting steering commands and vehicle discrete commands are sent to the booster over the guidance data link, decoded and input to the attitude control system. Radio guidance is not used for upper stages due to visibility constraints with ground radar tracking stations.

Ground-command guidance has been used for many years for commercial communications satellites. Here, a launch vehicle such as the Atlas-Centaur injects the spacecraft with an integral propulsion system into a transfer orbit to the desired geosynchronous orbit. Spin-stabilization provides the required attitude control. Ground radar tracking provides the necessary navigational data. Attitude data is provided by satellite sensors. These data are processed by a ground computer to generate any required attitude maneuver commands implemented by the satellite reaction control system and also to generate the required initiation time for the apogee kick motor. Such systems have proved quite reliable and are relatively inexpensive compared to inertial guidance systems.

Table 1 shows the general avionic functions required in an upper stage vehicle and also how these functions are implemented for both inertial guidance and the ground-command guidance alternatives. The ground-command, spin-stabilized alternative is clearly the simplest and least expensive with regard to on-board equipment. This is followed by the ground-command, three-axis stabilized case. The inertial guidance alternative requires the most on-board equipment.

Table 2 shows some examples of launch vehicles (boosters and upper stages) that have used both types of guidance alternatives. There are two noteworthy examples, DMSP and TIROS-N, where the upper stage functions are integral with the spacecraft avionics functions. These are truly integral propulsion systems. DMSP is a DoD meteorological satellite developed by RCA Corporation, Astro Electronics Division, for the USAF Space Division. TIROS-N is a similar type RCA spacecraft developed for NASA-GSFC.

DMSP is a current-day example of the use of spacecraft avionics to perform the upper stage guidance functions. The current block 5D-2 version of DMSP is launched by the radio-guided Atlas vehicle. After spacecraft separation, the final ΔV maneuver is performed by a solid rocket motor with inertial guidance, all integral to the spacecraft. The final orbit is approximately 450 n.m. in altitude, circular and sun-synchronous. Table 3 provides more detail concerning the avionics equipment. The propulsion system for the final ΔV maneuver is integral to the spacecraft. The guidance equipment, computer and inertial measurement unit, has three accelerometers that are not required for the spacecraft on-orbit operations. With this exception, the same inertial sensors and computer used to perform the upper stage function are used on orbit.

TABLE 1

UPPER STAGE GUIDANCE SYSTEM ALTERNATIVES

AVIONIC FUNCTION	GROUND COMMAND		INERTIAL GUIDANCE
	SPIN STABILIZED	THREE-AXIS STABILIZED	
GUIDANCE	• OPEN LOOP, COMPUTATIONS DONE ON GROUND	• OPEN LOOP, COMPUTATIONS DONE ON GROUND	• CLOSED LOOP, IN ON-BOARD COMPUTER
NAVIGATION	• GROUND-BASED RADAR TRACKING	• GROUND-BASED RADAR TRACKING	• IMU & ON-BOARD PROCESSING
CONTROL	• RCS TO PRECESS SPIN AXIS • PROCESSING DONE ON GROUND • EARTH SENSOR/ SUN SENSOR FOR ATTITUDE	• TVC SYSTEM FOR POWERED FLIGHT • RCS SYSTEM FOR COAST • ON-BOARD PROCESSING • ARU - ATTITUDE AND ATTITUDE RATES • STEERING COMMANDS GENERATED ON GROUND	• TVC & RCS SYSTEM WITH ON-BOARD PROCESSING
GROUND SUPPORT	• MAJOR SUPPORT NEEDED	• MAJOR SUPPORT NEEDED	• NO SUPPORT
TT & C	• UPLINK COMMAND AND DOWNLINK TLM	• UPLINK COMMAND AND DOWNLINK TLM	• TLM DESIRABLE

TABLE 2

UPPER STAGE GUIDANCE SYSTEM EXPERIENCE

EXPERIENCE	GROUND COMMAND		INERTIAL GUIDANCE	
	SPIN STABILIZED	THREE-AXIS STABILIZED	AVIONICS IN UPPER STAGE	AVIONICS IN SPACECRAFT
FLIGHT EXPERIENCE	• GPS • COMMERCIAL GEOSYNC MISSIONS • NATO • FLTSATCOM	• AGENA • BURNER II • INTERPLANETARY - MARINER - VIKING - VOYAGER	• TIIIC/TRANSTAGE • TIIIB/AGENA • ATLAS/CENTAUR • TIIID/CENTAUR • ATLAS/AGENA • THOR/DELTA • T34D/IUS • STS/IUS	• DMSP • TIROS-N
FUTURE MISSIONS	• SDS		• T34D/TRANSTAGE • STS/CENTAUR	

Table 3

INERTIAL GUIDANCE WITH SPACECRAFT AVIONICS - EXAMPLE

DMSP - 5D-2

-Launch Systems: Atlas booster (radio guidance) with upper stage inertially guided by spacecraft avionics (Integral propulsion system).

-Propulsion for Upper Stage Function: 1 SRM-STAR 37; 4 85# hydrazine thrusters for TVC, ΔV trim and Atlas separation; and 8 2# nitrogen thrusters for roll control during SRM burn and coast control.

-Avionics Required for On-Orbit Operations: 2 computers (RCA 234), one is back-up; primary attitude reference system - 4 gyros (3 + 1 skewed for redundancy) and star sensor; and a backup attitude reference system - horizon sensor and sun sensor.

-Avionics Required for Ascent Guidance: 1 computer; an inertial attitude reference system and a 3-axis accelerometer package.

-Advantage of Using Payload Avionics: precision attitude reference system and computer available; accelerometers are a low-cost addition and maximizes P/L deliver to orbit.

There are numerous examples of commercial communications satellites that have used a version of the ground-command guidance. Typically, the booster delivers the spacecraft with an integral propulsion system into a transfer orbit to the final geosynchronous orbit. As an example, consider FLTSATCOM, a DoD communications satellite that is launched by the Atlas-Centaur vehicle. The Atlas places the Centaur and spacecraft into a park orbit and the inertially-guided Centaur performs the perigee ΔV maneuver. After separation, the spacecraft with its integral propulsion system is spin-stabilized in a transfer orbit to the final geosynchronous orbit . Spacecraft sun and earth sensors provide the attitude data, ground tracking provides the navigational data, and all guidance processing (attitude determination, navigation, ΔV orientation and ignition time) is done on the ground. The spacecraft remains in the transfer orbit for 4-6 revolutions to provide good ephemeris information. FLTSATCOM has a three-axis stabilized, momentum-biased attitude control system using sun and earth sensors for attitude determination. An additional set of sun and earth sensors are added to the spacecraft to provide attitude data during the spin-stabilized transfer orbit phase. The on-orbit spacecraft RCS system is used during the transfer orbit for spin-up, attitude maneuvering to desired apogee ΔV maneuver, despin, and the final ΔV trim maneuver.

GUIDANCE ALTERNATIVE EVALUATION

The guidance system alternatives for the upper stage function described in Table 1 can be qualitatively compared. Table 4 compares four guidance alternatives; ground-command guidance in spinning and three-axis stabilized configurations, inertial guidance with

Table 4
Guidance Alternative Evaluation

1. Ground-Command, Spin-Stabilized

Advantages	Disadvantages
-Minimal Avionics: No TVC System, small RCS to precess spin axis, simple attitude sensing (earth & sun sensors) and no major processing requirement -Good injection accuracy available with trim burns -Low power requirements -Good for spin-stabilized spacecraft -Good experience base from commercial satellites using solid rocket apogee kick motor	-Active nutation control for unstable inertia configurations -Additional balancing requirement -Dynamics uncertainties due to fuel slosh in large liquid spinning propulsion systems -Some payloads not compatible with spinning (loads and configuration) -Lack of Autonomy -Ground support required Long injection time-line if multiple low thrust liquid burns are used for apogee ΔV

2. Ground-Command - Three-Axis Stabilized

Advantages	Disadvantages
-Moderate avionics requirement: no navigation system, or guidance computer, need TVC & RCS for control, attitude reference system, and processing for control -Experience Base: Agena and burner II. -Minimal on-board software requirements	-Less tolerant to propulsion dispersions and vehicle uncertainties -Lack of Autonomy -Long injection time-line if multiple low thrust liquid burns are used for apogee ΔV -Ground support required

3. Inertial Guidance - Avionics in Upper Stage

Advantages	Disadvantages
-Good experience base -Good injection accuracy -Overcomes propulsion dispersions and vehicle uncertainties (closed-loop guidance) -Low recurring software costs	-Avionics expensive: cost, power and weight

4. Inertial Guidance - Share Spacecraft Avionics

Advantages	Disadvantages
-Maximum use of common avionics for upper stage and spacecraft -Ideally suited for single satellite program with multiple launches	-Lack of commonality between different spacecraft avionics - computers & inertial sensors. -Problem in implementing common software requirements in different spacecraft computers -Vehicle testing more complex. -Integration more complex

conventional upper stage avionics and finally inertial guidance with some or all avionics in the spacecraft.

For certain DoD missions, autonomous operation of the launch vehicle after lift-off is a desirable and sometimes a required capability. In Table 4, autonomous operations is considered a desirable capability and as such, is a disadvantage for the ground-command guidance alternative. Autonomy, or the lack of it, should not be a consideration for commercial applications or for DoD missions that do not require autonomous operations.

The ground-command, spin stabilized alternative appears most cost effective for missions where; a solid rocket motor propulsion system can be used for the transfer orbit; autonomy is not a consideration; and the spacecraft configuration allows spin-stabilization. The added avionics equipment (earth and sun sensors), if any, is a minimal increase to the spacecraft set. Any attitude propulsion (spin-up, precession maneuvers, despin and ΔV trim) can be provided by the spacecraft reaction control system

For missions using an all liquid or hybrid propulsion system (solid rocket motor for the perigee burn and a liquid engine for the apogee burn), there is a certain design risk associated with large spinning liquid propulsion systems. One concern with large spinning liquid stages is the uncertainty in the dynamics due to fuel slosh. The effects are less predictable in spinning than non-spinning vehicles.

The ground-command, three-axis stabilized is a good alternative for missions that do not require autonomous operations. The design risk associated with large spinning liquid stages is no longer present.

There is, however, an increase in the avionics equipment. A thrust vector control system is required, implemented with a gimballed nozzle or with reaction control. Reaction control is also required during the coast portions of the transfer orbit. These attitude control systems for the coast and the powered flight phases are additional to what is available on the spacecraft. The spacecraft gyro attitude reference and computer could be used during the transfer orbit.

When autonomous operations, after lift-off or after deployment from the orbiter in STS launches, is a requirement, then the inertial guidance alternative is the best approach. The issue of injection accuracy is sometimes used as a requirement for inertial guidance. DoD spacecraft under development have an orbit maneuvering or station keeping capability so that any required orbit adjustment to eliminate injection errors can be achieved with the spacecraft system. With this assumption, injection accuracy is not an issue in choosing between ground-command and inertial guidance systems. The considerations in choosing between an upper stage with its own avionics and a system where spacecraft avionics are shared with an upper stage is discussed in the next section.

SPACECRAFT AVIONICS

Table 5 shows the general types of avionic subsystem requirements and how these requirements are implemented or mechanized on spacecraft and conventional inertially-guided upper stages such as the IUS, Centaur or Transtage vehicles. A comparison of the implementations shows which spacecraft avionic subsystems have application to the upper stage requirements. Generally, the spacecraft components having application to the upper stage function are; gyros, star sensors, sun sensors and earth sensors for attitude sensing; the spacecraft computers for data management; batteries for power and the TT&C system. The spacecraft reaction control system thrust level is generally too low and the system has the wrong configuration for use during the coast phase or for roll control during the powered-flight phase of the transfer orbit.

TABLE 5

AVIONICS REQUIREMENTS/IMPLEMENTATION COMPARISON

REQUIREMENTS	IMPLEMENTATION	
	UPPER STAGE	SPACECRAFT
NAVIGATION	GYROS & ACCELEROMETERS	
ATTITUDE CONTROL	RCS	REACTION WHEELS MOMENTUM WHEELS CMGs RCS GRAVITY GRADIENT MAGNETIC TORQUERS
THRUST VECTOR CONTROL	GIMBALLED NOZZLE - HYDRAULIC - ELECTROMECHANICAL RCS SECONDARY INJECTION	RCS
ATTITUDE SENSING	GYROS STAR SENSORS	GYROS EARTH SENSOR SUN SENSOR STAR SENSOR
DATA MANAGEMENT	GENERAL PURPOSE COMPUTER	SPECIAL PURPOSE COMPUTERS AND MICROPROCESSORS
COMMUNICATIONS	S-BAND	S-BAND & OTHERS
POWER	BATTERY	SOLAR + BATTERY

A survey was performed of the avionics equipment required in current DoD spacecraft under development and the results are summarized in Table 6. These spacecraft have a geosynchronous-class orbit. Also shown, is the equipment required by an inertially-guided upper stage to transfer the spacecraft from an Orbiter park orbit to the geosynchronous final orbit. Generally the spacecraft computer, gyro package and star sensor could be used by the upper stage. The addition of a three-axis accelerometer package is required by the spacecraft to perform the navigation function required by the upper stage. The reaction control system on the spacecraft is of no use during the transfer phase because its thrust level is too low. Figure 1 shows two possible configurations whereby the spacecraft avionics can be shared with the upper stage. Figure 1b) covers the case where the spacecraft data management system has insufficient capacity to perform the upper stage GN&C functions. In this case, the processing is distributed with a small microprocessor added to the upper stage to

TABLE 6

UPPER STAGE/SPACECRAFT AVIONICS EQUIPMENT MIX

HARDWARE COMPONENT	UPPER STAGE	SPACECRAFT	APPLICABILITY OF S/C EQUIPMENT TO UPPER STAGE REQMTS.
ATTITUDE SENSING	THREE-AXIS GYRO PACKAGE STAR SENSOR	GOOD QUALITY THREE-AXIS GYRO PACKAGE SUN SENSOR STAR SENSOR EARTH SENSOR	YES, WITH MODS NO YES NO
VELOCITY SENSING	3-AXIS ACCELEROMETER PACKAGE	NONE	NONE
ATTITUDE CONTROL	RCS	RCS - 1# MOMENTUM EXCHANGE DEVICES	NO NO
THRUST VECTOR CONTROL	GIMBALLED ENGINES	RCS - 1#	NO
ΔV CONTROL	RCS	RCS - 1#	POSSIBLE
COMPUTER	GENERAL PURPOSE OR DISTRIBUTED	CPU's FROM 16 BIT, 16K, 200 KOPS, FIXED-POINT TO 16 BIT, 64K, 360 KOPS FLOATING POINT	YES

perform all the control functions. The required attitude information and steering commands come from the spacecraft side of the interface. The concept of sharing of spacecraft avionics with the upper stage seems ideally suited to space programs requiring multiple launches. For these cases the nonrecurring costs, can be allocated to each launch. The nonrecurring costs include implementing the upper stage software in the spacecraft computer, testing the software, interfacing with and modifying the spacecraft gyro package, adding an accelerometer package, interfacing with the spacecraft star sensor and the spacecraft-to-upper stage integration. If, however, a common upper stage is used with one or two spacecraft from each of a number of satellite programs, the nonrecurring costs per program are significant. Implementing and testing the GN&C software into different spacecraft computers is a significant cost item. Each spacecraft gyro package will be different: some will have redundancy, some single degree-of-freedom instruments and some two degree-of-freedom gyros. The cost of interfacing the navigation equations with each gyro package and at best, a common accelerometer package, is a consideration.

SUMMARY

Upper stage guidance alternatives have been identified and the relative qualitative advantages and disadvantages of each addressed. The spacecraft avionics equipment available in current DoD space programs that is applicable to the upper stage GN&C function has been identified. Finally, the cost effectiveness of sharing spacecraft avionics with the upper stage is highly dependent of the number of launches in a particular space program and the available spacecraft avionics. While there are no major technical issues involved in the concept of sharing avionics between the spacecraft and upper stage, it is clear that trade studies of the costs are necessary to select the integrated GN&C approach.

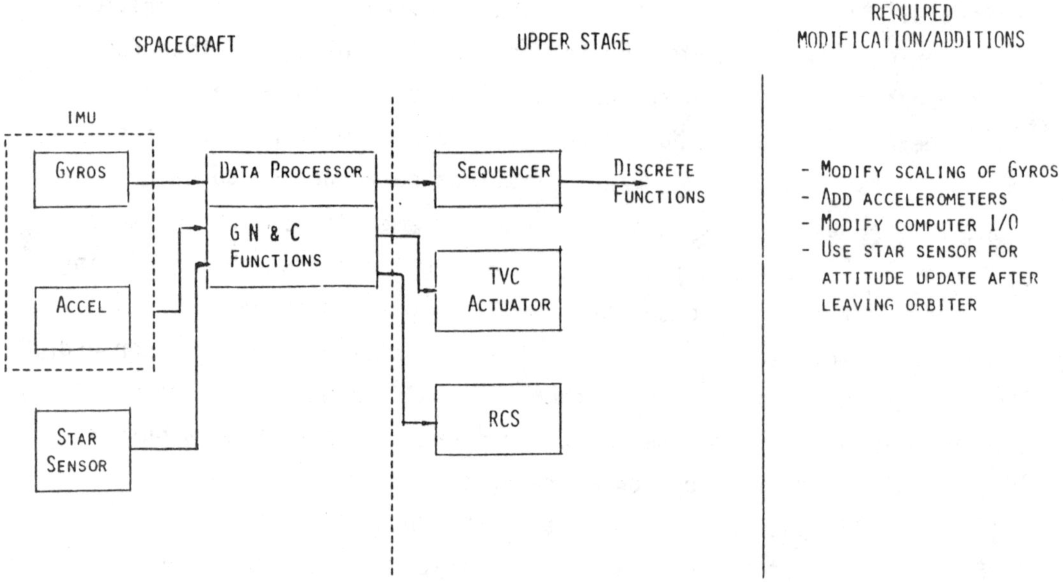

FIGURE 1A) INTEGRATED SPACECRAFT/UPPER STAGE AVIONICS
ALL PROCESSING ON SPACECRAFT

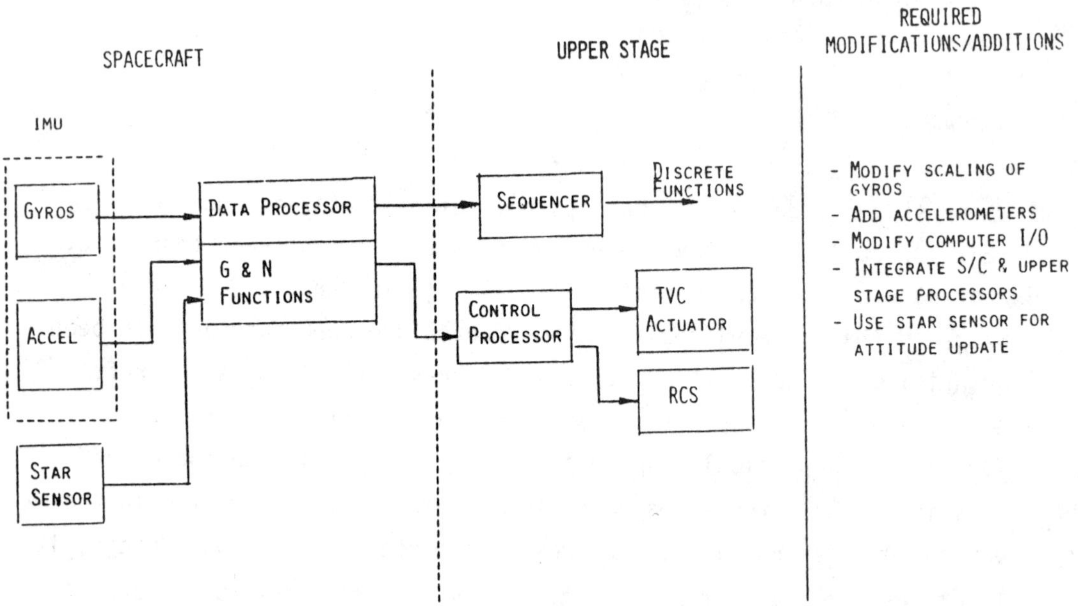

FIGURE 1B) INTEGRATED SPACECRAFT/UPPER STAGE AVIONICS -
DISTRIBUTED PROCESSING

Section V
RECENT EXPERIENCES

SESSION V

Chairman: Louis Herman
The Aerospace Corporation

Local Chairman: John C. Durrett
Martin Marietta Denver Aerospace

The following papers were not available for publication:

AAS 84-055 Shuttle Guidance and Control Lessons Learned, K. Cox, NASA Johnson Space Center, K. Daly, C.S. Draper Laboratories

AAS 84-057 High Frequency Jitter Effects in Control Scan Laser Tracking Systems, Major R.L. Van Allen, Air Force Weapons Laboratory

The following paper numbers were not assigned:

AAS 84-061 to -099

AAS 84-051

INERTIAL UPPER STAGE/TRACKING DATA RELAY SATELLITE (IUS/TDRS) MISSION POST-FLIGHT ANALYSIS

ANTHONY K. GOODFELLOW*, THEODORE R. ANDERSON AND MICHAEL T. OSHIMA+**

At approximately 83 seconds into a planned 107-second burn of the second stage solid rocket motor (SRM2) of the IUS/TDRS mission, a near-catastrophic anomaly occurred. The anomaly caused spin-up of the IUS-TDRS combination to 192 deg/sec. with the result that mission sequencing failed to continue. Subsequently, contingency commands were uplinked, enabling IUS/TDRS separation just prior to IUS battery depletion. This paper addresses the vehicle dynamics and control system analyses performed to determine the cause of the spin-up anomaly.

INTRODUCTION

The IUS, developed under contract to the Air Force Systems Command Space Division, is designed to carry spacecraft payloads from low earth orbit to geosynchronous or other higher-energy orbits. The IUS is boosted into low earth orbit by either the Air Force Titan 34D or the STS shuttle orbiter. The basic IUS vehicle is a two-stage SRM configuration with a mono-propellant hydrazine reaction control system (RCS) used for attitude control and to provide for velocity trim following SRM burns. During powered flight, two electromechanical actuators deflect the SRM nozzle for pitch and yaw control and the RCS is used for roll control.

The first IUS flight (October 30, 1982) utilized the Titan 34D booster and successfully placed a combined DSCS II/III payload into orbit. The flight performance for this flight is discussed in the reference. The second flight utilized the shuttle orbiter with the TDRS-A as payload. The anomaly which occurred during this flight on April 5, 1983 is the subject of discussion here.

* *Manager IUS Flight Technology, Boeing Aerospace Company, Seattle, Washington*

** *Principal Engineer, Boeing Aerospace Company, Seattle, Washington*

+ *Specialist Engineer, Boeing Aerospace Company, Seattle, Washington*

ANALYSIS OBJECTIVES

After a first look at the telemetered flight data, it appeared that the SRM nozzle had locked into a nearly fixed position at the time of the anomaly. A closer look at the data at the time of and subsequent to the anomaly established some characteristics that were initially inexplicable. These were:

- Body-axis linear acceleration data appeared incompatible with the perceived body-axis angular rate data as derived from saturated gyros and with the anticipated spin rate that was estimated and should have been achieved with the fixed nozzle position.
- Subsequent to SRM2 burnout, pitch and yaw axis nozzle motion in response to commands was regained.
- After initial nozzle motion that was not in response to commands, the pitch axis thrust vector control (TVC) appeared to jam while the yaw axis TVC showed small angle responses to changes of command direction.

The flight analysis objectives were thus twofold:

1. Establish the validity and compatibility of the flight data.
2. Establish, by computer simulation, failure scenarios that provided emulation of the flight data.

BACKGROUND

IUS Controls analyses use a standard right handed coordinate set illustrated in Fig. 1. Actuator positions and motions are such that pitch actuator retraction causes positive pitch motion and yaw actuator extension causes positive yaw motion.

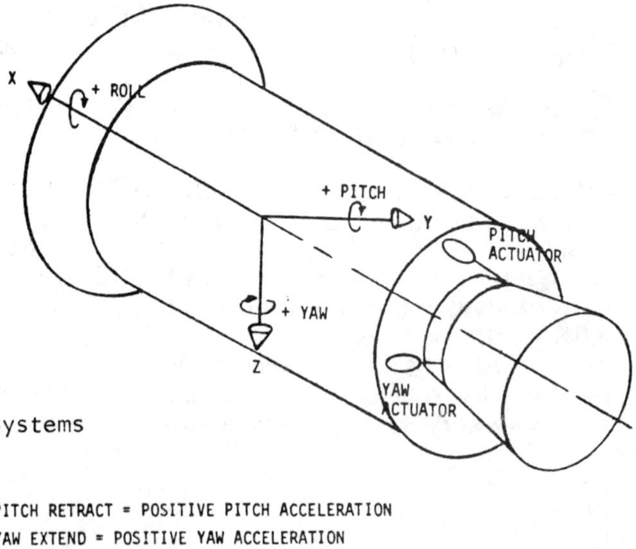

Figure 1
Powered Flight Coordinate Systems

PITCH RETRACT = POSITIVE PITCH ACCELERATION
YAW EXTEND = POSITIVE YAW ACCELERATION

The basic functioning of the Flight Control System is illustrated in the block diagram on Fig. 2. The Redundant Inertial Measurement Unit (RIMU) supplies rate gyro and accelerometer data to the navigation software where they are integrated to provide attitude, position and velocity. The RIMU contains 5 gyros and 5 accelerometers in skewed conical configurations. The gyro saturation thresholds are nominally ±30 deg/sec. and the accelerometers ±20 g's. No one instrument is aligned with an IUS body axis; thus body-axis outputs are obtained by transformations performed by the software. In the event of detected failures, different transformation matrices are required, depending upon the remaining "good" instrument configurations. The independent instrument outputs are not telemetered; however, the instrument cluster in use is identified.

During SRM burns, the guidance quaternion is "differenced" with the navigation quaternion to provide body-axis attitude errors. These, together with the body-axis attitude rates, are supplied to the pitch and yaw axis TVC algorithms and the roll-axis RCS algorithm. These data are then modified by appropriate gains, digital filtering and limiting processing, and the resulting outputs are then issued as commands through the Signal Conditioning Unit (SCU) to the TVC servo systems and roll control thrusters.

Fig. 3 identifies specifics of the TVC telemetry. The commands generated within the computer are fed directly to the telemetry formatter, constituting the most accurate representation of the issued values. The digital commands are converted to analog in the SCU using an 8-bit converter. The first element in the TVC controller is a buffer amplifier which echoes the command back to the computer through an A/D converter in the SCU. Other telemetry data processed through the SCU are the position monitor and axis currents. The position data is a value representing the combined outputs of a potentiometer co-linear with the actuator and another potentiometer positioned 180 deg from it around the periphery of the nozzle. The SCU telemetry data of each axis are all pre-formatted in the SCU to be within 3 milliseconds of each other. However, they lag the computed command by about 120 msec. in the raw data.

FLIGHT DATA

The flight data whose initial compatibility was in question consisted of the TVC and RIMU data described below:

TVC Servo System Data

Fig. 4 shows pitch and yaw commanded and actual TVC positions together with axis current at the anomaly event and shortly thereafter. Note that current magnitude only is telemetered. The pitch actuator moved from +0.16 in. to −0.17 in. in an uncommanded manner. The autopilot then generated correcting commands but the actuator failed to respond. Yaw position showed a drop to a negative peak of −0.26 in. from an original value of −0.025 in., then rapidly returned to −0.11 ins followed by a slower return to −0.042 in. The negative yaw peak preceded the abrupt stop of the pitch motion by 80 msec. and all rapid movement in both axes ceased after 200 msec. TVC axis-current data indicate the pitch actuators were at current limit in 120 msec. followed by the yaw actuator at 280 msec.

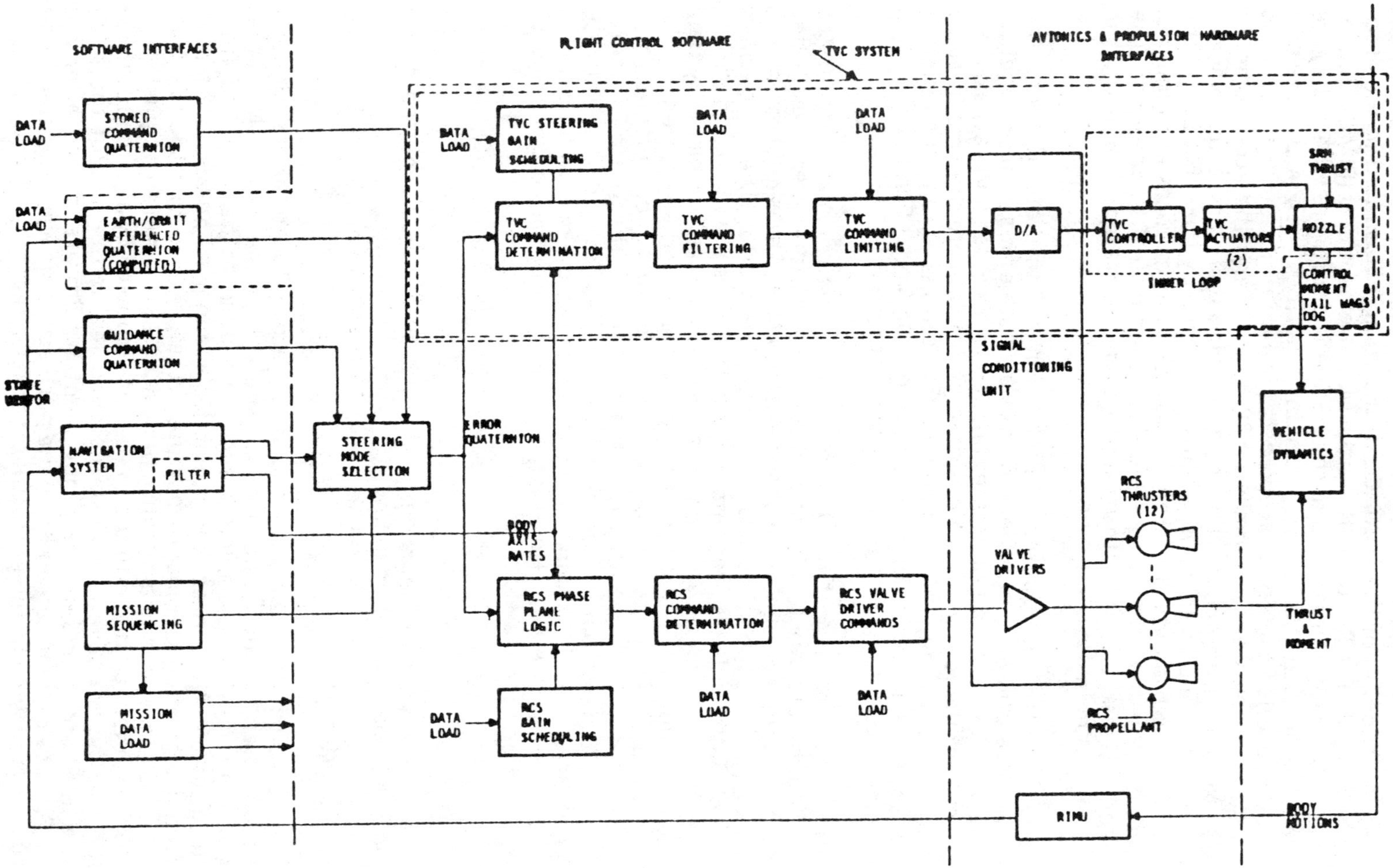

Figure 2 Flight Control System Functional Block Diagram

TVC CONTROL TELEMETRY

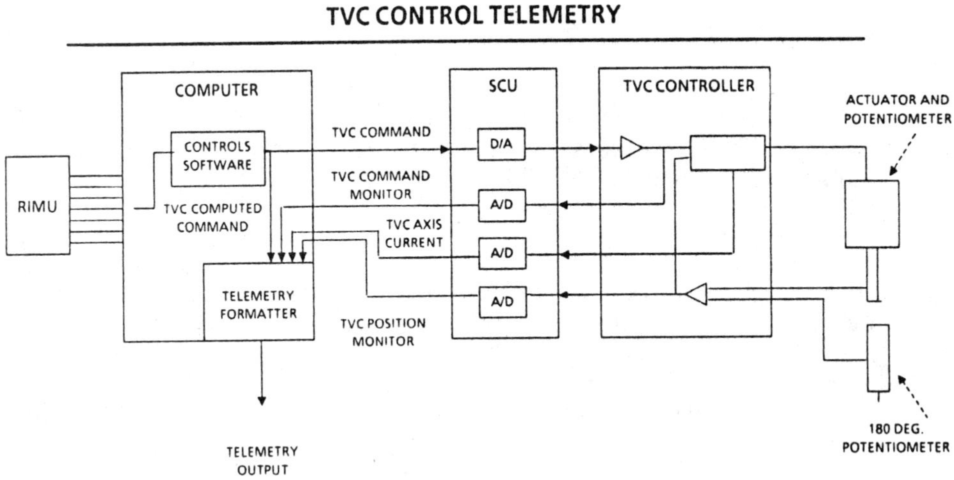

Figure 3

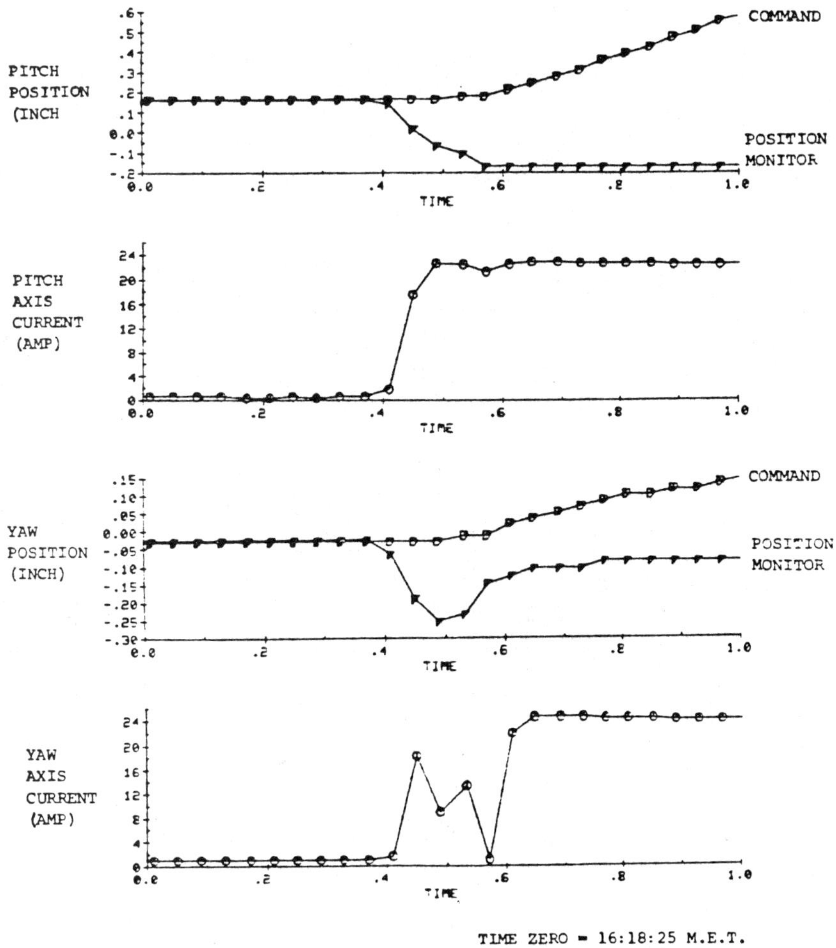

Figure 4 TVC Actuator Flight Data

Accelerometer Data

Fig. 5 illustrates accelerometer data resolved into body axes. The X-axis shows the combined effects of both external thrust and centripetally induced accelerations. The transverse axes show initial rotational acceleration and external side force effects, followed by peaks induced by centripetal effects.

Attitude Rate Data

Fig. 6 depicts the telemetered vehicle attitude rates. The thick lines depict data periods, the thin lines, data drop-outs due to signal loss resulting from the vehicle tumble. Because of the gyro saturation rate limits, these data rapidly become inaccurate as indicators of true vehicle rates, but continue to provide a signal based on a computation from the individual gyro outputs.

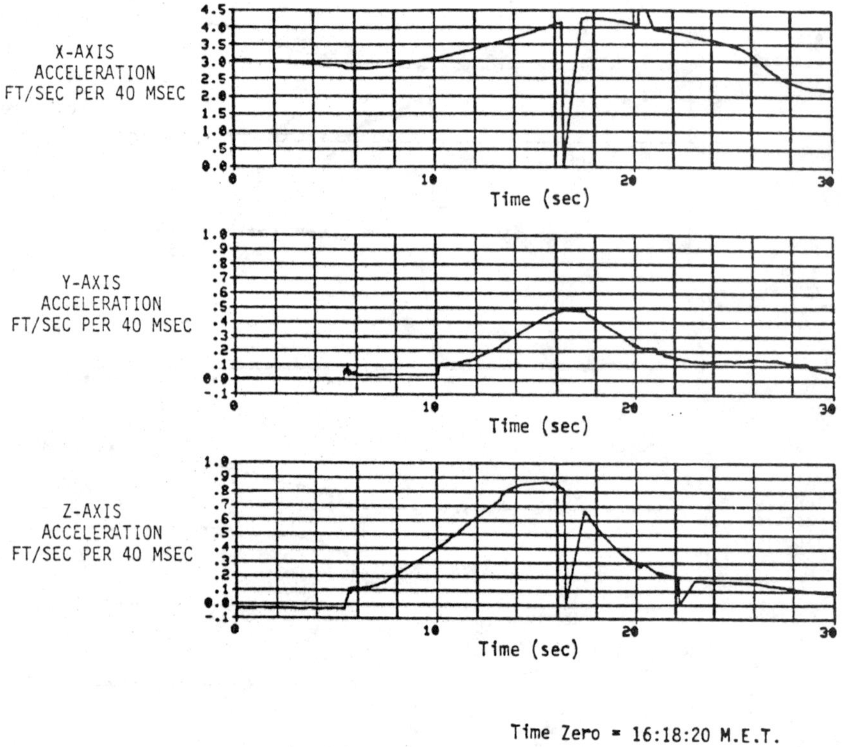

Figure 5 Telemetered Body Axis Accelerations

Also illustrated in Fig. 6 is the roll thruster "on count" trace. This shows the roll RCS activity in response to commands during the burnout period.

Additional data available to assist in the compatibility assessment were the accelerometer parity vector data and the ground station received-signal-strength plots. The accelerometer parity vector is a measure of the degree of discrepancy among the five instruments – both in differences among the instruments themselves and in differences resulting from the fact that they are not truly co-located. The high attitude rates associated with the spin-up resulted in centripetal acceleration differences among the accelerometers which showed up in the parity vector. The parity vector is normally calculated for RIMU fault detection and isolation (FDI) purposes.

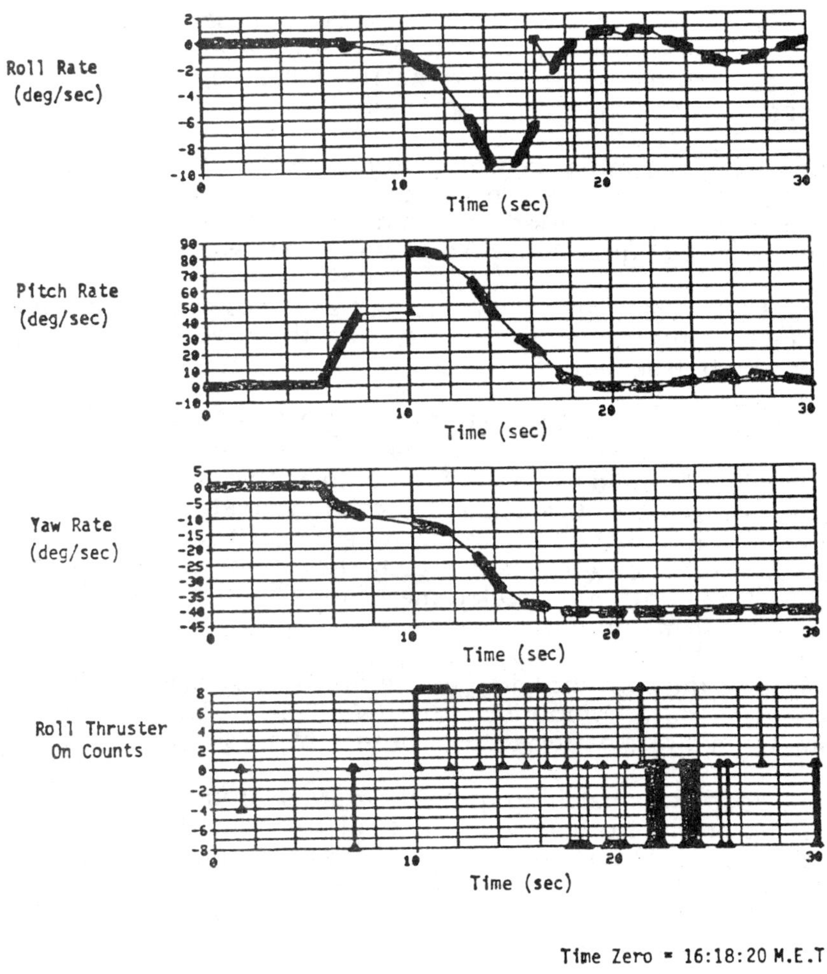

Figure 6 Telemetered Attitude Rates and Roll Thruster on Times

ANALYSIS APPROACH

Since the true vehicle angular rates were fundamental elements required in the data compatibility assessment, the prime objective was to obtain these parameters. Consequently, the analysis approach was broken down into three steps, each requiring the construction of specialized computational tools and with the objective of matching predicted and flight data at every step. These steps were:

(1) At selected time points, pass estimates of body rates through a simulation of the saturated gyros and transform to the telemetered body rates. Iterate the process until the predicted telemetry values match the flight data.

(2) Utilize the discrete rates determined in (1), together with the estimated angular accelerations due to displaced nozzle and roll thrusters, to obtain accelerometer data and parity vector data from kinematic relationships. Match predicted and flight data.

(3) Starting from the initial vehicle attitude at the start of the anomaly, integrate the body rate profiles determined in (1) and, utilizing a representation of the Medium Gain Antenna pattern, estimate the ground-received signal strength and match to mission data.

Steps (1) and (2) were eventually combined and a computer algorithm was used to minimize the squared error between the predicted and the flight data to obtain a better match.

VEHICLE DYNAMICS ANALYSIS RESULTS AND CONCLUSIONS

The rate profiles obtained using the foregoing analysis method are shown in Fig. 7. Data points were obtained at 2 second intervals through 20 seconds, then a match was obtained at steady state, i.e., 30 seconds. This profile produced the accelerometer data match shown in Fig. 8 by the crosses superimposed on the flight accelerometer data. All points obtained are within 5 percent in magnitude.

Fig. 9 represents the corresponding match of the body-axis rates to the flight data. The final data comparison of this phase of the analysis was the accelerometer mid-rate parity vector. These data are shown in Fig. 10.

Fig. 11 depicts spin up matching using the data drop-out periods as a source of the spin-up profile truth model. The small circles are flight data and have been curve-fitted as shown by the continuous line. The crosses are obtained from the rate profile data.

Finally the derived ground-station-received signal strength is shown as the continuous trace of Fig. 12. The crosses indicate the observed peaks of ground-received signal strength taken from the tracking station data. Magnitude and time-phasing correlation was deemed satisfactory.

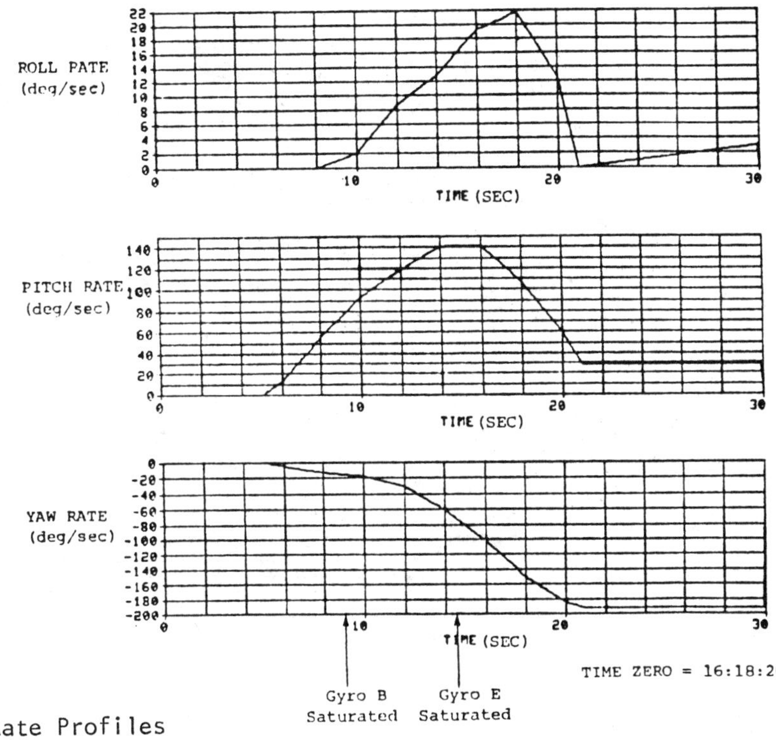

TIME ZERO = 16:18:20 M.E.T.

Figure 7
Vehicle Body Rate Profiles

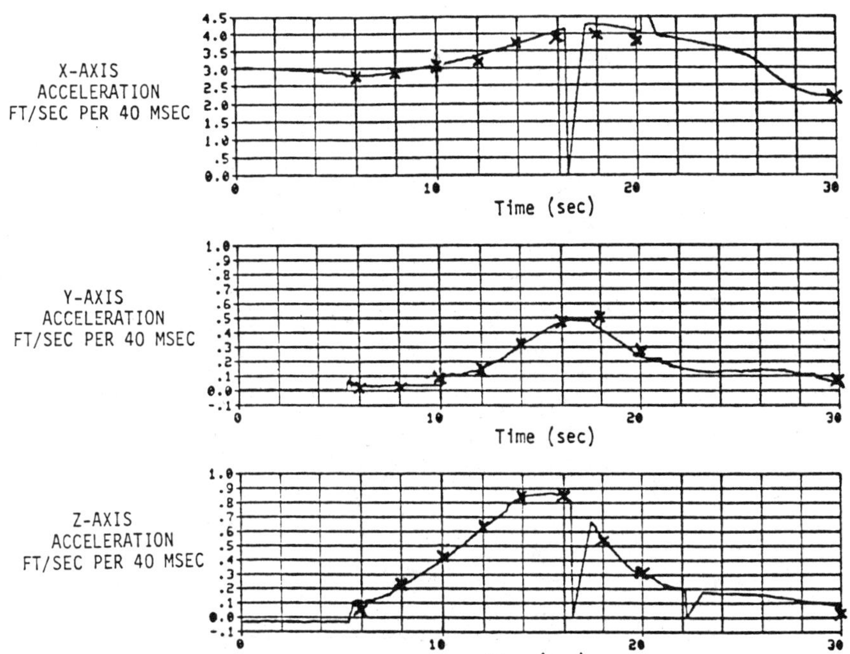

Figure 8
Acceleration Profile Comparison

Time Zero = 16:18:20 M.E.T.

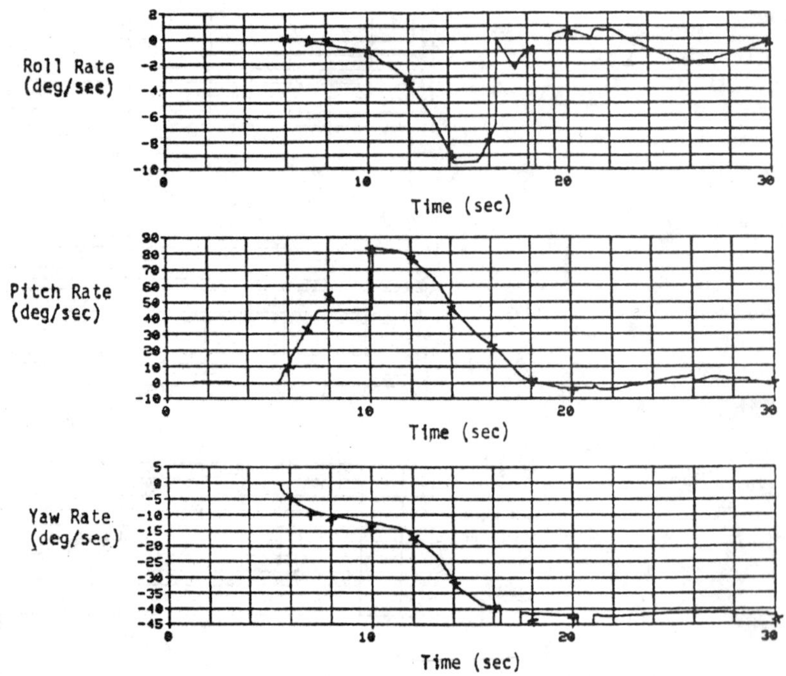

Figure 9

Rate Data Comparison

Time Zero = 16:18:20 M.E.T.

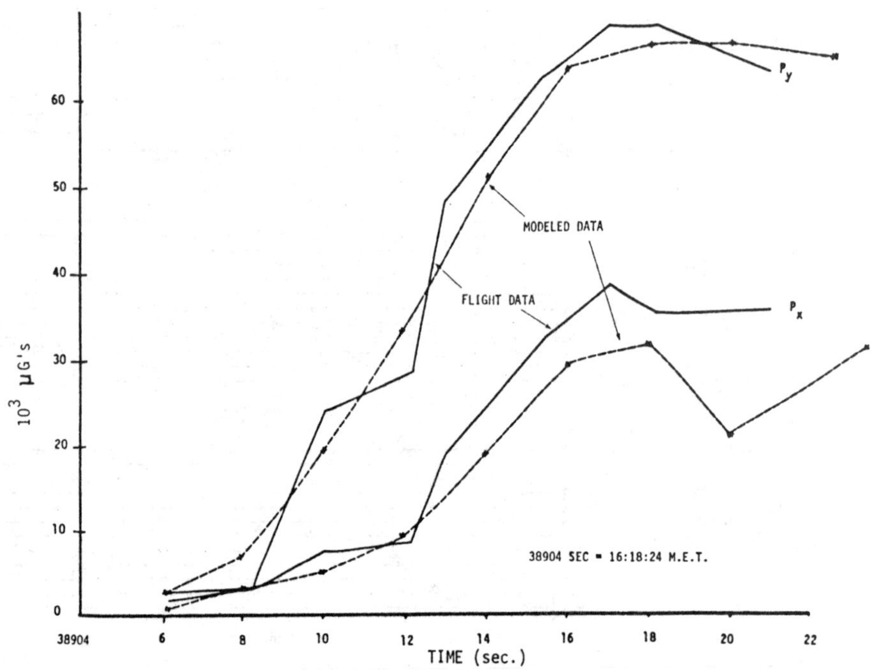

Figure 10 Accelerometer Mid-Rate Parity Vector Match

The significant factor in the rate profile analysis was that the pitch and yaw profiles correlated somewhat in phasing and polarity, if not in magnitude, as compared to the telemetered data. However, the roll-axis data did not compare in either magnitude or polarity. The roll thruster on-count trace gave the clue to this apparent anachronism. The roll control RCS control authority had a 10-to-1 margin over the TVC-induced roll disturbance throughout the period. Hence, since it was indicating positive roll thrusting for 7.5 plus seconds, the true roll rate was positive and not negative as indicated. This roll jet activity also accounted for the fact that although the vehicle initially started with significant "pitching" motion, it soon "rolled" over into a mostly "yawing" motion as a result of the gyroscopic cross-coupling from the combined roll and yaw rates. This accounted for a somewhat lower final transverse spin (tumble) rate than would have been expected if there was no roll rate due to RCS activity.

The final conclusion was that all of the telemetry data related to vehicle dynamics and kinematics, including the TVC position data, were compatible.

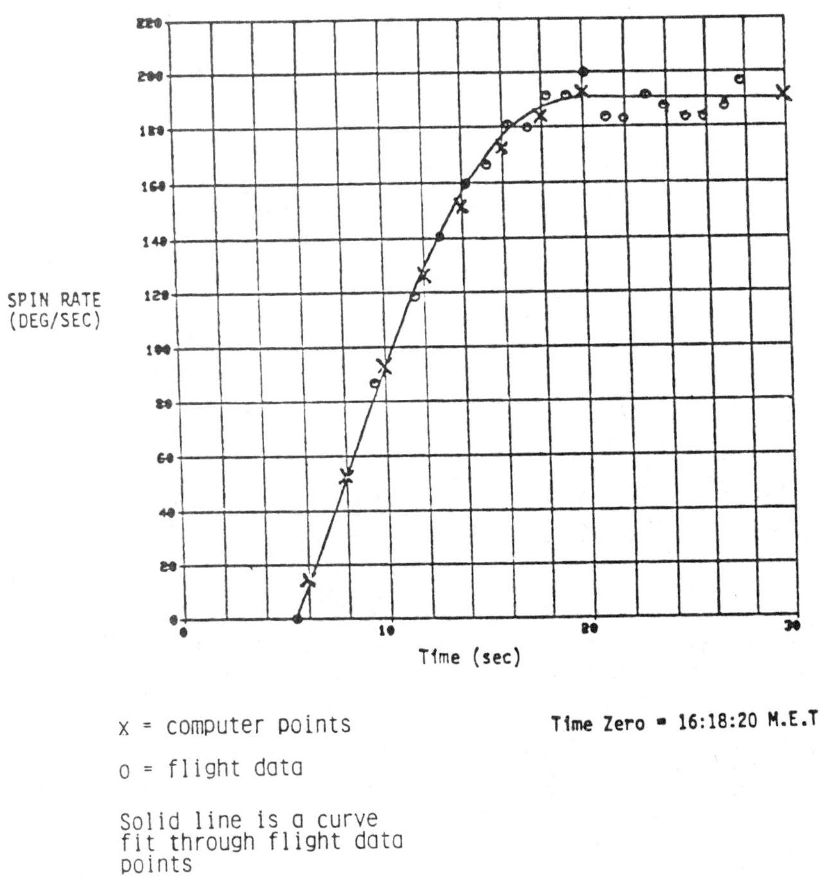

Figure 11 Telemetry Drop Out Spin Rate Comparison

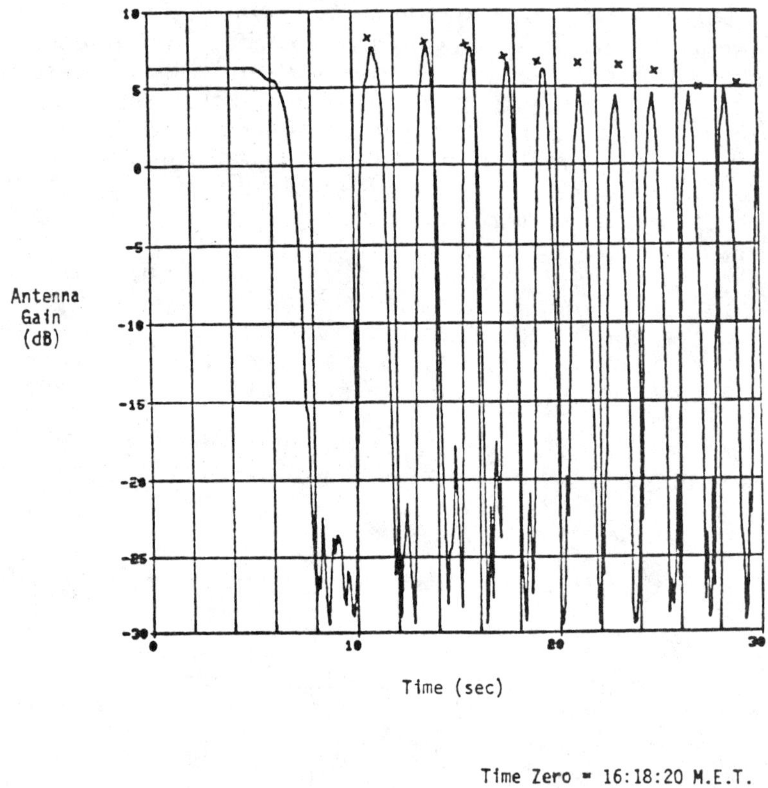

Figure 12 Ground Station Received Signal Strength Comparison

POST-BURNOUT TVC MOTION ANALYSIS

After SRM2 burnout, the TVC system appeared to regain control of the nozzle, so verification of correct command processing was required to exonerate the possibility of a software-induced anomaly.

Verification of the on-board algorithm processing of RIMU data to obtain TVC commands was performed by driving the Controls Software verification tool with inputs of the flight data. Fig. 13 shows the flight-reported computed commands and the simulated calculated commands which overlay the flight data in an acceptable fashion. Leading edges of data periods exhibit transient start up behavior as would be expected with the processing of the many difference equations. The flight data processing, of course, did not experience the data dropouts and thus did not exhibit the transients.

NOZZLE MOTION ANALYSIS

Having finally concluded that a hardware failure had caused the anomaly, the nozzle motion was analyzed to resolve the failure between the TVC servo system and the nozzle rotation mechanism known as the Techroll Joint®

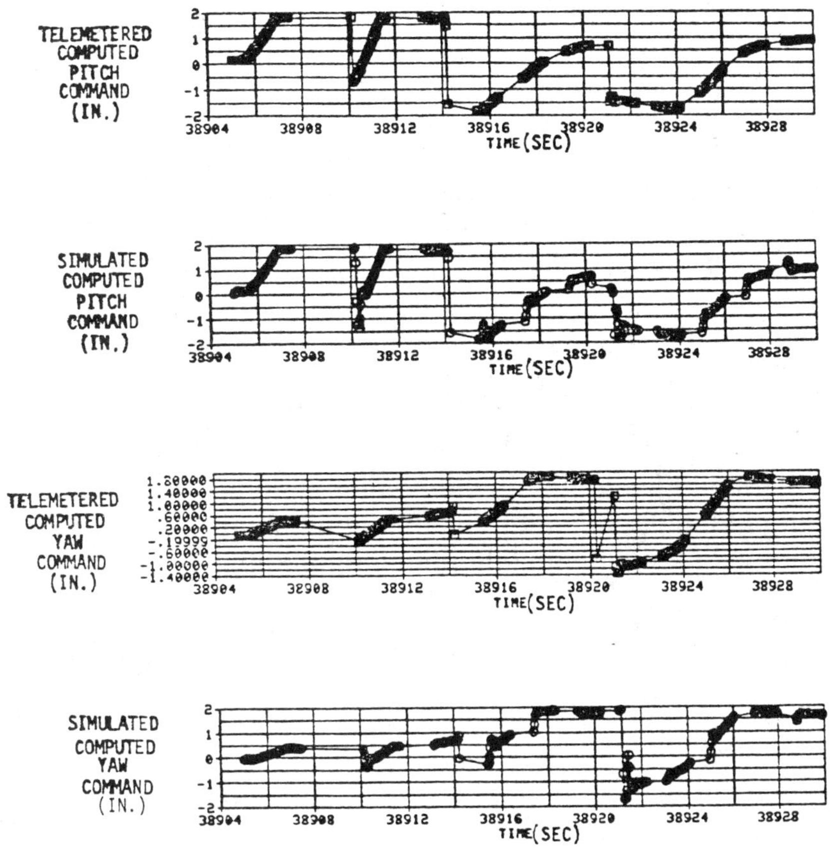

38904 SEC = 16:18:24 M.E.T.

Figure 13 TVC Command Verification

(TRJ). The TVC servo system installation and a detail of the TRJ are illustrated in Fig. 14.

The TRJ consists of a fixed nozzle ring, movable seal housing, retaining ring, gib ambient axial restraint, and a Techroll Seal (TRS). The TRS is composed of two-ply Kevlar wrapped in neoprene rubber. The movable, low-torque TRS is a constant-volume, silicone-oil-filled bearing that uses a seal with two rolling convolutes that permit omniaxial deflection of the nozzle assembly. The seal's outer convolute is protected by a thermal boot; the inner convolute is protected by a dust seal.

Significant effort was expended by the TVC servo system supplier in an attempt to implicate this system as the cause of the failure. All efforts in this direction failed and no viable failure mode was identified that could have resulted in the observed vehicle and nozzle motions. It was then concluded that a failure of the TRJ had occurred.

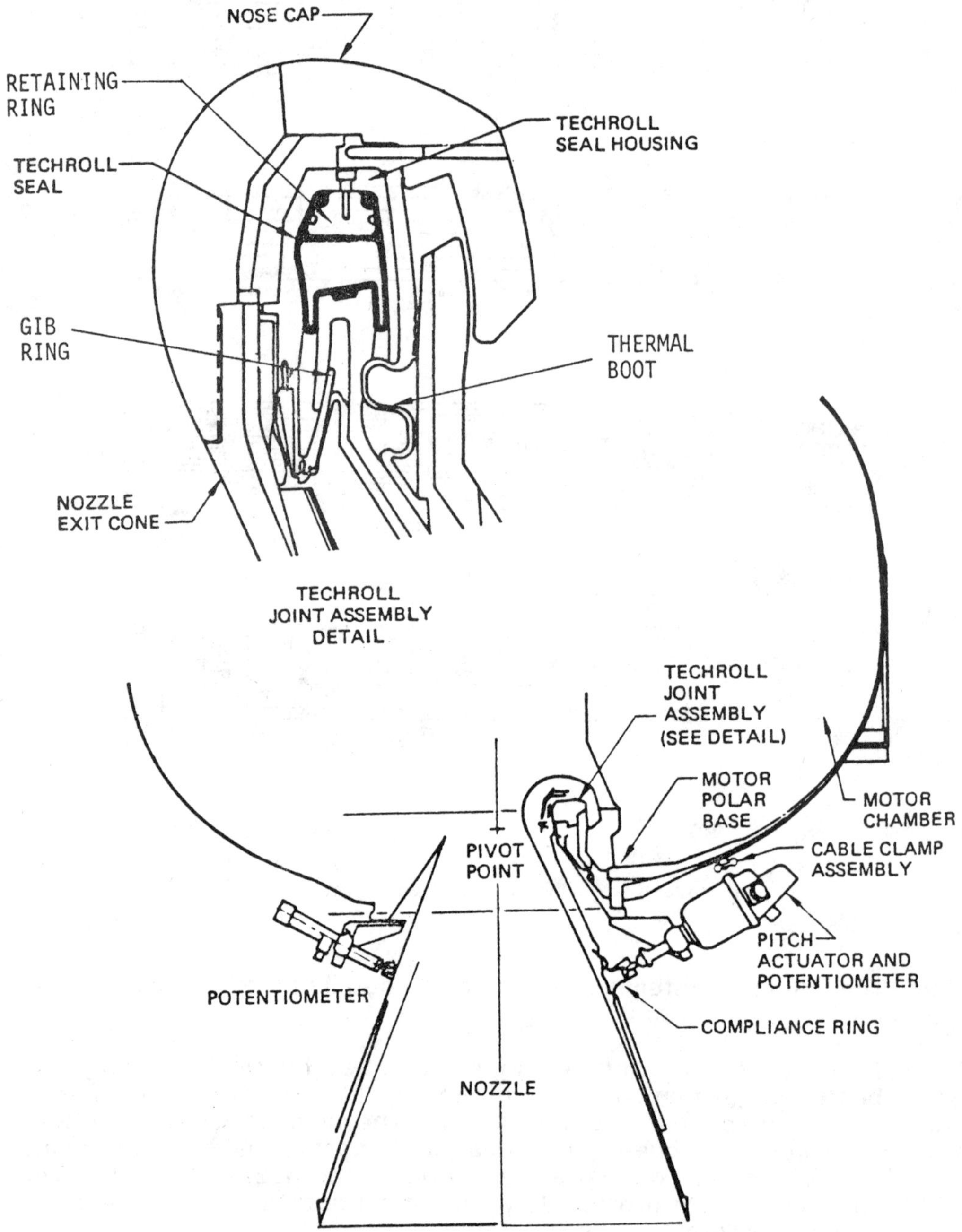

Figure 14 Techroll Joint Assembly (With Detail)

If the TRS were to leak silicone oil, the very high blowoff force (due to motor chamber pressure) would cause the TRS to collapse, allowing the nozzle to translate rearward until it bottomed out against a restraint. This restraint in combination with the blowoff force would create a significant moment which would tend to self-center the nozzle, with the potential for overpowering the TVC actuators in either a backdrive or overdrive condition.

In order to determine how the TRJ failed, a computer simulation analysis of the nozzle motion was conducted. The nozzle motion simulation models the TVC actuator and nozzle load characteristics in detail down to the functional level of the major electronic blocks. Nozzle translational kinematic relationships are included. The simulation was utilized to examine various TRS failure scenarios to determine their compatibility with the observed position and current monitor data during the anomaly.

The simulation was first used in an open loop mode and, using the reported position data as a forcing function, the resulting current signature was obtained and compared to the flight data. The forcing function was applied at the combined feedback point, representing pure rotation. The current signatures for pitch and yaw axes for this scenario are shown on Fig. 15.

It can be postulated that given the 40 msec. sample rate of the telemetry stream, a match could be made to the pitch-axis data. However, in no way does the yaw-axis current approach matching the dual spike characteristic of the flight data. Hence it was concluded that the failure mode must have involved translational motion of the TRS.

The second method of simulation utilization was in a closed loop mode, with nozzle translation, in order to simulate a collapsing techroll seal. Under this method of utilization, both TVC position and current data became discriminators.

A failure scenario matrix was devised covering both slow and fast seal leak rates and varying collapse strokes. The maximum collapse stroke possible is 1 in. at ambient pressure. However, the SRM chamber pressure at the time of the anomaly reduced the maximum value to about 0.8 in. Sample simulation results are shown in Figuress 16 and 17. As can be seen, the best match was obtained with a stroke of 0.8 in. at a collapse rate of 6.7 in/sec.

Results of the control system simulation analysis of the TVC system were construed to confirm that the nozzle assembly moved aft rapidly at the onset of the control anomaly. Such an aft motion can only occur when the TRS collapses, and the following scenario provides the most-likely description of events and the best match to the flight data.

The TRS collapsed at such a rapid rate that the actuators could not extend fast enough to compensate for the rapid aft motion of the nozzle because of inertia and friction effects of the actuator and the rate feedback component (due to aft motion) in the control loop. As a result, the nozzle pivoted about the clevis, where the actuator arm attaches to the compliance ring, causing the nozzle to lock in the position indicated by flight telemetry data.

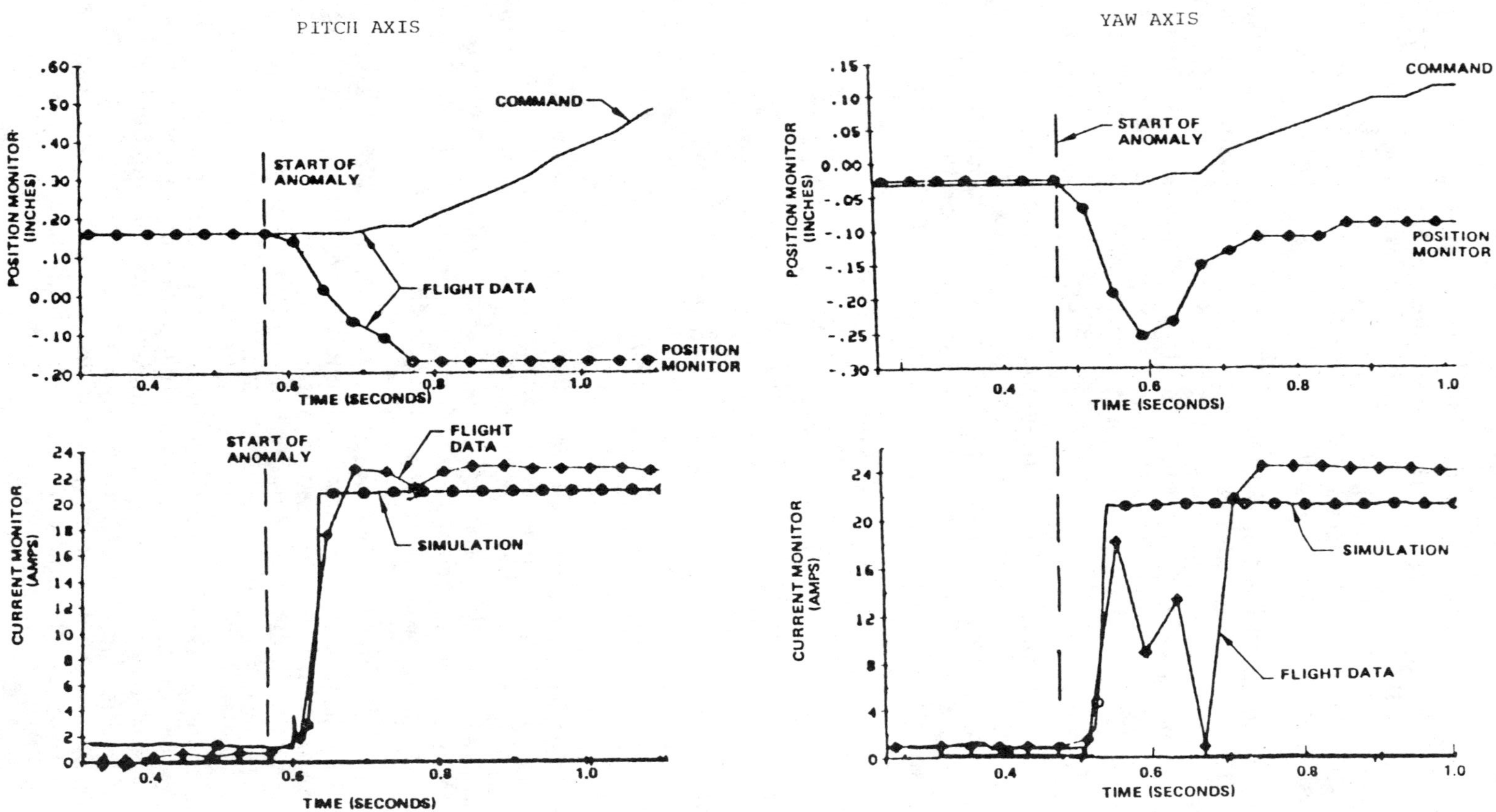

ROTATION ABOUT NORMAL PIVOT

Figure 15 Time Domain Simulation Results

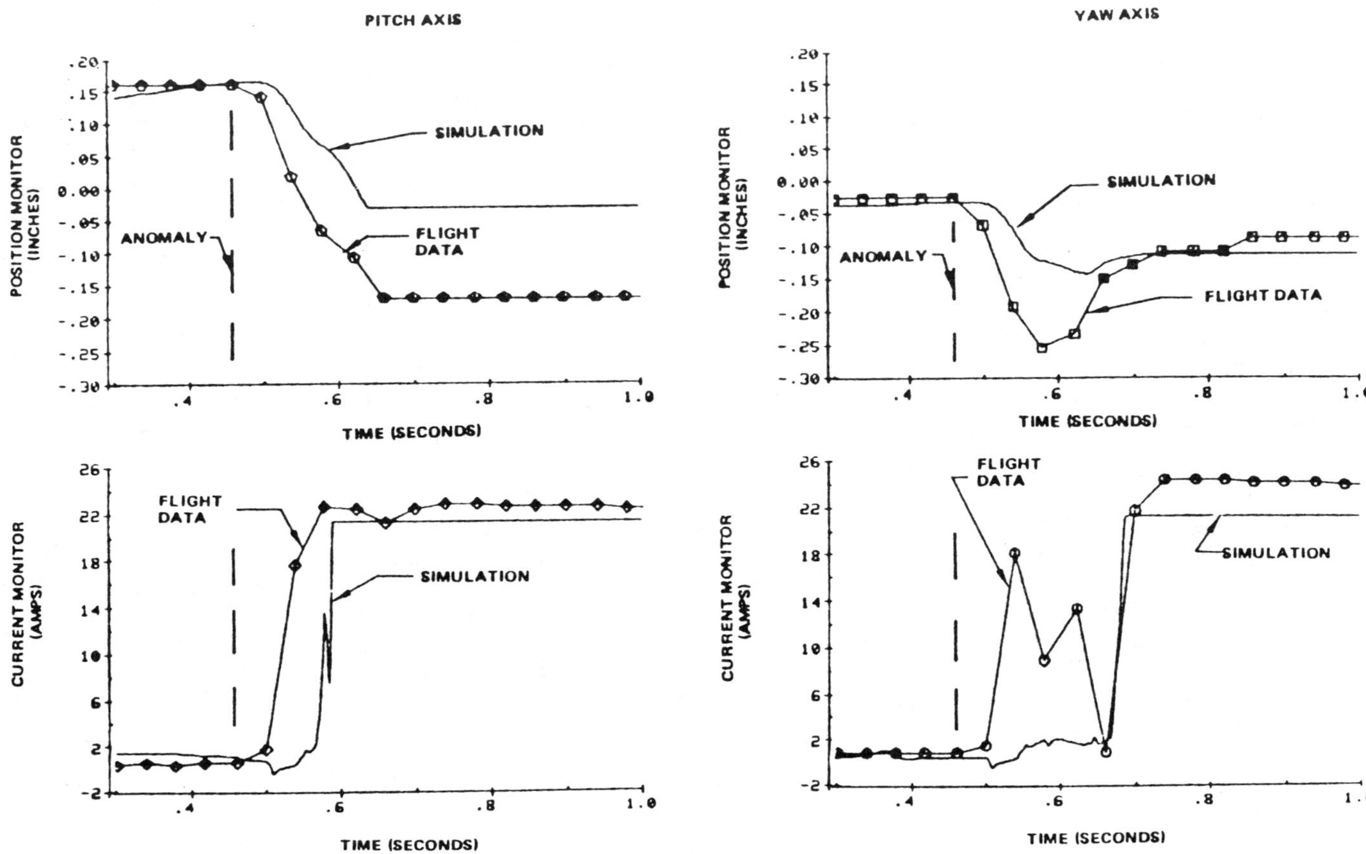

COLLAPSE RATE = 2.86 In/Sec; DISTANCE = 0.4 In

Figure 16 Time Domain Simulation Results

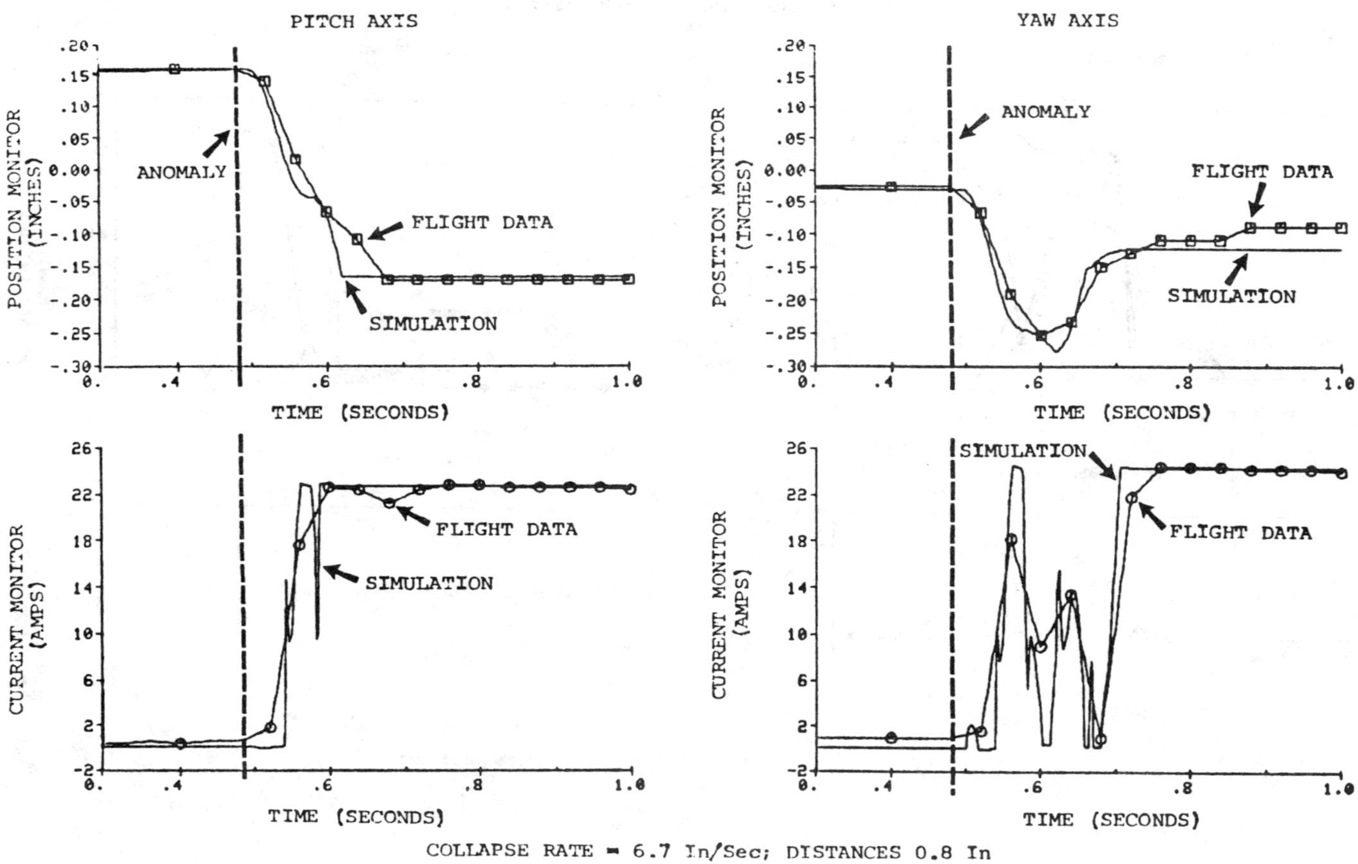

Figure 17 Time Domain Silulation Results

The aft motion of the nozzle drove the pitch actuator housing into the cable clamp and pad mounted on the motor dome (see Fig. 14). This caused a bending moment to be induced into the pitch actuator, which resulted in the actuator motor rotor coming into binding contact with the stator. The yaw actuator, having no obstructions behind it, recovered to near null and was held in that position by motor blowoff forces. Since the yaw actuator rotor was free to move, it responded slightly to commanded reversals consistent with nozzle and bracket compliance. Because the pitch actuator armature was not free to rotate, it was unable to respond to commanded reversals. The pitch position drifted slowly toward null as reduced motor pressure permitted relaxation of the binding forces.

The pitch and yaw actuators started to regain control authority just before or at motor burn completion. Actuator motion was now possible because the blowoff force was near zero; hence, moment induced into the pitch actuator had been removed. The TVC recovered to full control after two cycles, aided by the thermal boot spring force and the rocking action of the actuator, which freed the binding caused by the collapsed TRS.

CONCLUSIONS

The derived failure scenario described above very closely matches all of the pertinent flight data at the start, during, and subsequent to the anomaly. Subsequently, significant hardware quality analysis and testing has been conducted in an effort to pinpoint the cause of failure. To date, all indications are that a breakdown of the thermal protection system resulted in exposing the TRS material to extremely high temperatures, resulting in what essentially resembled a blow-out. Program recovery efforts are now directed at remedying the problem.

REFERENCE

A. K. Goodfellow and Captain Charles Vono, "First Flight Performance of the Control System of the Inertial Upper Stage," presented to the *American Astronautical Society Annual Rocky Mountain Guidance and Control Conference*, February 5-9, 1983.

AAS 84-052

INERTIAL UPPER STAGE
REDUNDANT INERTIAL MEASUREMENT UNIT SPACE PERFORMANCE

R.A. Baum*, G.E.S. Morrison**, R.C. Peters+

The Inertial Upper Stage (IUS) Redundant Inertial Measurement Unit (RIMU) is used in the Redundant Inertial Navigation System (RINS) of the IUS. The IUS was developed for use as an upper stage on the NASA Space Shuttle vehicle, and USAF Titan 34D space boosters. The RIMU contains five gyros and five accelerometers, and associated electronics, arranged in a redundant configuration such that no single component failure compromises system inertial performance. Failure Detection and Isolation (FDI) algorithms in the navigation computers automatically eliminate a failed sensor's data from the navigation computations, and the mission navigation function continues on uninterrupted.

To date, an IUS vehicle has been deployed once from a Titan 34D booster in October, 1982, and once from the Space Shuttle in April, 1983, to carry spacecraft to geosynchronous orbit. Telemetry data from these flights provide an indication of the RIMU's navigation and FDI performance in space. On the Space Shuttle flight, some gyro false alarm activity took place, during which the RIMU maintained continuous proper navigation function. Analyses of flight data, and subsequent laboratory tests, have attributed these false alarms to a gyro transcient effect, induced by the onset of specific angular rate profiles. The contribution of this transient effect to attitude error is insignificant. However, it can be quite noticeable when viewed through the FDI algorithms as an instantaneous error in rate, rather than attitude.

An improved methodology for the determination of FDI parameters has evolved from this space flight experience. It is influenced primarily by sensor attitude and velocity error criteria, rather than by instantaneous rate and acceleration error. Also, based on the confidence gained in on-board generated FDI data as a measure of RIMU health during flight, the capability of the FDI algorithm has been expanded to allow self-contained evaluation of, and recovery from, false alarms.

* *Hamilton Standard, Division of UTC, 1690 New Britain Avenue, Farmington, Connecticut 06032*

** *Boeing Aerospace Co., P.O. Box 3999, MS: 8K-85, Seattle, Washington 98124*

\+ *The Aerospace Corporation, Box 92957, Los Angeles, California 90009*

INTRODUCTION

The Inertial Upper Stage (IUS) was developed by the Boeing Aerospace Company under contract to the USAF Space Division for use as an upper stage on the National Aeronautics and Space Adminstration (NASA) Space Shuttle vehicle and USAF Titan 34D space boosters. Technical support is provided to the IUS program by the Aerospace Corporation. The Hamilton Standard Division of United Technologies Corporation developed the IUS Redundant Inertial Navigation System (RINS) in response to USAF/Boeing specifications.

The IUS program entered Full Scale Development (FSD) in January, 1978. The FSD program included the design, qualification, and production of the first nine flight systems. Component qualification was completed in 1980. First flight was aboard a Titan 34D booster in October, 1982. First flight aboard a Space Shuttle vehicle was in April, 1983. Both flights carried spacecraft to geosynchronous orbit.

HARDWARE DESIGN

The IUS RINS consists of a Hamilton Standard Redundant Inertial Measurement Unit (RIMU), and dual Delco Electronics Computer Units (CU), arranged in the redundant configuration shown in Figure 1.

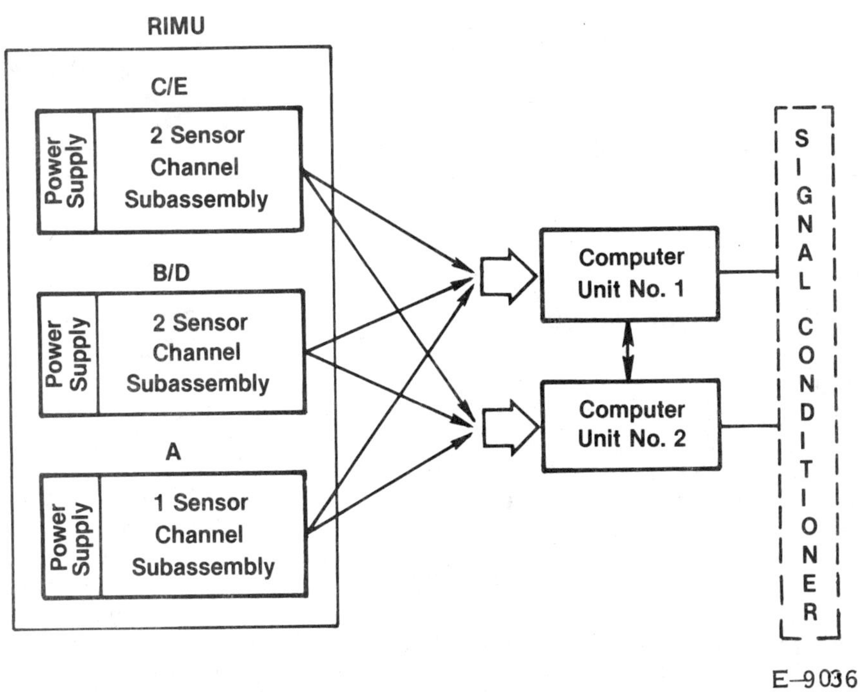

Fig. 1 IUS RINS Redundant Configuration

The RIMU uses five (5) Hamilton Standard RI-1010, pulse-torqued, single-degree-of-freedom, rate integrating, gas bearing gyros, and five (5) Kearfott 2401, pulse-torqued, single-axis, linear accelerometers, installed in a single cluster assembly. The gyro and accelerometer excitation, rebalance and output electronics are paired up, and partitioned into three, separately powered channels. A failure in any one power supply will still leave at least three gyro and accelerometer pairs in service.

The vector directions of the sensor input axes are arranged symmetrically about the surface of a cone whose half angle has its cosine equal to the square root of 1/3. This provides a globally uniform distribution of like sensor error statistics, as perceived in orthogonal vehicle axes. In addition, it takes maximum performance advantage afforded by the combination of data from five like sensors. The cone axis is tilted from the X-RIMU (IUS thrust) axis by a rotation of approximately 41.2 degrees. This orientation places two of the gyro's output axes parallel to the gravity vector on the launch pad, thus optimizing prelaunch azimuth alignment accuracy. It also places the output axes of these gyros parallel to the IUS vehicle thrust axis. In addition, all gyro spin axes and accelerometer pendulous axes are perpendicular to the thrust axis, thus optimizing inertial navigation performance during boost and IUS burns.

A "Kinematic Reversal" mechanism provides the capability of rotating the sensor cluster about a horizontal axis through an automatic sequence of four orientations, spaced 90 degrees apart. This is used during Earth prelaunch operation to perform a precise, self-contained alignment and calibration of the RIMU in order to provide a check of redundant system readiness, and to achieve the best possible mission accuracy.

Additional details regarding the RINS hardware and the IUS navigation and flight control mechanization may be found in References 1 and 2.

RIMU REDUNDANCY MANAGEMENT

The RIMU Failure Detection and Isolation (FDI) design satisfies the IUS reliability level and single failure tolerance criteria, while providing potential for additional levels of FDI capability. The computationally cross-strapped, five sensor array provides the ability to detect and isolate any first sensor failure using data comparisons alone. Computer software implemented data comparisons (i.e., "FDI Algorithms"), in conjunction with hardware Built-In-Test Equipment (BITE), are used to detect and isolate all RIMU subassembly failures. The key to successful RIMU operation is the ability to detect and isolate sensor performance degradation (i.e., "soft") failures. The magnitude of a soft failure is such that it tends to be masked by noise, but which ultimately degrades orbital injection accuracy if left to propagate throughout the mission. The FDI algorithms associated with the IUS RIMU solve this problem by means of a series of threshold checks upon the magnitude of "parity vectors" formed from the gyro and accelerometer data. The known relative placement of the inertial sensors permits the development of linear dependence equations for the sensor outputs. These linear dependence equations define the equivalence, or parity, among the sensor outputs and, as such, a failure in any sensor will cause a deviation from the expected parity.

For the pentad array of sensors, the associated parity space is two dimensional, and the parity algorithm is relatively simple. An illustration of the derived parity vector is presented in Figure 2. The amplitude of the parity vector is a measure of the inconsistency in the redundant sensor data. The direction of the parity vector is toward the placement of the failed sensor input axis in parity space.

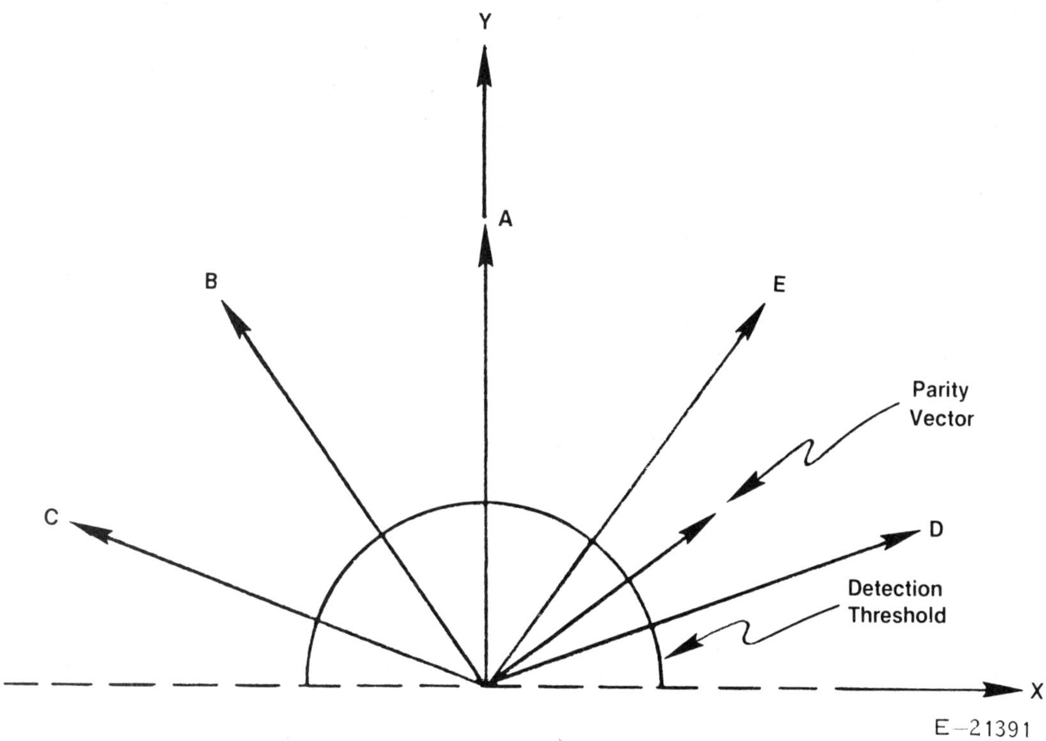

Fig. 2 RIMU FDI Via Mid-Value Selection in Parity Space

Failure detection and pair isolation is signaled when the parity vector amplitude exceeds a minimum value beyond which a failure is considered to have occurred. The strapdown navigation computations are then executed using only data from the failure-free triad (ABC in the example of Figure 2). Sensor isolation is signaled when the parity vector amplitude is sufficiently large and points in sufficient proximity to the failed sensor axis. Subsequent strapdown navigation computations are then executed using data from the remaining quartet of sensors. When no failures are indicated, the strapdown navigation computations gain the performance advantage provided by combining data from the full pentad array of sensors.

The parity vector checks are performed at three different computational rates: high-rate (50/sec), mid-rate (25/sec), and low-rate (2/sec), as illustrated in the block diagram of Figure 3. Adaptive thresholds are employed at each of the levels of parity checks. The thresholds are defined in terms of a constant plus a variable function of the sensed amplitude of the angular rate and linear acceleration environments. The magnitude of the constant term is programmable as a function of the flight phase, e,g., high during "dynamic" phases such as launch boost, IUS burns, staging and angular acceleration events; and low during "quiescent" phases such as parking orbit and transfer orbit coast. Parity vector filtering is performed at 50/sec for both the mid-rate and low-rate checks. The mid-rate filter is first-order to provide medium noise attenuation; the low-rate filter is second-order to allow high noise attenuation. The adaptive threshold forcing functions for these two checks receive the same filtering as their respective parity vectors.

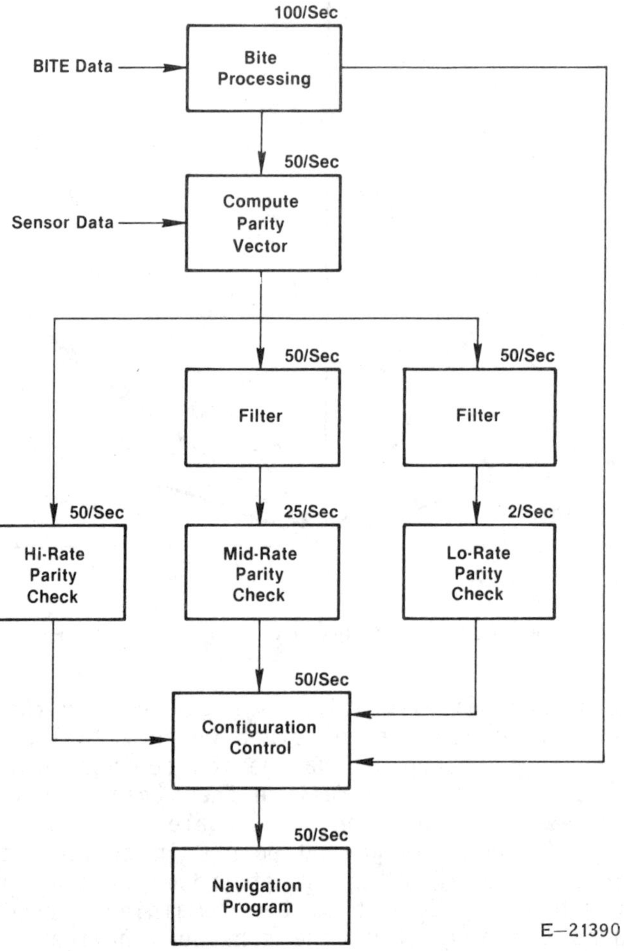

Fig. 3 IUS RIMU FDI

The high-rate check protects against hard failures; it is performed on the unfiltered sensor data before the data are passed on to the navigation and control algorithms. The mid-rate check protects against the accumulation of error due to a hard failure that just escapes the high-rate check. A parity vector which exceeds a threshold at the high- or mid-rate level will result in reconfiguration to one of five triads of sensors involving three adjacent input axes on the surface of the cone (also adjacent in parity space). No further high- or mid-rate checks are performed after reconfiguration. Low-rate parity checks are continued, however, in order to allow further isolation of the failure to a single sensor.

The low-rate parity check also protects against soft failures. A low-rate parity vector which exceeds its threshold will first result in reconfiguration to one of the five adjacent triads. The low-rate parity check continues, however, and if an isolation decision function is satisfied, reconfiguration to the remaining quartet occurs. The sensor isolation decision function is a combination of thresholds on both the amplitude and phase angle of the parity vector in parity space. After a low-rate parity vector check has caused reconfiguration to a quartet, no further parity checks are made unless the software is reset by an external command.

BITE processing monitors critical voltages, sensor loop closure discretes, gyro Spin Motor Rotation Detector (SMRD) discretes, and quality of data transmission from all three RIMU channels. Fresh values of the voltages and discretes are transmitted to the computers at a 100/sec rate. The BITE processing results in sensor BITE flags which are made available at the 50/sec rate to assist in the high-rate sensor configuration selection. A confirmation counter prevents spurious noise from initiating a permanent reconfiguration due to BITE.

The parity vector FDI can only isolate a single gyro and single accelerometer failure. If more than one gyro or accelerometer failure occurs, the FDI system relies on BITE for further failure isolation. If only three like sensors are in service, and one is flagged by BITE, the software will default to that configuration indicated by the first two confirmed BITE flags. Data from at least three like sensors is always processed by the navigation software.

Accuracy requirements are met by the RINS considering the total ensemble of possible redundant configurations. That is, the navigation error statistics are estimated based on a weighted combination of like statistics for each redundant configuration, where the weight factors are the derived probabilities of operating in that redundant configuration. Presentations of the analyses to estimate RIMU error budgets, determine the probabilities of component failure and FDI/BITE coverage, and combine these to estimate total ensemble mission performance can be found in References 3, 4 and 5.
The selection of threshold values was based upon a consideration of the estimated statistics of sensor errors and noise that might exist during a dynamic or quiescent mission phase. Simply put, the threshold-to-sensor statistic ratio must be high enough so that a good sensor is not

removed (false alarm), and navigation performance is degraded; and low enough so that a failed sensor is not missed (missed alarm), and navigation performance is unacceptable. RIMU parity vector data and sensor error statistics were accumulated from acceptance and qualification testing at Hamilton Standard, during periods of stationary (quiescent) operation, and during random vibration (dynamic) tests. Additional RIMU parity vector data were gathered in rocket sled tests at the Central Inertial Guidance Test Facility (CITGF) at Holloman Air Force Base, and in acoustic tests at Boeing. Evaluations of the effect of the RIMU error model as seen in parity space were conducted using a six-degree-of-freedom flight simulation in a scientific computer at Boeing.

A more detailed discussion of the analyses and tests which contributed to the RIMU parity vector threshold selection process is presented in Reference 6. But, as stated in Reference 6, a major FDI design process validation/ improvement would only result from a real flight, and its attendant real environmental information.

1ST FLIGHT PERFORMANCE

The first IUS mission placed two Defense Satellite Communication System spacecraft (DSCS II and DSCS III) into geosynchronous equatorial orbits, following deployment into a 80x95 nautical mile park orbit by a Titan T34D boost vehicle. Liftoff of the T34D occurred at 12:05 A.M., EDT, on October 30, 1982, from Launch Complex 40 at Cape Canaveral Air Force Station. The flight duration was approximately seven hours from launch to orbit insertion.

Telemetry data regarding the state of the RIMU were available up through T34D boost, first IUS Solid Rocket Motor (SRM-1) burn, and approximately one hour into the transfer coast to geosynchronous altitude (two hours into the flight). At this point, a telemetry problem developed; this prevented the receipt of RIMU state information for the rest of the flight. The available telemetry data indicated that no BITE flags were set. During the T34D boost and SRM-1 burn, gyro and accelerometer parity vectors were at least seven times lower than the thresholds used during these dynamic periods of flight. During the coast periods, the gyro parity vector was observed to undergo a transient coincident with the onset of attitude manuevers. The transients were at least two times smaller than the adaptive threshold used during these "quiescent" phases of flight. At the time, the source of the transients appeared to be consistent with known gyro dynamics under angular acceleration.

Despite the telemetry problem, the self-contained, inertially guided flight continued as programmed on board. At the scheduled time for payload deployment, the DSCS spacecraft were detected at their planned locations by radar. Final orbit parameters for the released spacecraft were obtained from tracking data. The orbital insertion errors for DSCS II and DSCS III are presented in Table I. Accuracies well within specification margins were achieved.
Further details regarding the IUS flight control system performance in the first flight may be found in Reference 7.

Table I

IUS FIRST FLIGHT SPACECRAFT PLACEMENT ACCURACY

Parameter	Error After Geosynchronous Insertion*		Allowable Error**
	DSCS II	DSCS III	
Semi-Major Axis (N.Mi.)	-13	-13	+120 -102
Eccentricity	+0.0002	+0.0006	+0.008 -0.001
Inclination (Deg.)	+0.009	+0.019	+0.085 -0.074

* Orbit determination from SCF
** Equivalent to STS-100 specification errors (in R-T-N coordinates)

2ND FLIGHT PERFORMANCE

The second IUS mission was designed to place the first Tracking Data Relay System (TDRS-A) spacecraft into geosynchronous orbit following deployment into a 153 nautical mile orbit by the Space Shuttle Challenger. Liftoff of the Space Shuttle from Launch Complex 39A at Cape Canaveral occurred at 1:30 P.M., EDT, on April 4, 1983. The flight duration was approximately seventeen hours from Space Shuttle launch to TDRS insertion.

Gyro and accelerometer parity vectors remained below thresholds, and there were no BITE flags set, during Space Shuttle boost into Earth orbit and subsequent Orbital Maneuvering System (OMS) burns. The thresholds, which had been set at dynamic levels during this time period, were reduced to quiescent levels shortly after OMS-2 for the expected on-orbit environment. During the scheduled in-bay star scan maneuvers, at 4:58 Mission Elapsed Time (MET), the gyro low-rate parity vector displayed a larger than expected transient response to the onset of the angular rate of the Space Shuttle Orbiter. The adaptive quiescent threshold was exceeded and the mid-triad ABC was selected, causing the data from gyros D and E to be excluded from processing by the navigation software. This gyro parity vector transient, in conjunction with the associated Orbiter rate profile, is depicted in Figure 4.

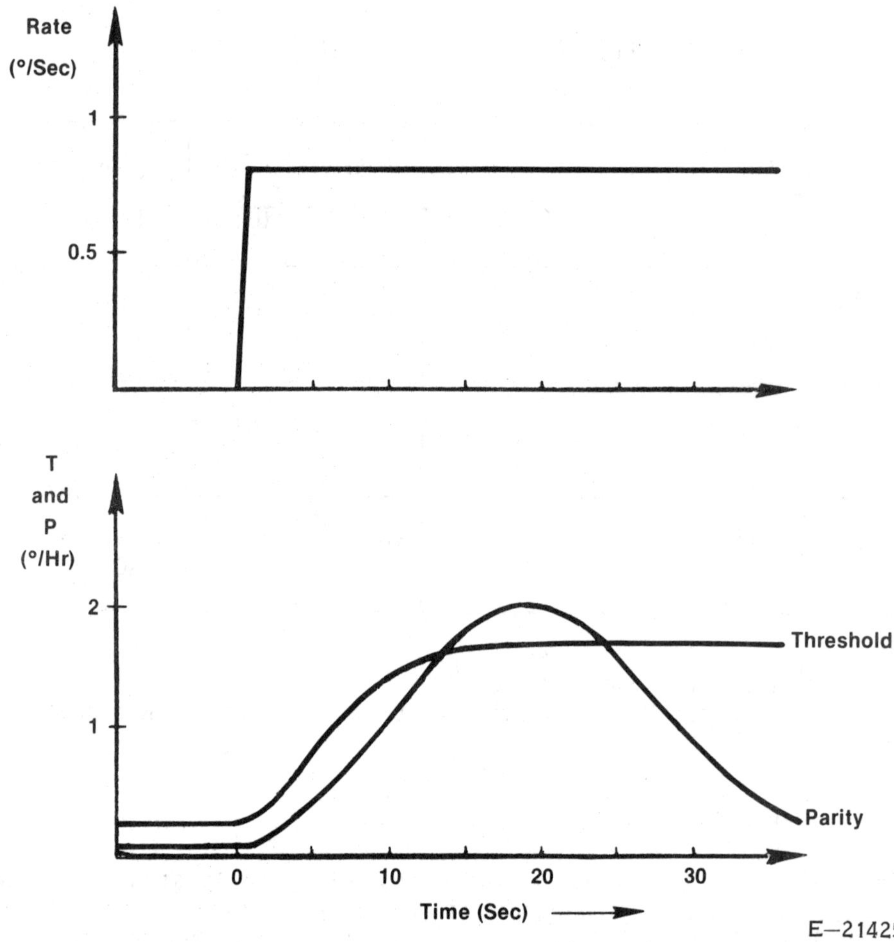

Fig. 4 Typical Gyro Parity Vector Transient in TDRS-A Mission

Subsequent telemetry data indicated that the gyro parity vector had returned to an acceptable level once the transient had passed. Accordingly, an unlink command was sent to IUS at 9:07 MET to restore the processing of data from all five gyros by the navigation software. The restoration was confirmed by telemetry.

Telemetry indicated at 10:07 MET that another similar gyro parity vector reconfiguration had taken place. It took place between IUS deployment from the Orbiter and IUS attitude control initiation. A dropout of telemetry during this period obscured the detailed events. Gyros D and E were again excluded from the navigation processing. And again, subsequent telemetry data indicated that the gyro parity vector had returned to an acceptable level. At 13:49 MET, another FDI reset command was uplinked to restore the processing of data from all five gyros by the navigation software.

A third gyro parity vector reconfiguration occurred at 16:08 MET during the star scan maneuver prior to the SRM-2 burn. The same transient characteristic was displayed, and the reconfiguration again resulted in the exclusion of data from gyros D and E by the navigation software. The SRM-2 burn was entered with the RIMU navigation processing in this state.

Approximately 80 seconds into the SRM-2 burn, a motor problem occurred, inducing a severe, 180 degree/second tumble of the IUS and its payload. The subsequent events, including the eventual separation of the TDRS-A spacecraft, and the use of its onboard fuel reserve to achieve geosynchronous orbit, are beyond the scope of this paper.

At this point it should be emphasized that during the entire mission up to the motor problem, RIMU BITE status was nominal and its navigation function performed satisfactorily. The gyro reconfigurations induced by the parity vector transients were handled properly by the FDI and navigation algorithms. The two uplinked FDI reset commands to restore the processing of all five gyros by the navigation software likely improved navigation accuracy somewhat. But even without those actions, the RIMU navigation accuracy would have been satisfactory. All three gyro reconfiguration events were coincident with the onset of attitude maneuvers, with the parity vector threshold at its quiescent level. The gyro parity vector was well below the quiescent threshold during truly quiescent environments, including the stationary time period just prior to the SRM-2 burn. Additional details may be found in Reference 8.

Nonetheless, the gyro false alarm events did cause considerable concern, and subsequent in-depth reviews of the subject by the IUS community raised the following key issues:

(1) Accuracy of the gyro error model.

(2) Sensor FDI based on instantaneous rate (acceleration) error rather than accumulated attitude (velocity) error.

(3) Self-contained transient recovery.

ACCURACY OF THE GYRO ERROR MODEL

Four available RIMU'S at Hamilton Standard and three at Boeing were subjected to a series of simulated mission manuevers (i.e., thermal rolls, star scans, and, dipouts) on Goerz three-axis, laboratory test stands. Gyro parity vector data from these tests displayed the same transient signature at the onset of attitude maneuvers as observed in the TDRS-A mission. As mentioned earlier, limited telemetry data from the first, DSCS mission also displayed a similar transient signature. A review of the RIMU gyro error model incorporated in the six-degree-of-freedom flight simulation revealed terms sensitive to angular acceleration. But these terms would only produce an impulse of error for a duration of about one second, and the subsequent low-rate gyro parity vector transient would peak in about four to five seconds, rather than the observed time to peak of about twenty seconds (see Figure 4).

A subsequent review of the dynamics of the gyro identified a previously unmodeled, transient gyro error source. An angular rate about the gyro Input Axis (IA) will result in a gyroscopic torque of the float about the gyro Output Axis (OA). That is the degree of freedom of the gyro which we depend upon; the sensed torqued about the OA is used as the measure of angular motion about the IA. However, the gyro has an infinitesimally small second degree of freedom. An angular rate about the OA will result in a gyroscopic torque of the float about the IA. The gyro float is suspended in fluid, and turns on pivots in bearing assemblies at both ends of the OA. The gyroscopic torque on the float about the IA will cause the float pivots to traverse the gaps represented by the clearance at either end of the float. This angular displacement of the float relative to the case about the IA is sensed by the gyroscope as if it was a vehicle motion. The maximum angular "cocking" of the float about the IA due to the nominal bearing clearance is only about 9 seconds of arc. This error is deterministic and depends upon the magnitude and sign of the rate about the OA. As the pivots traverse the gap, and the float pushes the viscous fluid out of its way, fluid turbine torques are developed about the OA. These error torques are random in sign and magnitude, being dependent upon the concentricity of the gaps between the float and the case. It has been determined from measurements on the gyro that it takes about 20 degrees of rotation about its OA to cause the pivots to traverse the entire gap. Thus, an angular rate of about 1 degree per second about the OA could result in an error torque about the OA for a time period of about 20 seconds. This is the characteristic indicated by the gyro parity vector transients at the onset of attitude maneuvers in flight, and confirmed in subsequent laboratory tests. The placement of the gyro OA's in the RIMU is such that roll rates (e.g., star scans and thermal rolls) will force this float cocking effect the most. Since the pivots could lie anywhere within the gap prior to the first rotation, the duration of the first transient is uncertain. But for periodic rotational reversals about the gyro OA of greater than 20 degrees, the effect will take on a corresponding periodic, and reversing characteristic.

Of the sample of RIMU's tested, the RIMU on the TDRS-A flight displayed the largest gyro parity vector transient response to the onset of rate. However, the float cocking transient error varies randomly in sign and magnitude from gyro to gyro. Therefore, the resultant transient in the RIMU gyro parity vector, being a combination of the transients on all five gyros, can be expected to vary widely from RIMU to RIMU. The TDRS-A flight RIMU's signature was within this expected variation.

This float cocking error does not significantly contribute to RIMU attitude error during rates, when compared to other rate sensitive errors such as gyro scale factor and IA misalignment. However, as we now know, it can produce a very noticeable effect when viewed through the parity vector algorithms as an instantaneous error in rate, rather than attitude.

The RIMU error model has been updated to include a gyro error term which is proportional to the magnitude of OA angular rate, while the gyro float cocks across the gap, and is zero otherwise. In addition, each RIMU will be subjected to the mission simulation test on the Goerz three axis test stand during Acceptance Tests at Hamilton Standard and Boeing, in order to accumulate a controlled data base with regard to parity vector transients.

SENSOR FDI BASED ON ATTITUDE AND VELOCITY ERROR

As has been noted, the FDI algorithm displayed a sensitivity to a gyro transient effect which contributed very little in terms of accumulated attitude error. This prompted a review of the FDI decision function as viewed in sensor attitude (velocity) space, rather than rate (acceleration) space. Given a failure, its isolation time is dependent upon the parity vector threshold and filter time constant. The accumulated attitude (velocity) error prior to isolation is equal to the integral of the failure magnitude over the isolation time. Given the gyro parity vector quiescent rate error thresholds and filter time constants, as baselined for the TDRS-A flight, the effective gyro attitude error threshold is a function of gyro failure magnitude as shown in curve A of Figure 5. Besides revealing the sensitivity to insignificant attitude error transients, another observation is of note from curve A. As the failure magnitude increases, the low-rate attitude error threshold becomes proportional to the square root of the failure magnitude. This has been found to be a property of the second order filter. It can be shown that, within sample time limitations, the attitude error threshold becomes independent of the failure magnitude, and proportional to the product of the rate error threshold and the filter time constant, for sufficiently large failures, if a first order filter is used. This is an important property of the first order filter, which can be exploited to allow the derivation of the rate (acceleration) error thresholds and filter time constants based on a specified, allowable attitude (velocity) error. As mentioned earlier, the selection of the baseline rate and acceleration error thresholds is substantiated by a large body of data and analyses. There is, however, room for adjustment in the filter time constants, provided the sensor noise is sufficiently attenuated. The effect, on total ensemble mission accuracy and reliability, of increasing the filter time constants, was evaluated using the Boeing NAV/FDI program. The best compromise between maintaining acceptable mission accuracy and reliability, and minimizing false alarms, was provided by the following adjustments to the parity vector filter time constants:

(1) Mid-rate filter time constant changed from 0.435 seconds to 1 second, and

(2) Low-rate filter time constants changed from a double at 3.86 seconds to 100 and 0.001 seconds, i.e., effectively changed to a first order filter with a time constant of 100 seconds.

The resulting adjustment in the gyro attitude error threshold is illustrated by curve B in Figure 5. A reduction in transient error sensitivity by as much as an order of magnitude is gained by this adjustment in the filter time constants. The same adjustment has been made in the accelerometer parity vector filters, with the same reduction in transient error sensitvity.

Tape recorded, 100/second, gyro and accelerometer pulse output data from two RIMU's subjected to the simulated mission test at Hamilton Standard were played through two versions of the FDI software: Baseline and Modified. Time histories of the gyro low-rate parity vector from one of these RIMU's during the thermal roll maneuver are shown in Figures 6A and 6B, for the Baseline and Modified filter time constants, respectively. The reduction in the transient sensitivity afforded by the time constant modification is quite noticeable.

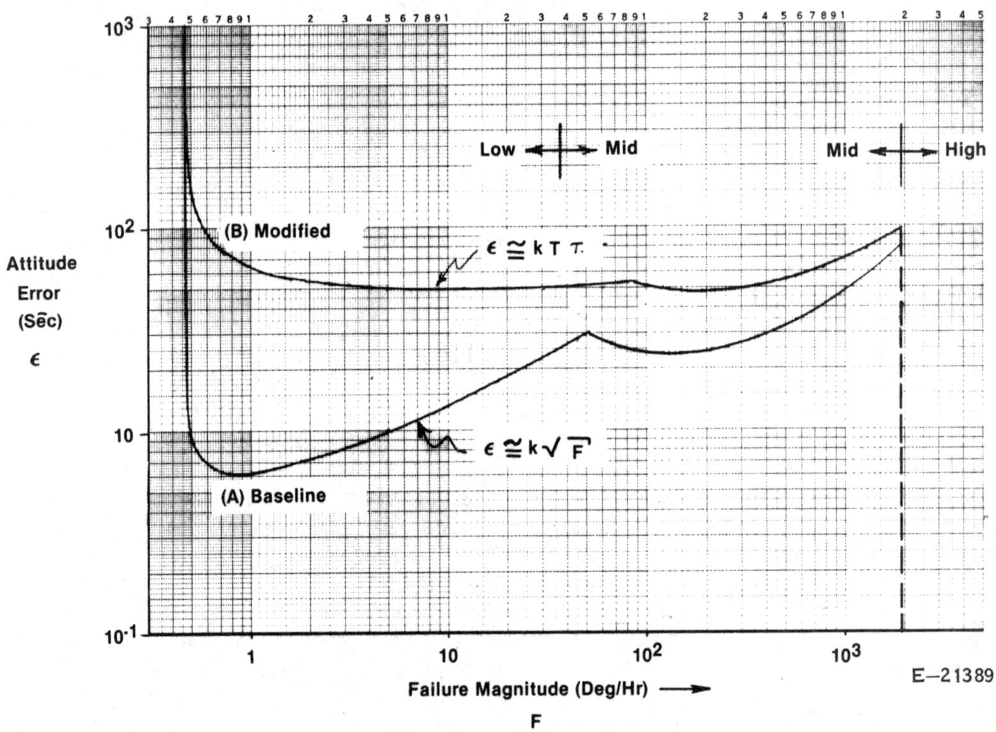

Fig. 5 RIMU Attitude Error Threshold Versus Gyro Failure Magnitude

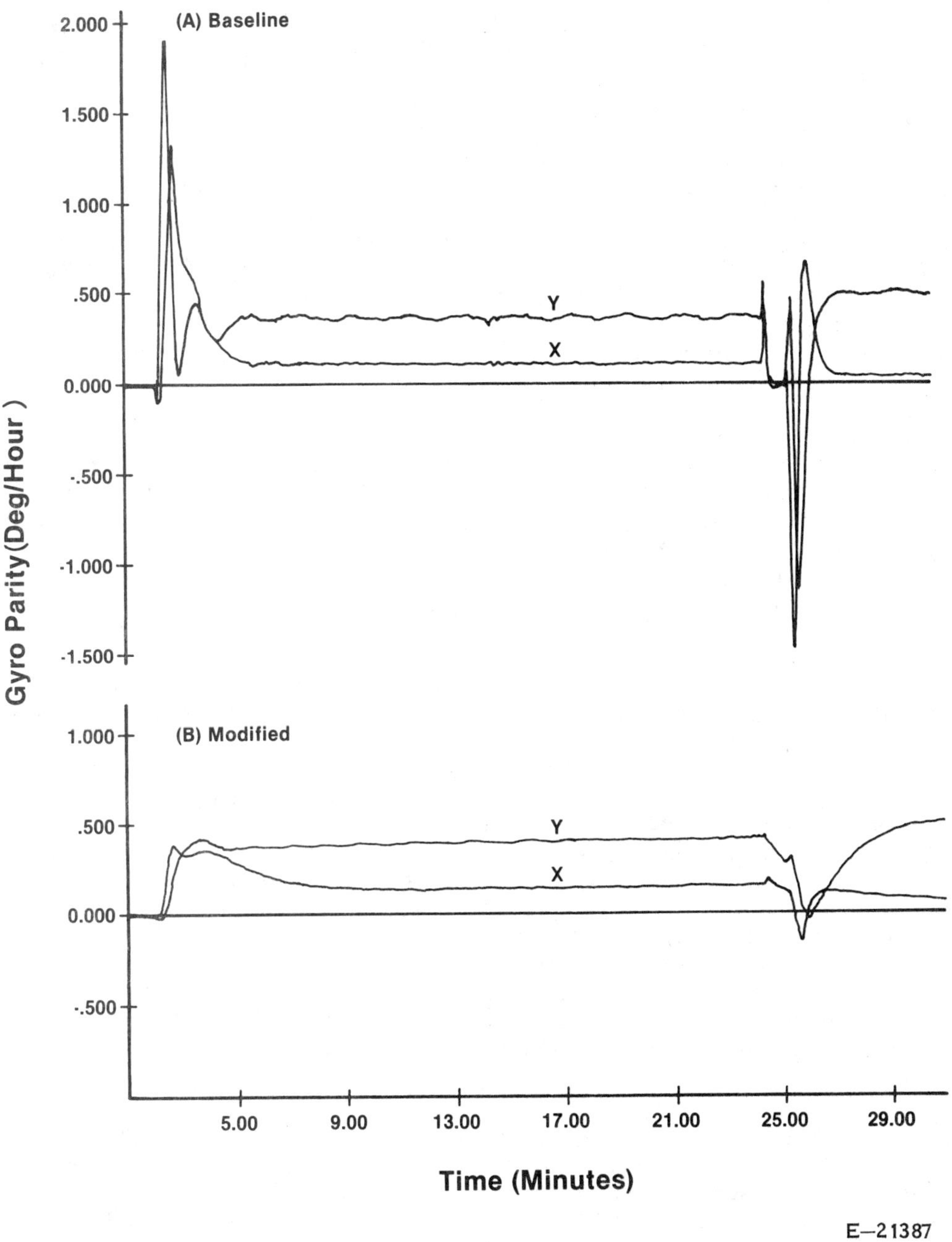

Fig. 6 RIMU 009 Gyro Low-Rate Parity Vector During Thermal Roll

SELF-CONTAINED TRANSIENT RECOVERY

The effectivity of the above improvement still depends upon prior knowledge of the range of mission environmental inputs, and the dynamic characteristics of the RIMU sensor responses. There could still exist a susceptability to unexpected phenomena which might be manifested by a transient in the filtered parity vector, yet contribute benignly to navigation accuracy. With the Baseline FDI algorithm, it took an uplink command to undo such a false reconfiguration in the TDRS-A flight. The decision to uplink an FDI reset command was based on a ground evaluation of telemetered parity vector and RIMU BITE data, i.e., no BITE flags set and all parity vectors returned to normal levels. It was noted that this same evaluation and decision process is readily within the real-time capability of the onboard FDI algorithm. Consequently, it was decided that the FDI algorithm should be modified to allow a graceful recovery after a transient reconfiguration, without external intervention.

As in the Baseline algorithm, reconfiguration is carried out upon the first exceedance of a high-rate, mid-rate or low-rate threshold. The reconfiguration is regarded, however, as a temporary, fault containment measure in the Modified algorithm. If some time later, the low-rate parity vector drops below the threshold, and stays below for a given "probation" time, the Modified algorithm reconfigures the system back to the pentad of sensors. This transient recovery capability is diagrammed in Figure 7, as seen in low-rate parity vector space.

When the parity vector exceeds the low-rate detection threshold, TD, the software reconfigures to an adjacent triad. At the same time, the probation counter, N, is initialized. When the parity vector is in the region above TD, but below the isolation threshold, TI, an adjacent triad is continually selected at a 2/sec rate. If the parity vector increases above TI, and satisfies the dot product test with a gyro IA in parity space, the software reconfigures from the triad to a quartet. If the dot product test is not satisfied, the software continues to select a triad. As long as the parity vector stays above TD, the software continues to select the triad or quartet at the 2/sec rate. If the parity vector drops below TD, the last selected triad is maintained and the probation counter starts incrementing. If the parity vector stays below until the probation counter times out, the software reconfigures back to the pentad. The probation timer has been baselined at 15 minutes.

The transient recovery capability of the Modified FDI algorithm is further illustrated by the RIMU sensor state transition diagram of Figure 8. The solid lines represent state transitions allowed in the Baseline FDI algorithm; dashed lines are added transitions allowed by the modification. The capability of the Modified algorithm to quickly react to a true sensor failure while in the midst of a transient reconfiguration, by reselecting a triad or quartet, is also indicated.

The FDI algorithm modifications to provide this transient recovery capability have been made to the flight software. The modified algorithm, including the revised parity vector filter time constants, has successfully passed validation tests at Boeing and Hamilton Standard.

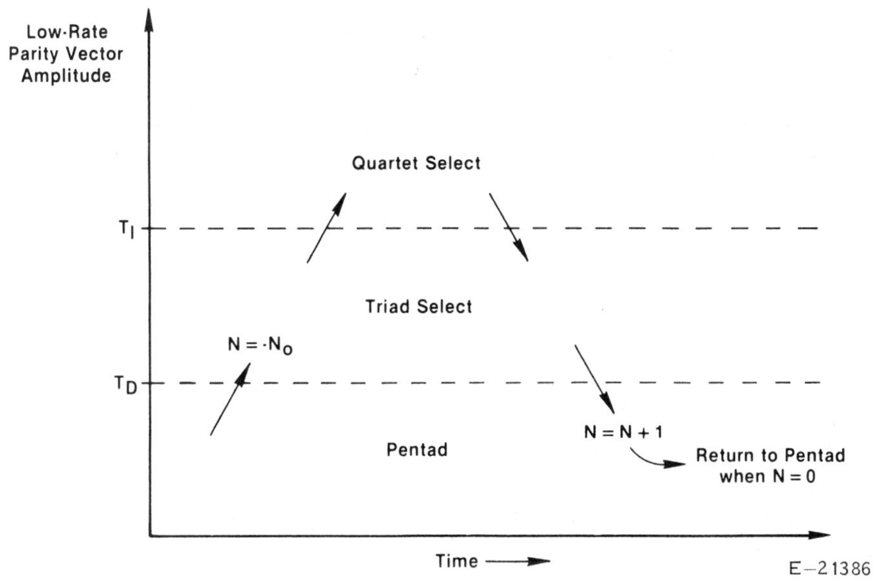

Fig. 7 RIMU FDI Transient Recovery

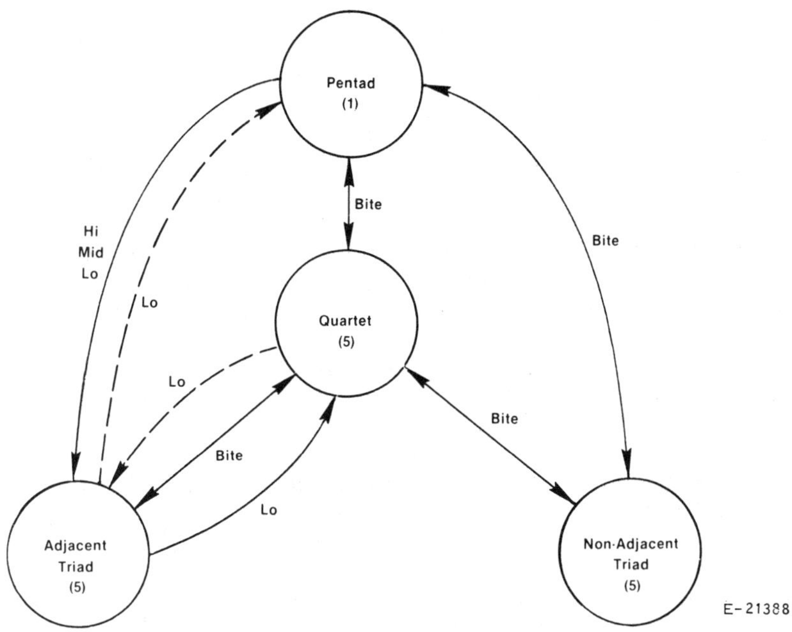

Fig. 8 RIMU Sensor State Transitions

CONCLUSIONS

There is no substitute for actual flight experience, just as there is no "perfect" flight simulation. From the initial RIMU space flight experience has evolved an improved methodology for the determination of FDI parameters, influenced primarily by attitude and velocity error criteria, rather than by instantaneous rate and acceleration error. Based on the confidence gained in parity vector and BITE data as measures of RIMU health during flight, the capability of the FDI logic has been expanded to allow self-contained evaluation of, and recovery from, false reconfiguration. In addition, the accuracy of the RIMU error model has been improved with regard to the simulation and test of known sources of gyro transient effects. These improvements provide a much higher level of transient tolerance of the RIMU FDI technique without compromising its fundamental FDI performance.

REFERENCES

1. Baum, R.A., Morrison, G.E.S. and Peters, R.C., "A Redundant Inertial Navigation System for IUS," Advances in Guidance and Control Systems using Digital Techniques, Paper No. 12, AGARD-CP-272, May, 1979.

2. Goodfellow, A.K., "Navigation and Flight Control of the Inertial Upper Stage," presented at the Annual Rocky Mountain Guidance and Control Conference, Paper No. AAS 79-034, February, 1979.

3. Goodstein, R., Joldersma, T. and Winkel, D.J, "Inertial Upper Stage Navigation Algorithms Evaluation," presented at the National Aerospace Meeting of the Institute of Navigation, March, 1982.

4. McArthur, W.G. and Baum, R.A., "Inertial Upper Stage Guidance System Mission Success Analysis," presented at the Annual Rocky Mountain Guidance and Control Conference," Paper No. AAS 79-035, February, 1979.

5. Daly, K.C., Gai, E. and Harrison, J.V., "Reliability and Accuracy Prediction for a Redundant Strapdown Navigator," presented at the AIAA Guidance and Control Conference, August, 1980.

6. Goodstein, R., Tse, B.K. and Winkel, D.J., "Failure Detection Technique for the Inertial Upper Stage Redundant Inertial Navigation System," presented at the AIAA Guidance and Control Conference, Paper No. AIAA-82-1622-CP, August, 1982.

7. Goodfellow, A.K., "First Flight Performance of the Control System of the Inertial Upper Stage," presented at the Annual Rocky Mountain Guidance and Control Conference, Paper No. AAS 83-086, February, 1983.

8. Goodstein, R., Tse, B.K. and Winkel, D.J., "Flight Test Results of the Inertial Upper Stage Redundant Inertial Measurement Unit Redundancy Management Technique," presented at the AIAA 22nd Aerospace Sciences Meeting, Paper No. AIAA-84-0568, January, 1984.

AAS 84-053

IN-FLIGHT RESCUE OF STRANDED TDRS-1 SPACECRAFT

H. Schmeichel and B.J. Ehlers*

On April 4, 1983, the first Tracking and Data Relay Satellite (TDRS-1), one of six to be built by TRW for SPACECOM and NASA, was launched by the space shuttle Challenger. During the ascent to geosynchronous orbit, the Boeing-built Inertial Upper Stage (IUS) booster malfunctioned, leaving TDRS stranded in a low, elliptical orbit.

This report decribes the amazing satellite recovery from a wild tumble and the dramatic three-month rescue effort to move TDRS into synchronous orbit. It took a total burn time of 44 hours from a pair of tiny one-pound thrusters, about 800 pounds of propellant, an adaptable, robust attitude control system and much resourcefulness from ground operators to accomplish this feat. TDRS-1 reached its proper earth orbit on June 29, 1983 and then began its ten-year mission as a data relay satellite.

INRODUCTION

After a flawless start for space shuttle Mission 6, the TDRS flight support team at the White Sands Ground Terminal (WSGT) in New Mexico settled into routine monitoring activities. The TRW-built payload, the first Tracking and Data Relay Satellite (TDRS-1), was not to be deployed from the Challenger's cargo bay until about 10 hours after liftoff. While the orbiter was circling the globe every 90 minutes, the Attitude Control Subsystem (ACS) specialists followed the shuttle motion by observing the changing TDRS gyro telemetry. The four gyros and the control processor had been turned on several hours before liftoff. They are normally not needed, however, until TDRS reaches its final destination in synchronous orbit.

Transferring TDRS from the low shuttle parking orbit to synchronous orbit is the job of the Inertial Upper Stage (IUS). This powerful two-stage rocket only needs two short burns to raise the 5,000 pound TDRS to an altitude of 19,323 nautical miles. The nominal mission plan includes the following sequence of events. Approximately one hour after leaving the orbiter bay, the first stage solid rocket motor (SRM-1) burns for

* *TRW Space and Technology Group, One Space Park, Redondo Beach, California 90278*

2.5 minutes. The resulting elliptical transfer orbit is then circularized by firing the second stage solid rocket motor (SRM-2) for 1.8 minutes about 5 hours later. Before separating from the IUS, TDRS is given a chance to deploy its solar panels, the C-Band antenna and the space-to-ground link (SGL) antenna. During this 30-minute deployment period, the IUS provides full three-axis stabilization in inertial space. Once TDRS is ejected from the IUS, it relies on its own attitude control system for stabilization. After separation, the two single-access antennas are deployed and TDRS is in its final on-orbit configuration as shown in Figure 1.

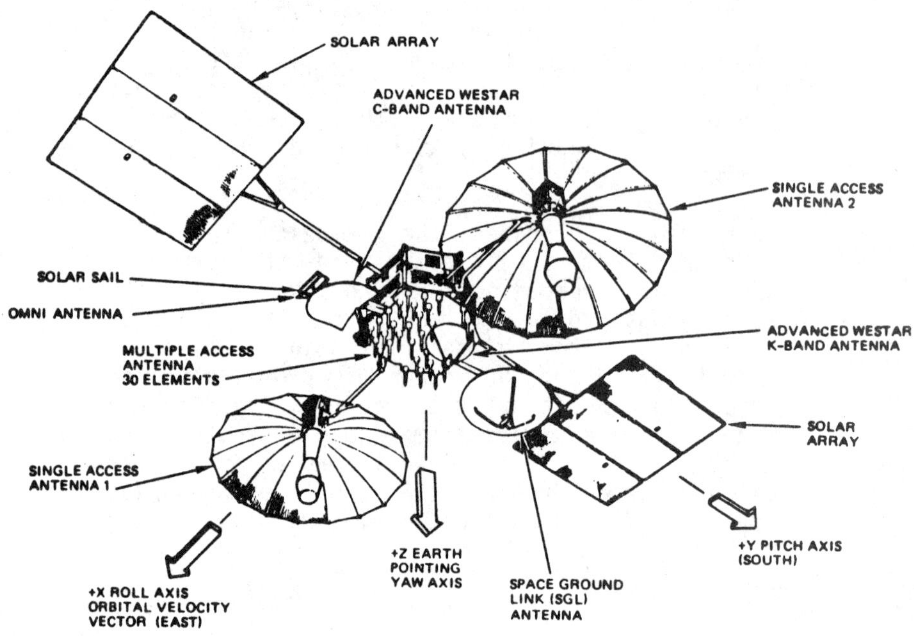

Figure 1. TDRS Deployed Configuration

On April 5, 1983, after the shuttle orbiter had circled the earth six times, it was time to deploy the IUS-TDRS payload. The astronauts opened the cargo bay and prepared the IUS for its first operational flight. Everything went according to plan. When a safe distance had been established between the orbiter and its released payload, the IUS fired its first stage rocket motor, sending TDRS on its way to its high orbit.

By now, activities at WSGT had picked up considerably. As the IUS went through its pre-programmed sequence of maneuvers, it paused at one-hour intervals to accept TDRS commands from WSGT. There were five such command opportunities, called dipout maneuvers, before the SRM-2 burn. Between dipouts, the IUS performed thermal roll maneuvers to keep TDRS temperatures within design limits.

Just before the first dipout, a RAM checksum error ocurred in the TDRS control processor. Since the control processor was needed right after separation, this problem had to be corrected in transfer orbit. During dipout 2, a RAM dump was performed. ACS specialists analyzed the octal dump data and determined that a single bit in the gyro transformation matrix had been flipped. The corrective RAM load was quickly put together and successfully sent during dipout 4. The last TDRS configuration commands were also sent at this time. No further WSGT commands were planned until the busy appendage deployment period.

The IUS second stage burn started right on schedule. ACS specialists at WSGT, monitoring the strip chart recorders, observed the rate jitter induced by the 17,000 pound thrust of the solid rocket motor. Suddenly, about 80 seconds into the burn, the TDRS gyro-sensed rates diverged and telemetry lock was lost.

A check with the IUS flight support team at Sunnyvale, California confirmed the suspicion that the IUS had lost control and was tumbling. Whether or not TDRS had separated from the booster could not be determined because of the intermittent telemetry lock. In the WSGT control room, the strip chart recorder pens were oscillating wildly. It seemed like the TDRS launch mission had come to a premature end and nothing could be done about it. But instead, it turned out to be the beginning of a long and dramatic rescue mission.

TUMBLE AND DESPIN

Intermittent telemetry hampered efforts to ascertain TDRS' condition. Immediately after the IUS malfunction, a team of engineers convened at WSGT to assess the situation and to determine what could be done to salvage the mission. The first and most urgent task was to somehow bring about IUS-TDRS separation.

Timeline of Events

Approximately ten minutes after the anomalous SRM-2 burn, the ground station at Goldstone reported that the signal strength was fluctuating at 2 second intervals indicating that TDRS was tumbling at 180 deg/sec. The first WSGT action was to configure the TDRS downlink for its best signal quality. Unfortunately, these efforts failed to provide the needed improvement and the TDRS telemetry remained intermittent and incoherent. Knowing that the IUS batteries, which are required for separation, were rapidly being drained, the WSGT crew requested that the Sunnyvale control center continuously command separation. Although the IUS sequencer, the internal timer that controls the launch events, may have been operational, it was critical that TDRS separate from the IUS in order to unblock the TDRS roll and pitch thrusters.

Following two anxious hours of waiting for either stabilization under IUS control or an indication of TDRS separation, WSGT began sending commands to enable the yaw channel of the Inertial Mode control system.

The roll and pitch channels were not enabled because the IUS partially blocked the TDRS roll and pitch thrusters. Five minutes after initiating Inertial Mode commanding from WSGT, Goldstone reported loss of all IUS data. Time for separation had run out because the IUS batteries were dead. Suddenly, the crew at White Sands faced the possibility of flying TDRS-1 with the spent SRM-2 stage attached to it.

Knowing that the IUS batteries were dead, WSGT started transmitting commands to enable Inertial Mode outputs in all three axes, in the hope that the TDRS thrusting would either despin the TDRS/IUS stack or the separated TDRS. Then, almost three hours after the wild ride had begun, Goldstone reported that their signal fluctuations were slowing and WSGT began locking on telemetry. Three hours and fifteen minutes after the ill-fated SRM-2 burn, TDRS stabilized with Inertial Mode enabled and separation indicated. From all initial indications, TDRS had despun from 30 RPM with its own Inertial Mode control system. However the question still remained; what had happened during the three-hour tumble?

TDRS Inertial Mode Control System

Prior to analyzing the despin, some background information on the Inertial Mode control system is necessary. Figure 2 presents a block diagram of the Inertial Mode control system. The TDRS attitude control subsystem has a versatile, "Chinese menu" design where three different thruster control modes can be enabled on a per-axis basis. One of these control modes, that provides three-axis inertially-stabilized attitude control, is the gyro-based Inertial Mode.

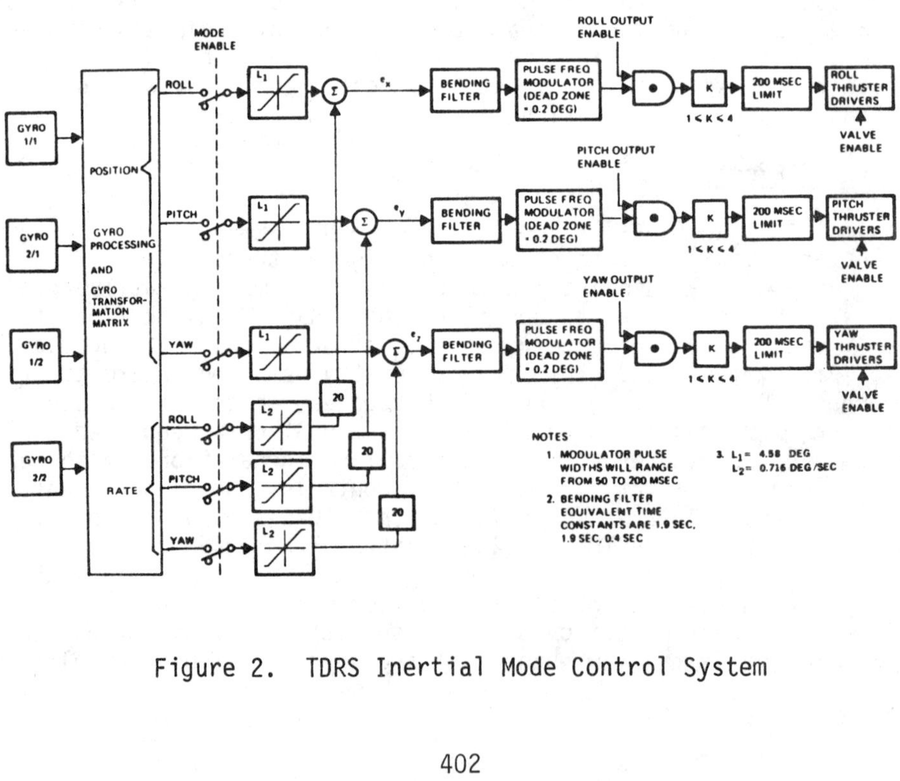

Figure 2. TDRS Inertial Mode Control System

Nominally, the Inertial Mode provides control during the two hours from separation to earth acquisition. To accomplish this, an interface automatically sets the gyro reference to the IUS separation attitude and enables Inertial Mode upon separation from the IUS. The maximum specified separation rates are 0.5, 0.5 and 1.0 deg/sec for roll, pitch and yaw, respectively.

The Inertial Mode derives rate and position information from the four gyros mounted to the spacecraft body. The gyros' sensitive axes are arranged in a pyramid, as shown in Figure 3, such that any three gyros can provide three-axis attitude information. The gyro transformation matrix contained in the control processor RAM, allows the ground operator to choose any combination of gyros for use in the control computations.

The Inertial Mode design is such that the rate feedback dominates when the position and rate errors are limited. For large initial rates, this system effectively acts like a rate control system. The Inertial Mode controller will command the thrusters to reduce the spin in a stable manner. Six thrusters, two per axis, are used to provide the control torques.

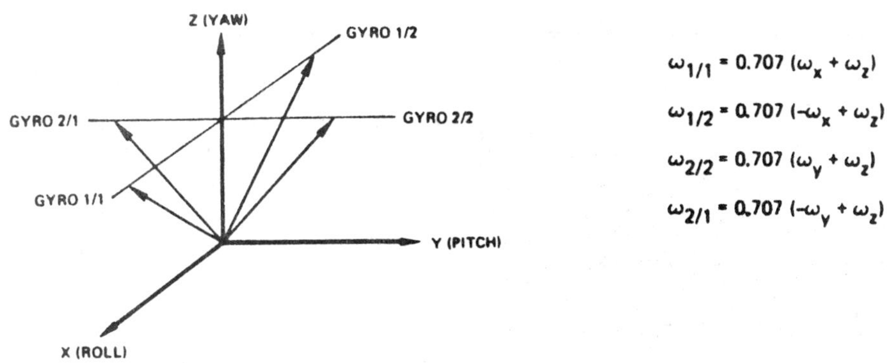

Figure 3. Gyro Input Axes in TDRS Body Reference

Prior to launch a study into the Inertial Mode's capabilities for separation in the stowed configuration was performed. This study showed well-behaved responses for the nominal cases with initial rates up to 2 deg/sec. As the flight data has shown, this design is robust enough to despin far beyond its design limits.

Post-Flight Investigation

Although telemetry was intermittent, careful review of the spotty telemetry as a whole enabled the anomaly investigation team to determine the sequence of events during the despin. Post-flight analysis of the Goldstone and sun sensor data provides a history, pictured in Figure 4, of the spin rate during the tumble. The Inertial Mode was enabled and despin began at 12:51 GMT when Sunnyvale apparently was able to command separation. Concurrently, WSGT started transmitting Inertial Mode enables for yaw only, not knowing separation had occurred. One and a half minutes after separation one of these commands got through and had the effect of disabling roll and pitch. Since WSGT had neither command nor telemetry verification, they continued commanding and shortly thereafter the Inertial Mode was disabled for unexplained reasons. In the following minutes, the IUS downlink was lost due to the drained IUS batteries. Subsequent commanding was set up to enable Inertial Mode outputs in all three axes, but Inertial Mode commands were not successfully implemented until 13:39 GMT.

Once enabled, the Inertial Mode control system with one-pound thrusters was able to despin the spacecraft in less than 25 minutes. Simulation of the despin using a three-axis rigid body siumulation predicts despin taking 39 percent longer. The difference can be accounted for by considering the damping effects of the 1318 pounds of hydrazine on board which aligned the spin vector with the maximum inertia axis. This axis lies between the TDRS roll and pitch axes. With the dominant rates about the roll and pitch axes, the saturated gyros will always give the correct control error polarity and the spacecraft will despin in a stable manner.

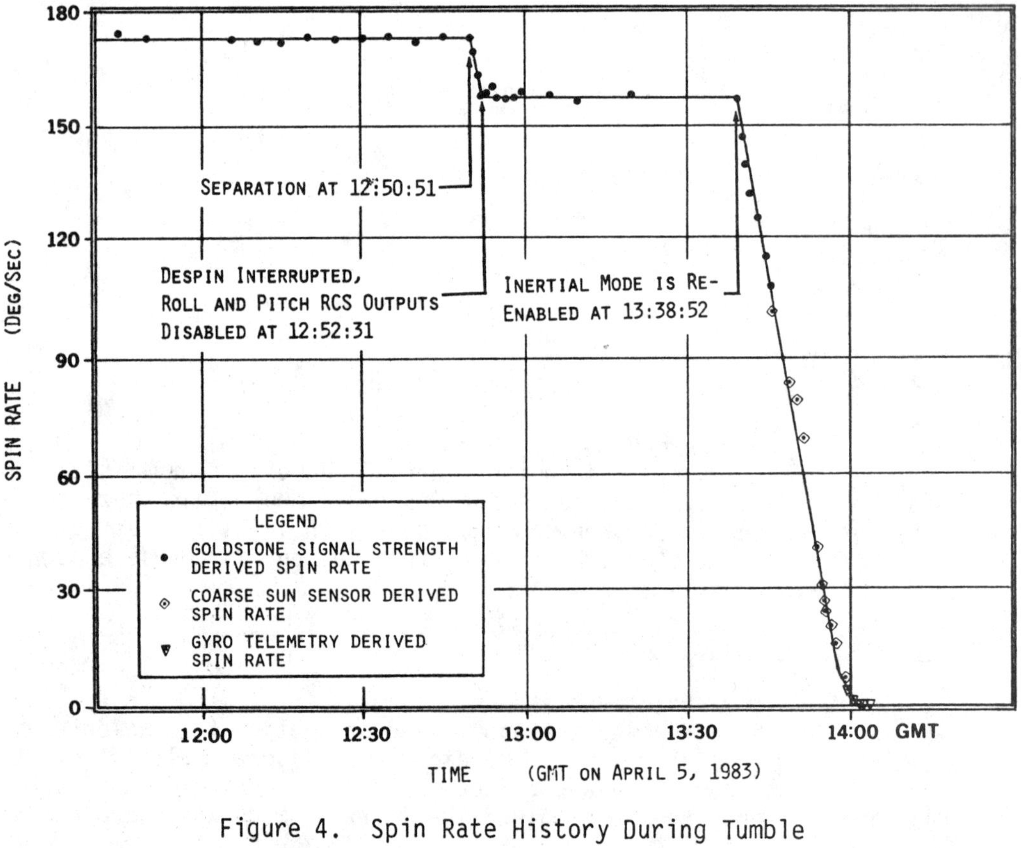

Figure 4. Spin Rate History During Tumble

HYDRAZINE DETONATION AND ROLL THRUSTER LOSS

Six hours after the IUS booster failure, TDRS was fully deployed and three-axis stabilized in Inertial Mode. At this time all spacecraft systems appeared to be functioning properly. But a sluggish attitude recovery from the second SA antenna deployment heightened suspicion that not all was well. Had TDRS suffered some damage from its rough ride on the IUS? It was decided to give the spacecraft an earth-pointed orientation where ground operators could continuously monitor its state of health.

1013 Anomaly

As viewed from TDRS, the earth appeared as a 17 degree disk at apogee and as a 26 degree disk at perigee. The onboard earth sensor processing logic is only designed for operation at synchronous altitude. Therefore, the changing earth size in this elliptical orbit produced a slowly varying pointing offset from nadir. When TDRS began Earth Mode control about 4 hours before reaching perigee, its yaw axis was pointed 4 degrees south of nadir.

In monitoring the Earth Mode limit cycle behavior, excessive roll thruster firings were noted. Figure 5 shows that the roll thrusters are located on the aft end of the spacecraft where the IUS was attached. Processed telemetry in the WSGT control room indicated roll thruster valve temperatures of 240 degrees Fahrenheit (F) because the ground software limited all valve temperatures to this maximum value. A later inspection of the raw telemetry revealed that the roll thruster temperatures were actually as high as 300 degrees F at this time.

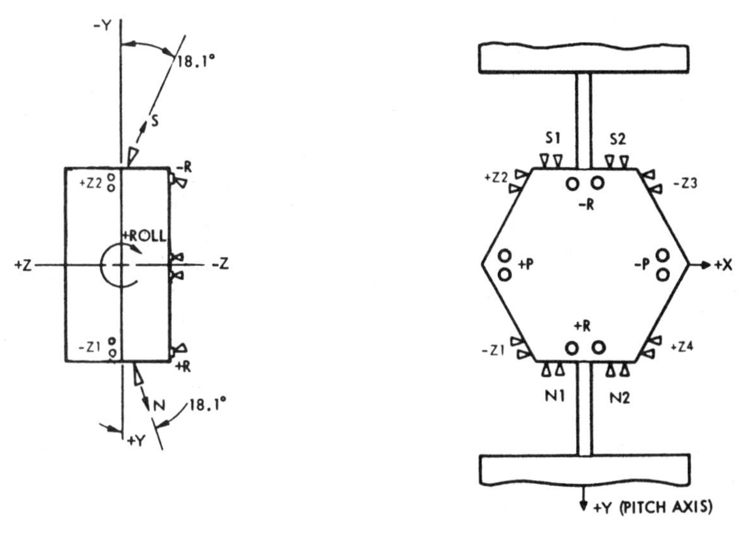

Figure 5. TDRS Thruster Layout

The TDRS propulsion subsystem, depicted in Figures 6, was in its nominal configuration according to the flight plan. Both propellant isolation valves AA and BA were open and the Bank A thrusters were enabled. Pressurization gas valves between the two tanks were closed with telemetry indicating pressures of 328 and 326 psi for Tank A and B, respectively.

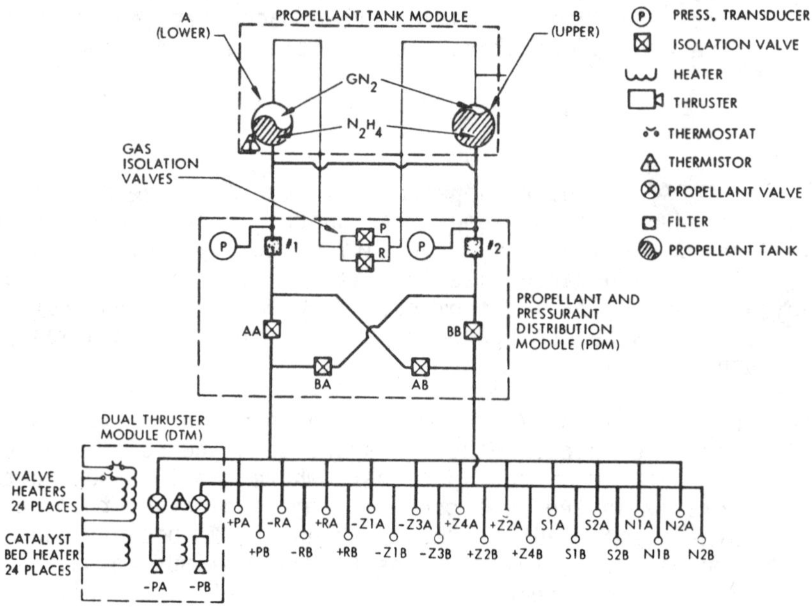

Figure 6. TDRS Propulsion Subsystem

At 10:13 GMT on April 5, 1983, the spacecraft roll error exceeded the positive deadzone and the -RA thruster was commanded to fire. Instead of a small corrective pulse, a huge impulse of about 25 ft-lb-sec was delivered! The -RA and -RB thruster temperatures plunged from about 300 degrees F to near zero indicating a propellant leak near this dual thruster module (DTM). At these low temperatures, some of the propellant in the lines froze and thus slowed down the propellant flow. In the meantime, the spacecraft rolled off the earth in a southerly direction.

Attitude Recovery

Figure 7 presents a plot of the spacecraft angular momentum over the next two hours. Four minutes after the hydrazine detonation, ground operators switched from Earth Mode to Inertial Mode control. This recovery attempt was unsuccessful. Thruster firing commands issued by the controller had no perceivable dynamic effect. In fact, for the next 20 minutes, the spacecraft roll rate continued to grow slowly to about 0.3 deg/sec. The gradual rate increase was probably due to continued hydrazine leakage near the negative roll thruster module.

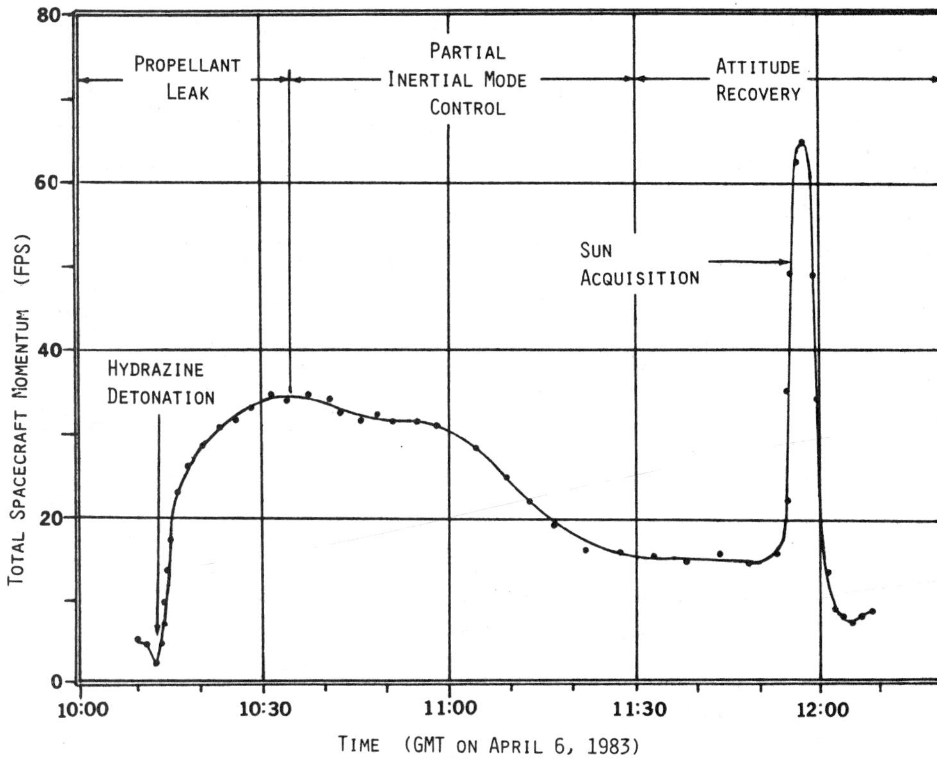

Figure 7. Propellant Leak and Attitude Recovery

About 20 minutes after the 1013 detonation, the isolation valves were closed. But gyro rate data indicates that some Inertial Mode control was taking place. This control action was probably due to residual propellant in the thruster manifold. When the manifold had been drained empty, the spacecraft rates leveled off. As Figure 7 indicates, the spacecraft momentum was reduced from a peak value of 33 to 16 ft-lb-sec.

At 11:30 GMT preparations were made to switch to Sun Mode. Since the sun sensors are mounted on the solar panels, the normal to the panel surface has to be aligned with the TDRS roll axis first. After disabling all ACS control, a 6 minute slew positioned the solar arrays for a +X axis sun acquisition. Simultaneously, the primary thrusters on Bank A were selected for active control.

When isolation valve BA was opened, tank pressures equalized and the Sun Mode pitch and yaw controllers converged. The Sun Mode pitch/yaw convergence required approximately 10 minutes. TDRS was now slowly rotating about the sunline with the arrays pointed at the sun. Attitude recovery to a power-safe condition had been achieved and it was time to assess the damage from the 1013 anomaly.

Which Roll Thrusters Survived?

After the successful sun acquisition, the RCS configuration was identical to that used just before the 1013 event. Isolation valves AA and BA were open and six Bank A thrusters were enabled. The tank pressures had only dropped by 1 or 2 psi indicating that very little propellant had been lost. Since the hydrazine detonation probably damaged the -RA thruster, it was decided to close the isolation valves for the A-side manifold and operate with B-side thrusters exclusively. By isolating the damaged A-side manifold, the possibility of a propellant leak through a ruptured line or valve was eliminated. RCS performance with this new configuration was nominal with one exception; the -RB thruster did not work.

The events following the 1013 anomaly have been reviewed here with the benefit of having scrutinized a lot of the pertinent flight data afterwards. Obviously, the real-time decisions had to be made on the basis of much less information. During this time the full extent of the damage suffered from the the 1013 detonation was not known. As a matter of fact, it might never be known! At that point in the rescue mission, it was only known that the long journey to geosynchronous orbit had to be made without the help of negative roll thrusters.

SEARCH FOR SAFE-HAVEN MODE

In geosynchronous orbit, TDRS uses an earth sensor and two speed-biased wheels for its long-term ACS operating mode known as Normal Mode. But the current orbit was not suitable for Normal Mode operation. The Sun Mode control system, shown in Figure 8, appeared to be the best candidate for an alternative long-term operating mode although it was not designed for this function. One of its obvious limitations is that it cannot be used during eclipses. Since TDRS experienced eclipses every 18 hours in its current orbit, the most immediate concern was a safe ACS mode for eclipse transitions.

Evolution of Eclipse Mode

TDRS had gone through its first two eclipses in Inertial Mode. For the first eclipse transition, it was fully three-axis stabilized on gyros. The second eclipse transition occurred after the 1013 anomaly and again all three axes were switched to Inertial Mode. But the spacecraft continued to roll at about 0.2 deg/sec because the negative roll thruster did not work. Attitude control was maintained in spite of saturated gyro position accumulators.

For the third eclipse the Sun Mode pitch and yaw channels were disabled while retaining roll rate control. When the spacecraft spin axis began to drift faster than expected, all ACS outputs were disabled. After the eclipse period, the sun-pointed attitude was recovered by switching back to Sun Mode. The acquisition transient produced a roll rate of about 0.4 deg/sec which eventually decayed to 0.23 deg/sec.

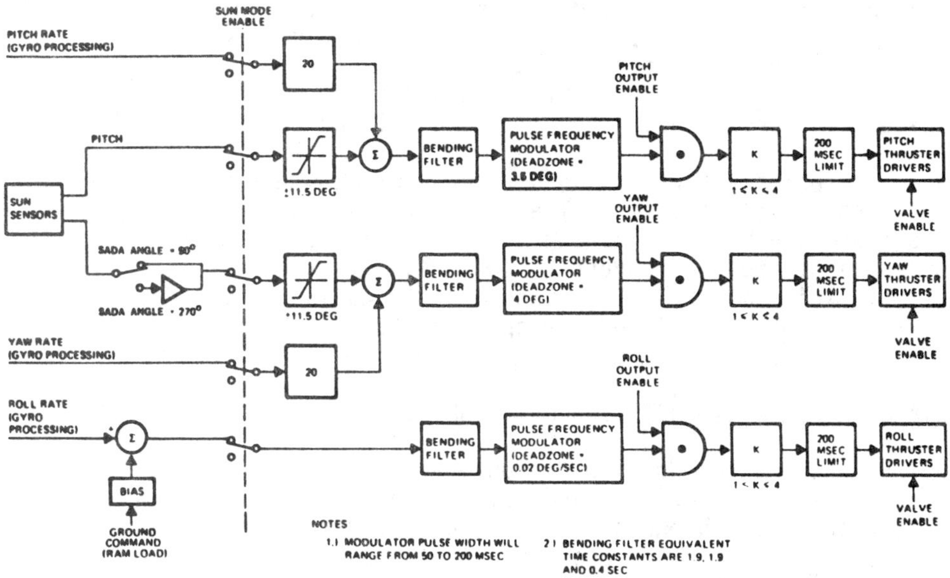

Figure 8. TDRS Sun Mode Control System

It was not desirable to go through such a sun reacquisition after each eclipse because this maneuver wasted propellant and also heated up the thrusters. Reviewing the second eclipse transition, however, it was realized that stable control is possible with saturated gyro accumulators. Rate information is not lost for any axis because it is derived from the unsaturated gyro counts.

A test for this "Eclipse Mode" was scheduled two hours before entering the fifth eclipse. All gyro accumulators were nulled and the pitch and yaw channels were switched to Inertial Mode. The roll axis remained in Sun Mode. About 3 minutes after this switch, the gyro 1/1 and 1/2 accumulators saturated but the spacecraft attitude remained stable. Because Eclipse Mode maintained basically the same attitude motion as Sun Mode, the switch back to Sun Mode caused only a minor control transient. The actual eclipse transition, a little later, was even smoother. At last, an acceptable Eclipse Mode had been found!

Sun Mode Control with One Thruster

After Eclipse Mode had been discovered, orbital operations became quite routine for the next 10 days. Outside of eclipse transitions, TDRS spent all of its time in Sun Mode, rolling about the sunline once every 24 minutes. Whenever the roll rate dropped below 0.23 deg/sec, the positive roll thruster fired. Sun pointing control required a positive pitch thruster duty cycle of about 0.12% and a positive yaw thruster (Z2) duty cycle of only 0.015%. This mode of operation consumed approximately half a pound of propellant each day.

A few days before the first delta V maneuver, the roll rate command was reduced from 0.25 to 0.12 deg/sec. As expected, the number of pitch thruster firings decreased substantially. But the surprising part was that roll and yaw thruster firings had stopped altogether. A single thruster thus provided sun-pointing control and roll rate control!

Since the spacecraft was spinning about an intermediate inertia axis, it has a tendency to tip over. The roll rate will naturally decay unless it is controlled. It turns out that the positive pitch thruster has a small positive roll component which stabilizes the roll rate near 0.12 deg/sec. This way a single pitch thruster was able to maintain a very delicate dynamic balance of the spacecraft.

DELTA V PHASE I - THRUSTING ALONG SUNLINE

For four weeks a team of TRW design specialists scrutinized flight data, ran computer simulations and evaluated various methods for boosting TDRS-1 into geosynchronous orbit. The satellite was in an orbit with a perigee of 11,804 and an apogee of 19,106 nautical miles and drifting 120 degrees per day relative to the earth. TDRS needed a total velocity change of about 1,000 feet per second to achieve synchronous orbit. Luckily, there was sufficient hydrazine fuel on board to do this job. TDRS was carrying about 1,300 pounds of propellant although only 200 to 300 pounds were required for the NASA mission. The extra propellant was originally intended for north/south stationkeeping in an alternate commercial mission known as Advanced Westar. This mission requirement was deleted but the extra propellant was purposely left on board because the IUS had already been programmed to carry that weight. In retrospect, this decision made the rescue mission possible.

Figure 9 illustrates the initial delta V plan. About three hours before each apogee burn, the roll motion was stopped in Inertial Mode with the spacecraft Z-axis 10 degrees from south-pointing. This attitude assures good antenna coverage and also put the effective delta V thrust vector near the orbit plane. The three-hour burn was approximately centered about apogee. At the end of each burn, the roll axis was switched back to Sun Mode and the spacecraft resumed its slow rotation about the sunline.

Stopping Spacecraft Roll Motion

Two problems had to be solved for stopping the rolling spacecraft at the proper burn attitude. First of all, the time for stopping the roll motion had to be predicted accurately. Secondly, a combination of thrusters had to be found that could do the job of the disabled negative roll thruster.

The process of estimating the roll stop time started about 8 hours before each burn. The roll reference was established when the earth sensor field-of-view crossed the earth near 6 pm local satellite time. Next the roll rate was measured each spacecraft rotation from gyro telemetry and then integrated to obtain the roll angle. Based on post-flight analysis, the achieved accuracy was usually within 10 degrees of the desired attitude. A key factor in this success was the stable and predictable spacecraft attitude dynamics in the new safe-haven mode.

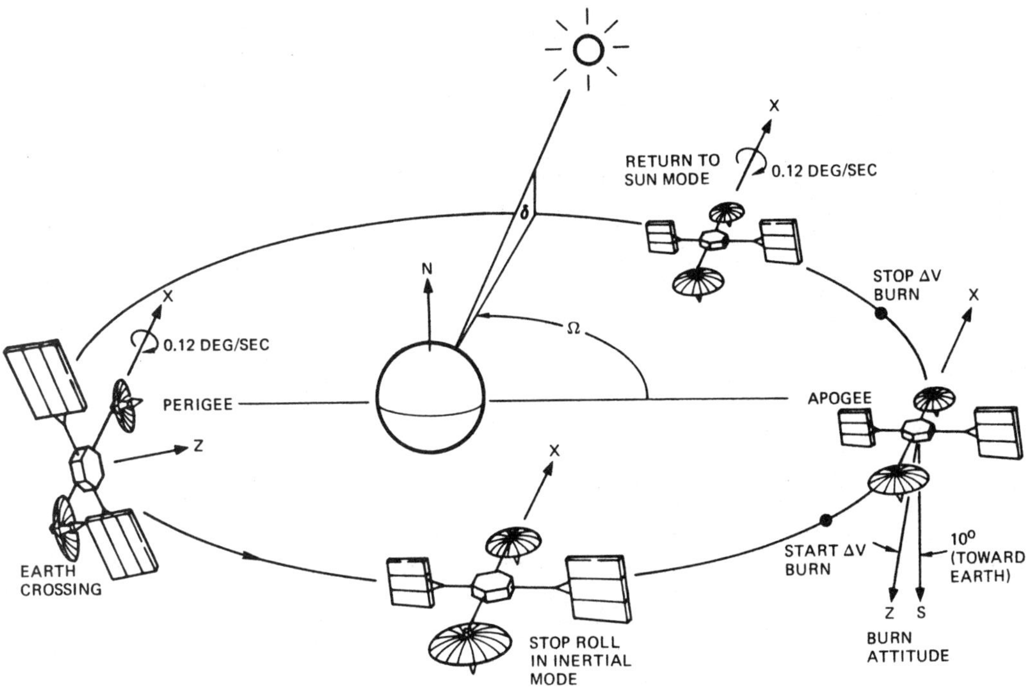

Figure 9. Maneuvers for Thrusting Along Sunline

The necessary negative roll torque was generated by alternately firing two yaw thrusters (Z2 and Z3). There are four yaw thruster modules mounted on the spacecraft as illustrated by Figure 10. Thruster pairs Z1-Z2 and Z3-Z4 are used for delta V burns. For yaw control, Z2 and Z4 produce positive torques and Z1 and Z3 produce negative torques. Each yaw thruster is canted by 17 degrees in order to minimize the pitch disturbance during a delta V burn. This 17-degree cant angle gives rise to a sizable roll torque component for each thruster. By ground commanding alternating pulses from Z2 and Z3, the net effect is a negative roll torque and thus, a substitute for the negative roll thruster was found.

It took a total firing time of about 10 seconds from the Z2-Z3 thruster combination to stop the TDRS roll motion. When the roll error was nulled, the onboard processor took action by firing the positive roll thruster. This cycle of onboard and ground-commanded thruster pulses continued until it was time to start the burn. A different method was employed during the burn as explained next.

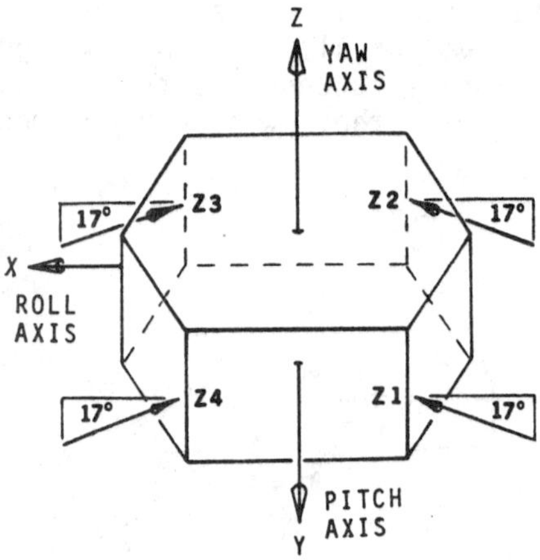

YAW THRUSTER TORQUE COMPONENTS (FT-LBS)

	X	Y	Z
Z1	1	0	-3
Z2	-1	0	3
Z3	-1	0	-3
Z4	1	0	3

Figure 10. Yaw Thruster Geometry

Roll Control During Burn

The TDRS ACS design includes the capability to activate a constant thruster pulse train during a delta V maneuver. This logic, referred to as disturbance compensation logic, was introduced to counteract known disturbances due to the delta V thruster pair. By partially cancelling these disturbances, it is possible to achieve tighter spacecraft attitude control during a burn and in this way improve antenna pointing performance.

For the current task at hand, this delta V compensation logic provided the means to achieve automatic roll control during a burn. In effect, this logic was used to introduce a large negative roll disturbance which requires only a positive roll thruster for control. But this roll disturbance had to be generated indirectly by pulsing a yaw thruster. The size of the roll disturbance can be adjusted by varying the yaw thruster duty cycle.

All delta V burns were performed with the thruster pair Z1-Z2. The Z2 thruster was pulsed with the onboard logic to produce a negative roll disturbance and a positive yaw disturbance. The yaw disturbance was controlled by the couple Z1-Z3 while the roll disturbance was handled by the +R thruster. Thus, three yaw thrusters were involved which served three interrelated functions. Z1 and Z2 provided the delta V thrust. Z1 and Z3 supplied the negative yaw torque. Z2 produced the positive yaw torque and also a negative roll disturbance.

Adjusting Thruster Duty Cycles

The first test burn occurred on May 2, 1983. It was scheduled to last for one hour but it had to be terminated after 42 minutes due to a hot roll thruster. A redline valve temperature of 240 degrees F had been established as a shutdown criterion for the burn. When the +R thruster reached this temperature, the burn was stopped.

After four burns, it was becoming apparent that steady-state duty cycles had to be below 0.5% or above 3% in order to keep the thruster valve temperatures at acceptable levels. Since these duty cycles are determined by the delta V disturbances, which varied significantly from burn to burn, there was little hope of consistently achieving low duty cycles. High duty cycles appeared more promising.

Burns 5 through 10 were performed with the high Z2 duty cycle of 20%. After two successful 3-hour burns, Burn 7 was terminated after 70 minutes because the -P thruster experienced reduced thrust. The remaining burns were cut short because the roll thruster got hot. Starting with the 10'th burn, the Z2 duty cycle was increased to 25%. As expected, this change increased the Z3 and +R duty cycles. However, there was no corresponding increase in the burn times. Figure 11 summarizes the Phase I burn durations.

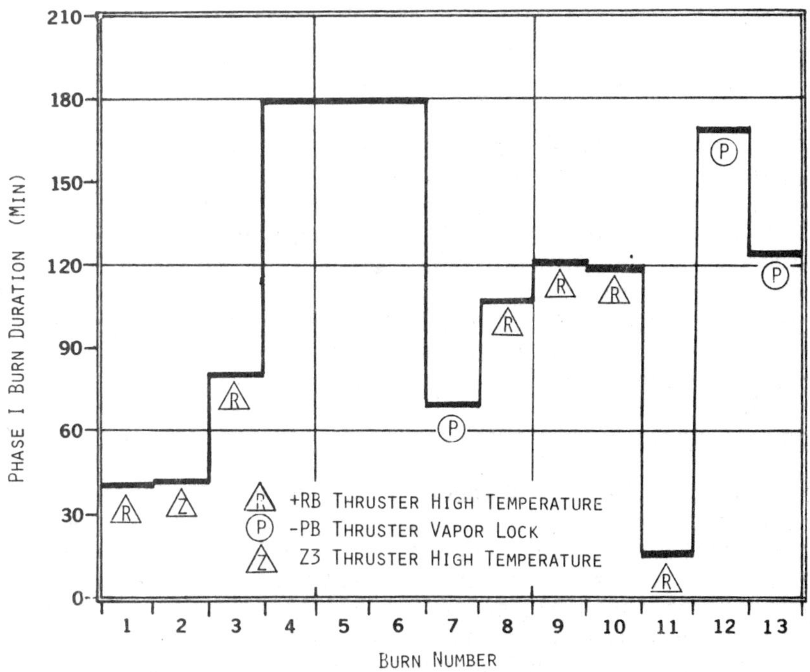

Figure 11. Phase I Burn Durations

The thruster duty cycle was only one of several critical factors which determined the valve temperature rise during a burn. The initial valve and line temperatures and the hydrazine flow rate all affected the thruster heating rate. As the tank pressures dropped, higher duty cycles were required to maintain the same flow rate, for example. Another phenomenom which exhibited itself more frequently was vapor lock. This condition occurs when warm hydrazine vaporizes before reaching the catalyst bed in the ignition chamber. It usually causes a sudden increase in the firing commands because the thruster impulse becomes drastically reduced.

Summary of Phase I Burns

In spite of all these difficulties, TDRS was slowly but surely approaching its final orbit. The perigee altitude had been raised from 11,800 nm to 15,850 nm. In doing this job, the tiny thrusters had expelled about 475 lbs of hydrazine into space over a cumulative burn time of 24 hours. There was still an adequate supply of fuel left but time had run out for the Phase I burns.

All Phase I burns were executed by thrusting along the sunline. The ideal time for this type of delta V was April 24 when the sunline was nearly lined up with the TDRS velocity vector at apogee. Phase I burns were spread over the time period from May 2 until May 19. For the last burn the sunline was 25 degrees away from the ideal thrust direction. It now became necessary to change to Phase II burns in order to improve the delta V efficiency.

DELTA V PHASE II - THRUSTING ALONG THE VELOCITY VECTOR

The nominal TDRS delta V mode, Earth Mode, uses an earth sensor and one gyro for three-axis control. However, it was decided not to use Earth Mode for the Phase II burns because of the small linear range of the earth sensor error signals. In Earth Mode minor attitude problems could possibly lead to an earth pointing loss and would require time-consuming attitude recovery procedures. Inertial Mode was considered to be the safest ACS mode for the Phase II burns. It had already proven itself as a robust and versatile control mode. In addition, Inertial Mode offers the option to adjust the burn direction, if necessary.

Maneuvers to Reach Burn Attitude

Figure 12 illustrates the spacecraft maneuvers required to reach the burn attitude. Starting out in Sun Mode, the first action was to stop the roll motion about four hours before the burn. This maneuver was the same as for the Phase I burn except that the spacecraft was stopped with the yaw axis in the orbit plane. The next maneuver was a yaw rotation to bring the roll axis into the orbit plane. Finally, a pitch maneuver swept the earth sensors across the earth and was terminated at the burn attitude.

Using the onboard logic normally used to compensate for known gyro drifts, a method was devised to make TDRS rotate about its pitch axis at 2 deg/min. It required a RAM load to start and stop the maneuver. When the earth came into the field-of-view of the earth sensors, preparations were made to stop the pitch motion. About 5 minutes after the nadir crossing, all gyro rate biases were removed, leaving the spacecraft inertially pointed. Since the desired yaw axis orientation was about 15 degrees east of nadir, the final part of the pitch maneuver was done at earth rate of 0.25 deg/min. When the burn attitude was reached, an earth rate bias was loaded into the gyro processing logic to keep the spacecraft attitude fixed relative to the earth.

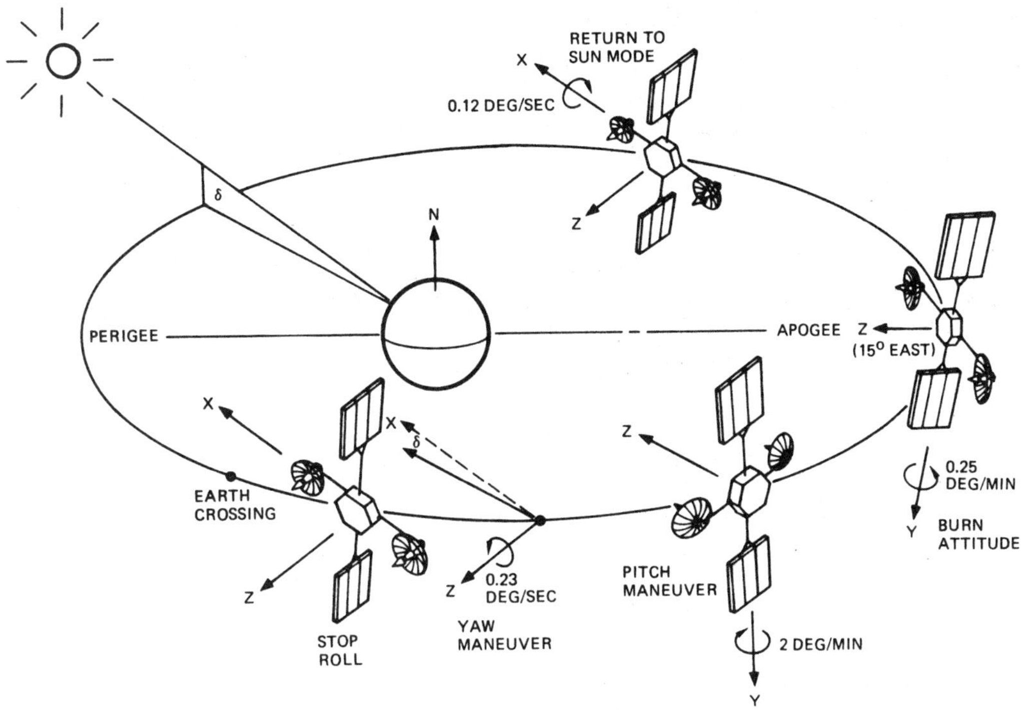

Figure 12. Maneuvers for Thrusting Along Velocity Vector

Phase II Test Burns

Two Inertial Mode test burns were scheduled for May 25 and 26, 1983 with the objective of checking out the new delta V procedure. The first Phase II test consisted of a 20 minute burn using a high duty cycle mode. Control performance during this short Inertial Mode burn was virtually identical to that of the earlier Sun Mode burns. For the second test, a new low duty cycle mode was tried in an effort to find a way around the recurring vapor lock conditions. The scheduled burn time was 1.5 hours, but the burn was stopped after 24 minutes because the Z3 thruster got too hot. Based on this experience, it was concluded that low duty cycle operation was too unpredictable. All subsequent Phase II burns were performed with a high Z2 duty cycle of 25%.

Both Inertial Mode test burns were executed with the Phase I maneuver sequence, for simplicity. Subsequently, a test run was made through the new maneuver sequence illustrated by Figure 12. The maneuvering went smoothly and closely followed the predicted timeline. There were a few anxious moments when the earth appeared about two minutes later than expected during the pitch maneuver. But this kind of deviation was well within the design tolerances of the maneuver timeline.

Coping with Thruster Vapor Lock

Figure 13 summarizes the Phase II burn durations and identifies the burns which were terminated due to a vapor lock occurrence. Vapor lock occurred in thrusters with initial valve temperatures as low as 180 degrees F. The negative pitch thruster, whose duty cycle generally fell in the range of 3.5-4.0%, seemed to be most susceptible to vapor lock. As the vehicle center-of-mass dropped due to propellant depletion, the pitch thruster duty cycle decreased proprtionately. After Burn 20, a pulse train of 0.5% was introduced for the +P thruster. This change effectively raised the -P thruster duty cycle by 0.5% and prevented pitch thruster vapor lock during all remaining burns.

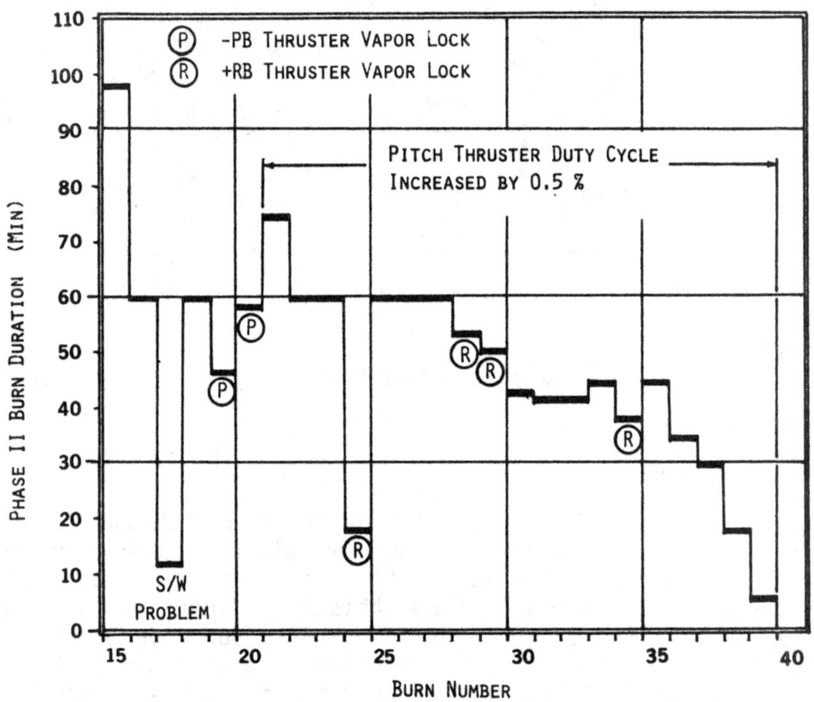

Figure 13. Phase II Burn Durations

From Burn 21 through 27 almost everything seemed to be going according to plan. One exception was Burn 24 which experienced a +R vapor lock after a 17-minute burn. This occurrence was due to poor catalyst bed heater management. The initial roll thruster temperature was excessively high because the catalyst bed heater had not been turned off early enough before the burn.

Trouble started again when the +R thruster experienced vapor lock on two successive burns, Burns 28 and 29. Not much could be done to change the +R thruster duty cycle. The only alternative at this time appeared to be shorter burns. Therefore, the one-hour burns were reduced to 45-minute or shorter burns. This plan worked quite well. Only one more vapor lock condition occurred near the end of Burn 34. But this minor perturbation did not delay the scheduled TDRS arrival time in synchronous orbit.

Final Burn to Geosynchronous Orbit

A new delta V procedure was developed for the final burn to geosynchronous orbit. After Burn 38, the spacecraft was not returned to Sun Mode but remained earth-pointed until the last burn. The pitch axis was switched to Earth Mode while the roll and yaw axes remained in Inertial Mode using an earth rate bias to keep the yaw axis earth-pointed. For the next 36 hours, the positive roll attitude error was controlled from the ground. By firing an occasional Z2-Z3 thruster pair, the spacecraft yaw axis was kept within one degree of nadir.

Figure 14 epitimizes TDRS' long and eventful climb to synchronous orbit. Starting from the low perigee altitude of 11,800 nautical miles on May 2, 1983, the 2 1/2 ton spacecraft was boosted to an altitude of 19,323 nautical miles using its tiny one-pound thrusters. This herculean task caused the thrusters to overheat and choke at times but they never quit. After 44 hours of thrusting over a two-month period, TDRS-1 arrived at its final destination on June 29, 1983.

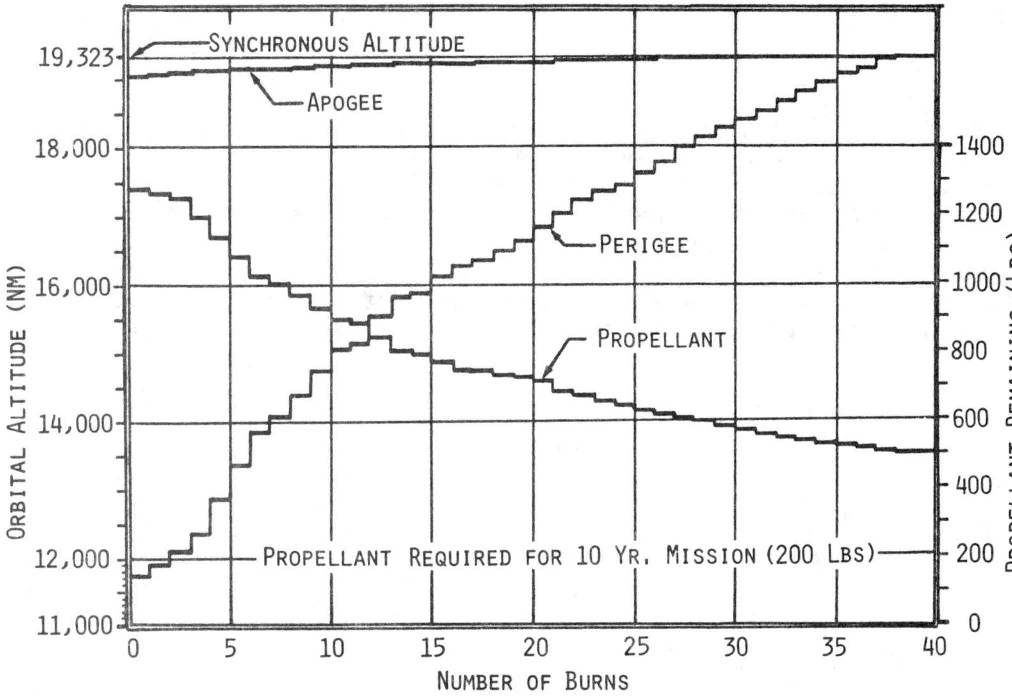

Figure 14. TDRS-1 Orbit Correction History

TDRS-1 FLIGHT PERFORMANCE SUMMARY

Over the three-month period the TDRS-1 propulsion and attitude control subsystems have experienced more activity than they would over a normal ten-year lifetime. Both systems were pushed far beyond their design limits. The successful conclusion of the rescue operation clearly demonstrates system capabilities that exceed mission requirements. Without these capabilities, the TDRS-1 rescue would not have been possible.

Designers of the TDRS attitude control subsystem incorporated considerable operational flexibility into their microprocessor-based system. Operating in an elliptical orbit with disabled negative roll thrusters, several new control modes were devised in flight to adapt to this new environment. The ability to select the TDRS control modes on a per-axis basis, known as a "Chinese menu" design, played a key role in all reconfigurations. Another design feature which turned out to be very useful during delta V maneuvers, was the onboard thruster pulse train logic.

Faultless operation of the coarse sun sensors, gyros and the Sun Mode control processing provided a safe haven throughout the rescue mission. Precise attitude control during the 39 delta V maneuvers resulted in efficient TDRS orbit adjustments. In summary, the TDRS attitude control subsystem performed extremely well under severe contingency conditions.

ACKNOWLEDGMENT

The successful rescue of the stranded Tracking and Data Relay Satellite required extraordinary efforts from many talented people at TRW, NASA and SPACECOM. After the rescue mission, TRW issued a comprehensive anomaly investigation report which was sponsored by the National Aeronautics and Space Administration (NASA). This paper is based on the findings of the anomaly investigation team. In writing the story of the TDRS-1 rescue, the authors especially appreciate the careful review and helpful suggestions provided by R.J.Moore, Dr. H.L.Mork and J.C.Sarina, all of TRW.

AAS 84-054

AUTOTRACKING FROM SPACE: THE TDRSS APPROACH

R.E. Spearing*, W.R. Harper**

The TDRSS will provide telecommunications support to near-earth orbiting satellites through the 1980s and into the 1990s. The system incorporates two operational satellites at geostationary altitude and a single ground station at White Sands, New Mexico. Of the many tasks facing the engineering team in development of this system, one of the most challenging was K-band autotrack. An approach not previously attempted placed the error detection, processing, and feedback elements for automatic control of the TDR satellite antennas on the ground. This approach offered several advantages to the designers but posed a number of interesting questions during the development program. The autotrack system design and its test program are described with emphasis given to areas of special interest in developing a working K-band service.

INTRODUCTION

The Tracking and Data Relay Satellite System (TDRSS) is being built to provide NASA with its next generation tracking and data acquisition capability for support of near-earth orbiting user spacecraft. The concept is novel in that only two geostationary satellites are used in place of many ground stations located around the earth. As depicted in Figure 1, these two satellites and in-orbit spare provide nearly global coverage to users of the system. Establishing and maintaining the TDRSS communication links is a complex task requiring careful review and selection of available design concepts and equally careful attention to implementation of the end product. This task becomes most interesting for the K-band links which require autotrack.

AUTOTRACK DESIGN

Each TDR satellite provides a focal point for transmission and reception of signals from users. The many services available can be evidenced by

* *TDRSS Ground Segment Manager, NASA Goddard Space Flight Center, Greenbelt, Maryland 20771*

** *Deputy of TDRSS System Engineering, TRW Space and Technology Group, One Space Park, Redondo Beach, California 90278*

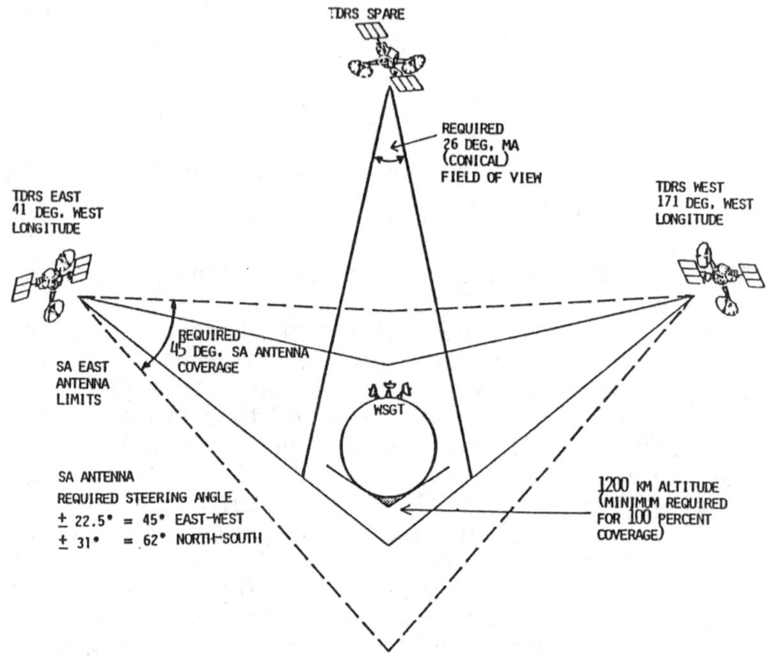

Figure 1. TDRSS System Configuration

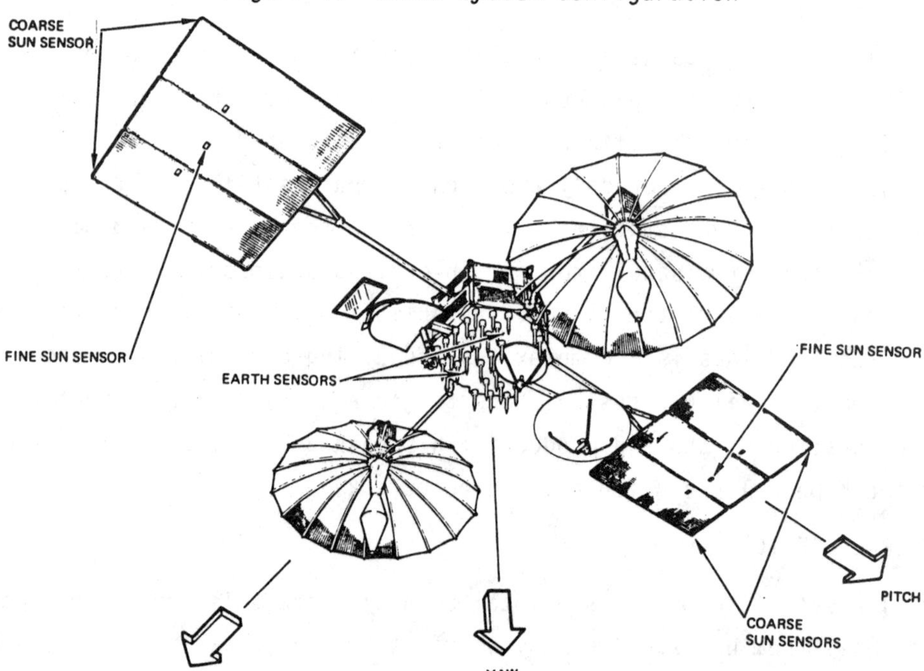

Figure 2. The Tracking and Data Relay Satellite

the many antennae illustrated in Figure 2. Of particular interest for this paper are the two 16-foot (4.9 meter) mechanically steerable single access (SA) antennas. The platform to which the antenna gimbals are attached is stabilized in three axes using momentum wheels in combination with earth sensors. Hydrazine thrusters are employed to periodically unload wheel momentum to maintain stabilization for long term operations.

The beamwidth of the SA antenna at K-band is approximately 0.28 degree. In order to provide satisfactory communication signal margin an overall pointing uncertainty of 0.06 degree is necessary. Two sources of error dominate the open loop pointing budget and resulted in the decision to incorporate autotrack as a means of meeting the accuracy requirement. The first is user satellite position uncertainty. This error is the result of non-ideal orbit determination by NASA and degradation in the predicted orbit with time, which can amount to as much as ± 9 seconds principally in the along-track direction. Conversion of this time value to angular uncertainty yields ± 0.14 degree.

The second significant error source is introduced by the TDRS attitude determination process. The satellite attitude control system utilizes earth sensors to maintain pitch and roll attitude to an accuracy of 0.15 degree. Ground processing of the earth sensor outputs is utilized to determine the actual satellite attitude for antenna pointing purposes to an accuracy of 0.1 degree. Spacecraft yaw attitude is controlled passively to an accuracy of ± 1 degree. The sun sensor provides a highly accurate measure of yaw attitude for several hours centered on 6:00 A.M. and 6:00 P.M. local satellite time. At other times yaw attitude must be estimated using reaction wheel telemetry and disturbance torque predictions. The uncertainty of the yaw estimate is ± 0.25 degree resulting in a maximum antena pointing error from this source alone of ± 0.13 degree.

Additional smaller contributions to pointing uncertainty (see Table 1) will not be addressed here but when considered with the sources noted above result in a maximum error of 0.24 degree. Thus, the job for autotrack is to acquire the user satellite signal at a maximum angular offset of 0.24 degree and reduce the nominal tracking offset to less than 0.06 degree for optimum data recovery.

Table 1

K-BAND SINGLE ACCESS OPEN-LOOP POINTING BUDGET

	One-Dimensional Errors				
	Fixed (Max) (deg)	Slow Variations (3σ) N-S (deg)	E-W (deg)	Yaw (deg)	Fast Variations (3σ) (deg)
1. Software					0.026
Antenna/Boom					
2. Thermal (boom)		-	0.021	0.013	
3. Thermal (antenna)		0.0	0.06	-	
4. Gimbal to antenna mechanical axis misalignment	1.0				
Misalignment measurement uncertainty	0.066				
Antenna mechanical to electrical boresight misalignment	0.2				
Earth Sensor					
5. Null alignment	0.05				
6. Mech alignment	0.05				
7. Bias errors		0.0500	0.0429	-	
8. Power supply		0.0100	0.0086	-	
9. Radiance uncertainty		0.0150	0.0129	-	
10. Noise and quantization					0.03
Control System					
11. Step size and dynamic oscillations					0.06
12. Gimbal drive position error					0.04
13. Drive axis wobble					0.05
14. Yaw coupling error		-	-	0.129	
15. User ephemeris error		-	-	0.122	
16. TDRS orbit determination		0.014	0.014	-	
17. Fixed error after calibration N-S / E-W	0.039 / 0.049				

Table 1

K-BAND SINGLE ACCESS OPEN-LOOP POINTING BUDGET (Continued)

	One-Dimensional Errors				
	Fixed (Max) (deg)	Slow Variations (3σ) N-S (deg)	E-W (deg)	Yaw (deg)	Fast Variations (3σ) (deg)
18. RSS of the slow varying errors		0.0550	0.0795	0.1776	
19. RSS of the fast varying errors					0.0963
20. Uncertainty ellipse major axis (3σ)	0.217				
21. Uncertainty ellipse minor axis (3σ)	0.111				
22. Probability of pointing within 0.24 degree	0.994				
23. Worst-case pointing loss (dB)	9.6				

The overall autotrack subsystem design is depicted in Figure 3. The design is unique in that most of the autotrack equipment is on the ground rather than in the TDRS. This approach enables selection of optimum processing parameters for each user, saves weight and complexity in the spacecraft, and improves reliability. Complex elements are on the ground where they can be repaired. This approach to the design made sense for the TDRSS because of the relatively low angular rates and angular accellerations encountered in tracking low earth orbiting satellites from geostationary altitude.

The K-band feed of the SA antenna is a central horn with four separate error horns on the perimeter of the central horn. The five-horn feed was selected to provide maximum data channel performance while maintaining adequate autotrack error channel signal performance. A photograph of the feed horn configuration is shown in Figure 4. The radio frequency (RF) signals received by the perimeter horns are processed by a microwave combiner which produces elevation (roll) and azimuth (pitch)

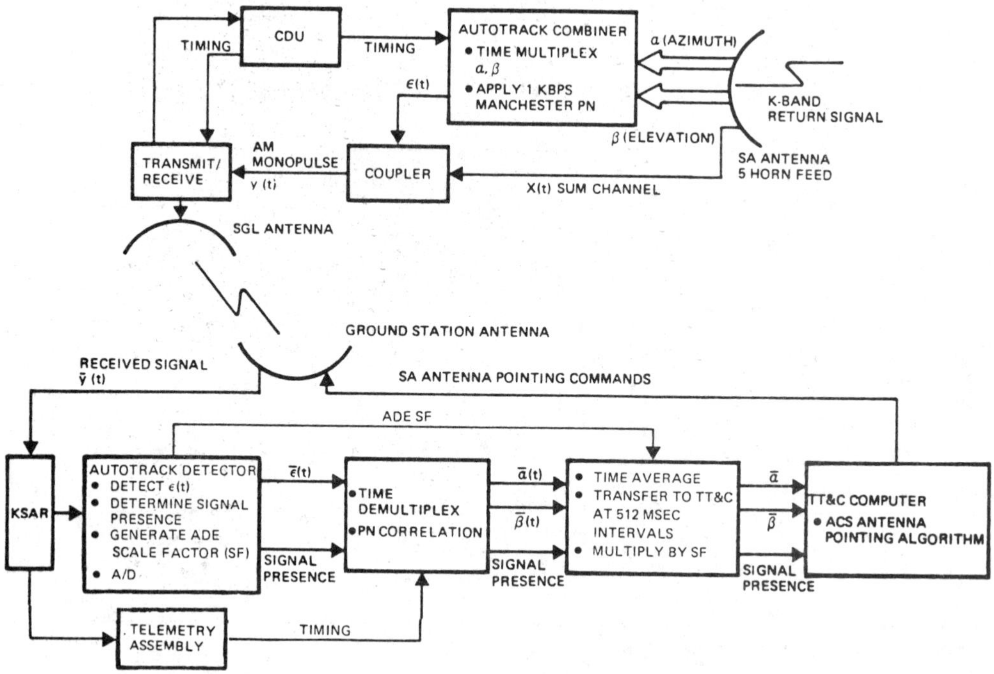

Figure 3. Angle Tracking - Autotrack System

Figure 4. Single-Access Antenna Feed

error signals. The resulting RF error signals are time division multiplexed and PN coded prior to being algebraically summed with the RF output of the central horn. The resultant is an amplitude modulated "monoscan" composite signal which is then transmitted to the ground. The multiplexing technique assures a high degree of electrical isolation between the azimuth and elevation error signals and the PN coding reduces the sensitivity to interference and spurious signals introduced in the transmission path. The technique has been demonstrated to produce negligible degradation to the user signals for all constant envelope modulation formats under even the most adverse tracking conditions. The amplitude modulation index is proportional in amplitude to the angular off set between the antenna electrical boresight and the line of sight to the user. The sense of the amplitude modulation (whether the signal amplitude is increased or decreased in any given time slot) is a function of the quadrant in which the user resides.

When received on the ground the composite signal is split into two paths. In one path the signal is demodulated to recover the user data. In this path the autotrack AM is stripped off since it does not contain data originating at the user. In the second path the AM signal is recovered using a square law detector. In this path the user data is ignored since it contains no pointing information. Because of the nature of the detection process, and AM signal is suppressed at low signal to noise ratios reducing the apparent scale factor of the autotrack system at a time when high signal gain would be desirable. The ADE utilizes a second detector preceded by a filter outside the normal user signal bandwidth to derive an estimate of signal-to-noise power ratio (SNR) permitting control of the gain as a function of SNR.

The ADE AM detector output and the SNR estimate are digitized for input to the Autotrack Interface Processor (AIP). The AIP is a general purpose digital computer that performs a number of numerical and logical operations on the two digitized data streams. The AM detector output is correlated with a locally generated version of the autotrack PN code. The effect of the correlation process is that the PN code is "stripped off" the signal and spurious signals introduced after the autotrack combiner are reduced in amplitude. This is an important feature because of the low AM modulation index during normal operation. The resultant is a digital data stream representing the time multiplexed azimuth and elevation error signals corupted by system noise. A demultiplexing opertion is performed which produces separate azimuth and elevation error signals that are subsequently integrated to reduce the magnitude of the system noise.

The SNR estimate is integrated and compared with a preset threshold to establish signal presence. Prior to exceeding the threshold (signal absence) the AIP error signal outputs are artificially set to zero preventing system noise from perturbating the open loop pointing process. These "default zero" conditions prevail until signal presence is declared. The AIP provides azimuth and elevation error signals to the TT&C computer whose magnitude is adjusted as a function of input SNR.

The processed error signals are input to the TT&C computer which performs the final pointing computations. The TT&C computer receives open loop steering data from the ground segment central computer which is derived from the user and TDRS orbits. It receives antenna alignment data from the spacecraft data base also resident in the central computer. Antenna position data is derived in the TT&C computer from spacecraft telemetry and a running log of antenna position commands. The computer then combines the open loop steering inputs with the antenna position data, attitude data, and autotrack error signals to produce a stream of antenna step commands for transmission to the spacecraft.

The spacecraft structural resonances must be carefully avoided in commanding the antennas. "Deadbeat" logic is incorporated into the tracking algorithm so that step commands are always sent in pairs timed so that they minimize structural perturbations.

With this knowledge of the autotrack loop design in hand, the pull-in and autotrack process can be described. Initial pointing of the antenna is based solely upon a users predicted ephemeris. Stepping commands are sent to the TDRS and the antenna is moved to the expected signal acquisition coordinates. Prior to the expected time of acquisition,a self calibration process is performed. The TT&C computer issues commands which terminate the input to the KSA return channel. This action allows the ADE to establish a system noise reference level for signal presence detection. As noted earlier, signal presence detection will start the autotrack process. Should signal presence not be determined, the SA antena will continue to be pointed toward the user by stepping commands generated on the ground at a rate based upon the users predicted ephemeris.

Once signal presence is declared, a useable error signal is transferred from the AIP to the TT&C computer which changes control modes from open-loop pointing to closed-loop. Figure 5 illustrates the transitions from open-loop pointing through pull-in to autotrack. Initially, only the polarity of the error signal is used, which establishes the direction in which to move the antenna. A sequence of rate bias stepping commands are sent to the TDRS SA antenna to drive it toward the user at a constant rate of 0.02 degree per second. Early designs used signal amplitude as well as polarity, but large variations in scale factor resulted in a reconsideration of the approach.

The pull-in process continues until a change in the polarity of each error signal (pitch and roll) is detected indicating that the SA antenna has passed through boresight. At this time the autotrack logic switches to its fine pointing mode and maintains antenna pointing. Figure 6 illustrates one side of the error signal amplitude response as a function of angle off boresight and is annotated to show major decision points for achieving autotrack.

AUTOTRACK PERFORMANCE VERIFICATION

The verification program for the autotrack subsystem is unique in that there was no end-to-end autotrack acquisition or tracking testing performed on the ground prior to launch. Instead, extensive use was made of a computer simulation which substituted for a ground based system test bed. The autotrack simulation program was developed during the initial design phase as an optimization tool to aid in the complex process of flowing performance requirements to the many affected spacecraft and ground segment equipments and software.

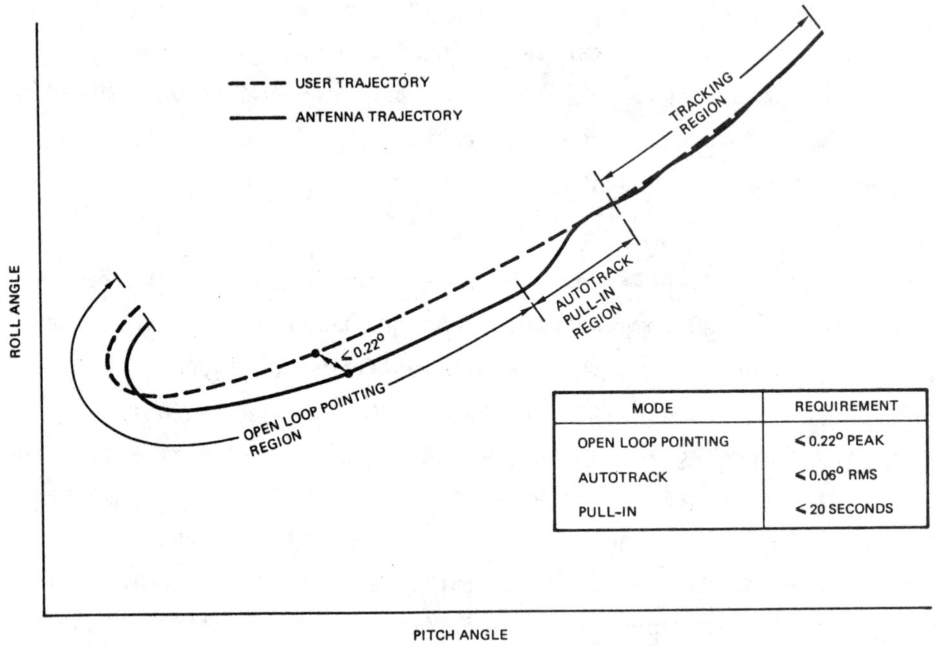

Figure 5. Autotrack Operating Regions

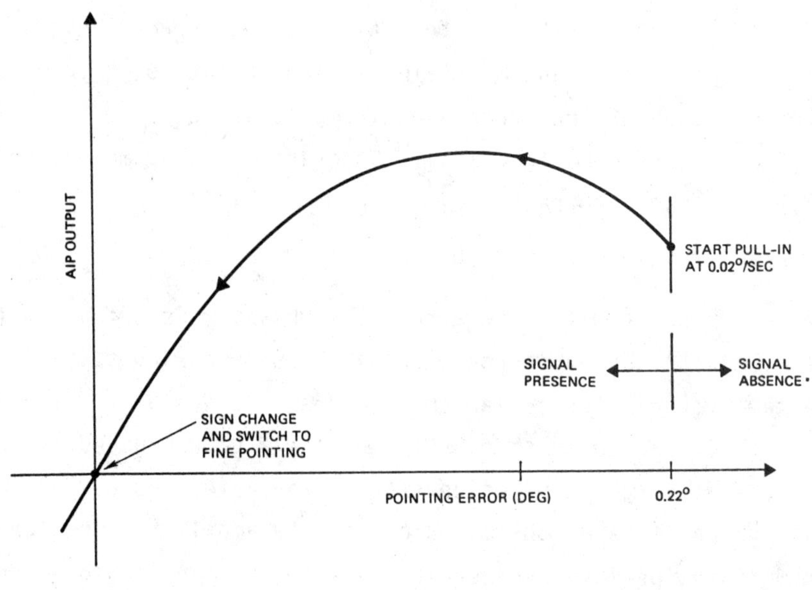

Figure 6. Autotrack Error Magnitude and Decision Points

The SA antenna characteristics were modeled as were the characteristics of the spacecraft autotrack combiner, the KSA communication channel including automatic gain control, and the ground segment Autotrack Detection Equipment (ADE). The digital signal processing operations performed in the Autotrack Interface Processor (AIP) were simulated as were the acquisition logic and tracking algorithms implemented in the Telemetry, Tracking and Command (TT&C) computer. The long loop delays introduced by the transmission path to and from the satellite on geostationary orbit and the computer processing and signal processing dalays were estimated and incorporated into the simulation.

For each equipment, a model reflecting the worst case specified limits of performance was developed as was a model representing the expected performance. As equipment was built and tested, the estimates were revised to reflect measured performance.

At each stage in the design evolution the effects of revising specification limits to accommodate developmental problems were evaluated using the simulation. At one point in the design, substantive changes in the ADE performance requirements were requested by a vendor when it was discovered that certain of the requirements imposed conflicting design constraints on critical components. It was necessary to completely revise the ADE simulation model and redo much of the early simulation work. The new simulations revealed that the revised ADE characteristics degraded tracking performance slightly, bringing performance closer to specified limits, but acquisition performance did not meet requirements. As a result, the acquisition logic was revised to correct the deficiency.

The simulation of the vehicle started with a relatively simple three body model consisting of the two SA antennas and a rigid body representing the rest of the spacecraft. It evolved into a six-body simulation model which includes the effects of the solar arrays and booms, the spacecraft structure and the mechanically steerable space to ground link (SGL) antena. The control characteristics of the attitude control system were modeled and incorporated into the simulation. The simulation has the ability to evaluate the acquisition and tracking performance of one SA antenna while slewing the other.

After the six-body model was incorporated into the simulation the required computer run time dramatically increased. It became apparent that a complete evaluation of autotrack performance using this version of the simulation was not economically practical. Instead, the three-body model was used for each of the dozens of user modulation formats, data rates and limiting signal levels to identify problem areas and areas of potentially marginal performance. For those cases, the six-body model was used to establish the actual performance.

As part of their own internal verification process, NASA contracted with the Applied Physics Laboratory (APL) of Johns Hopkins University to independently develop a simulation tool. The TDRSS contractor team cooperated with APL in the effort by defining the performance characteristics of the various system elements. The APL personnel assigned to the task quickly assembled a working computer simulation from routines used on another system and routines developed for this application. The results of the APL and TRW simulations agreed remarkably well from the first. The differences were quickly traced to correctable weaknesses in the two simulations. This proved to be a mutually rewarding effort since the NASA evaluators quickly built confidence in the merit of the overall verification approach and the TDRSS contractor team incorporated graphical capabilities into their simulation which the APL program demonstrated had great functional utility in optimization and trouble-shooting. Compliance with system acquisition and tracking performance requirements is now demonstrated using the simulation in concert with user on orbit service demonstrations.

The first TDRS was launched on 4 April 1983. After an extended rescue mission which is the subject of numerous technical papers, some of which are being presented at this conference, the autotrack system was finally tested on orbit. A comprehensive autotrack test series was performed consisting of antenna alignment calibrations, static and dynamic acquisition and tracking tests, and repeated acquisitions of user satellites including Landsat, STS-8 and STS-9/SL-1. Only two problems were encountered in the testing, both of which were outside the equipment and software included in the simulation: initial gimbal position determination and yaw estimation.

RECOMMENDATIONS

The autotrack subsystem in TDRSS is proving to be a robust design in day-to-day service. The long loop design which placed most critical elements on the ground has proven to be practical and cost effective. While the complexity of some elements is greater than would be required for a short loop design implemented on the satellite, ADE complexity was dictated primarily by the need to accommodate the wide range of data rates (1 kb/s to 300 Mb/s) and modulation formats (dozens) required by the system level specification. It is unlikely that a satellite borne ADE could have been as nearly optimized for all users as the ground based unit. Additionally, the attitude data processing required to achieve the small residual attitude error would have substantially increased satellite data processing requirements. As a result, the long loop approach to the design selected for TDRSS is recommended for any application where autonomy is not a major consideration.

The two areas where problems were encountered on orbit were amenable to simulation, but no simulations were developed. The software component which commands the antenna in such a way that its gimbal angles are unambiguously defined suffered from several subtle internal timing and logic errors. The process of developing a suitable simulation would have concentrated more attention on these areas of the design. The simulation would have provided a software test bed for the actual code which would have caught implementation errors as well as design errors. A ground test was performed using the actual satellite in the assembly area as a test bed, but it only highlighted the most obvious problems. Because of schedule time constraints associated with the satellite hardware it was not possible to repeat the test after the problems were fixed thus more subtle problems remained.

The second problem involved the yaw estimator. The yaw estimator software exhibited a higher degree of sensitivity to missed telemetry data points than proved to be tolerable. Additionally, the restart rules in case of computer failure introduced large transients into the system. An end-to-end ground test was impractical for this software.

A simulation employing the six-body satellite model could have been developed to exercise all aspects of the software design. The recommendation arising from both of these problems is to identify all critical elements of each subsystem and devise suitable methods of verification prior to first use on orbit.

ACKNOWLEDGMENTS

The authors wish to thank a number of people who have contributed a part of themselves to this paper. To Jim Ransom of Westinghouse and Glenn Preston of TRW whose comments, sage advice and reference materials were invaluable to assembling the paper. To Nadia Habash who suffered through the typing of endless rough drafts. To Fred Beisel and Austin Cullen of Space Communications Company (SPACECOM)* who reviewed the manuscript in record time and to Ed Coy, Ron Browning and Nelson Pixley, the SPACECOM, NASA and TRW TDRSS program managers, respectively, who actively encouraged us to tell this story, our many thanks.

*A partnership affiliated with Continental Telecom, Inc. and Fairchild Industries, Inc.

AAS 84-056

COSMIC RAYS, SINGLE EVENT UPSETS AND THINGS THAT GO BUMP IN THE NIGHT

Stephen S. Cunningham*

Bi-stable devices in outer space can undergo anomalous state changes which can damage or make inoperative their electrical circuitry and even entire space systems. Some of these state changes are due to electrostatic discharges. Recently, however, more and more of these changes are the result of the little understood condition of cosmic-ray-induced single event upsets. And as satellites become more complex, so will the rate of cosmic-ray-induced changes. These upsets are not the result of solar illumination, witness that transient (soft) and permanent (hard) upsets due to cosmic rays can occur at any time during orbit, day or night. Thus "things that go bump in the night" is a good clue that the event was induced by cosmic rays and not electrostatic discharge.

This paper reviews the cosmic ray environment in space and shows how and to what extent it upsets bi-stable devices. Some of the spacecraft currently experiencing these upsets, such as the Tracking and Data Relay Satellite System, Landsat 4 and Galileo, are surveyed to illustrate the severity and frequency of the problem. Hardware and software techniques to detect and correct upsets and ways to reduce a system's susceptibility to them are presented.

Finally, an outline of a prototype specification for the design and verification of cosmic ray hardened electronics is provided to aid Guidance and Control engineers.

* *Physicist, General Electric Space Systems Division, P.O. Box 8555, Room 8036 Building 8, Philadelphia, Pennsylvania 19101*

INTRODUCTION

Upset and even permanent damage of spaceborne electronics may be caused by the cosmic ray environment of outer space. As the speed and sophistication of spaceborne computers and memories improve, their susceptibility to cosmic ray damage increases. Actually, spacecraft are even now experiencing serious -- and expensive -- problems because no protection against cosmic-ray-induced upsets was included in their electronic designs.

This paper presents a review of the problem of Single Event Upsets and two of its main causes, Electrostatic Discharge (ESD) and Galactic Cosmic Rays. ESD is so well understood that systems to eliminate it have been implemented on many satellites. The cosmic ray problem is relatively new however, and the guidance and control designer, since the implementation of on-board processors and memory is generally handled by him, is often the first to become acquainted with it. Thus, this paper will briefly describe ESD and then concentrate on the cosmic ray problem. A chronology of the discovery of cosmic rays is provided, while a quick look at the problem being experienced by spaceborne systems will illustrate both the nature of cosmic rays and the hazards they pose to unprotected systems. Protection schemes, and examples of current systems that have successfully implemented them while maintaining required system performance, are presented.

Finally, an outline for a specification is provided to help Guidance and Control engineers to develop and provide a system which is fully protected against cosmic-ray-induced single event upsets.

WHAT IS A SINGLE EVENT UPSET?

This term describes a situation in which a bi-stable device is forced to change state through a process which is outside the normal read-write process for that device. This single bit flip, which is random across a group of bits, usually occurs without permanent damage to the device. Typically, a bit which has undergone a flip will remain in that state until the processor re-writes it with new data (References 1 and 2). Once a bit has been re-written, all evidence of the upset is erased in that device. The problem to the spacecraft designer is the possible effects of the erroneous data which has been introduced into memory by the flip (see Figure 1).

Depending on the operation of the particular satellite system, and on the ability of either the satellite or the ground controller to detect and correct an error and to restore normal operation, the effects of a single event upset may vary widely. Serious damage is, of course, difficult to quantify and the occurrence rate may often be the deciding factor in determining whether a problem is an IRRITANT or a SHOW STOPPER.

- AN ANAMOLOUS CHANGE TO A BI-STABLE LOGIC DEVICE

- SEUs IN MODERN ELECTRONICS IN SPACE ARE CAUSED BY:
  - COSMIC RAYS
  - ELECTROSTATIC DISCHARGE
  - TRAPPED PROTONS

- THE SYSTEM OR SUBSYSTEM DAMAGE CAUSED BY SEU MAY VARY:

	EFFECTS
- TRANSIENT MEMORY UPSET - UNNOTICED	BENIGN
- TRANSIENT MEMORY UPSET	↓
- PROCESSOR UPSET	
- LATCH-UP	SEVERE
- PERMANENT DEVICE DAMAGE	
- FALSE COMMAND GENERATION	↓
- SYSTEM HARDWARE DAMAGE	
- TOTAL LOSS OF SATELLITE	CATASTROPHIC

Figure 1. What is a Single Event Upset?

Although single bit anomalies may occur for various reasons, the SEUs of concern here are due to electrostatic discharge or galactic cosmic rays. Because these two causes may be confused, and since their "fixes" are radically different, both causes will be explained.

ELECTROSTATIC DISCHARGE (ESD)

Both local electrical and electrostatic discharges can wreak havoc on sensitive satellite electronics and memory devices, even destroying some materials. Even a relatively small discharge may release sufficient energy to cause transient damage, i.e., upsets in bi-stable devices. Although a bit-flip looks the same regardless of the cause, ESD can be identified sometimes because the discharge occurs at certain times in the satellite's solar day (Reference 3). Electrostatic build-up between parts of a satellite may be triggered (discharged) by an attitude change, such as a change in exposure to the sun, or power switching on the vehicle.

The two main contributors to electrostatic build-up are solar illumination and plasma bombardment. Since illumination via the photoelectric effect liberates electrons, it causes a charge build-up. Inhomogeneities in the surface, holes, crevices, protusions, and different materials, as well as different solar aspect angles due to geometry, also lead to the build-up of charge differences between different regions on the body. Plasma bombardment varies because of geomagnetic disturbances, particularly such solar activity as flares and storms. Again, the magnitude of the potential differences depends on vehicle geometry and construction.

Measurements made on the SCATHA Satellite and ATS-6 uncovered static surface potentials as high as 20,000 volts (Reference 4). Clearly, careful attention to grounding and bonding is a key to controlling and eliminating charge build-up.

Given a highly charged condition, a change in the plasma environment or the solar incidence could initiate a discharge. This release, like a local lightening bolt, can cause physical damage or, depending on its location, cause a disruption or upset in a sensitive semi-conductor circuit. Some flight measurements and experience suggest a strong correlation between ESD and satellite local time. Discharges, for example, often occur soon after sunrise, probably due to the sudden change caused by the photoelectric mechanism. Correlation of upsets with solar storm activity or satellite local time is, therefore, often a clue that the cause of the disturbance is electrostatic discharge.

COSMIC-RAY-INDUCED SINGLE EVENT UPSET

What is a Cosmic Ray?

The current model defines the cosmic ray as ionizing radiation of extraterrestrial origin, including high-energy electrons and protons. Particle energies vary, but very high energy particles have been identified with all elements. The Galaxies are the main source of cosmic rays, not the sun. Some solar flares, however, are ion rich and therefore occasionally contribute to the cosmic ray environment. The near Earth cosmic ray environment decreases during periods of high solar activity. This effect is thought to be due to the disruption of the Earth's magnetic field and, therefore, some additional deflection of incoming charged particles causes a reduced cosmic ray environment in the near Earth environment. This variation of cosmic ray flux and solar activity closely follows the eleven-year sunspot cycle.

On a day-by-day basis, however, the galactic cosmic ray flux strikes a satellite from all directions, at all hours, and is not related to sun incidence angle or satellite local time. Cosmic-ray-induced upsets are equally likely at any point or any time in the orbit, thus "things that go bump in the night" may be a strong indication that the cause of the upset is the cosmic ray.

Cosmic ray research began when scientists observed the unexplained discharge of electroscopes and began searching for the cause of the mysterious ionizing radiation. Experiments conducted by Hess and Millikan established that the radiation was extraterrestrial; Millikan named it "Cosmic Ray." V.F. Hess, 24 years after the event (Reference 5), received the Nobel Prize in 1936 for the initial discovery of cosmic rays. Figure 2 presents a chronology of cosmic ray research.

1912:	V.F. HESS	- DISCOVERS THAT IONIZING RADIATION IS EXTRATERRESTRIAL
1913:	C.T.R. WILSON AND OTHERS	- SUGGESTS SOURCE IS ELECTRICAL STORMS OR AIRBORNE RADIOACTIVITY
1926:	R.A. MILLIKAN	- PROVES RADIATION ORIGINATES ABOVE THE EARTH'S ATMOSPHERE; NAMES RADIATION COSMIC RAYS
1928:	R.A. MILLIKAN	- POSTULATES THAT COSMIC RAYS ARE VERY HIGH ENERGY GAMMA RAYS -- PHOTONS
1929:	GEIGER & MUELLER	- GEIGER COUNTER INVENTED
1929:	BOTHE & KOHLHORSTER	- "DAS WESEN DER HOEHENSTRAHLUNG" - PROVES THAT COSMIC RAYS ARE PARTICLES NOT PHOTONS
1931:	B. ROSSI ET AL	- PROVES PARTICLE ENERGIES ARE EQUAL TO BILLIONS OF ELECTRON VOLTS
1930-1950:		- COSMIC RAY INVESTIGATORS DISCOVER EXISTENCE OF NEUTRON, NEUTRINO, PI-MESON, MUON, ETC.
1958:	J. VAN ALLEN	- COSMIC RAY INSTRUMENTS ON SPUTNIK II AND EXPLORER I & II CONFIRM EXISTENCE OF THE VAN ALLEN RADIATION BELTS
1962:	WALLMARK & MARCUS	- PREDICT COSMIC RAY EFFECTS ON MICRO-ELECTRONICS
1970:	MAY & WOODS	- OBSERVE UPSETS IN DYNAMIC RAMS
1974:	BINDER,SMITH,HOLMAN	- OBSERVE UPSETS IN RAMS IN SPACE
1983:		- MANY SATELLITES ON-ORBIT SUFFER COSMIC RAY UPSETS
1986:		- COMBINED RADIATION AND RELEASE SATELLITE (CRRES) TO BE LAUNCHED

Figure 2. Cosmic Ray Chronology.

The work of Bothe and Kohlhorster, and Rossi et al, proved that cosmic rays were particles with fantastic penetrating capability due to energies of millions to billions of electron volts. Measurements of this magnitude eliminated from consideration many natural processes and sources since even nuclear fission and fusion reactions failed to produce particles with these energies. Researchers began seeking for an understanding of the origin of cosmic rays and the mechanism by which these high energies are generated. Between 1930-1950 these investigations led to the discovery of sub-nuclear physics and the identification of the neutron, neutrion, Pi-Meson, muon, etc. Several Nobel prizes were awarded for this theoretical and experimental effort, but, although the physics of the particles became clear, the question of cosmic ray origin remained unanswered.

Cosmic ray measuring equipment was aboard the earliest orbiting satellites, Sputnik I and II and Explorer I and II. These vehicles discovered the intense radiation belts around the Earth, which were named for the leading American experimenter, Van Allen, who had earlier predicted their existence and described the belts and the strange satellite data obtained from them.

The elliptical orbits of the satellites through the belts resulted in varying and interesting geiger-counter readings, i.e., the maxima of signals did not occur with maximum altitude. This discovery was the beginning of the effort to understand the radiation effects on satellite electronics. While early instruments aboard Sputnik II and Explorer I experienced intermittent operation due to the extreme variations in particle counts, later devices successfully measured particle densities versus altitude and, thus, the belts were charted.

Since the launch of the first Sputnik in 1957, predictions were made by Wallmark and Marcus, 1962 (Reference 6) that single event upsets exist in spaceborne electronics. In the early 1970s, upsets were observed in laboratory conditions; upsets in satellite random access memories were observed in 1975 (Reference 7). Since then, there have been an increasing number of cosmic-ray-induced problems in operational satellites. As the number of on-board computers and their memories increase, there is every expectation that the frequency of reported upsets and their system impact will also increase. This growing concern is exemplified by an increase in both the calculations and measurements of cosmic ray effects. A major cosmic ray experiment will be flown in 1986 on the Combined Radiation and Release Satellite (CRRES). This satellite will carry a micro-electronic measurements package which will obtain cosmic ray upset data on many types of parts (Reference 8).

The Mechanism for Cosmic-Ray-Induced Single Event Upset

The accepted model for the cosmic ray is a particle (an ion) with exceptionally high energy of the order of hundreds to thousands of MEV. When a high energy particle passes through matter, the particle loses energy by ionization which creates electron-hole pairs in the material. As shown in Figure 3, the deposition of 3.6 electron volts in silicon creates a single pair; increasing numbers of pairs result from greater energy deposition. If this charge is deposited near particular regions in a semi-conductor device, an accumulation of charges occurs sufficient to flip a bit. The actual charge deposition and circuit interaction is well characterized for most circuit types, such as bi-polar, CMOS, CMOS/SOS, NMOS, etc. (See Reference 9, Pickel and Blandford, for a complete explanation.) For the purpose of this report, it is enough to know that the process occurs in nearly every type of semi-conductor circuit.

The charge required to flip a bit is the critical charge (Qc). The magnitude of this charge is a function of circuit parameters, particularly the size of the sensitive region. The critical charge parameter is determined by experimental measurements of upset rates and their comparison with data collected at large particle accelerators where active devices are tested. If a group of devices is evaluated under similar test conditions, the critical charge may be viewed as a screening parameter to compare the upset sensitivities of various parts. The crux of the dilemma for modern electronics is that the critical charge, which is the threshold for unwanted upsets,

- A PARTICLE PASSING THROUGH MATERIAL LOSES ENERGY BY IONIZING THE MATERIAL AND CREATING PAIRS OF ELECTRONS AND HOLES

  IN SILICON: 3.6 eV PRODUCES A SINGLE e^-, e^+ PAIR
  3.6 MeV = 10^6 ELECTRONS = 0.16 PICO COULOMBS (PC)

- THIS CHARGE NEAR A DEPLETION REGION IN A DEVICE RESULTS IN CHARGE COLLECTION
- THE CHARGE REQUIRED TO FLIP A BIT IS THE <u>CRITICAL CHARGE</u> (QC)
- THE CRITICAL CHARGE ASSOCIATED WITH MODERN HIGH-SPEED CIRCUITS IS EQUAL TO 0.01 TO 1.0 PICO COULOMBS

Figure 3. The Mechanism for Cosmic-Ray-Induced Single Event Upset

decreases rapidly as the speed of devices increases and feature size decreases. Current devices exhibit critical charge values of 0.01 to 1.0 pico-coulombs. Figure 4 shows the amount of charge deposited in a 1 micron slice of silicon as a function of particle energy. Several different ions are shown in order to illustrate the different charge depositions expected from the most common cosmic ray particles. It is clear that cosmic iron ions deposit a great deal of charge per micron, and that the charge deposited in the sensitive region of a device (which is several microns thick) is well above the critical charge values which have been measured for many semi-conductor parts.

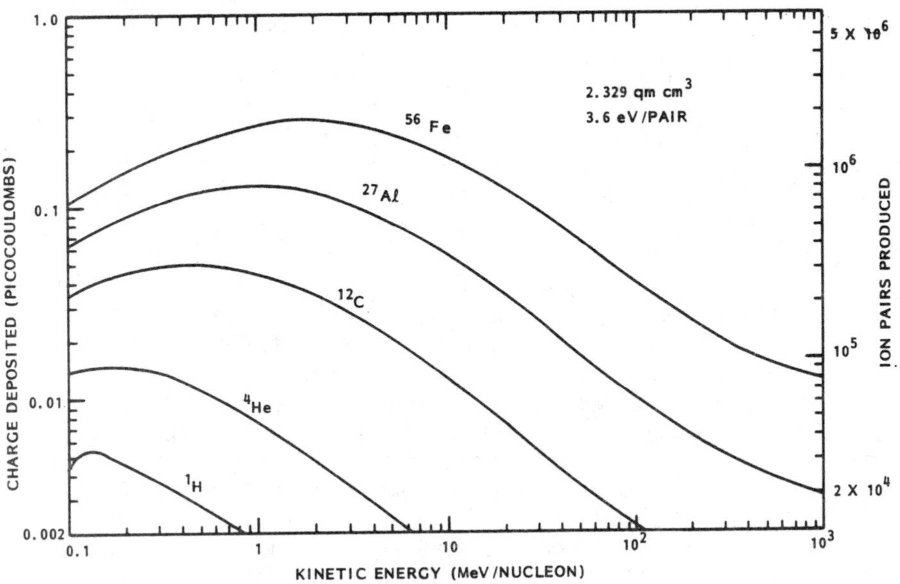

Figure 4. Charge Deposited in 1 Micron of Silicon

COSMIC-RAY-INDUCED SEU IN TODAY'S SPACE PROGRAMS

Considering current operational space systems which are having problems, the severity of cosmic-ray-induced problems become clear. In all cases, cosmic rays are thought to be the cause of the observed upsets. Each of the problems have been studied by the design groups and operations teams who discovered them and who, eventually, worked out ways to deal with them. The following classic characteristics are shared by each problem and help to identify their causes:

- Upsets are transient
- Upsets are removed by re-writing the particular cells
- Upsets occur frequently without correlation to solar activity
- Random bits are upset
- Problems occur in chips which are known to be susceptible (20-20 Hindsight)

The rates being observed -- one upset every few days -- are troublesome and cannot be ignored or passed off as one-time anomalies.

Figure 5 presents upset information on the GPS Receiver on LANDSAT D, Satellite A, and the Tracking and Data Relay Satellite (TDRS).

The Receiver/Processor Assembly (R/PA) carried as an experiment on Landsat D and D-prime is an electronic processor subsystem which receives the transmissions from the Global Positioning Satellite (GPS) and determines host satellite postion; it is not part of the satellite attitude and control subsystem. The package operates independently of the ACS subsystem and its output, i.e., the satellite location, is inserted into the downlink telemetry for evaluation on the ground. The R/PA includes a processor and a memory equal to 56,000 16-bit words. The memory is regularly checked by the processor using a check sum routine. If an error is detected, a telemetry flag is set and the subsystem processor ceases operation and goes into a dormant (safe) mode. Upon ground command, the entire memory contents can be downlinked for analysis. After studying many upsets, it was established that the bit flips were randomly distributed through the memory and that the memory could be rewritten following an upset. In one case a single permanent (hard) failure involving a single bit was observed. It could have been caused by a random part failure or a cosmic-ray-related hard failure. The frequency of upsets and the effort involved in re-starting it finally caused ground controllers to de-activate the R/PA experiment.

- LANDSAT 4/LANDSAT D PRIME
  - GPS RECEIVER - RECEIVER PROCESSOR ASSEMBLY (R/PA)
  - 56K X 16 BITS HR-6508 RAM CHIPS
  - 1 UPSET PER 4-5 DAYS
  - RECOVERY VIA SAFE MODE AND GROUND RE-BOOT
- SATELLITE A
  - MEMORY IN COMMAND SUBSYSTEM
  - REGULAR UPSETS CAUSE INCORRECT COMPONENT SWITCHING
  - WORK AROUND = REGULAR REGISTER RE-WRITES/RESETS SEVERAL PER DAY INSTEAD OF 1 PER 5 DAYS
- TDRS
  - ATTITUDE CONTROL COMPUTER MEMORY 93L422 RAM CHIPS
  - 1 UPSET PER 1 1/2 DAYS
  - NO PROTECTION AGAINST A "BAD" COMMAND

Figure 5. Current Events.

The upsets observed in Satellite A reflect cosmic ray problems in a memory implemented with HR 6508 chips. Here a processor-controlled command sequencer uses the memory for the storage of future commands. A serious system problem exists since inadvertent power switching could lead to electrical power or thermal problems that could endanger the entire satellite. A work-around scheme has been devised to regularly reload critical registers to reduce the probability of incorrect register data. This scheme is working, but it forces the ground system operations to send many command loads every day rather than the one load every few days that was the design goal.

The attitude control computer problems on the TDRS are potentially very damaging, since upsets in this subsystem can quite easily become major problems due to critical parameter storage in susceptible RAM devices. Since this satellite is in a geosynchronous orbit, electrostatic discharge (ESD) was initially suspected. When upsets did happen, memory contents were downlinked for analysis and ESD was eliminated as the cause. Investigators determined that "things were going bump in the night" and that these upsets were not related to solar activity or to local satellite time. Once the source of the difficulties was identified, ground controllers were able to prepare reload/re-boost procedures for contingency use. After this, many upsets were observed with no serious consequence. At least one potentially serious problem did occur on the TDRS when a bias momentum wheel reference speed register was upset. A reference value, stored in RAM, is used by the wheel drive electronics to maintain a constant wheel speed. Upon upset, the wheel began to slow down, a situation which could have led to loss of the satellite. Fortunately, the problem was detected, corrective action was taken from the ground station, the reference was reset and the system resumed normal function. Inadvertant thruster firings, loss of earth presence, and processor interrupts are other potential problems which could happen to the TDRS (Reference 11). Figure 6 illustrates three instances of dealing with cosmic-ray-upset problems following either real or projected occurrences.

- GALILEO VEHICLE AND PROBES
  - COMPUTER SUSCEPTIBLE TO GALACTIC COSMIC RAYS AND HIGH ENERGY IONS FROM JOVIAN MOONS
  - COMPUTER REDESIGN AND PARTS CHANGE WILL PRECLUDE PROBLEMS
- SPACEBORNE FLIGHT COMPUTER
  - UPSET IN RAM CHIP CAUSES SWITCH OVER TO REDUNDANT COMPUTER
  - "FIX" - DUAL RAMS AND "OR" LOGIC IMPLEMENTED ON SUBSEQUENT UNITS
- SOLAR MAX MISSION
  - CENTRAL UNIT IN COMMAND SUBSYSTEM USED 93L422 CHIPS
  - SOFT FAILURES IN FLIGHT UNITS CAUSED PART CHANGE IN CU USED ON LANDSAT 4 AND LANDSAT D'

Figure 6. Current Events (Continued)

The Galileo vehicle, which has not been launched at the time of this report, will avoid cosmic ray problems based on recent data received from inter-planetary probes. Specifically, the environment around Jupiter, which is interesting because of the trapped particles and high energy ions emitted from eruptions on the moon Io, was measured by another probe; this environmental data was used to make predictions of upsets for both of the two processors on the Galileo probe. Upsets can be expected during the final and most critical phase of the flight, and the single-shot nature of the mission made a computer outage an unacceptable risk. This is the reason for the recent announcement that the on-board processor would be re-designed to eliminate the chance of a single event upset (Reference 11). It has been established, however, that, during the transit portion of the mission, the vehicle could experience upsets due to galactic cosmic rays, but recovery through on-board or ground-assisted means would be sufficeint to maintain the mission.

The upsets in the Spaceborne Flight Computer and the Solar Max Central Unit, illustrate the problem in two more semi-conductor part types. Since both the computer and the CU, which is used in the Command and Data Handling module, are built as a number of units, hardware fixes could be used to prevent further occurences of upset-related problems.

Single Event Upsets In Modern Semi-Conductors

The rate of SEU occurrences in electronic parts is determined by the cosmic ray environment and the device's parameters. Both factors can be controlled or modified within limits. But, in fact, when considering real parts and real spacecraft missions, many of the variables are really not controllable by the subsystem designer (see Figure 7).

- DEVICE PARAMETERS - CRITICAL CHARGE, BIT DENSITY, FEATURE SIZE
- COSMIC RAY/PROTON ENVIRONMENT:
  - SOLAR ACTIVITY (11 YEAR CYCLE)
  - SOLAR FLARES THAT ARE ION RICH
  - SATELLITE ALTITUDE VS. RADIATION BELTS
  - SATELLITE ALTITUDE VS. EARTH'S MAG FIELD
  - PASSAGE THROUGH SOUTH ATLANTIC ANAMOLY
- MODIFICATION OF ENVIRONMENT THROUGH SHIELDING CANNOT BE DONE EFFICIENTLY.

CONCLUSION: LEARN TO LIVE WITH SINGLE EVENT UPSET

Figure 7. SEU Rate Determining Factors

The important device parameters influncing upset rate are the critical charge, the device feature size, and the density of bit or cell packing. The critical charge parameter, discussed previously, is the key parameter used to compare the susceptibility of various part types to an upset caused by cosmic rays or any other type of disturbing energy. The magnitude of the critical charge varies greatly with different part classes, e.g., CMOS, CMOS/SOS, Bi-Polar. Some progress toward increasing the critical charge for a certain part during manufacturing and through circuit design techniques has been made. This is an important step in SEU protection because increasing the critical charge decreases the susceptibility of the part to upset.

Unfortunately, the critical charge depends on the circuit feature size with a dependence of $1/(\text{Feature}^2)$. Many desirable circuit features such as speed, bits per chip, and low power consumption are achieved with ultramodern small feature sizes. Three micron circuits are available today, while submicron sizes are coming soon. These advanced circuits, VHSIC in particular, will undoubtedly experience serious problems with SEU.

Finally, bit density is a factor since the chances of an upset depend on the probability of a chip being "hit" by a cosmic ray. In short, as the need for more on-board memory increases, the density of bits per chip -- which are vulnerable to cosmic rays -- increases dramatically. To fully understand this bit density problem, consider that a spaceborne computer designed with mid-seventies technology used memory chips with 10 micron feature size and provided only a few hundred words of memory. Based on calculations and flight data, an upset rate of less than one per year was forecast. On the other hand, the TDRS computer, with modern chips (93L422) and 50 kilobits of memory, is suffering upsets at the rate of one every day and a half. Unprotected designs built with even more advanced technology can expect even higher upset rates.

The Effect Of Shielding On Galactic Cosmic Rays

A first survivability concept to consider would be to eliminate the problem by shielding the sensitive components from the incoming cosmic rays. Aluminum or Tantalum shielding is often used effectively to protect against the damage expected from electrons trapped by the earth's magnetic field. In these cases, particle energies are in the range of several kilo-electron volts and 0.25-0.50 cm (0.1-0.2 inches) provide a significant reduction to the flux incident upon electronics. Unfortunately, cosmic rays have energies of hundreds to thousands of kilo-electron volts and the effectiveness of absorptive shielding decreases with incident particle energy.

Figure 8 depicts the cosmic ray flux as a function of energy deposited in silicon. The deposition curve is plotted for several thicknesses of shielding. Notice that shielding of 20 g/cm^2 reduces the flux by less than a factor of 10 compared to the standard satellite shielding of 2 g/cm^2 provided by a typical skin plus electronics box. Clearly, from Figure 8, a factor of 10 decreases in occurrence is not a great victory since an upset every 15 days is not much better than one every 1.5 days in the life of a 10-year system. Also, the weight and volume of 20 g/cm^2 shielding, 3.0 inches of aluminum, is not a very reasonable solution.

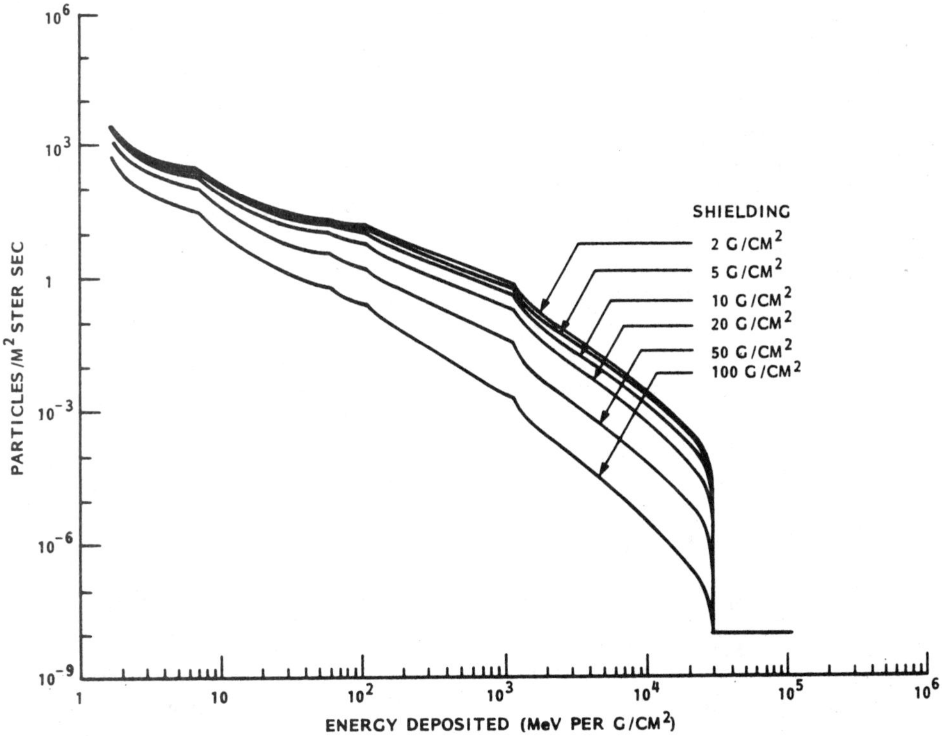

Figure 8. Cosmic Ray Flux as a Function of Shielding.

Upset Rates for Electronic Parts

Since shielding cannot make the problem go away, the next concept is to understand the sensitivity of various parts and to strive to make a design using upset proof parts. The desire for upset immunity must be balanced against other requirements, however; and the demands for speed, low power consumption, and small size limit the choice of acceptable parts. The growing concern for this problem has stimulated several organizations to begin to gather parts data for the processors and the memory devices that are used in satellite electronics.

The susceptibility of parts to cosmic-ray-induced upset can be measured by direct testing. The usual source for generating cosmic rays for testing is a high energy accelerator such as the cyclotron at the University of California, Berkeley. Such a machine can produce rare gas ions, principally Krypton, that have the appropriate energy levels, 150 Mev. The device under test is set up in an active mode and illuminated with ions. The number and rate of upsets is determined by reading and writing the bi-stable cells. This test data allows the characterization of parts and a determination of parameters such as the critical charge, Qc. If the critical charge parameter in the same reference environment is derived for each part type, the critical charge may be used to compare the upset sensitivity of various parts.

High energy particle testing is also used to screen parts for latch-up. Again, devices under test must be operating. Following their exposure the device's parts are evaluated. Latch-up has occurred when a part is stuck in one state and does not respond to write operation. A latched part is usually restored to normal function by powering down. Upon re-applying power, the device resumes normal operation. Although the latch-up phenomena is similar in appearance to x-ray induced latch-up, thresholds and susceptibilities are not always identical. Parts should be screened using cosmic rays of the correct energy to determine if the latch-up phenomena occurs.

The result of the testing is the expected upset rate as a function of energy. A plot of this rate for several different device volumes is shown in Figure 9 (Reference 1). Each particular part would have a separate data line if actual part data was plotted.

The upset rates shown here are the results of a calculation which depends on measured data for parts and a standard environment. The environment is specified as the 10% environment (Reference 2) meaning that the actual environment is worse than this only 10% of the time. The regions labeled CMOS-BULK and CMOS-EPI reflect the test data. This chart shows the strong dependence of upset rate on both feature size, stated as the collection volume, and the critical charge. Not shown here is data for bi-polar devices which have an upset rate greater than CMOS-BULK, and CMOS-SOS, which has an upset rate lower than even the CMOS-EPI. The mechanisms in these circuits and the differences in sensitivity are clearly explained by Pickel in Reference 9.

The most realistic data is that gathered in space by actual operating components. Clearly, a great deal is being learned from the problems that are occurring. A carefully controlled measurements program will be conducted on the CRRES which will be launched in 1986. Its Microelectronics Package will contain many parts which will be operating with regular read-write cycles. A very-high-speed telemetry system will collect data for downlinking and analysis. This experiment will greatly increase the upset rate data base.

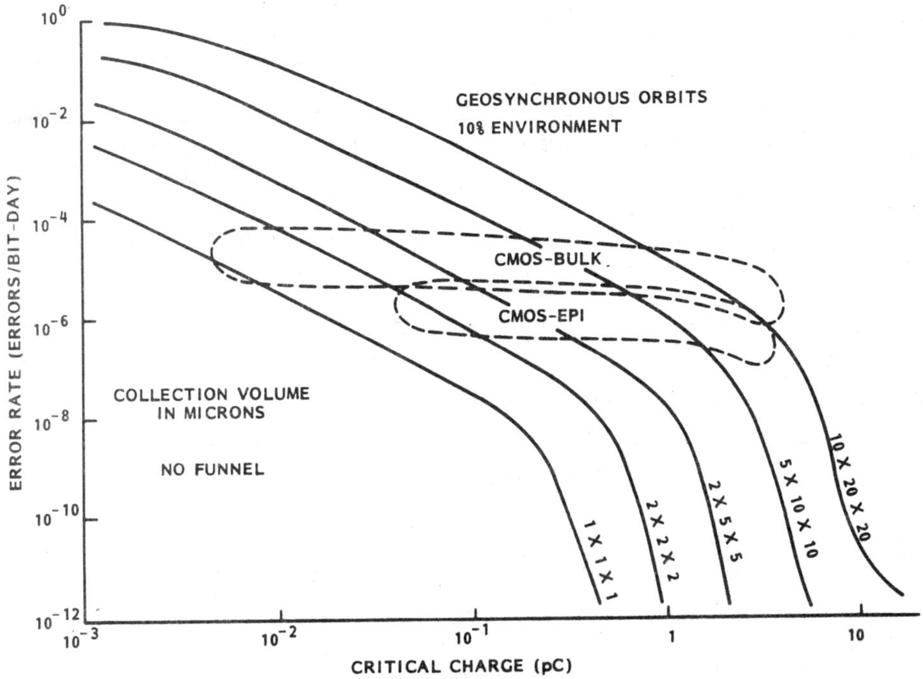

Figure 9. Cosmic Ray Upset Rates.

Note that, although the indicated rates are low for a single bit, about 10^{-6} errors/bit-day, the expected error rate for a memory will be much greater. Modern spacecraft computers are using total memory of hundreds of kilobits while newer designs have several megabits of sensitive memory. For example, the Galileo Orbiter has 2.5×10^6 bits (Reference 13). Thus upset rates of the order of one per day are to be expected with some devices. This number will continue to grow as the amount of on-board memory increases. When one considers advanced designs for autonomous satellites with memories of 10^6-10^8 bits, the upset rates quickly become unbearable. It should be very clear by now that methods to deal with single event upset must be developed and implemented on modern spaceborne computers.

METHODS TO DEAL WITH SINGLE EVENT UPSET

The four methods shown in Figure 10 represent techniques that are in use or planned for use on satellite electronics. The imposition of a cosmic ray SEU specification forces computer designers to evaluate these methods, to trade-off the penalties, and to develop a protection scheme while still meeting the system performance requirements.

Clearly avoiding the problem through parts selection looks attractive. In fact, many older technologies such as plated wire, or core or solid state memory with large feature size elements (>10 microns) can be used to achieve a small, or zero, upset rate. The problem here is usually operating speed and power consumption, with speed after being the most formidable requirement. The parts may not meet design or system performance goals. Various investigators have designed and built upset-resistant parts while maintaining circuit performance. Andrews et al (Reference 10) built 2 Kbit CMOS RAM chips based on the Harris HS 6508RH chip. These parts have been tested at 140 Mev Krypton; they demonstrated significantly lower upset rates than the standard part.

- AVOID THE PROBLEM VIA PARTS SELECTION

  PENALTY - PARTS MAY NOT MEET DESIGN REQUIREMENTS OR SYSTEM PERFORMANCE GOALS

- TRIPLE REDUNDANT SYSTEMS (CPU AND MEMORY) WITH MAJORITY VOTING

  PENALTY - WEIGHT, POWER, REAL ESTATE
VOTING EQUIPMENT MUST BE PROTECTED

- ERROR DETECTION AND CORRECTION TO INITIATE SYSTEM INTERRUPT OR EVEN AUTONOMOUS CORRECTION

  PENALTY - WEIGHT, POWER, REAL ESTATE
I.E., 6 BITS TO PROTECT 16 BITS
PROTECTION OF PROCESSOR STILL NEEDED

- REGULAR RESET/RELOAD VIA GROUND COMMAND

  PENALTY - DEPENDS ON RATE AND SEVERITY

Figure 10. Methods for Dealing with Single Event Upset.

Triple memory redundancy can beat the speed problem as well as provide a high degree of fault tolerance from other causes, such as parts failures, etc. The bad news is, additional weight and power is required for the additional memory. Protection must also be provided for the common "voting" electonics.

Error detection and correction (EDAC) offers the advantage of a single memory, but with additional bits stored to support the Hamming code or whatever scheme is used to support the EDAC. Also, the processor and software must perform EDAC on some sort of regular basis to insure a correct memory. As upset rates increase, one could quickly reach the point where most of the time is spent doing EDAC rather than any productive work. One current design has a predicted upset rate of slightly less than one per day. In this case, since EDAC is planned for several times per minute, the chances favor quickly detecting and correcting the upset.

The final method to deal with SEU, regular reset, would be used on a system in orbit. Regular commands may be the only alternative and the only way to protect against the occurrence of an event that could place the total system in jeopardy. Computer systems which can be reconfigured or reprogrammed from the ground might allow other alternatives. This method is usually unsatisfactory, unless an occasional reset is acceptable within the system performance requirements.

SPACE SYSTEMS WITH COSMIC RAY PROTECTION

Several systems have incorporated cosmic ray protection in their design and operation. Protection has been accomplished using some or all of the general methods which were discussed previously. Details of these protected systems are presented in Figure 11.

DSCS III, in orbit for about 18 months, has achieved a great deal of single event protection through the use of parts selection and command subsystem design. The main computer memory is implemented with RCA CD-4061 chips which exhibit a very low susceptability to SEU. These chips have 10-micron feature size and exhibit a critical charge value which is greater than the amount of energy deposited by a typical cosmic ray. Upsets in the processor could result in a logical or arithmetic error, however, the design of the system provides protection against system damage as a result of an upset. A logical error would trigger a computer error which would, in turn, initiate a reset of the processor that would have no effect on the computer operation. An arithmetic error could lead to an erroneous or unintentional command. The system is protected from this, however, since all critical commands require an enable/execute sequence which could not result from a single event. Furthermore, the thruster subsystem's design protects it by allowing only a maximum of 1-second thruster firing with a single command. Thus, a disastrous thruster firing would require a string of commands instead of a single bad command.

Error Detection and Correction (EDAC) is an attractive scheme since the memory and processor run in the normal way. During processor operations, the EDAC program and the additional memory bits are transparent to the system. EDAC is invoked periodically, with the optimum design running EDAC during an idle period of the processor.

- DSCS III (LAUNCHED 1982)
  - MEMORY USES CD-4061 CHIPS - NOT SUSCEPTIBLE
  - PROCESSOR ERRORS HANDLED BY SOFTWARE/HARDWARE DESIGN
  - NUCLEAR SURVIVABILITY DESIGN ADDS SOME PROTECTION
- DSCS III (FUTURE)
  - EXPECT PROBLEMS IN NEW MEMORY, WILL ADD EDAC
- FLIGHT COMPUTER (1983)
  - ERROR DETECTION CAUSES SWITCH TO REDUNDANT COMPUTER
- RECEIVER PROCESSOR ASSEMBLY (1984)
  - NEW SOFTWARE TO PROTECT PART OF MEMORY VIA EDAC
  - CPU IS NOT PROTECTED BY THIS SCHEME
- TDRSS AND SYSTEM A (1983)
  - RESET/RELOAD VIA GROUND COMMAND - REGULARLY

Figure 11. Systems with Cosmic Ray SEU Protection.

The trade here is the projected upset rate versus the frequency of running EDAC. If error rates are equal to one per day, many EDAC cycles per day will insure an error free processor/memory.

Redundant computers offer protection but at a high cost in terms of money, power, and weight. Also, their voting and redundancy logic requires additional processor support, especially if the system is a fully autonomous one. At present levels of cost and processor complexity, EDAC appears to be a better choice than redundancy.

The reset/reload scheme is illustrated by the TDRS and System A. This scheme is surely not the method of choice in either case; but it is the only available choice when the problem is discovered after launch. A paper presented at this meeting by H. Nakano (Reference 11) discusses the TDRS operation fully. This mode of operation can be successful, but it requires constant attention on the part of the ground control system, plus good luck.

PROPOSED SPECIFICATION FOR COSMIC RAY PROTECTION

The examples described above have established two things: cosmic-ray-induced SEU is a serious problem; viable solutions exist. The key ingredient is a clear specification which could be used to drive the design of processor electronics beginning with the earliest requirements. If the cosmic ray environment and the requirement for protection is a part of the original equipment specification, the hardening and protective schemes can be accommodated with the minimum perturbation to the program and at minimum cost. The most successful protection schemes begin at the concept definition stage of development. Add-on "Band-Aids" solutions are the least desirable way to try for protection.

The key points of a proposed specification are presented in Figure 12. The first item is a description of the environment. This description provides the design goal for the system design as well as the test levels needed for both development and qualification testing. The environments shown in Figures 13a through 13d are the most significant contributors to the SEU problem (Reference 2).

The next two points in the specification suggest specific prohibitions on phenomena that could easily lead to serious system damage. Both points may already exist for other reasons, such as good engineering, but these points should be emphasized in a specification. The final point allows for the occurrence of some upsets but with a frequency that can be tolerated by the entire system and its mission requirements. Here, the outage rate and the time allowed for recovery must be carefully considered so that some design flexibility exists while meeting mission performance requirements. This item can easily become the major cost driver to the protected design, especially if the recovery is required rapidly and/or autonomously without benefit of ground command intervention. This is the trickiest part of the specification process.

- DESCRIBE COSMIC RAY ENVIRONMENT (SEE FIGURES 13a THRU 13d)
- LATCH-UP SHALL BE PREVENTED EITHER BY SCREENING OUT SUSCEPTIBLE PARTS OR BY SPECIFIC DESIGNS WHICH INCORPORATE PROTECTIVE MEASURES
- SUBSYSTEM OR HOST SYSTEM DAMAGE DUE TO AN UPSET-GENERATED COMMAND OR CONDITION SHALL NOT BE ALLOWED
- SUBSYSTEM OUTAGE DUE TO SINGLE EVENT UPSET SHALL BE MINIMIZED AND THE TIME ALLOWED FOR RECOVERY WILL BE SPECIFIED

Figure 12. Key Points for Proposed Specification.

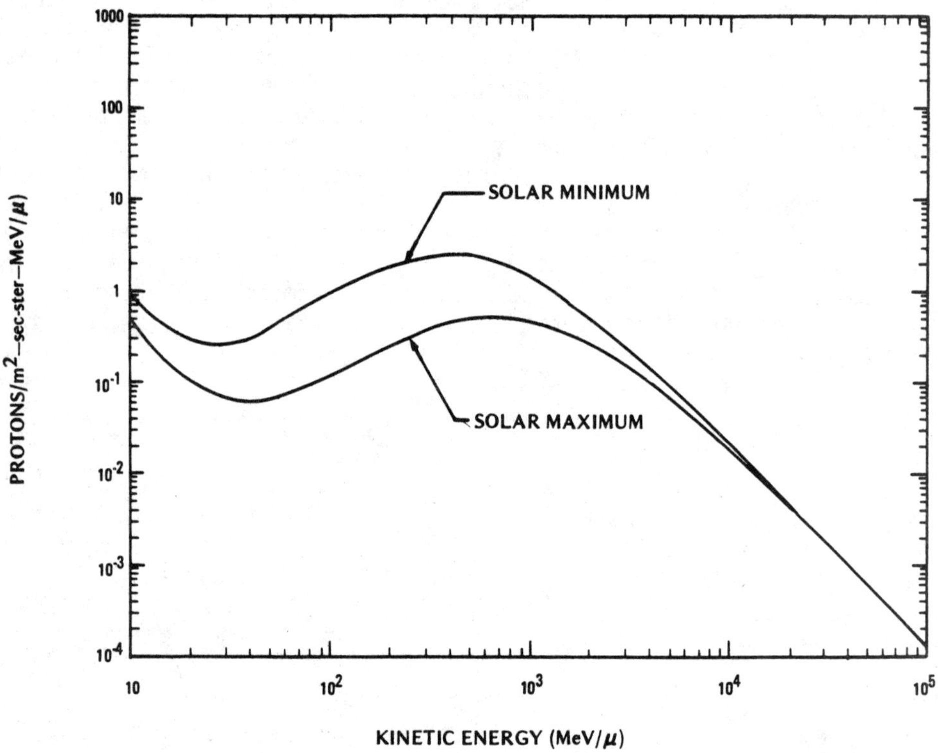

Figure 13a. Comic Ray Environment-Hydrogen

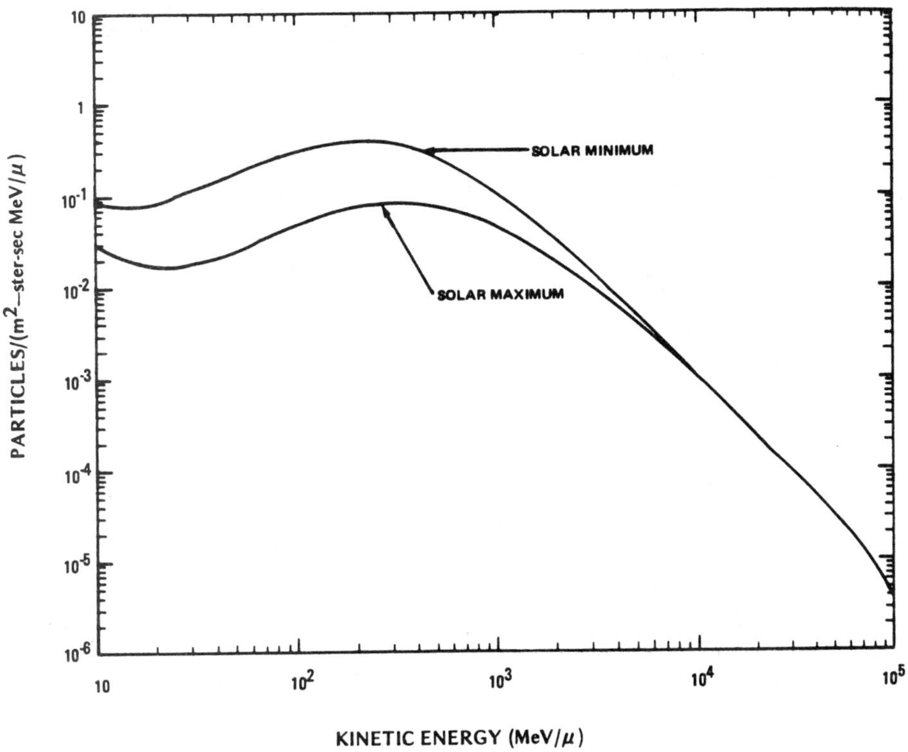

Figure 13b. Cosmic Ray Environment-Helium

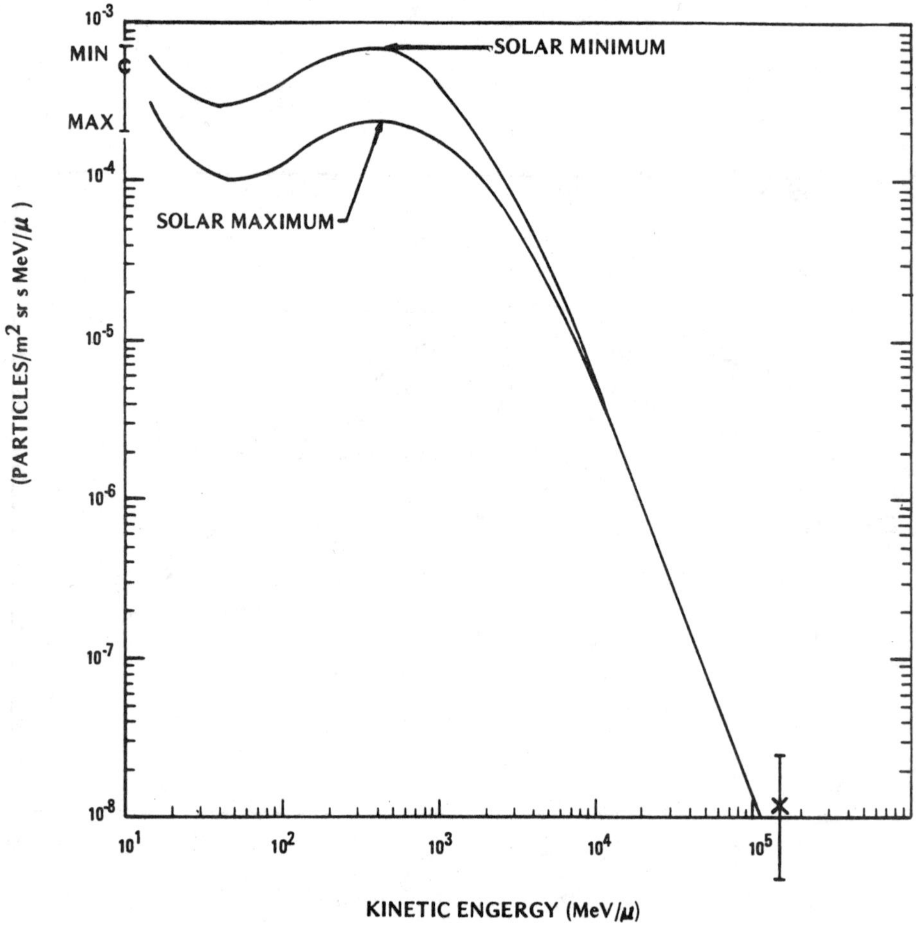

Figure 13c. Cosmic Ray Environment - Iron

CHARGE GROUP	ABUNDANCE/ALPHA PARTICLES	REPRESENTATIVE ELEMENT	FLUX (PARTICLES/CM^2-SEC)
Z = 1	8/1	PROTONS (Z=1)	4
Z = 2	1/1	ALPHA PARTICLES (Z=2)	0.5
3 Z 5	1/48	-----	-----
6 Z 9	1/16	OXYGEN (Z=8)	3 X 10^{-2}
10≤Z≤14	1/75	SILICON (Z=14)	7 X 10^{-3}
15≤Z≤19	1/600	-----	-----
20≤Z≤30	1/200	IRON (Z=26)	3 X 10^{-3}
Z>30	1/(8 X 10^5)	-----	-----

Figure 13d. Relative Abundance of Heavier Nuclei in the Cosmic Ray Environment

CONCLUSIONS

The existence of cosmic rays as a natural environment has been established by the work of many investigators since 1912. The susceptibility of modern spaceborne electronic to cosmic-ray-induced single event upsets has been demonstrated on several space systems. The potential for serious problems exists in critical systems, such as TDRSS and Galileo, on future systems.

Several systems have been flown with workable upset protection schemes. The use of non-susceptible parts or protective and error-tolerant techniques, have provided SEU protection without degrading the performance requirements of each system.

The proposed specification suggests a system level approach that requires performance when operating in the cosmic ray environment and prevents permanent damage to the host system while an acceptable rate and duration of transient upset is allowed. The equipment supplier is free to choose among various design options and penalties in order to provide the required level of performance. Verification of the protected system is accomplished through a combination of gathering test data on parts and analyzing that data to predict upset rates for a complete processor/memory system. Further analysis and hardware tests are used to demonstrate recovery schemes and the ability of the system to prevent damaging responses to the transient situations caused by the cosmic ray upsets.

- GALACTIC COSMIC RAYS CAN CAUSE UPSETS AND THUS DAMAGE CRITICAL MISSION FUNCTIONS OF ORBITING SYSTEMS
- SEU CAN BE CONTROLLED THROUGH PARTS SELECTION OR THE IMPLEMENTATION OF PROTECTIVE MEASURES
- THE PROPOSED SPECIFICATION ALLOWS EQUIPMENT SUPPLIERS DESIGN LATITUDE WHILE STILL MEETING THEIR SYSTEM PERFORMANCE REQUIREMENTS

REFERENCES

1. E. L. Peterson "Single Event Upsets in Space: Basic Concepts," IEEE 1983 Radiation Effects Conference, July 1983.

2. J. H. Adams, et al "Cosmic Ray Effects on Microelectronics Part 1,": The Near-earth Particle Environment NRL Report No. 4506 August 25, 1981.

3. Solar Cell Radiation Handbook, Jet Propulsion Laboratory, publication 77-56, November 1977.

4. R. E. Townsend, et al "Source Book of the Solar Geophysical Environment," USAF Air Weather Service AFGWC-TN-82-002, August 1982.

5. B. Rossi, "Cosmic Rays," McGraw-Hill, New York 1964

6. J. T. Wallmark and S. M. Marcus "Minimum Size and Maximum Packing Density of Non-Redundant Semiconductior Devices," Proceedings of the IRE, 286, (1962).

7. D. Binder, E. C. Smith, A.B. Holman, Satellite Anomalies from Galactic Cosmic Rays, IEEE Trans. Nuclear Sciences Vol. NS-22, No. 6, December 1975.

8. Aviation Week and Space Technology, March 14, 1983.

9. J. C. Pickel "Single Event Upset Mechanisms and Predictions" Tutorial Short Course, IEEE, 1983 Nuclear and Space Radiation Effects Conference, July 1983.

10. J. L. Andrews, J. E. Schroeder, B. L. Gingerich, W. A. Kolasinski, R. Koga, S. E. Diehl, Single Event Error Inmune CMOS RAM, IEEE Trans. on Nuclear Sciences Vol. NS-29, No. 6, December 1982.

11. H. Nakano, TRW Inc., 1984 AAS Rocky Mountain Conference, February 1984 (to be published)

12. Aviation Week and Space Technology, October 31, 1983.

13. C. P. Jones, and M. R. Landano, The Galileo Spacecraft System Design, AIAA 21ST Aerospace Sciences Meeting, January 1983.

AAS 84-058

A PROTECTIVE METHOD AGAINST RAM UPSETS DUE TO COSMIC PARTICLES*

Hiroshi Nakano and Harry Schmeichel**

A variety of attitude control functions are performed on the TDRS satellite using microprocessor-based electronics. This paper describes the recent experience of RAM bit upsets by cosmic particles and shows how a ground procedure has been implemented to correct some upset conditions which might otherwise lead to divergence of earth pointing. All TDRS system requirements are met with this ground procedure. A modification approach is also suggested, should there be a mission requirement in the future for an autonomous operation.

INTRODUCTION

Since its recent launch, the Tracking and Data Relay Satellite (TDRS) has encountered numerous unaccounted intermittent bit state changes in the random access memory (RAM) of the control processor electronics (CPE), which performs most of attitude control computations and logic executions. Subsequent analyses attribute this phenomenon to cosmic particles. As the nature of RAM upsets became clear, a study was initiated to investigate the potential effects of RAM upsets on satellite operations and develop a corrective ground procedure. What follows are the results of these studies. Since the discussion focuses on RAM utilization by the control processor, the basic CPE architecture and control laws are described only to the extent relevant to the topic.

---

* *Work sponsored by Space Communications Company under Contract No. 76159.*

** *TRW Electronics and Defense Sector, One Space Park, Redondo Beach, California 90278*

RAM UTILIZATION ON TDRS

The CPE used on the TDRS is an 8-bit special purpose computer to perform the computations and logic operations required for attitude control as well as reaction wheel control. As shown in Fig. 1, it is structured around two AMD 2901 4-bit slice-type LSI microprocessor chips which provide the basic two's complement arithmetic computation capability. Three AMD 2909 microsequencer LSI chips and a 1024-byte x 32-bit read-only-memory (ROM) are used to implement the microprogrammed set of 108 instructions. The control program (list of macro-instructions) is stored in another ROM which is directly addressed by a 13-bit program counter. Though the CPE has the capability to handle 8192 bytes of ROM, 6144 bytes are wired in, and 5494 bytes are actually used for the program. The RAM is made up of 256 bytes by 4-bit 93L422 memory chips and is organized as four pages of 256 8-bit bytes each. The program uses a total of 527 byte locations distributed over the first three pages as shown in Figs. 2, 3 and 4. The last is a blank spare page. 244 bytes on the first page (Page 0) are used for a variety of functions such as stack registers and scratch pads for the micro-program, telemetry buffers and memory buffer. The memory buffer provides temporary locations for new RAM values sent from the ground before they are shifted to the desired RAM locations. The rest of the RAM on this page consists of firmware configuration flags and variables associated with sensor data processing. On Page 1, 4 bytes are constants and 141 bytes are allocated for control variables which are updated every computational cycle. All 137 bytes on Page 2 are constants used in various parts of the control laws. These are changed only by ground commands. This page is the only page where a checksum is performed.

The ACS operation is divided into Inertial, Earth, Normal and Sun Modes, and the RAM is accessed to varying degrees in different modes.

RAM HITS ON ORBIT

From the first day of orbit, single state changes of RAM bits have been observed on the TDRS, approximately at a rate of once a day. After reviewing a number of possible causes, such as surface electrostatic discharges, high energy electrons, inner belt protons, and high energy solar protons, cosmic particles (iron ions) are currently considered the most probable mechanism for these RAM bit upsets. The exact extent of the RAM upsets is somewhat difficult to assess, since this requires checking of the entire RAM on a regular basis. This is particularly true for Page 1 where most words are continually updated by the control program. On the other hand, any upset on Page 2 is more easily identified because all the words are constant and any change is signaled by a memory checksum failure in the telemetry. The blank page also provides an easy assessment since the bits should be all zeros here. During evaluation of the RAM upset phenomena these pages were monitored. However, the blank page is no longer checked on a regular basis. During

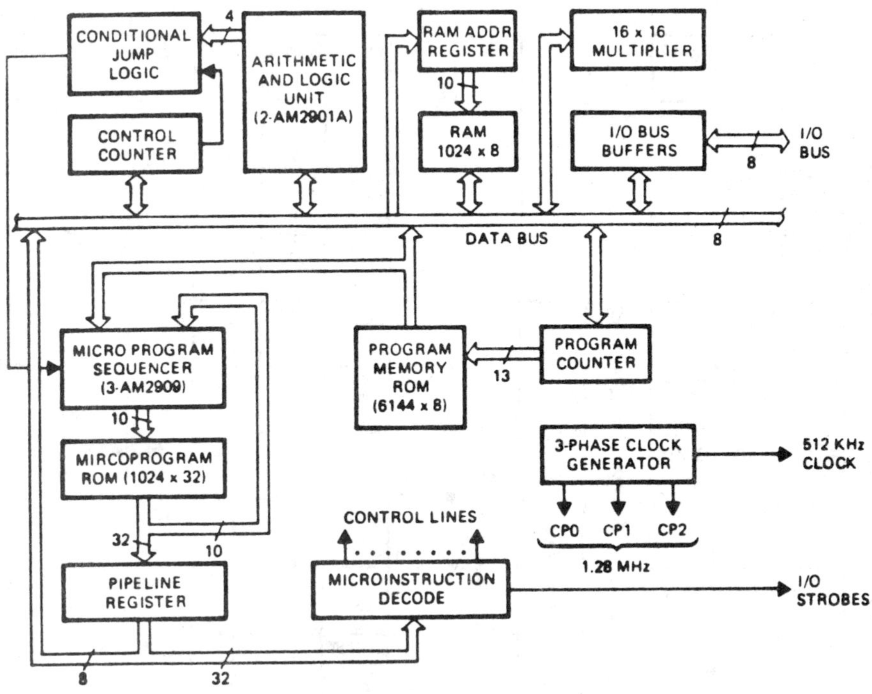

Fig. 1. CPE Block Diagram

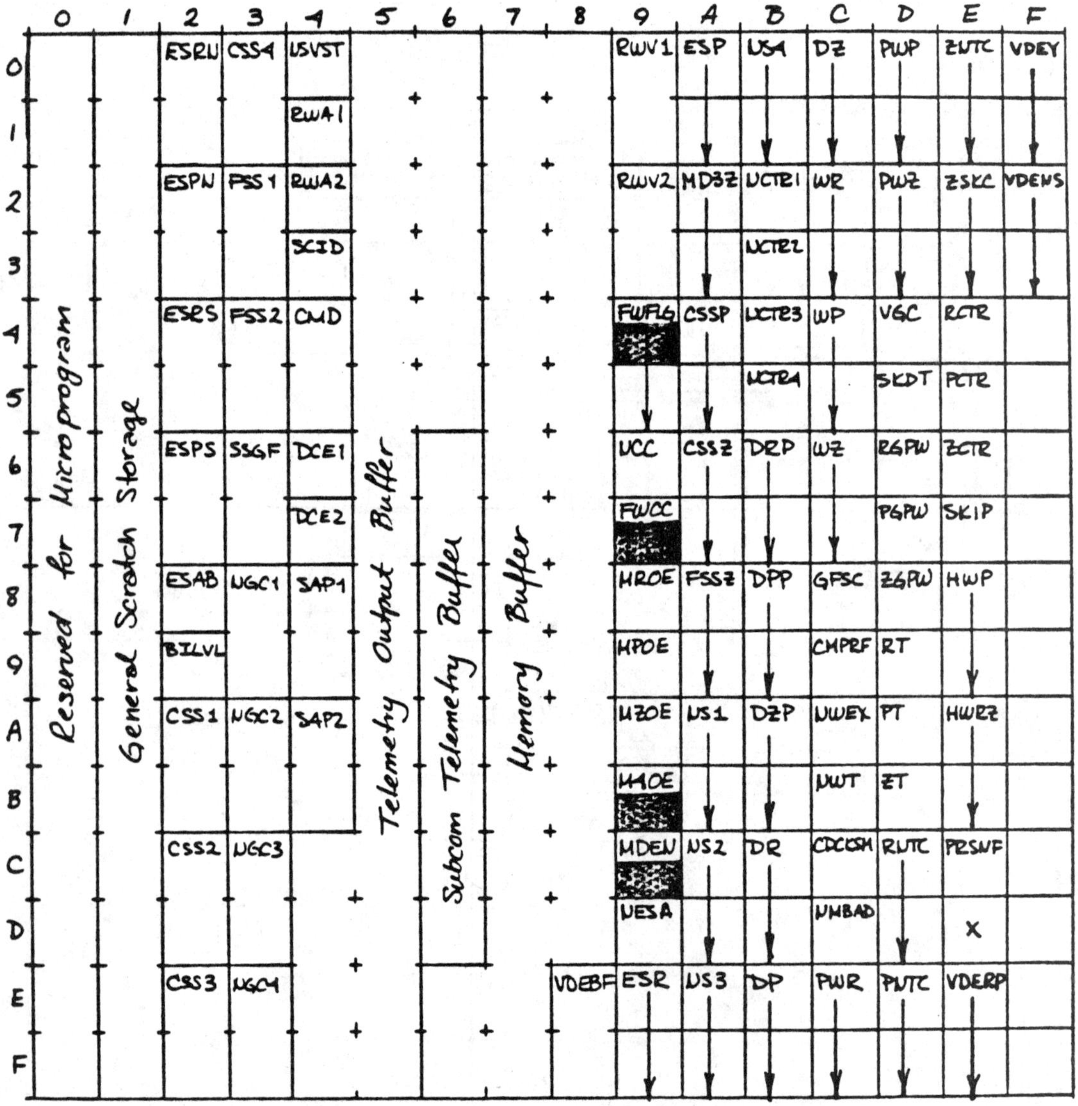

Fig. 2. RAM Assignments - Page 0

	0	1	2	3	4	5	6	7	8	9	A	B	C	D	E	F
0	E2R1	E2P4	M3PWR	M3-ZBS2	ESPP	M4ZA3	E5R1	E5P4	RW10A2	MVDEC						
1	↓	↓	↓	↓	↓	↓	↓	↓	↓	MRCDTS						
2	E2R2	E2P5	M3PWP	M3-RIS	M4FZ	M4ZA4	E5R2	E5P5	RW20A1							
3	↓	↓	↓	↓	↓	↓	↓	↓	↓							
4	E2R3	E2Z1	M3PWZ	M3-PIS	M4IZ	M4ZA5	E5R3	E5Z1	RW20A2							
5	↓	↓	↓	↓	↓	↓	↓	↓	↓							
6	E2R4	E2Z2	M3-RBS1	M3-ZIS	M4PF	M4ZA6	E5R4	E5Z2	RWI1							
7	↓	↓	↓	↓	↓	↓	↓	↓	↓							
8	E2R5	E2Z3	M3-RBS2	M3-RFS	M4IP	M4ZA7	E5R5	E5Z3	RWI2							
9	↓	↓	↓	↓	↓	↓	↓	↓	↓							
A	E2P1	E2Z4	M3-PBS1	M3-PFS	M4ZA0	M4ZA8	E5P1	E5Z4	MDCRA							
B	↓	↓	↓	↓	↓	↓	↓	↓	↓							
C	E2P2	E2Z5	M3-PBS2	M3-ZFS	M4ZA1	M4ZA9	E5P2	E5Z5	MDSTA							
D	↓	↓	↓	↓	↓	↓	↓	↓	↓							
E	E2P3	DVIS2	M3-ZBS1	ESRP	M4ZA2	M4ZJ	E5P3	RW10A1	MDFIU							
F	↓	↓	↓	↓	↓	×	↓	↓	↓							

Fig. 3. RAM Assignments - Page 1

	0	1	2	3	4	5	6	7	8	9	A	B	C	D	E	F
0	ESBR	TAU2	TAU10	KHLMP	M3RFP	M3ZIK	M3ZK1	M4FKP	RWVCB							
1							x									
2	ESBP	TAU3	TAU11	KHLMZ	M3RFZ	M3ZFG	M3RIF	M4FP	KPWR							
3							M3PIF									
4	FSSBZ	TAU4	TAU12	DVIG2	M3RMG	M3ZFP	M3ZIF	M4FZP	KPWP							
5							M3RFF									
6	NCN1	TAU5	KGP	GFBR2	M3PIK	M3ZFZ	M3PFF	M4KIP	KPWZ							
7	NCN2						M3ZFF									
8	NCN3	TAU6	KR	GFBP2	M3PFG	M3ZMG	M4KDZ	WRC	CKSMX							
9	NCN4															
A	NPB1	TAU7	MOML	GFBZ2	M3PFP	M3ZK2	M4FB	RWB1								
B	NPB2															
C	NPB3	TAU8	GFSL	M3RIK	M3PFZ	M3ZK3	M4FA	RWB2								
D	NPB4		x													
E	TAU1	TAU9	KHLMR	M3RFG	M3PMG	M3ZK4	M4KIZ	RWVCL								
F																

Fig. 4. RAM Assignments - Page 2

the first 203 days of operation between April 5 and October 22, 1983, 56 RAM upsets were observed (8, 10, 4, 8, 9, 9 and 8 on the monthly basis). All but one were observed on Page 2. The attitude control operation is mostly in Normal Mode and any RAM bit upset associated with this mode is of a particular concern. There have been six of these, but only two occurred after the operation entered Normal Mode on June 29, 1983.

The most serious upset so far occurred on October 3, 1983 when the reaction wheel speed bias was changed on Wheel #1 from 2782.5 to 2542.5 rpm, initiating wheel speed run down. Fortunately, this anomaly was noted by the ground personnel immediately and the bias was corrected within 1.5 minutes. A small transient attitude error resulted for several minutes. Peak errors of -0.11 deg in roll and -0.65 deg in pitch were observed. All other upsets were of fairly benign nature and no significant anomalies resulted in the attitude control operation.

POTENTIAL EFFECTS AND REQUIRED CORRECTIVE ACTIONS

All the RAM bit assignments have been reviewed and their effects on the Normal Mode operation have been assessed. The study has identified 40 bits in 18 parameters listed in Table 1 to be the most critical ones. These are critical in the sense that an upset of any one of the 40 bits will lead to an earth pointing loss without ground operator intervention. Five of the critical bits are located on Page 0, and the rest are located on Page 2. The RAM locations containing these critical bits are shaded in Figs. 2 and 4. Two additional locations on Page 1 are also shaded. Although any bit upset in these bytes will not lead to an earth pointing loss, it can cause a large pointing disturbance. All the parameters identified here are of the following types:

- o ACS configuration flags
- o Normal Mode integrator states
- o Normal Mode gains
- o Normal Mode filter parameters
- o Earth sensor biases
- o Reaction wheel speed biases
- o Reaction wheel voltage command limits

The control system's response to each of critical bit upsets is briefly described in Table 1. When a bit is upset, it takes one to several minutes before earth pointing is lost, as indicated in the table.

Since most of the critical bits are located on Page 2, any upset here will be indicated by a memory checksum failure. The ground operator is alerted by an alarm whenever the memory checksum fails. If he rapidly resets all the Page 2 Normal Mode parameters to their nominal values, the control system will recover and no further action is required. Of the five critical bits on Page 0, three are telemetered and can be monitored. The other two bits are not telemetered, but upset of either can be detected through telemetry. In one case, the wheel

Table 1. Critical RAM Bits for Normal Mode Operation

Parameter	Location	Bit	T(Min:Sec)	Spacecraft Response
M4ØE*	09B	0	> 5:00	Roll output disabled
		1	> 5:00	Pitch output disabled
MDEN*	09C	3	> 5:00	Normal Mode processing disabled
FWCC	097	1	3:00	ESA selection changed—will trigger fail-safe response
FWFLG(LSB)	094	5	3:00	Wheel voltage commands set to zero—wheels will run down
RWVCL(MSB)	27F	7	3:00	Wheel voltage commands set to zero—wheels will run down
RWVCB(MSB)	281	7	3:00	Wheel voltage commands set to zero—wheels will run down
M4KDZ(MSB)	269	7	> 5:00	Roll loop unstable
M4KIZ(MSB)	26F	7	> 5:00	Roll loop unstable
M4FA(MSB)	26D	2	> 5:00	Roll loop unstable
		3	> 5:00	
		4	> 5:00	
		5	> 5:00	
		6	> 5:00	
		7	> 5:00	
M4FB(MSB)	26B	3	> 5:00	Roll loop unstable
		4	> 5:00	
		5	> 5:00	
		6	> 5:00	
		7	> 5:00	
M4FKP(MSB)	271	7	> 5:00	Pitch loop unstable
M4KIP(MSB)	277	7	> 5:00	Pitch loop unstable
M4FZP(MSB)	275	7	> 5:00	Pitch loop unstable
M4FP(MSB)	273	5	> 5:00	Pitch loop unstable
RWB1(MSB)*	27B	1	1:55	Loss of earth presence—will trigger fail-safe response
		2	1:06	
		3	1:43	
		4	1:43	
		5	1:43	
		6	1:43	
		7	1:06	
RWB2(MSB)*	27D	1	1:55	Loss of earth presence—will trigger fail-safe response
		2	1:06	
		3	1:43	
		4	1:43	
		5	1:43	
		6	1:43	
		7	1:06	
ESBR(MSB)*	201	7	> 5:00	
ESBP(MSB)*	203	7	1:47	

Notes: T is the time from RAM hit to earth presence loss. Parameters with asterisk can be checked directly using Normal Mode telemetry.

command voltage is set to zero. In the other case, the fail-safe logic is triggered. In either case, the wheels begin to run down. This can be stopped by immediately turning off the fail-safe logic and thus bringing the wheels back to their bias speeds. By enabling the Normal Mode pitch and roll outputs in a preferred sequence, the earth pointing operation can be recovered in less than an hour in most cases. This preferred sequence has been developed in anticipation of the critical bit upset and it is shown in Fig. 5. If a quick recovery is not successful, the earth must be reacquired via sun acquisition. This can take 6 to 12 hours.

On October 22 an operational procedure was implemented to reset all Page 2 Normal Mode parameters on every checksum failure. During the following 43 days (up to December 6), 16 upsets have been observed including 7 Normal Mode parameters. In every case of the Normal Mode parameter upsets, the Page 2 parameters were reset within 30 seconds and the earth pointing operation was in no way affected. Thus the RAM bit upset problem appears to be fully under control.

POSSIBLE MODIFICATIONS FOR AUTONOMY

Although the TDRS mission does not require autonomous operation and the current method for protecting the Normal Mode operation from RAM upsets meets all system requirements, some alternate methods were considered as product improvements minimizing or eliminating the need for this ground procedure. Of the several approaches reviewed, those listed below are relatively simple and compatible with the current CPE architecture.

1) Majority Rule Logic All the critical parameters are stored in three separate locations and these are read and compared whenever needed. Since the probability of two locations being upset within a very short time is extremely low, the control logic is designed to ignore one value if it is not the same as the other two. As the critical parameters are shown to be few, additional RAM locations required for this would be minimal.

2) Protected Latches All constants in RAM are relocated into ROM or special latches, each of which is protected from spurious state change by an RC network. The time constant of this network is selected such that only a pulse of relatively long duration generated by a ground command can change the latch state. An effort will be directed toward minimizing the number of latches without sacrificing the system flexibility. For example, parameters such as control loop gains and wheel bias voltages are stored in ROM as groups of selectable values representing the intended variational ranges. The selections would be done by means of ground commandable flags residing in the protected latches. Or, these parameters can be stored in the latches using larger quantizations. In either way the required latch number will be considerably less than the RAM bit number currently utilized. The bending mode

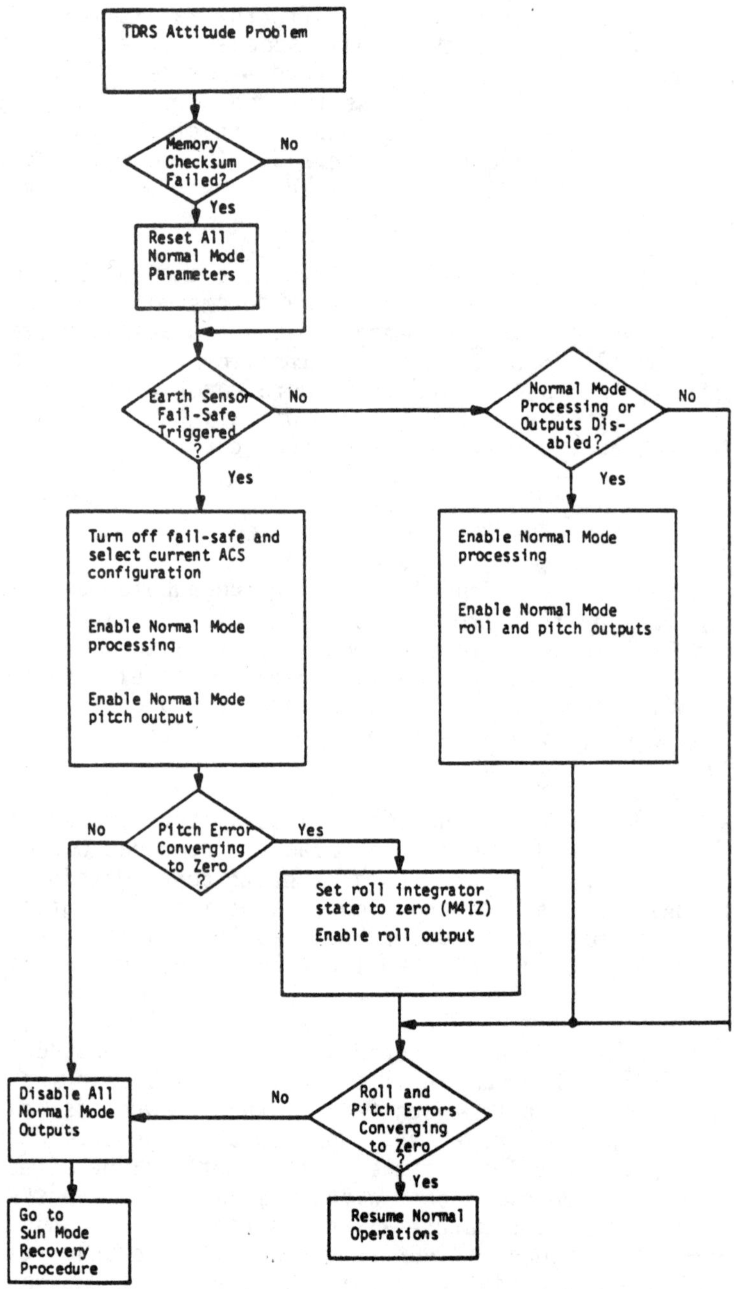

Fig. 5. Operator Response to Critical RAM Hits

filter coefficients can be stored in ROM in normalized formats and only the normalizing constants will be stored in the protected section. If all the Normal Mode constants are removed from the RAM in this way, the operation will be quite impervious to RAM upsets. A quick examination shows that no more than 40 latches are required for the current controller. These will probably fit on a 2x5 or 2x6 inch circuit board. Since the CPE is housed in a unit with other electronics, addition of such a board should be a relatively minor modification.

3) Automatic Refreshing Logic If computational speed requirements preclude reading of the protected latches every time these values are needed in the computation, they can be temporarily stored in the RAM and refreshed at some convenient intervals. An appropriate logic criteria can be also selected to execute refreshment of these parameters. An example is detection of a computed control command far exceeding the level expected during the normal operation.

4) Limited Variable Range Since the possible range of a byte in the RAM does not necessarily correspond to the range of the variable stored there, a bit upset may cause the variable to exceed the designed limit. Such an anomaly can be prevented by masking out the unwanted significant bits, checking against a predetermined limit, or taking the absolute value if the variable is not supposed to change sign.

Probably the best approach is to mix all the above techniques in some fashion.

CONCLUSIONS

In summary, the TDRS attitude control system was found to be susceptible to random RAM bit upsets by cosmic particles. However, the basic controller design and the RAM usage allowed a relatively minor ground procedure to be developed to circumvent this problem. Experience up to date indicates this procedure to be effective. If some form of autonomy is required in a future mission, the necessary changes can be easily accommodated into the basic structure of the control system.

ACKNOWLEDGEMENT

The simulation study made by J.H. Turner to determine the potential effects of RAM upsets and the helpful suggestions provided by R.J. Moore during the preparation of this paper are gratefully acknowledged.

APPENDIX

AAS 84-059

LECTURE SERIES: VOLUME I - ARTIFICIAL INTELLIGENCE

AN OVERVIEW OF ARTIFICIAL INTELLIGENCE AND ROBOTICS*

William B. Gevarter+

Artificial Intelligence (AI) is an emerging technology that has recently attracted considerable attention. Many applications are now under development. The goal of Artificial Intelligence is focused on developing computational approaches to intelligent behavior. This goal is so broad - covering virtually all aspects of human cognitive activity - that substantial confusion has arisen as to the actual nature of AI, its current status and its future capability. This volume, the first in a series of NBS/NASA reports on the subject, attempts to address these concerns. Thus, this report endeavors to clarify what AI is, the foundations on which it rests, the techniques utilized, applications, and participants and, finally, AI's state-of-the-art and future trends. A glossary and numerous sources for further information on AI are given at the end of the report.

It is anticipated that this report will prove useful to government and private engineering and research managers, potential users, and others who will be affected by this field as it unfolds.

---

* *This paper appears in full in the microfiche supplement to this volume, AAS Microfiche Series, Volume 48.*

+ *NASA Ames Research Center, 244-7, Moffett Field, California 94035*

AAS 84-060

LECTURE SERIES: VOLUME II - ROBOTICS

AN OVERVIEW OF ARTIFICIAL INTELLIGENCE AND ROBOTICS*

William B. Gevarter[+]

This is the second volume of a two volume NBS/NASA report on an *Overview of Artificial Intelligence and Robotics*. This report is intended to be read by the interested technically-oriented layman, such as engineering managers, government funding officers, and others who desire an overall perspective of the field but are unable to obtain it from the highly technical and unintegrated literature in the field, or from the more flamboyant but non-technical feature articles in the popular press.

The report incorporates a summary of the basic concepts utilized in each of the many technical areas, a review of the state-of-the-art, and indication of the organizations involved together with their funding sources and principal activities, research requirements, and a 5-10 year forecast of the emerging technology.

As robotics is a very rapidly changing field - with new companies, products and researchers entering the field each month - a report like this can only be regarded as an imperfect snapshot in time, and would benefit from periodic updating. Thus, quantitative information shown in this report may perhaps best be regarded as indicative rather than authoritative. Despite such shortcomings, this report should be helpful in providing a base, for organizations and individuals in and out of the government, for planning future robotic research and development, applications and utilizations.

The following topics are covered in the report: Primitive functions accomplished by a robot, manipulator arm configurations, actuators, mobility, end effectors, control, sensor-controlled robots, robot vision, programming a robot, kinematics and dynamics, research requirements, commercially available robots and components and vision systems for robots, current robot research, applications - now and in the future, the expanding robot population, and future directions. Numerous references are cited at the end of the report.

---

* *This paper appears in full in the microfiche supplement to this volume, AAS Microfiche Series, Volume 48.*

+ *NASA Ames Research Center, 244-7, Moffett Field, California 94035*

THE 1985 ANNUAL ROCKY MOUNTAIN GUIDANCE AND CONTROL CONFERENCE WILL BE HELD AT KEYSTONE

February 2 - February 6, 1985

Contacts for the 1985 Annual
Rocky Mountain Guidance and Control Conference

CHAIRMAN:	Charles Cullian Ball Aerospace Systems P.O. Box 1062 Boulder, Colorado 80306 (303) 441-4392
CO-CHAIRMAN:	Parker Stafford Martin Marietta Aerospace P.O. Box 179 Denver, Colorado 80201 (303) 977-3226
SECTION CHAIRMAN:	Don Parsons Martin Marietta Aerospace P.O. Box 179 Denver, Colorado 80201 (303) 977-7109

Reservations should be made directly with:

KEYSTONE LODGE
Box 38
Keystone, Colorado 80435
(303) 468-2316

ANNUAL ROCKY MOUNTAIN GUIDANCE & CONTROL CONFERENCE
1984 REGISTRANTS

Rex Agler
Martin Marietta
P. O. Box 179
Denver CO 80201

Dr. Harrison B. Albert
Ball Aerospace
P. O. Box 1062
Boulder CO 80306

Pam Alstott
Martin Marietta
P. O. Box 179; MS: D6072
Denver CO 80201

Neal T. Anderson
Ball Aerospace
P. O. Box 1062
Boulder CO 80306

Erwing Anisman
Aerospace Corp
2350 E. El Segundo Blvd
El Segundo CA 90009

Paul Arlton
University of Colorado
Boulder CO 80309

Max Bachmann
TRW
One Space Park, MS: 82/2425
Redondo Beach CA 90278

Frank Barbera
Rockwell International
12214 Lakewood Blvd
Downey CA 90241

Robert B. Barnes
Barnes Engineering Co
44 Commerce Road
Stamford CT 06904

David Barrows
MDAC, MS: 14-3
5301 Bolsa Ave
Huntington Beach CA 92647

Richard L. Bates
General Electric Co
Elec Park Bldg 3 Rm 33
Syracuse NY 13221

Robert A. Baum
Hamilton Standard 1690
New Britain Ave
Farmington CT 06032

Edward J. Bauman
Department of Electric Engr.
University of Colorado
Colorado Springs CO 80907

Steve Beach
Lockheed Missile & Space Co.
P. O. Box 504
Sunnyvale CA 94086

Robert L. Berry
Martin Marietta
P. O. Box 179
Denver CO 80201

Fred Bikle
Martin Marietta
P. O. Box 179 MS: L8232
Denver CO 80201

Charles E. Bond
Martin Marietta
P. O. Box 179
Denver CO 80201

Richard Bortins
Lockheed Missile & Space Co.
P. O. Box 504 0/62-11 B/104
Sunnyvale CA 94086

Walt Boyne
9134 Continental
Alexander VA 22309

Larry Brandon
NASA/MSFC
Marshal Space Flight Center
Huntsville, AL 35812

George S. Briggs
Vought Corporation MSF-29
Box 225907
Dallas TX 75265

Mick Burton
British Aerospace Dynamics Group
Argyle Way, Stevenage
Herts, England

Joseph G. Callahan
Northrop Precision Products Div
100 Morse St.
Norwood Mass 02062

James Cheng
RCA Astro
P. O. Box 800
Princeton NJ 08540

Chia-Chuan Chow
University of Colorado
Boulder CO 80309

Harold E. Cook
Martin Marietta
P. O. Box 179
Denver CO 80201

D. E. Cornick
Martin Marietta
P. O. Box 179
Denver CO 80201

Kenneth J. Coughlin
Martin Marietta Aerospace
7000 E. Quincy #311
Denver CO 80237

Paul S. Cramer
Martin Marietta
P. O. Box 179
Denver CO 80201

Charles A. Cullian
Ball Aerospace
P. O. Box 1062
Boulder CO 80306

Robert D. Culp
University of Colorado
Campus Box 429
Boulder CO 80309

David Cunningham
Sperry
5540 E. Charter Oak
Scottsdale AZ 85254

Steve Cunningham
General Electric
P. O. Box 8555, Rm 8036
Philadelphia PA 19101

Rodd Cusic
University of Colorado
Boulder CO 80309

David J. Cwynar
Martin Marietta
P. O. Box 179
Denver CO 80201

Phil Dahl
The Aerospace Corp.
El Segundo CA 90009

Phyllis L. Dahl
TRW
One Space Park
Redondo Beach CA 90278

Scott Dahl
Martin Marietta
273 S. Humboldt St.
Denver CO 80209

Kevin C. Daly
C. S. Drapin Lab
555 Tech Sq
Cambridge MA 02139

Daniel B. DeBra
Stanford University
Aero & Astro
Stanford CA 94305

Jorge H. Decanini
TRW/STG
One Space Park
Redondo Beach CA 90278

R. H. DeFrees
TRW, Electronics & Defense Sector
One Space Park
Redondo Beach CA 90278

Robert L. DelBoca
CLS Laser Systems Inc
5 Jeffery Dr. P. O. Box 767
South Windsor CT 06058

Michael Devirian
JPL-NASA
EI-5, NASA Headquarters
Washington DC 20546

Ed Diffendal
Hamilton St...card
8939 S. Sepulveda Blvd
Los Angeles CA 90045

Ed Dorroh
Martin Marietta Corp
P. O. Box 179
Denver CO 80201

John C. Durrett
Martin Marietta
P. O. Box 179
Denver CO 80201

Brian Edmister
BEI Electronics
1701 W. Euless Blvd #425
Euless TX 76021

Bruce J. Ehlers
TRW
Redondo Beach CA 90278

Zubin Emsley
Ball Aerospace Systems Divsion
Boulder CO 80306

Patrick G. Engstrom
Ball Aerospace
P. O. Box 1062, BLDG EO-2
Boulder CO 80306

Harry Erwin
NASA - JSC - EE6
NASA Road One
Houston TX 77058

Edward A. Euler
Martin Marietta
P. O. Box 179
Denver CO 80201

C. W. Francis Everitt
Stanford University
Varian Physics Bldg.
Stanford CA 94305

James Farler
Martin Marietta
P. O. Box 179
Denver CO 80201

Tom Fenner
Martin Marietta
P. O. Box 179
Denver CO 80201

Jack R. Ferguson Jr.
USAF Academy
Colorado Springs CO 80840

Alan Fleming
TRW/STG
One Space Park 03/2661
Redondo Beach CA 90278

Bill Frazier
Ball Aerospace
MP4
Boulder CO 80306

Randy W. Frey
Martin Marietta
P. O. Box 5837 MP-15
Orlando FL 32855

John E. Gaiser
Space Applications Corp
440 Persian Dr
Sunnyvale CA 94089

Bud Gates
Martin Marietta
P. O. Box 179
Denver CO 80201

Lawrence M. Germann
Ball Aerospace
P. O. Box 1062
Boulder CO 80306

William B. Gervarter
NASA Ames
244-7
Moffett Field CA 94035

James Gilchrist
Aerospace Corp
El Segundo CA 90009

Rond S. Goetz
Aerospace Corp
P. O. Box 91055
Los Angeles CA 90009

Stewart J. Golding
Singer-Kearfott
150 Totowa Road
Wayne NJ 07470

A. K. Goodfellow
Boeing Aerospace Co.
2404 128th Ave S. E.
Bellevue WA 98005

Dr. Andrew L. Gordon
NAS/LERC MS 500-120
21000 Brookpark road
Cleveland OH 44135

Richard Gran
Grumman Aerospace
MS: A08-35
Bethpage NY 11714

Greg Griffin
Hamilton Standard
8939 S. Sepulveda
Los Angeles CA 90045

Michael D. Griffin
Johns Hopkins Univeristy
Applied Physics Lab
Johns Hopkin Road
Laurel MD 20707

Morgan A. Gunst Jr.
Quantic Industries Inc
990 Commercial St.
San Carlos CA 94070

Dennis Haley
Martin Marietta
P. O. Box 179
Denver CO 80201

Mike Haley
Hamilton Standard
1690 New Britain Ave
Farmington CT 06010

Grover W. Hall, Jr.
Martin Marietta
P. O. Box 179
Denver CO 80201

R. J. Hamann
Fokker B. V.
P. O. Box 7600
1117 ZJ Schiphol, The Netherlands

John C. Harcinske
Martin Marietta
P. O. Box 179
Denver CO 80201

Bruce Hardy
System Analysis and Control
P. O. Box 1381
Ridgecrest CA 93555

William R. Harper
TRW Space & Technology Group
One Space Park
Redondo Beach CA 90278

Tom Harrington
TRW Federal Systems Division
One Space Park
Redondo Beach CA 90278

Richard Harris
BEI Electronics
1221 Pearl St. #13
Boulder CO 80302

R. S. Harris
British Aerospace Dynamics Group
Briston Division FPC 641
Fitton Bristol BS12 7QW England

Michael W. Hendrix
Martin Marietta
P. O. Box 179
Denver CO 80201

Carl Henrikson
Ball Aerospace Systems
Box 1062
Boulder CO 80306

Lou Herman
Aerospace Co.
El Segundo CA 90009

James L. Hieatt
TRW Space & Technology Group
One Space Park BLDG R5 Rm 1281
Redondo Beach CA 90278

Edwin S. Jeris
NASA Lewis Research Center
21000 Brook Park Rd
Cleveland OH 44120

Randy Jones
University of Colorado
Boulder CO 80306

William C. Jones
Honeywell Inc
13350 US Hwy 19
Clearwater FL 33516

Marshall H. Kaplan
Spacetech Inc
P. O. Box 1109
State College PA 16804

Terry J. Kelly
Ball Aerospace Systems
P. O. Box 1062 MS1 BE-6
Boulder CO 80306

Richard Klabunde
Naval Weapons Center
Code 3912
China Lake CA 93555

Jerry Kollodge
Ball Aerospace Systems Division
Boulder CO 80306

Marvin A. Kollodge
Honeywell Inc
2600 Ridgeway Parkway
Minneapolis MN 55413

John M. Kolvek
Bendix GSD
28 Wierimus Road
Hillsdale NJ 07642

Mark R. Kozemchak
US Air Force
Kirtland AFB NM 87117
AFWL/ARAA

Randall Kuhns
Boeing Aerospace
3407 61st Ave S. W.
Seattle WA 98116

E. E. Lademann
Singer-Kearfott
150 Totowa Road
Wayne NJ 07470

Carol Lane
TRW
One Space Park
Redondo Beach CA 90278

Tim Larson
Hughes Aircraft Co.
P. O. Box 92919
Los Angeles CA 90009

Wiley J. Larson
USAF
US Air Force Academy CO 80840

James M. Lecher
Lockheed Engr & Mgmt Services
1830 NASA Rd 1
Houston TX 77058

D. A. Levinson
Lockheed Missile & Space
3251 Hanover St.
0/52-57, B/205
Palo Alto CA 94304

Joel Levinthal
BENDIX
Teterboro NJ 07608

Lynn R. Lewis
Ball Aerospace
P. O. Box 1062
Boulder CO 80306

Richard A. Lucal
USAF
USAF Academy CO 80840

Edward B. Madigan
Martin Marietta
P. O. Box 179
Denver CO 80201

T. R. Mahoney
Singer Kearfott
1150 McBride Ave
Little Falls NJ 07424

Paul Makus
Bendix
88 Deerfield Road
Wayne NJ 07470

Dr. Chris Markland
European Space Agency
Estec
Noordwyk Netherlands

Willard Marquis
Martin Marietta
P. O. Box 179
Denver CO 80201

William McArthur
Aerospace Corp
El Segundo CA 90009

L. F. McGlinchey
Jet Propulsion Lab
4800 Oak Grove Drive MS:198-326
Pasadena CA 91109

John W. McIntosh
Martin Marietta
P. O. Box 179 MS: 4010
Denver CO 80201

W. S. McMath
Communications Research Center
P. O. Box 11490 Station H
Ottawa Ontario Canada

James McQuerry
Ball Aerospace
P. O. Box 1062
Boulder CO 80306

Phil Meinhardt
USAF 4400 Dauphine St.
CAP - USAF NSA (E)
New Orleans LA 70116

Daniel E. Mercier
USAF
USAF Academy CO 80840

Paul H. Merritt
Hughes Aircraft Co.
P. O. Box 9301
Albuquerque NM 87119

James D. Micheal
NASA/MSFC
NASA Marshall Space Flight Center
Huntsville AL 35812

M. S. Misra
Martin Marietta Aerospace
P. O. Box 179
Denver CO 80201

F. F. Mobley
Applied Physics Lab
Laurel MD 20707

John Montgomery
Martin Marietta
P. O. Box 179 MS: 1130
Denver CO 80201

David G. Morgenthaler
Martin Marietta
P. O. Box 179
Denver CO 80201

Dr. Hal Mork
TRW
One Space Park
Redondo Beach CA 90278

Lou Morine - TO 510
Martin Marietta
P. O. Box 179
Denver CO 80201

Gerald E. Morrison
Boeing Aerospace Co.
P. O. Box 3999 MS: 8K-85
Seattle WA 98124

Noboru Muranaka
NEC Corp
4035 IKEBE - CHO MIDORI - KU
Yokohoma 226 Japan

Eiji Nakagawa
NEC Corp
4035 IKEBE - CHO MIDORI - KU
Yokohama 226 Japan

Hiroshi Nakano
TRW
One Space Park
Redondo Beach CA 90278

Arun Nayak
HR Textron
2485 NcCaber Way
Irvine CA 92714

R. D. Neal
Martin Marietta
P. O. Box 179
Denver CO 80201

Doug New
Ball Aerospace
Boulder CO 80306

Dr. Keiken Ninomiya
Institude of Space & Ast. Sci.
4-6-1 Komaba
Tokyo 153 Japan

Hal Nylander
Martin Marietta
P. O. Box 179 MS: D6072
Denver CO 80201

Norman A. Osborne
Martin Marietta
P. O. Box 179 MS: T0510
Denver CO 80201

Donald H. Parsons
Martin Marietta
P. O. Box 179 MS: D6110
Denver CO 80201

Robert C. Peters
Aerospace Corp
4114 Michelle Dr
Torrance CA 90503

Vincent L. Pisacane
JHU/Applied Physics Lab
Laurel MD 20709

John Pitre
Aerospace Corp
2250 El Segundo
El Segundo CA 90009

R. Stephen Price
Martin Marietta
P. O. Box 179 MS S0510
Denver CO 80201

R. L. Proffit
Quantic Industries
990 Commercial St.
San Carlos CA 94070

Glenn Quasius
Honeywell Inc.
13350 U.S. 19 South
Clearwater FL 33516

Alberto Ramos
COMSAT
Clarksburg MD 20879

Robert B. Rice Jr.
Martin Marietta
P. O. Box 179
Denver CO 80201

John W. Ritter
Teledyne Systems Co.
19601 Nordhoff St
Northridge CA 91324

David A. Rogers
Ball Aerospace Systems
P. O. Box 1062
Boulder CO 80306

Dr. Richard E. Rose
TRW Space & Technology Group
One Space Park
Redondo Beach CA 90278

Susan Ross
Ball Aerospace Systems Division
P. O. Box 1062
Boulder CO 80306

Michael Z. Rutkowski
Ithaco Inc
735 W. Clinton St.
Ithica NY 14851

James Sanborn
Martin Marietta
P. O. Box 179
Denver CO 80201

John Sand
Ball Aerospace
Boulder CO 80306

David Sanders
Honeywell Inc.
555 Denver Tech Center
Parkway #5
Englewood, CO 80111

Roger T. Schappell
Martin Marietta
P. O. Box 179 MS 0570
Denver CO 80201

Robert Schein
Martin Marietta Aerospace
P. O. Box 179
Denver CO 80201

Harry Schmeichel
TRW
Redondo Beach CA 90278

R. Schroer
Martin marietta
P. O. Box 179 MS: G6420
Denver CO 80201

Roy M. Scott
Ball Aerospace
P. O. Box 1062
Boulder CO 80306

S. R. Schrock
Martin Marietta
P. O. Box 179
Denver CO 80201

Eric Schwarm
Martin Marietta
P. O. Box 179
Denver CO 80201

Eric Seale
University of Colorado
Boulder CO 80306

Paul L. Shattude
Martin Marietta
P. O. Box 179 MS: T0510
Denver CO 80201

Paul S. Shirley
Air Force Weapons Lab
ARWL/ARLB
Kirtland AFB NM 87117

Thomas J. Sisk
Martin Marietta
5650 W. Rowland Ave
Littelton CO 80123

S. A. Skaron
TRW Inc Electronics & Defense Sys
One Space Park
Redondo Beach CA 90278

Harvey Sokoloff
The Aerospace Corp
9145 Gerald Ave
Sepulveda CA 91343

Robert Spahr
Northrop Corp
100 Morse St.
Norwood MASS 02062

Thomas P. Sparn
Univeristy of Colorado LASP
5525 Central Ave
Boulder CO 80301

Robert E. Spearing
Goddard Space Flight CTR
Greenbelt MD 20771

Parker Stafford
Martin Marietta
P. O. Box 179
Denver CO 80201

Kirk Stewart
Ball Aerospace
P. O. Box 1062
Boulder CO 80306

Douglas w. Stoner
University of Colorado
Engineering Center Campus Box 429
Boulder CO 80309

W. R. Sturgeon
Systems Analysis and Control
P. O. Box 1381
Ridgecrest CA 93555

C. O. Swanson
Singer-Kearfott
150 Totowa Road MC 10B16
Wayne NJ 07470

Stephen Z. Szirmay
JPL
4800 Oak Grove Dr
Pasadena CA 91109

Ralph H. Tate Jr.
USAF
USAF Academy CO 80840

Hal Taylor
Hamilton Standard
1690 New Britain Ave
Farmington CT 06110

A. J. Throckmorton
Martin Marietta
P. O. Box 179
Denver CO 80201

Maj Robert L. Van Allen
AF Weapons Lab/ARLB
Kirtland AFB NM 87117

Don Vanlandingham
Ball Aerospace
P. O. Box 1062
Boulder CO 80306

Richard VanRiper
Sperry
Phoenix AZ 85001

Robert H. Van Vooren
TRW S & TG
One Space Park R4/1098
Redondo Beach CA 90278

D. D. Van Winkle
McDonnell Douglas
6751 Red Coach Drive
Huntington Beach CA 92647

Dominique Vilain
MATRA-ESPACE
Velizy 78146 Cedex
France

Robert L. Wasley
Northrop Corp
100 Morse St
Norwood MASS 02062

Jack F. Wade
Ball Aerospace
Boulder CO 80306

Leslie Walker Jr.
Orbital Systems LTD
8HFO P. O. Box 700
Lanaham MD 20706

Mary S. Walker
NASA/GSFC
Greenbelt MD 20771

James Wambold
Penn State
Research Bldg B
University Park PA 16802

Frank W. Wargocki
Ball Aerospace Systems Division
P. O. Box 1062
Boulder CO 80306

Robert L. Wasley
Northrop Corp.
100 Morse St.
Norwood, MASS 02062

Frank D. Watson
R.C.A.
2424 N. Tejon
Colo Sprngs CO 80907

August Wessels
NASA GSFC
Code 407
Greenbelt MD 20771

Robert Wickwire
Martin Marietta
2041 Rosecrans Ave
El Segundo CA 90245

R. F. Wilkinson
TRW
One Space Park
Redondo Beach CA 90274

Ivan Joe Williams
TRW
One Space Park
Redondo Beach CA 90274

W. E. Williams
Rockwell Space Division
12214 Lakewood blvd
Downey CA 90241

Robert K. Williamson
The Aerospace Corp
2350 El Segundo
El Segundo CA 90245

John M. Wilson
Martin Marietta
P. O. Box 179
Denver CO 80201

John W. Wokak
Bendix GSD
Teterboro NJ 07608

David L. Younkin
USAF
HQ Space Division L.A.
3912 Denwood Ave
Los Alamitos CA 90720

Richard T. Yoshida
Martin Marietta
P. O. Box 179 MS: G6420
Denver CO 80201

Proceedings of the Rocky Mountain
Guidance and Control Conferences are available as follows:

Guidance and Control 1984, Volume 55,
Advances in the Astronautical Sciences;
Microfiche Supplement: Volume 48, AAS Microfiche Series

Guidance and Control 1983, Volume 51,
Advances in the Astronautical Sciences;
Microfiche Supplement: Volume 44, AAS Microfiche Series

Guidance and Control 1982, Volume 48,
Advances in the Astronautical Sciences;
Microfiche Supplement: Volume 38, AAS Microfiche Series

Guidance and Control 1981, Volume 45,
Advances in the Astronautical Sciences;
Microfiche Supplement: Volume 36, AAS Microfiche Series

Guidance and Control 1980, Volume 42,
Advances in the Astronautical Sciences;

Guidance and Control 1979, Volume 39,
Advances in the Astronautical Sciences;
Microfiche Supplement: Volume 31, AAS Microfiche Series

Guidance and Control 1978, Volume 29,
AAS Microfiche Series
(Includes one paper from earlier conference)

Order from Univelt, Inc., P.O. Box 28130, San Diego, CA 92128

INDEX

NUMERICAL INDEX

VOLUME 55 ADVANCES IN THE ASTRONAUTICAL SCIENCES, GUIDANCE AND CONTROL 1984

VOLUME 48 AAS MICROFICHE SERIES, Supplement to Volume 55, Advances, 1984

(Annual Rocky Mountain Guidance and Control Conference, February 4-8, 1984, Keystone, Colorado)

*AAS 84-001 Development of the Attitude and Orbit Control Subsystem for the OLYMPUS Satellite, M. Burton

AAS 84-002 EUROSTAR Multimission Platform Attitude and Orbit Control Subsystem, R. Oskian, J.F. Poussin, B. Govin

AAS 84-003 Marine Observation Satellite-1 System and Control Concepts, Y. Ishizawa, M. Kusanagi, T. Shimamura, G. Shirako, E. Nakagawa

AAS 84-004 Attitude Control of the Infrared Space Observatory and its Predecessors, R.J. Hamann

AAS 84-005 Attitude Determination and Control of the Hipparcos Satellite, D.P. Vilain, R.S. Harris

AAS 84-006 PLANET-A Attitude and Orbit Control Subsystem, K. Ninomiya, K. Uesugi, H. Hara, H. Yamamoto, N. Muranaka, A. Katoh

AAS 84-007 to -010 Not assigned

AAS 84-011 Direct Quaternion Determination from BARS Measurements, S.A. Skaron

AAS 84-012 Digital Standard Star Tracker, J.P. McQuerry, Jr.

AAS 84-013 Interferometric Angle Sensor, H.B. Albert, L.M. Germann (v48 Micro)

AAS 84-014 Not available

---

* *Unless otherwise indicated all papers appear in Volume 55, Advances in the Astronautical Sciences.*

AAS 84-015 Retroreflector Field Tracker, F.E. Wargocki, S.E. Ross (v48 Micro)

AAS 84-016 Airborne Laser Lab: Description and Experimental Results, P.S. Shirley (v48 Micro)

AAS 84-017 Robotics in Engineering Education, E.J. Bauman, D.W. Erickson

AAS 84-018 Artificial Intelligence and Computer Vision for Advanced Manipulator Systems, D.C. Haley, D.G. Morgenthaler, J.C. Sanborn

AAS 84-019 Not available

AAS 84-020 Strapdown Inertial Guidance Performance in Space, M.J. Haley, H.B. Taylor

AAS 84-021 ISTP - NASA's Next International Cooperative Program (summary) R.S. Tatum (v48 Micro)

AAS 84-022 to -030 Not assigned

AAS 84-031 Not available

AAS 84-032 The First Role of the Manned Maneuvering Unit in Space, J.C. Harcinske, R.A. Schein, A.M. Ray, D.J. Cwynar

AAS 84-033 Laser Docking System, H.O. Erwin

AAS 84-034 Rendezvous and Docking with Remote Piloted Vehicles, J.D. Micheal

AAS 84-035 Control of a Flexible Robot for Space Applications, R. Gran

AAS 84-036 Docking of a Spacecraft with an Unrestrained Orbiting Structure (Abstract) D.A. Levinson, T.R. Kane

AAS 84-037 to -040 Not assigned

AAS 84-041 IUS Propulsion, Guidance and Control - An Integrated Design, H. Sokoloff

AAS 84-042 Gamma Guidance - Design and Results for Two Flights, R.H. Kuhns, G.C. Coomer

AAS 84-043 Centaur D-1A Guidance/Software System, A.L. Gordan

AAS 84-044 Navigation & Control Considerations for Space Based Orbital Maneuvering Systems, L. Brandon

AAS 84-045 Guidance and Control for the Transfer Orbit Stage (TOS), R.B. Schroer

AAS 84-046 Integrated Launch Vehicle and Spacecraft Avionics - A Search for a Feasible Option, J.D. Gilchrist

AAS 84-047 to -050 Not assigned

AAS 84-051 Inertial Upper Stage/Tracking Data Relay Satellite (IUS/TDRS) Mission Post-Flight Analysis, A.K. Goodfellow, T.R. Anderson, M.T. Oshima

AAS 84-052 Inertial Upper Stage Redundant Inertial Measurement Unit Space Performance, R.A. Baum, G.E.S. Morrison, R.C. Peters

AAS 84-053 In-Flight Rescue of Stranded TDRS-1 Spacecraft, H. Schmeichel, B.J. Ehlers

AAS 84-054 Autotracking from Space: The TDRSS Approach, R.E. Spearing, W.R. Harper

AAS 84-055 Not available

AAS 84-056 Cosmic Rays, Single Event Upsets and Things that Go Bump in the Night, S.S. Cunningham

AAS 84-057 Not available

AAS 84-058 A Protective Method Against RAM Upsets Due to Cosmic Particles, H. Nakano, H. Schmeichel

AAS 84-059 Lecture Series: Volume I - Artificial Intelligence - An Overview of Artificial Intelligence and Robotics, W.B. Gevarter (v48 Micro)

AAS 84-060 Lecture Series: Volume II - Robotics - An Overview of Artificial Intelligence and Robotics, W.B. Gevarter (v48 Micro)

AAS 84-061 to -099 Not assigned

AUTHOR INDEX*

* *For each author the paper number is given. The page numbers refer to Volume 55, Advances in the Astronautical Sciences. "Mic" refers to the AAS Microfiche Series, a supplement to Advances in the Astronautical Sciences.*